AF334587

Protein Kinase C

Protein Kinase C

Edited by

J. F. Kuo

New York   Oxford
OXFORD UNIVERSITY PRESS
1994

Oxford University Press

Oxford   New York   Toronto
Delhi   Bombay   Calcutta   Madras   Karachi
Kuala Lumpur   Singapore   Hong Kong   Tokyo
Nairobi   Dar es Salaam   Cape Town
Melbourne   Auckland   Madrid

and associated companies in
Berlin   Ibadan

Library of Congress Cataloging-in-PublicationData
Protein kinase C / [edited by] J. F. Kuo.
p.   cm.   Includes bibliographical references and index.
ISBN 0-19-508101-3
1. Protein kinase.
I. Kuo, J. F. (Jyh-Fa), 1933–
[DNLM: 1. Protein Kinase C—metabolism.
WL 104 P9665 1993]   QP606.P76P734   1933
612'.01516—dc20   DNLM/DLC   for Library of Congress   93-24246

9 8 7 6 5 4 3 2 1

Printed in the United States of America
on acid-free paper

Preface

Since its discovery in 1979 by Professor Yasutomi Nishizuka and his associates, protein kinase C (PKC) has generated a level of research interest seldom seen for any single biological system. After more than a decade of exponential and continuing growth, owing largely to its unique biological roles inherent in its molecular and regulatory diversities, our knowledge of PKC has reached sufficient maturity that a comprehensive survey of the field seems warranted. Although this book cannot be exhaustive in its coverage, it places special emphasis on areas of PKC research where major contributions have already been made and future directions indicated. Accordingly, the book begins with a chapter on history and perspectives on PKC and continues with a consideration of basic principles (molecular and catalytic properties, activity regulation, signal transduction) and selective key target tissues and processes (cancer, nervous system, smooth muscle, heart, exocytosis, and apoptosis). It concludes with a discussion of protein phosphatases, the other half of the enzyme system responsible for maintenance of the steady state of protein phosphorylation.

I am deeply indebted to many individuals, active and major investigators in their respective areas of research interests, who have graciously contributed chapters to the book. And, I also want to thank Professor Nishizuka for friendship and inspiration and for the wonderful gift of PKC, for which we all share the joys (and often agonies) in the course of uncovering its never-ending secrets.

Atlanta
August 1993   J.F.K.

Contents

Contributors

Robert M. Bell
Department of Biochemistry, Duke University Medical Center, Durham, N.C.

Mary A. Bittner
Department of Pharmacology, University of Michigan Medical School, Ann Arbor, Mich.

Alton L. Boynton
Department of Cell and Molecular Biology, Pacific Northwest Research Foundation, Seattle, Wash.

Joan Heller Brown
Department of Pharmacology, University of California School of Medicine, La Jolla, Calif.

P. Jeffrey Conn
Department of Pharmacology, Emory University School of Medicine, Atlanta, Ga.

Robert I. Glazer
Department of Pharmacology, Georgetown University School of Medicine, Rockville, Md.

Hiroyoshi Hidaka
Department of Pharmacology, Nagoya University School of Medicine, Nagoya, Japan

Ronald W. Holz
Department of Pharmacology, University of Michigan Medical School, Ann Arbor, Mich.

Richard E. Honkanen
Department of Cell and Molecular Biology, Pacific Northwest Research Foundation, Seattle, Wash.

Kuo-Ping Huang
Section on Metabolic Regulation, Endocrinology and Reproductive Research, Research Branch, National Institute of Child Health and Human Development, National Institutes of Health, Bethesda, Md.

J. F. Kuo
Department of Pharmacology, Emory University School of Medicine, Atlanta, Ga.

J. David Lambeth
Department of Biochemistry, Emory University School of Medicine, Atlanta, Ga.

Michael J. McCabe, Jr.
Institute of Chemical Toxicology, Wayne State University, Detroit, Mich.

Charles W. Mahoney
Section on Metabolic Regulation, Endocrinology and Reproductive Research, Research Branch, National Institute of Child Health and Human Development, National Institutes of Health, Bethesda, Md.

Masakatsu Nishikawa
The Second Department of Internal Medicine, Mie University School of Medicine, Tsu, Japan

Catherine A. O'Brian
Department of Cell Biology, The University of Texas M. D. Anderson Cancer Center, Houston, Tex.

Sten Orrenius
Department of Toxicology, Karolinska Institute, Stockholm, Sweden

Peter J. Parker
Protein Phosphorylation Laboratory, Imperial Cancer Research Fund, London, U.K.

Michel Pucéat
Laboratoire Physiologie Cellulaire Cardiaque, INSERM U-21, Université Paris-Sud, Orsay Cedex, France

Andrew F. G. Quest
Department of Biochemistry, Duke University Medical Center, Durham, N.C.

J. David Sweatt
Division of Neuroscience, Baylor College of Medicine, Houston, Tex.

Protein Kinase C

1

Protein Kinase C: History and Perspectives

PETER J. PARKER

AN HISTORICAL VIEW

Defining the Pathway

The now classical phosphorylation cascade at the center of the signal transduction pathway, which lies between β-adrenergic agonists and glycogenolysis, has greatly influenced concepts of cell signaling. In particular the broad-specificity, second-messenger-dependent protein kinase, cyclic AMP-dependent protein kinase (PKA), has served as a benchmark for protein kinases and one for which the greatest structural detail is now known (Knighton et al., 1991).

In the late 1960s, PKA was one of only two members of this now unwieldy protein kinase superfamily. Its position within the glycogenolytic signaling pathway led to the expectation that distinct broad-specificity protein kinases might play similar roles in other pathways. The search for these has led to the identification of a number of second-messenger-responsive protein kinases including Ca^{2+}-calmodulin-dependent (Hanson and Schulman, 1992), cyclic GMP-dependent (Lincoln and Corbin, 1983), and diacylglycerol (DAG)-dependent activities (Nishizuka, 1984). It is this last group of proteins that, broadly speaking, constitutes the protein kinase C (PKC) family.

Like phosphorylase kinase, PKC was first identified functionally in a proteolytically activated form (Inoue et al., 1977; Takai et al., 1977). It was subsequently shown to be reversibly activated by Ca^{2+}/phospholipid and DAG (Takai et al., 1979). This DAG sensitivity proved to be a keystone to the signaling pathway, providing a specific consequence to the previously documented agonist-induced phosphatidylinositol (PtdIns) ''turnover'' (Hokin and Hokin, 1958; Hokin, 1985). This lipid pathway had already been drawn into sharp focus in the mid-1970s (Michell, 1975), but the key players were missing. The identification of PKC and

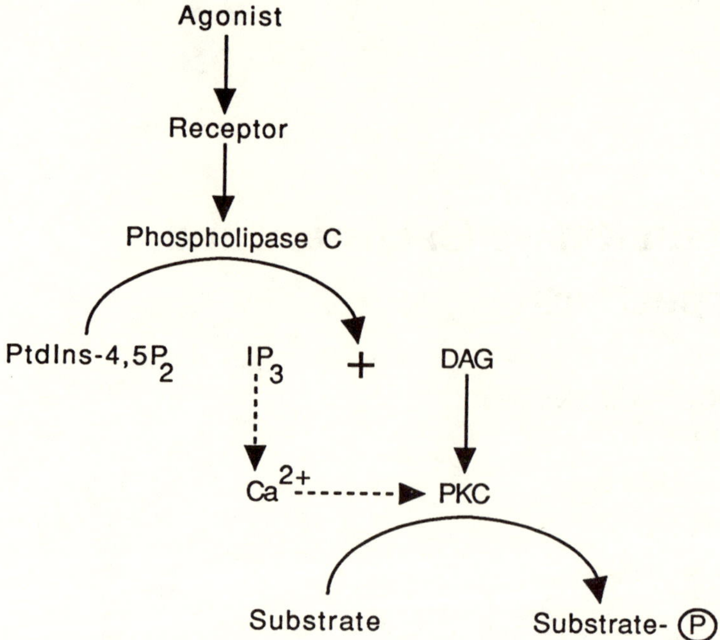

Figure 1.1 The classical inositol lipid pathway. The signal transduction pathway as illustrated depicts the action of an extracellular agonist on its receptor with consequent activation of a phospholipase C specific for the inositol lipids. The breakdown of phosphatidylinositol-4,5 bisphosphate (PtdIns-4,5P$_2$) yields inositol 1,4,5 trisphosphate (IP$_3$) and diacylglycerol (DAG). The former causes release of Ca^{2+} from intracellular stores, while the latter activates protein kinase C (PKC), leading to phosphorylation of protein substrates.

its DAG dependence produced a flurry of activity that was also spurred on by the issue of Ca^{2+} mobilization; the consequence was the pathway outlined in Figure 1.1. Agonists known to induce PtdIns turnover (acutely through PtdIns 4,5 bisphosphate hydrolysis) would increase Ins 1,4,5 trisphosphate (and hence Ca^{2+}) and coincidentally elevate DAG. The latter would activate PKC and so effect alterations to cell function.

Although the assembly of events shown in Figure 1.1 was compelling, direct evidence for such a pathway was required. In essence, the ''proof'' of involvement of a kinase in a particular phosphorylation event is only one step away from the proof for a pathway:

1. Demonstration of phosphorylation of a protein on the same site(s) *in vivo* as that phosphorylated by the kinase in question *in vitro*.
2. Evidence that activation of the kinase *in vivo* leads to phosphorylation of the target protein on the correct site.
3. Evidence that physiological agonists produce effectors and activation of the kinase.

These criteria were met through an elegant series of studies, particularly in platelets where the phosphorylation of p40 (now termed pleckstrin) was studied. Of great value in this work was the use of a membrane-permeant diacylglycerol, 1-oleyl-2-acetylglycerol (OAG) (Kaibuchi et al., 1983). Exposure of platelets to OAG was shown to induce p40 phosphorylation, and peptide mapping showed that *in vitro,* purified PKC phosphorylation of p40 yielded an identical phospho-peptide map. These studies were further supported following the identification of PKC as a target for the phorbol ester class of tumor promoters (Castagna et al., 1982). Thus, for example, 12-0-tetradecanoylphorbol-13-acetate (TPA) was shown to elicit responses similar to OAG in platelets (Sano et al., 1983). In addition, direct evidence of agonist-induced accumulation of diacylglycerol in platelets was provided (Kawahara et al., 1980). Part or all of these types of anal-ysis have been carried out in many different cell types, providing an enormous weight of evidence for the operation of this pathway.

As a consequence of the mechanism of activation of PKC, one further element can be included in the implication of this kinase in signal transduction, and that is the phenomenon of translocation. On activation, cytosolic[1] PKC forms a mem-brane-associated complex that is only slowly dissociated on extraction in chelat-ing agents at 4°C; this provides an assay based upon the cytosol/particulate dis-tribution of PKC that is considered to reflect activation. Such acutely induced translocations have been documented in many different contexts, suggesting acti-vation of PKC.

Messenger Generation

The number of agonists that will stimulate inositol lipid degradation is substantial (see Meldrum et al., 1991), indicating that the Ca^{2+}/PKC pathways play important roles in feedforward or feedback controls in many situations. The spectrum of agonists that will trigger this pathway encompasses more than one receptor super-family and it is now clear that distinct members of the phosphatidylinositol-specific phospholipase C (PtdIns-PLC) gene family enable coupling by different receptor types. Thus receptor tyrosine kinases (or receptors with associated tyro-sine kinases) can couple through the PtdIns-PLCγ subclass (see Wahl and Car-penter, 1992). By contrast, seven transmembrane receptors coupling through G-proteins do so either via Gqα-PtdIns-PLCβ interactions (Taylor et al., 1991; Taylor and Exton, 1991) or via Gβγ-PtdIns-PLCβ interactions (Camps et al., 1992; Carozzi et al., 1993; Katz et al., 1992). These two G-protein mechanisms appear to define pertussis toxin insensitive (Gqα) and sensitive (Gβγ) coupling, respectively (Katz et al., 1992).

It appears then that there are multiple inputs into the simplistic pathway out-lined in Figure 1.1. Given that individual agonists triggering a PtdIns response can have distinct effects even upon the same cell, it is evident that the timing and/or context of the response is critical to the cell's interpretation.

The PKC Family

The rather singular output from the PtdIns-PLC family is not retained at the level of PKC since it is now clear that there exists a large family of PKC-related gene

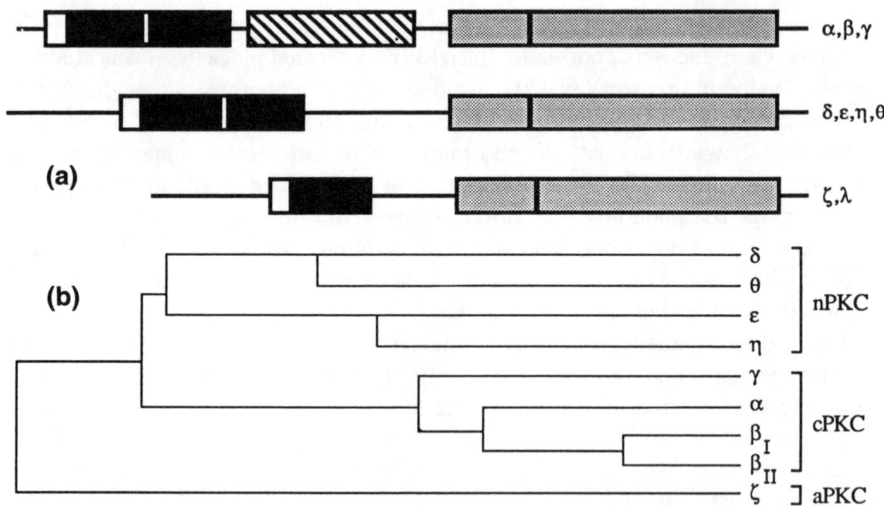

Figure 1.2 Structure and relationship of members of the PKC gene family.

(a) General domain structure: inhibitory pseudosubstrate site (open box), cysteine-rich effector binding domain (black box), Ca_2^+/phospholipid binding domain (hatched box), and c-terminal kinase domain (mottled box).

(b) Dendrogram showing the predicted evolutionary relationship between members. The sequences used for the alignment were: rat δ (Ono et al., 1988); murine θ (Osada et al., 1992); murine ε (Schaap et al., 1989); rat γ (Knopf et al., 1986); bovine α (Parker et al., 1986); human $β_I$ (Coussens et al., 1987); human $β_{II}$ (Coussens et al., 1986); rat ζ (Ono et al., 1989).

products (Figure 1.2). Within mammals there is clear conservation of PKC isotypes such that human PKC α is more closely related to bovine PKC α than it is to human PKC β. This suggests a conservation of selective functions that defies the more trivial explanation of redundancy. Many but not all of these PKC isotypes have been shown to be activated by DAG (see below) and as such are likely to respond to DAG *in vivo*. For these, specification of function probably relates to substrate specificity (see below).

Genes encoding PKC family members have been isolated from higher organisms (mammals) (see for example legend to Figure 1.2) to the simplest eukaryotes (yeast) (Levin et al., 1990; T. Toda, personal communication 1993) and there have even been reports of a prokaryotic PKC-like activity (Norris et al., 1991). The genes that have been identified fall broadly speaking into three groups, the calcium-dependent cPKCs, the calcium-independent nPKCs, and the atypical aPKCs (Nishizuka, 1992). All of the cDNAs isolated to date from yeast to humans appear to fall into these categories on the basis of domain structure and in many cases on biochemical evidence. Not surprisingly, when objective sequence alignments are carried out for these proteins, again they fall into three groups that correlate with the c, n, and aPKC nomenclature (see Figure 1.2)—the exceptions are the yeast PKC-1, pck1, and pck2 gene products, which all retain an extended aminoterminus and thus score poorly on such an alignment (not shown). The aPKC seem to have diverged earlier than the n and cPKC groups. This is reflected in functional differences (see below) and questions the assignment of this class of kinases as PKCs.

While the aPKC isotypes might be considered "outside" the PKC family, it is nevertheless pertinent to ask what the extent of PKC diversity is. In *Drosophila melanogaster,* three PKC genes have been identified to date at chromosomal locations 53B, 53E, and 98F (Rosenthal et al., 1987; Schaeffer et al., 1989). One of these, 53E, is expressed only in the adult compound eye and in no other tissue throughout development. Mutations in this gene clearly demonstrate a specialized role in light adaptation, consistent with this unique pattern of expression (Smith et al., 1991). It is likely that a distinct species of PKC is also expressed in mammalian retina, a prediction yet to be verified but for which some biochemical evidence has been presented (Fujisawa et al., 1992). Although this speculation remains just that, it is now established that distinct PKC species are highly expressed in tissues outside the central nervous system (the source of the first generation of PKC isotypes) and that an expanding number of PKC (-related) mammalian genes have now been identified (see legend to Figure 1.2; unpublished data). The extent to which these and other PKC isotypes are tissue/cell type restricted is not yet clear, but on current evidence it is quite possible that anyone's favorite tissue will express a novel PKC gene product and in the long run it is unlikely that the Greek alphabet will be able to cope!

GENERAL PERSPECTIVE

Extending the Signaling Pathway

The functional differences between members of the PKC gene family provide direct evidence that the context in which they can become activated is varied. This is most obvious in comparing the cPKC and nPKC subfamilies. Thus *in vitro* cPKCs require Ca^{2+} for full activity, whereas nPKCs are activated by DAG/ phospholipid alone (see reviews by Stabel and Parker, 1991; Nishizuka, 1992). In the classical pathway, the hydrolysis of PtdIns $4,5P_2$ *in vivo* produces both a Ca^{2+} stimulus [via inositol-trisphosphate (IP_3)] and DAG: both nPKCs and cPKCs would be expected to become activated. Indeed, this appears to be the case (e.g., Kiley et al., 1991). By contrast there is accumulating evidence that DAG from noninositol lipid sources (reviewed by Exton, 1990) can also activate PKC *in vivo* but that this is selective for the Ca^{2+}-independent nPKC species (Kiley et al., 1991; Olivier and Parker, in preparation). The implication is that DAG produced in the absence of a coincident rise in Ca^{2+} is "available" to certain PKCs and not to others. Whether in fact this is a spatial constraint on the PKC or is due to the dynamics of the membrane association event (i.e., the regulatory influence of Ca^{2+}) is yet to be resolved.

There is published evidence that fatty acids will activate PKC directly; it is not clear that this is a reversible, regulated phenomenon (discussed in Parker et al., 1989). However, like Ca^{2+}, fatty acids have been shown to act synergistically with DAG in the activation of PKC *in vitro* (Shinomura et al., 1991) and may thus serve a cooperative role *in vivo*. There are a number of possible sources of such fatty acids (e.g., phospholipase A_2, DAG-lipase); however, their role in the physiological regulation of PKC *in vivo* has yet to be established.

While the context of DAG may be important in controlling activation, at least one isotype (PKC ζ) appears insensitive to DAG (or phorbol esters) in a variety of contexts (Gschwendt et al., 1992; Ono et al., 1989; Ways et al., 1992). This enzyme does show lipid-dependent activity (Ono et al., 1989; Nakanishi and Exton, 1992) and has also been shown to be activated by PtdIns-3,4,5P$_3$ and PtdIns-3,4P$_2$ (Nakanishi et al., 1993). Activation of PKCs by inositol lipids has a significant history (see Chauhan et al., 1991), and although this might reflect an element of nonspecificity (with respect to isotypes), the activation of PKC ζ by the polyphosphorylated isomer PtdIns-3,4,5P$_3$ appears to be a high-affinity event ($K_a \sim 50$ nM) suggesting a specific effect (Nakanishi et al., 1993). If PKC ζ turns out to represent one of the primary targets for this recently discovered and assumed bioactive lipid, PtdIns-3,4,5P$_3$ (reviewed by Downes and Carter, 1991; Parker and Waterfield, 1992), it would provide an intriguing bifurcation of signaling events operating on the common precursor PtdIns-4,5P$_2$ (Figure 1.3).

The role of distinct lipid products in the activation of PKC isotypes substantially muddies the hitherto clear waters of DAG-dependent PKC activation. Whereas the evidence for the dependence of physiological responses on DAG accumulation is well supported by a wealth of data (see above), the same is not (yet) true for fatty acid contributions to DAG activation; the *in vivo* evidence on PtdIns-3,4,5P$_3$ is to date nonexistent. Thus the actual physiological contributions of these effectors has yet to be established, and this is an area of some importance in establishing PKC involvement in the context of particular signaling scenarios.

Establishing Roles for PKC Isotypes

There is a significant problem in determining the action or inaction of PKC, specifically with respect to particular PKC gene products. While evidence of substrate phosphorylation is indicative, as yet no physiological substrates substantiate the action of a specific PKC isotype. In this situation the only effective attack on the problem is an immunological one. Studies on the expression and subcellular distribution of individual PKC isotypes provide evidence for selective regulation. In its simplest form this can be described as an increase in the membrane association of a particular isotype (Kiley et al., 1991). At higher resolution, immunofluorescence can provide a more rigorous appraisal of distribution (e.g., Ito et al., 1991)—although this may not necessarily reflect activation.[2]

The subcellular distribution of specific PKC isotypes revealed by immunofluorescence indicates distinct localizations within a particular cell type (Kiley, unpublished). How this is achieved/maintained physiologically (sorting processes/binding proteins; see Mochlyrosen et al., 1991) and how it relates to the ability of particular PKC species to perceive DAG (or other lipid effectors) produced at specific locations have yet to be established.

It is quite possible that generalizations with respect to localization will not be valid, but that each cellular context will display its own idiosyncrasies. It is per-

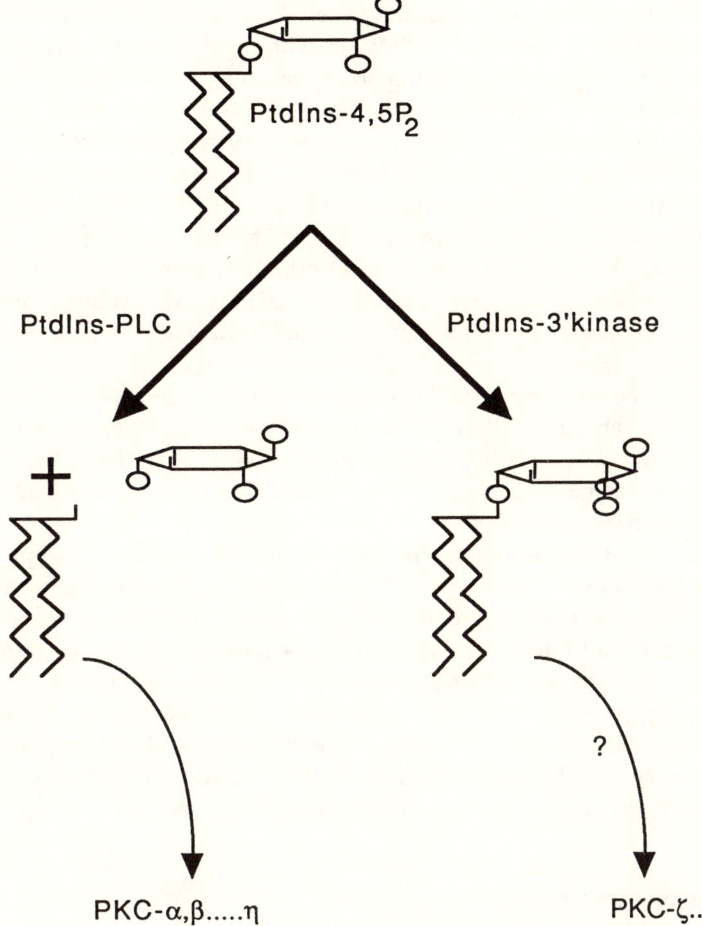

Figure 1.3 Bifurcation at PtdIns-4,5P$_2$. The polyphosphoinositide PtdIns-4,5P$_2$ is shown undergoing one of two fates through the action of a phospholipase C (PtdIns-PLC) or an inositol lipid kinase specific for the 3′-OH position (PtdIns-3′kinase). The products DAG and PtdIns-3,4,5P$_3$ may be involved in the selective activation of distinct PKC species (see text).

tinent that chapters within this volume deal with specific biological problems and do not apply too broad a brush to these issues.

Effector-Independent Control of PKC

There are two aspects of the control of PKC that are not enmeshed in effector dependence—namely, transcriptional regulation and phosphorylation. The control of PKC gene expression remains poorly understood; only limited information is available on the acute or long-term regulation of expression. Similarly, the promoters of only two PKC genes have been described to date (Chen et al., 1990; Niino et al., 1992; Obeid et al., 1992). Notwithstanding this paucity of information, it would seem that this is an important element in the overall control of PKC action with respect to both developmental changes in expression (e.g., the postnatal PKC-γ up-regulation in the central nervous system; Hashimoto et al., 1988) and more acute agonist-induced changes (e.g., induction of PKC-α on differentiation of HL60 cells; Edashige et al., 1992).

Phosphorylation of PKC-α has been shown to be a necessary event for its latent function—that is, in an unphosphorylated form there is no effector-dependent (or -independent) kinase activity (Pears et al., 1992). The region necessary for this phosphorylation has been mapped recently, and in view of its conservation, phosphorylation is likely to be necessary for all PKC isotypes (Cazaubon and Parker, submitted). There is circumstantial evidence that this ''permissive'' phosphorylation is not carried out through an autophosphorylation mechanism but that another protein kinase is responsible (Pears et al., 1992). The identity of this kinase and the acute/developmental regulation of this permissive phosphorylation have not yet been established.

The regulation of transcription/translation of PKC genes and the phosphorylation of PKC are in a sense necessary permissive events in priming a cell for responsiveness to effector production. The extent to which either of these is acutely regulated to alter the nature of a physiological response is not known. However, there is a significant literature describing acute changes in extractable PKC activity that may well reflect alterations in steady-state level (transcription/translation) and/or in specific activity (altered phosphorylation?). A number of these observations show rapid alterations (e.g., Salari et al., 1990) that may be more readily accounted for by a posttranslational mechanism.

FUTURE PERSPECTIVES

Even leaving aside transcription, translation, and posttranslational modification, the physiological activation of PKC isotypes remains a complex problem. Although it is clear that exogenously added membrane-permeant DAG can activate a number of isotypes to produce cellular responses, the necessary penetrating properties and extracellular origin of these DAGs mean that the compartmentalized aspect of agonist-induced DAG production is lost. Furthermore, even though such exogenous DAG is metabolized, the length of an induced activation may not reflect the acute production of DAG induced by agonists, and indeed the concentrations routinely employed with exogenous DAG may override the need for other associated changes, such as elevated Ca^{2+}, that are relevant to physiological responses. These problems are generally accepted in the use of the poorly metabolized cell-penetrating phorbol esters, but are only now being appreciated in the use of permeant DAGs.

In attempting to come to a fully integrated understanding of PKC activation it will be necessary to define the location of each PKC isotype and likewise the site(s) at which agonists induce DAG (and other effectors) production. Progress on the former issue has clearly been made through the use of isotype-specific antibodies; the latter problem will be less tractable. It is possible that a direct immunological assessment of activation could be made through the use of conformationally dependent monoclonal antibodies. These have been described for PKC γ where antibody binding has been shown to be inversely correlated with activation *in vitro* (Cazaubon et al., 1990). Whether this specificity can be preserved on fixation for an immunofluorescent analysis of activation status is not known, but would prove invaluable in a comprehensive analysis of PKC activation.

It is implicit in the above that activation of PKC leads to the phosphorylation of specific target proteins. In fact, many proteins will serve as substrates for PKC *in vitro,* but only a few have been clearly established *in vivo* (see review by Woodgett et al., 1987). Of these, a number are membrane-associated proteins including receptors, c-src, and the cortical cytoskeleton-associated myristoylated alanine-rich C kinase substrate (MARCKS) protein. The localization of these proteins may be important in their ability to serve as PKC substrates, although mutation of both c-src and MARCKS to block myristolylation and membrane association has an inhibitory effect or no effect on PKC-dependent phosphorylation (Buss et al., 1986; Graff et al., 1989). This suggests that membrane localization of potential substrates is not a necessity and implies that the diffusability of some substrates may influence susceptibility to phosphorylation. The existence of PKC-controlled kinase cascades may imply indirect action in some instances, as clearly established for S6 phosphorylation via S6 kinase (see Kozma et al., 1990). This has an interesting compartmental consequence to it in that active PKC is apparently membrane associated; phosphorylation of other diffusible kinases provides a strategy for the propagation of signals through the cell. Elucidation of the spectrum of direct PKC substrates in a cell and the ability of expressed PKC isotypes to phosphorylate them are critical to the ultimate biochemical definition of these signaling pathways.

NOTES

1. There are forms of PKC that are tightly associated with the particulate fraction prior to agonist stimulation and indeed forms that are not neutral detergent extractable (i.e., membrane associated) both before and following agonist stimulation. However, a discussion of these is beyond the scope of this chapter.

2. The ''translocation'' empirically observed for activated PKC is discussed in the text and in effect represents a stabilization event with respect to extraction. Fixation and permeabilization for immunofluorescence may provide evidence of translocation; however, in principle an individual PKC may be membrane associated but inactive prior to agonist action and may not alter significantly as a consequence of agonist action.

REFERENCES

Buss, J. E., M. P. Kamps, K. Gould, and B. M. Sefton. 1986. The absence of myristic acid decreases membrane binding of p60[src] but does not affect tyrosine kinase activity. *J. Virol.* 58:468–474.

Camps, M., A. Carozzi, P. Schnabel, A. Scheer, P. J. Parker, and P. Gierschik. 1992. Isozyme-selective stimulation of phospholipase C-β_2 by G-protein γ-subunits. *Nature* 360:684–686.

Carozzi, A., M. Camps, P. Gierschik, and P. Parker. 1993. Activation of phosphatidylinositol lipid specific phospholipase C-β_3 by G-protein $\beta\gamma$ subunits. *FEBS Lett.* 315: 340–342.

Castagna, M., Y. Takai, K. Kaibuchi, K. Sano, U. Kikkawa, and Y. Nishizuka. 1982. Direct activation of calcium-activated, phospholipid-dependent protein kinase by tumor-promoting phorbol esters. *J. Biol. Chem.* 257:7847–7851.

Cazaubon, S., C. Webster, L. Camoin, A. D. Strosberg, and P. J. Parker. 1990. Effector dependent conformational changes in protein kinase Cγ through epitope mapping with inhibitory monoclonal antibodies. *Eur. J. Biochem.* 194:799–804.

Chauhan, A. H. Brockerhoff, H. M. Wisniewski, and V. P. S. Chauhan. 1991. Interaction of protein-kinase-C with phosphoinositides. *Arch. Biochem. Biophys.* 287:283–287.

Chen, K.-H., S. Widen, S. G. Wilson, and K. -P. Huang. 1990. Characterization of the 5'-flanking region of the rat protein kinase Cγ gene. *J. Biol. Chem.* 265:19961–19965.

Coussens, L., P. J. Parker, L. Rhee, T. L. Yang-Feng, E. Chen, M. D. Waterfield, U. Francke, and A. Ullrich. 1986. Multiple, distinct forms of bovine and human protein kinase C suggest diversity in cellular signalling pathways. *Science* 233:859–866.

Coussens, L., L. Rhee, P. J. Parker, and A. Ullrich. 1987. Alternative splicing increases the diversity of the human protein kinase C family. *DNA* 6:389–394.

Downes, C. P., and A. N. Carter. 1991. Phosphoinositide 3-kinase: a new effector in signal transduction? *Cell. Signal.* 3:501–513.

Edashige, K., E. F. Sato, K. Akimaru, M. Kasai, and K. Utsumi. 1992. Differentiation of HL-60 cells by phorbol ester is correlated with upregulation of protein kinase C-α. *Arch. Biochem. Biophys.* 299:200–205.

Exton, J. 1990. Signalling through phosphatidylcholine breakdown. *J. Biol. Chem.* 265: 1–4.

Fujisawa, N., K. Ogita, N. Saito, and Y. Nishizuka. 1992. Expression of protein kinase C subspecies in rat retina. *FEBS Lett.* 309:409–412.

Graff, J. M., J. I. Gordon, and P. J. Blackshear. 1989. Myristoylated and nonmyristoylated forms of a protein are phosphorylated by protein kinase C. *Science* 246:503–506.

Gschwendt, M., H. Leibersperger, W. Kittstein, and F. Marks. 1992. Protein kinase Cζ and η in murine epidermis. *FEBS Lett.* 307:151–155.

Hanson, P. I., and H. Schulman. 1992. Neuronal Ca^{2+}/calmodulin-dependent protein kinases. *Annu. Rev. Biochem.* 61:559–601.

Hashimoto, T., K. Ase, S. Sawamura, U. Kikkawa, N. Saito, C. Tanaka, and Y. Nishizuka. 1988. Postnatal development of a brain-specific subspecies of protein kinase C in rat. *J. Neurosci.* 8:1678–1683.

Hokin, L. E. 1985. Receptors and phosphoinositide-generated second messengers. *Annu. Rev. Biochem.* 54:205–236.

Hokin, L. E., and M. R. Hokin. 1958. Phosphoinositides and protein secretion in pancreas slices. *J. Biol. Chem.* 233:805–810.

Inoue, M., A. Kishimoto, Y. Takai, and Y. Nishizuka. 1977. Studies on a cyclic nucleotide-independent protein kinase and its proenzyme in mammalian tissues II. *J. Biol. Chem.* 252:7610–7616.

Ito, T. O. Kozawa, H. Hidaka, and H. Saito. 1991. Thrombin-induced translocation of protein-kinase-C in human megakaryoblastic leukemia-cells (meg-01). *Arch. Biochem. Biophys.* 291:218–224.

Kaibuchi, K., Y. Takai, M. Sawamura, M. Hoshijima, T. Fujikura, and Y. Nishizuka. 1983. Synergistic functions of protein phosphorylation and calcium mobilisation in platelet activation. *J. Biol. Chem.* 258:6701–6704.

Katz, A., D. Q. Wu, and M. I. Simon. 1992. Subunits-βγ of heterotrimeric g-protein activate β₂ isoform of phospholipase-c. *Nature* 360:686–689.

Kawahara, Y., Y. Takai, R. Minakuchi, K. Sano, and Y. Nishizuka. 1980. Phospholipid turnover as a possible transmembrane signal for protein phosphorylation during human platelet activation by thrombin. *Biochem. Biophys. Res. Commun.* 97:309–317.

Kiley, S., P. J. Parker, D. Fabbro, and S. Jaken. 1991. Differential regulation of protein kinase C isozymes by thyrotropin-releasing hormone in GH$_4$C$_1$ cells. *J. Biol. Chem.* 266:23761–23768.

Knighton, DR, JH Zheng, LF Teneyck, VA Ashford, NH Xuong, SS Taylor and JM Sowadski. 1991. Crystal-structure of the catalytic subunit of cyclic adenosine-monophosphate dependent protein-kinase. *Science* 253:407–414.

Knopf, J. L., M.-H. Lee, L. A. Sultzman, R. W. Kriz, C. R. Loomis, R. M. Hewick, and R. M. Bell. 1986. Cloning and expression of multiple protein kinase C cDNAs. *Cell* 46:491–502.

Kozma, S., S. Ferrari, P. Bassand, M. Siegmann, N. Totty, and G. Thomas. 1990. Cloning of the mitogen-activated S6 kinase from rat liver reveals an enzyme of the second messenger subfamily. *Proc. Natl. Acad. Sci. U.S.A.* 87:7365–7369.

Levin, D. E., F. O. Fields, R. Kunisawa, J. M. Bishop, and J. Thorner. 1990. A candidate protein kinase C gene, PKC1, is required for the *S. cerevisiae* cell cycle. *Cell* 62: 213–224.

Lincoln, T. M., and J. D. Corbin. 1983. Characterization and biological role of the cGMP-dependent protein kinase. *Adv. Cyclic. Nucl. Res.* 15:139–192.

Meldrum, E., P. J. Parker, and A. Carozzi. 1991. The PtdIns-PLC superfamily and signal transduction. *Biochim. Biophys. Acta* 1092:49–71.

Michell, R. H. 1975. Inositol phospholipids and cell surface receptor function. *Biochim. Biophys. Acta* 415:81–147.

Mochlyrosen, D., H. Khaner, and J. Lopez. 1991. Identification of intracellular receptor proteins for activated protein-kinase-C. *Proc. Natl. Acad. Sci. U.S.A.* 88:3997–4000.

Nakanishi, H., K. A. Brewer, and J. H. Exton. 1993. Activation of the ζ isozyme of protein kinase C by phosphatidylinositol 3,4,5-trisphosphate. *J. Biol. Chem.* 266:13–16.

Nakanishi, H., and J. H. Exton. 1992. Purification and characterization of the ζ isoform of protein kinase C from bovine kidney. *J. Biol. Chem.* 267:16347–16354.

Niino, Y. S., S. Ohno, and K. Suzuki. 1992. Positive and negative regulation of the transcription of the human protein kinase C β gene. *J. Biol. Chem.* 267:6158–6163.

Nishizuka, Y. 1984. The role of protein kinase C in cell surface signal transduction and tumor promotion. *Nature* 308:693–695.

Nishizuka, Y. 1992. Intracellular signalling by hydrolysis of phospholipids and activation of protein kinase C. *Science* 258:607–614.

Norris, V., T. J. Baldwin, S. T. Sweeney, P. H. Williams, and K. L. Leach. 1991. A protein kinase-C-like activity in *Escherichia coli. Molec. Microbiol.* 5:2977–2981.

Obeid, L. M., G. C. Blobe, L. A. Karolak, and Y. A. Hannun. 1992. Cloning and characterization of the major promoter of the human protein kinase C β gene. *J. Biol. Chem.* 267:20804–20810.

Ono, Y., T. Fujii, K. Ogita, U. Kikkawa, K. Igarashi, and Y. Nishizuka. 1989. Protein kinase C ζ subspecies from rat brain: Its structure, expression and properties. *Proc. Natl. Acad. Sci. U.S.A.* 86:3099–3103.

Osada, S.-I., K. Mizuno, T. C. Saido, K. Suzuki, T. Kuroki, and S. Ohno. 1992. A new member of the protein kinase C family, nPKCθ, predominantly expressed in skeletal muscle. *Molec. Cell. Biol.* 12:3930–3938.

Parker, P. J., and M. D. Waterfield. 1992. Phosphatidylinositol 3-kinase: A novel effector. *Cell Growth Diff.* 3:747–752.

Parker, P. J., L. Coussens, N. Totty, L. Rhee, S. Young, E. Chen, S. Stabel, M. D. Waterfield, and A. Ullrich. 1986. The complete primary structure of protein kinase C—the major phorbol ester receptor. *Science* 233:853–859.

Parker, P. J., G. Kour, R. M. Marais, F. Mitchell, C. J. Pears, D. Schaap, S. Stabel, and C. Webster. 1989. Protein kinase C—a family affair. *Molec. Cell. Endocrinol.* 65: 1–11.

Pears, C., S. Stabel, S. Cazaubon, and P. J. Parker. 1992. Studies on the phosphorylation of protein kinase C-α. *Biochem. J.* 283:515–518.

Rosenthal, A., L. Rhee, R. Yadegari, R. Paro, A. Ullrich, and D. V. Goeddel. 1987. Structure and nucleotide sequence of a *Drosophila melanogaster* protein kinase C gene. *EMBO J.* 6:433–441.

Salari, H., V. Duronio, S. Howard, M. Demos, and S. L. Pelech. 1990. Translocation-independent activation of protein kinase C by platelet-activating factor, thrombin and prostacyclin. *Biochem. J.* 267:689–626.

Sano, K., Y. Takai, J. Yamanishi, and Y. Nishizuka. 1983. A role of calcium-activated phospholipid-dependent protein kinase in human platelet activation. *J. Biol. Chem.* 258:2010–2013.

Schaap, D., P. J. Parker, A. Bristol, R. Kriz and J. Knopf. 1989. Unique substrate specificity and regulatory properties of PKC-ε: A rationale for diversity. *FEBS Lett.* 243:351–357.

Schaeffer, E., D. Smith, G. Mardon, W. Quinn, and C. Zuker. 1989. Isolation and characterization of two new *Drosophila* protein kinase C genes, including one specifically expressed in photoreceptor cells. *Cell* 57:403–412.

Shinomura, T., Y. Asaoka, M. Oka, K. Yoshida, and Y. Nishizuka. 1991. Synergistic action of diacylglycerol and unsaturated fatty acid for protein kinase C activation: Its possible implications. *Proc. Natl. Acad. Sci. U.S.A.* 88:5149–5153.

Smith, D. P., R. Ranganathan, R. W. Hardy, J. Marx, T. Tsuchida, and C. S. Zuker. 1991. Photoreceptor deactivation and retinal degeneration mediated by a photoreceptor-specific protein kinase C. *Science* 254:1478–1484.

Stabel, S., and P. J. Parker. 1991. Protein kinase C. *Pharm. Therap.* 51:71–95.

Takai, Y., A. Kishimoto, M. Inoue, and Y. Nishizuka. 1977. Studies on a cyclic nucleotide-independent protein kinase and its proenzyme in mammalian tissues I. Purification and characterization of an active enzyme from bovine cerebellum. *J. Biol. Chem.* 252:7603–7609.

Takai, Y., A. Kishimoto, U. Kikkawa, T. Mori, and Y. Nishizuka. 1979. Unsaturated diacylglycerol as a possible messenger for the activation of calcium-activated, phospholipid-dependent protein kinase system. *Biochem. Biophys. Res. Commun.* 91:1218–1224.

Taylor, S. J., and J. H. Exton. 1991. Two α subunits of the G_q class of G proteins stimulate phosphoinositide phospholipase C-β_1 activity. *FEBS Lett.* 286:214–216.

Taylor, S. J., H. Z. Chae, S. G. Rhee, and J. H. Exton. 1991. Activation of the β_1 isozyme of phospholipase C by α subunits of the G_q class of G proteins. *Nature* 350:516–518.

Wahl, M., and G. Carpenter. 1992. Selective phospholipase C activation. *Bioassays* 13:107–113.

Ways, D. K., P. P. Cook, C. Webster, and P. J. Parker. 1992. Effect of phorbol esters on KC-ζ. *J. Biol. Chem.* 267:4799–4805.

Woodgett, J. R., T. Hunter, and K. L. Gould. 1987. Protein kinase C and its role in cell growth. In: *Cell Membranes: Methods and Reviews,* edited by E. L. Elson, W. A. Frazier, and L. Glaser. Plenum, New York, pp. 215–340.

2

Molecular and Catalytic Properties of Protein Kinase C

CHARLES W. MAHONEY
KUO-PING HUANG

Members of the Ca^{2+}/phospholipid/diacylglycerol-dependent serine/threonine protein kinase (PKC) gene family are highly conserved during evolution. These enzymes are believed to be involved in the signaling pathway that utilizes second messengers generated from the turnover of membrane phospholipids. Extracellular ligands, including hormones, growth factors, neurotransmitters, and antigens, bind to their respective cell surface receptors and, through a coupling to G-proteins and activation of the various phospholipases, a plethora of metabolites are generated. Many of these, such as diacylglycerol (DAG), inositol-trisphosphate (IP_3), lysophospholipids, free fatty acids, and phosphatidate, can affect PKC activity directly or indirectly by altering $[Ca^{2+}]_i$ or the local membrane environment. The stimulated PKCs phosphorylate the target substrates at discrete cellular locations, which results in cellular responses. Frequently, phosphorylation of certain PKC substrates further amplifies the input signal by affecting other signaling pathways. In addition to the phosphorylation-mediated responses, physical association of the activated PKCs with other proteins can potentially regulate certain cellular functions. Members of this enzyme family, at least ten isozymes from recent count, are believed to play important roles in signaling mechanisms involved in both normal physiological responses and pathological development (K.-P. Huang 1989, 1990; Nishizuka, 1986, 1988, 1992). However, in spite of a broad implication of PKCs in cellular regulation derived from studies with pharmacological agents, detailed mechanisms for the functioning of each of these enzymes *in vivo* are largely unknown.

MOLECULAR STRUCTURE OF PKC FAMILY MEMBERS

The mammalian PKC enzyme family consists of at least ten members, α, βI, βII, γ, δ, ϵ, ζ, η, θ, and λ, coded for by nine genes (for review see Nishizuka, 1992).

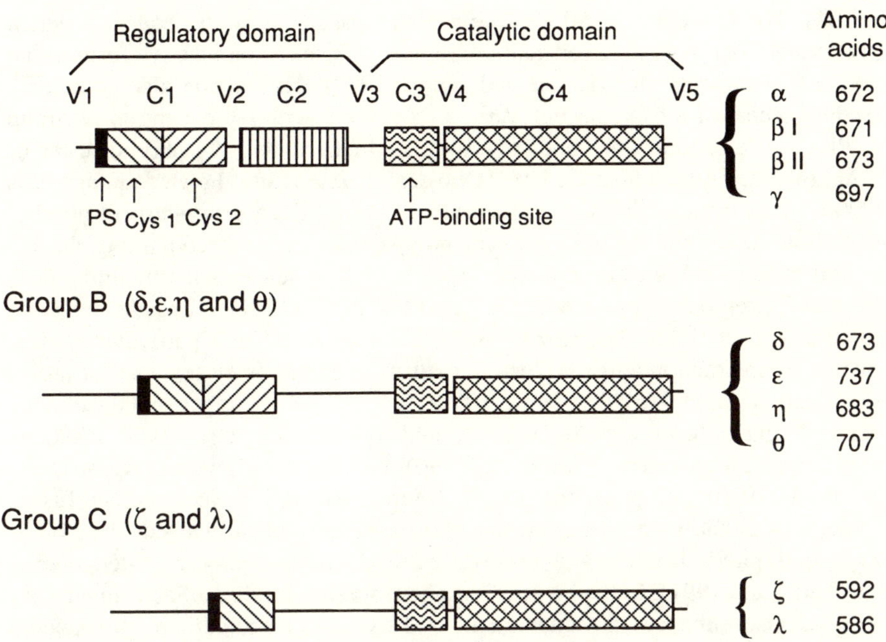

Figure 2.1 Schematic structure of PKC members. Group A (conventional) PKCs contain conserved C1–C4 and variable V1–V5 regions, two Cys-rich Zn^{2+} fingers, and C2, the putative Ca^{2+}-binding domain. Group B (novel) PKCs also contain two Cys-rich fingers yet lack C2, and Group C (atypical) PKCs have a single Cys-rich finger and lack C2. All members contain pseudosubstrate (PS) and C3/C4 (catalytic) domains. Group B PKCs contain an extended N-terminal V1 domain that restricts substrate specificity.

PKC βI and βII are derived from a single gene by alternative splicing of the 3′ exons (Ono et al., 1987). Additional PKC family members resulting from alternative splicing of these genes are likely to be uncovered. These enzymes can be subdivided into group A (cPKC) (α, βI, βII, and γ), group B (nPKC) (δ, ε, η, and θ), and group C (aPKC) (ζ and λ) on the basis of their common structural features (Figure 2.1). The genes for PKC α, β, and γ are known to reside on different human chromosomes. All these enzymes consist of a single polypeptide chain having several regions of highly conserved sequences. The carboxy-terminal halves of all these PKC molecules are the catalytic domains containing the ATP- and substrate-binding sites, and the amino-terminal halves are the regulatory domains containing the Ca^{2+}, Zn^{2+}, phospholipid, and DAG/phorbol ester-binding sites. The groups B and C PKCs are different from those of group A in the regulatory domain in that a putative Ca^{2+}-binding C2 region is absent; these two groups of enzymes are insensitive to Ca^{2+}. The group C PKCs, in contrast to groups A and B PKCs, have a single cysteine-rich Zn^{2+} finger motif in the C1 region.

The primary structure of the PKCs, on the basis of amino acid sequence homology, consists of variable regions, V1–V5, interspersed by the conserved domains, C1–C4. The C1 region of all PKCs contains a pseudosubstrate sequence, which presumably is responsible for maintaining the enzyme in an inactive form in the absence of the activator (House and Kemp, 1987). This region also contains a tandem repeat of a cysteine-rich Zn^{2+} finger motif, with the exception of group C PKCs that contain only one such motif, which confers the phorbol ester or DAG binding (Kaibuchi et al., 1989; Ono et al., 1989a). Site-directed mutagenesis of two of the cysteine residues to serine in rat brain PKC γ causes loss of phorbol ester-binding activity. A recombinant polypeptide fragment containing the C1 region binds phorbol ester in the presence of (PS), whereas that containing both C1 and C2 regions requires both Ca^{2+} and PS. Purified PKC βI has been shown to contain four tightly bound Zn^{2+} ions that may stabilize a particular conformation by interaction with one histidine nitrogen and three cysteine sulfur atoms (Hubbard et al., 1991; Quest et al., 1992). A similar Zn^{2+} finger sequence motif is also found in the protooncogenes c-*raf* and A-*raf* (Bonner et al., 1986; Ishikawa et al., 1986), n-chimaerin (Hall et al., 1990), 80-kDa diacylglycerol kinase (Sakane et al., 1990), and phospholipase A_2 (Maragnore, 1987). Recombinant fusion proteins of glutathione S-transferase and the cysteine-rich domains of PKC binds Zn^{2+} and phorbol ester, and removal of Zn^{2+} inhibit phorbol ester-binding (Ahmed et al., 1991). In the presence of phorbol ester, binding of additional Zn^{2+} to PKC also enhances the attachment of PKC to the membrane cytoskeleton (Zalewski et al., 1990). The cysteine-rich Zn^{2+} finger motif of PKC—$HX_{12}CX_2CX_{10-14}CX_2CX_4HX_2CX_7$ C (C_6H_2)—is similar to, yet apparently unique relative to, those of transcriptional factor IIIA (Berg, 1990) and GAL4 (Pan and Coleman, 1989), and human glucocorticoid (Luisi et al., 1991) and estrogen receptors (Schwabe and Rhodes, 1991). However, there is no evidence at present of sequence-specific DNA binding by PKC. Both cysteine-rich regions of PKC when expressed individually in insect cells exhibit [^{3}H]PDBu-binding activity with reduced affinity (Burns and Bell, 1991). However, binding of more than one phorbol ester per PKC has not been demonstrated and PKC ζ, which contains only one cysteine-rich region, is not activated by phorbol esters and does not translocate or down-regulate in response to phorbol esters or membrane-permeant DAG (Ways et al., 1992).

Zn^{2+} fingers, which have been found in over 200 proteins, appear to consist of at least three different classes; the transcription factor TFIIIA (C_2H_2 with three conserved hydrophobic residues), the estrogen hormone receptor type ($C_4 + C_4$), and the PKC type (C_6H_2) (Rhodes and Klug, 1993). TFIIIA contains nine similar C_2H_2 Zn^{2+} fingers, each of which appears to be able to bind in a base-specific manner to the major groove of DNA. X-ray crystallography of Zif268, a Zn^{2+} finger containing transcription factor in the TFIIIA class, indicates that the Zn^{2+} finger of this transcription factor consists of a β sheet hairpin containing the two Cys residues followed by an α helix containing the two His residues, with the four Cys residues being coordinated by a single Zn^{2+} molecule (Pavletich and Pabo, 1991). The three conserved hydrophobic residues of the Zif268 Zn^{2+} finger form a hydrophobic pocket at the tip of the finger between the β sheet hairpin and the α helix and presumably stabilize the Zn^{2+} finger. In contrast, the estrogen hormone receptor type of Zn^{2+} finger contains a tandem repeat of non-identical functioning Zn^{2+} fingers. In this case, the first Zn^{2+} finger has four conserved Cys residues that complex a single Zn^{2+} molecule and the sequence ZZCXZ at the fourth Cys, where Z residues are critical and X is not for DNA-binding specificity. The second Zn^{2+} finger of the estrogen receptor also has four conserved Cys residues that complex a single Zn^{2+} molecule, yet unlike the first finger, the residues surrounding the fourth Cys are not involved with DNA-binding specificity; amino acid residues between the first and second conserved Cys residues are. In addition, amino acid residues in the second finger are located at the dimerization interface (Luisi et al., 1991). The secondary structure of the tandemly repeated Zn^{2+} fingers of the estrogen receptor, in contrast to that of the TFIIIA type, consists of random coil containing the first two conserved Cys residues followed by an α helix containing the third and fourth conserved Cys residues (Schwabe and Rhodes, 1991). The second Zn^{2+} finger of the estrogen receptor, which is involved with dimerization, also appears to be important in providing the proper spacing between the two first DNA-binding Zn^{2+} fingers of the dimer to allow for occupancy of both half-sites on the DNA (Luisi et al., 1991). The PKC family appears to belong to a third class of Zn^{2+} fingers, each of which contains six Cys and two His residues that appear to bind two molecules of Zn^{2+} (Figure 2.2). Although there are no three-dimensional data available for the PKC-type Zn^{2+} fingers, differences between the phorbol ester-binding Zn^{2+} fingers of groups A and B PKCs, and between chimaerin and the non-phorbol ester-binding PKC ζ and DAG kinase, may be due in part to differences in hydrophobicity. The first Zn^{2+} finger of group A PKCs and chimaerin have a strongly hydrophobic region around the third and fourth conserved Cys residues. In contrast, the first and single Zn^{2+} fingers of the non-phorbol ester-binding DAG kinase and PKC ζ, respectively, are weakly hydrophobic in the third and fourth conserved Cys region and are strongly hydrophilic in the N-terminal region of the fingers (Figure 2.2). PKC δ, ϵ, η, and θ have reduced hydrophobicity at the third to fourth conserved Cys region of their first Zn^{2+} fingers relative to those of group A PKC and chimaerin. The second Zn^{2+} fingers of PKC α, β, and γ, and δ, ϵ, and η, are weakly hydrophobic to not hydrophobic at their third to fourth conserved Cys region, yet have moderate hydrophobicity at the mid-portion of their sequences. Because PKC deletion constructs containing one or the other Zn^{2+} finger, expressed in insect cells, bind [^{3}H]PDBu tightly (K_d = 21–41 nM), but 10- to 20-fold less than native enzyme (Burns and Bell, 1991), and because individual Zn^{2+} fingers from PKC can bind phorbol ester, its binding may be controlled by the overall hydrophobicity of each finger. The binding of a single phorbol ester molecule to native PKCs may sterically block a second phorbol ester binding site on the second Zn^{2+} finger. The possibility of having two nonequivalent second-messenger binding sites in groups A and B PKCs has been hypothesized to explain the differential effects of several PKC-activating tumor promoters (Kraft et al., 1988).

The C_2 region is believed to be the Ca^{2+}-binding domain because the groups B and C PKCs, which lack this region, are insensitive to Ca^{2+}. This region does

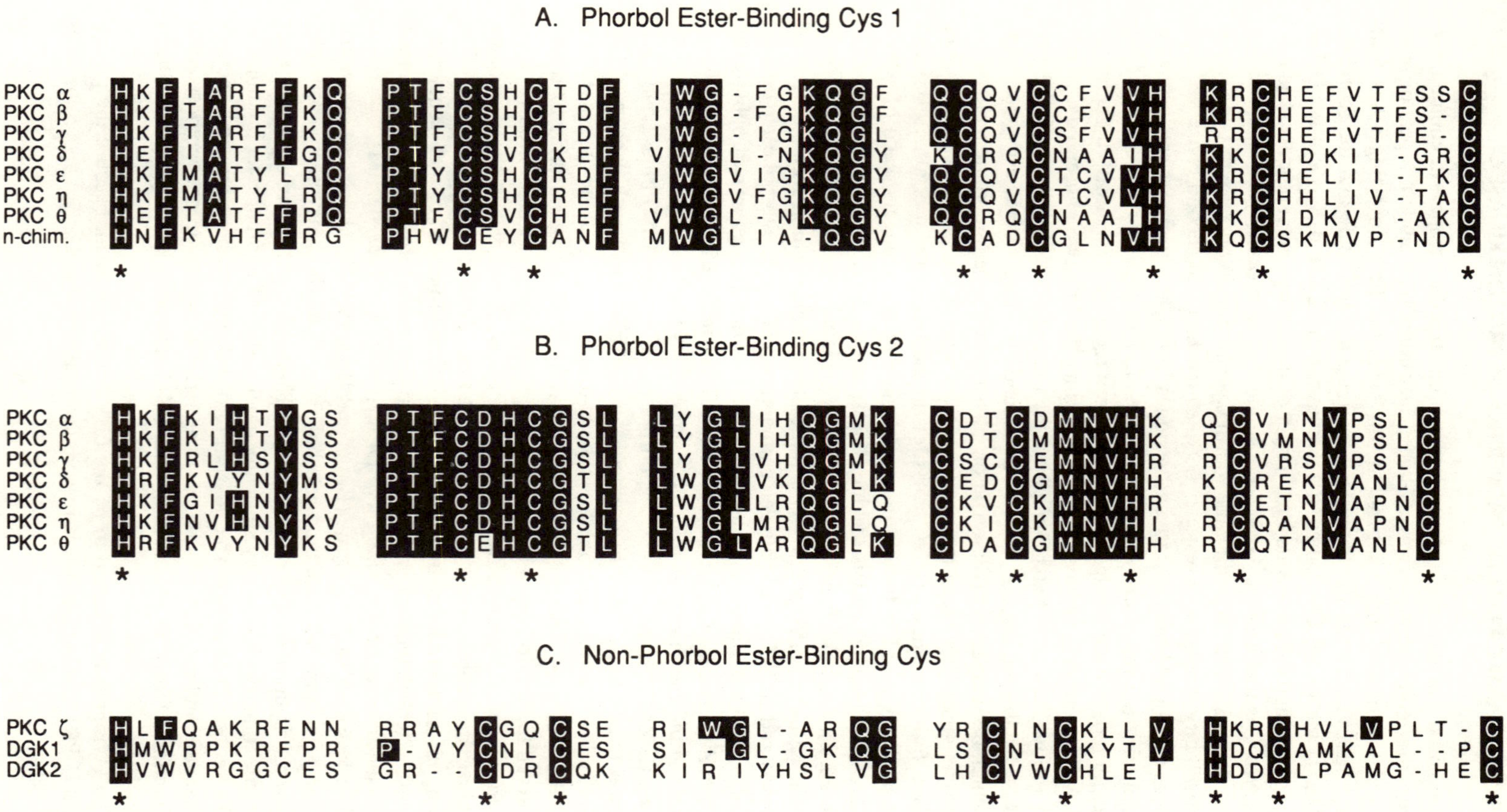

Figure 2.2 Amino acid sequence homology of Cys-rich Zn^{2+} fingers of PKC and non-phorbol ester-binding related proteins. Alignment of (A) phorbol ester-binding Cys 1; (B) phorbol ester-binding Cys 2; and (C) non-phorbol ester-binding Cys domains. Amino acid sequence identity is highlighted.

not contain a typical calmodulin Ca^{2+} binding site, and the high-affinity Ca^{2+} binding site might form through appropriately spaced amino acid side chains or through coordination with bound PS (Bell and Burns, 1991). Sequence segments homologous to the PKC C2 region have also been identified in a synaptic vesicle-specific Ca^{2+}/PS-binding protein, synaptotagmin (p65) (Perin et al., 1990), cytosolic phospholipase A_2 (Clark et al., 1991), phospolipase C γ (Stahl et al., 1988), and GAP (Vogel et al., 1988). A sequence motif for these Ca^{2+}-binding proteins corresponding to PKC C2 is shown in Figure 2.3 (cf. Clark et al., 1991). The C3 region is the putative ATP-binding domain that includes a Gly X Gly XX Gly X_{16}Lys consensus sequence for ATP binding and C4 is part of the catalytic domain that may participate in the recognition of the substrate.

The V1 region of group A PKCs is relatively short compared with that of group B and C enzymes; for the latter groups, the V1 region may confer substrate specificity of these enzymes. The hinge region (V3) between the catalytic and regulatory domains contains the protease-sensitive sites, which on cleavage by Ca^{2+}-activated calpain or by trypsin generate a phorbol ester-binding and a protein kinase fragment. This region appears to be fairly flexible to permit dislodging of the pseudosubstrate region from the catalytic domain on binding of the activators. In addition, these enzymes undergo intramolecular autophosphorylation at three well separated regions located in both the catalytic and regulatory domains of PKC (K.-P. Huang et al., 1986a; Flint et al., 1990). It has also been suggested that the V3 and C3 regions of the various PKCs may interact with specific receptors, thus targeting the activated enzymes to specific locations within the cells (James and Olson, 1992). Selective translocation or localization of PKCs may provide a mean of regulating substrate specificities and differential susceptibilities to proteolysis.

TISSUE, CELLULAR, AND SUBCELLULAR DISTRIBUTION OF PKCs

PKC isozymes show distinct tissue, cellular, and subcellular distribution, and frequently more than one isoform is expressed in a cell type. PKC α, β, γ, δ, ϵ, and ζ are highly expressed in brain (Nishizuka, 1988). PKC γ is exclusively expressed in brain and spinal cord (Yoshida et al., 1988; Nishizuka, 1988, 1992); PKC η, θ, and λ are most abundant in skin and lung (Osada et al., 1990; Bacher et al., 1991), skeletal muscle (Osada et al., 1992), and ovary and testis (Akimoto et al., in preparation; see Nishizuka, 1992), respectively. Because of the preferential abundance of PKC γ, η, θ, and λ in their respective tissues, it is expected that these isoforms should have tissue-specific functions. In contrast, PKC α, δ, and ζ are expressed in most tissues. This ubiquitous expression suggests that all three groups are essential to general cell function.

The various PKCs are differentially expressed in various regions and cell types of the brain. Immunological analysis with isozyme-specific antibodies indicates that in cerebellum PKC γ predominates in the dendrites, cell bodies, and axons of Purkinje cells. Hence it is likely that PKC γ is of importance in both presynaptic and postsynaptic functions (F. L. Huang et al., 1987; Kose et al., 1988). PKC β is localized mainly in cerebellar granule cells and PKC α in granule and Purkinje

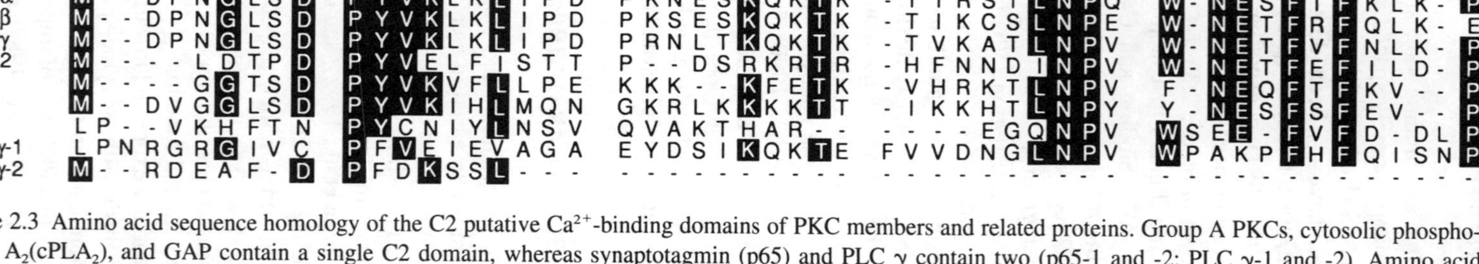

Figure 2.3 Amino acid sequence homology of the C2 putative Ca^{2+}-binding domains of PKC members and related proteins. Group A PKCs, cytosolic phospholipase A$_2$(cPLA$_2$), and GAP contain a single C2 domain, whereas synaptotagmin (p65) and PLC γ contain two (p65-1 and -2; PLC γ-1 and -2). Amino acid sequence identity is highlighted.

cells. Most other neurons in the brain contain PKC α, β, and γ. Differential distributions of PKC βI and βII in rat brain (Hosoda et al., 1989). PKC α, β, and γ in the visual pathway of monkey brain (F. L. Huang et al., 1989b) and in the neurons of rat striatum and substantia nigra (Yoshihara et al., 1991) are evident. *In situ* hybridization histochemical analysis with probes against PKC α, β, γ, and ϵ also reveals distinctive expression patterns of these respective mRNAs in the various brain regions (Young, 1988). In addition to the prominent presence of these PKCs in the neurons of CNS, glial cells also express selective types of PKCs, in particular PKC α and βII (Masliah et al., 1991).

Various PKC subtypes have been found to be associated with the nucleus, some of which can translocate to this organelle depending on the cell type and activator. PKC β has been found in liver nuclei (Rogue et al., 1990), PKC α and β in neutrophil nuclei (Perletti et al., 1991), PKC α and γ in brain nuclei (Buchner et al., 1992), and PKC η exclusively in nuclei (Greif et al., 1992). Phorbol esters translocate PKC α to 3T3 fibroblast nuclei (Thomas et al., 1988; Leach et al., 1989) and to multidrug-resistant breast carcinoma nuclei (Lee et al., 1992), whereas PKC β is induced to translocate to nuclei in HL60 cells by this agent (Hocevar and Fields, 1991). Ia-binding ligands and cAMP translocate PKC to B cell nuclei (Cambier et al., 1987) and α-thrombin treatment of fibroblasts translocates PKC α to the nucleus (Leach et al., 1992). IGF-1 treatment of 3T3 fibroblasts translocates group A PKC to the nucleus (Divecha et al., 1991). Constructed PKC α proteins consisting of deletion of the regulatory domain or deletion of C-terminal parts of the catalytic domain have been shown to localize in the nucleus of transfected COS-1 cells (James and Olsen, 1992). It was suggested that a nuclear targeting sequence in the hinge and N-terminal part of the catalytic domain is present in PKC α. Phorbol ester treatment of PKC α-transfected COS-1 cells resulted in translocation of PKC α to the nucleus, which suggests that the binding of activators to PKC induces a conformational change in the protein, thereby exposing the nuclear targeting sequence (James and Olsen, 1992). Alessenko and colleagues (1992) have demonstrated that during liver regeneration, nuclear PKC α levels declined, whereas nuclear PKC δ levels increased. The presence of both groups A and B PKCs in the nucleus suggests that both the Ca^{2+}-dependent and -independent activation of PKC in the nucleus may be critical for the regulation of gene expression.

With a commonly used method of extraction of PKCs with buffer-containing metal chelators, group A PKCs, in general, have higher levels in the cytosolic than in the membrane-associated fractions (Ashendel et al., 1983). On cell stimulation by a variety of extracellular ligands, the membrane-associated enzyme increases at the expense of the cytosolic one (Kraft et al., 1982; Kraft and Anderson, 1983). A rise in either internal Ca^{2+} or DAG or addition of phorbol ester can cause this cytosolic-membrane redistribution/translocation (Melloni et al., 1985; Wolf et al., 1985a; Gopalakrishna et al., 1986). In contrast, PKC δ is predominantly membrane associated in rat brain and 3Y1 cells (Mizuno et al., 1991; Ogita et al., 1992). PKC δ expressed in transfected COS-1 cells, however, is distributed about 50 percent cytosolic and about 50 percent membrane associated (Ogita et al., 1992). PKC ϵ and PKC ζ, similar to group A PKCs, are predominantly cytosolic in brain (Gschwendt et al., 1992; Koide et al., 1992; Saido et al., 1992a), PKC η, most abundant in skin and lung, is predominantly associated with the particulate fraction, like PKC δ (Gschwendt et al., 1992). PKC ζ, which is predominantly cytosolic, has a single Cys-rich Zn^{2+} finger, does not respond to PMA, and does not translocate or down-regulate (Greif et al., 1992; Gschwendt et al., 1992).

Group A PKCs have been found to associate with the cytoskeleton (Werth and Pastan, 1984; Jaken et al., 1989) and myofilament (Liu et al., 1989). PKC α in rat embryo fibroblasts is associated with focal contacts of the cytoskeleton and co-localizes with vinculin and talin (Jaken et al., 1989). Activation of PKC by phorbol ester treatment of REF cells results in actin filament depolymerization and reorganization of vinculin (Jaken et al., 1989). Many cytoskeletal proteins are PKC substrates, for example, vinculin (Werth and Pastan, 1984), filamin (Kawamoto and Hidaka, 1984), profilin (Hansson et al., 1988), desmin (Inagaki et al., 1988), MARCKS (Hartwig et al., 1992), and troponin T, troponin I, and C-protein associated with cardiac myofibrils (Venema and Kuo, 1993).

Thyrotropin-releasing hormone (TRH) treatment of GH_4C_1 pituitary cells results in translocation of PKC α, β, δ, and ε to the cytoskeletal fraction (Kiley et al., 1992). Stimulation of heart cells with α-adrenergic agonist results in the association of PKC with myofibrils (Mochly-Rosen et al., 1990). A 40-kDa PKC ε-related activity was associated with and phosphorylated cytokeratins 8 and 18 (CK8, CK18), cytoskeletal proteins found primarily in epithelial cells (Omary et al., 1992). A 40-kDa tryptic fragment of purified PKC ε bound purified CK8 and CK18, and the ε pseudosubstrate peptide inhibited phosphorylation of CK8/18 in immunoprecipitates (Omary et al., 1992). Peptide maps of CK8 and CK18 phosphorylated by either purified PKC ε or by associated kinase in immunoprecipitates were similar, and antibodies to the catalytic fragment of PKC ε and CK8/18 co-localized with use of immunofluorescent staining. These data indicate the presence of a constitutively active 40-kDa PKC ε in the cytoskeletal fraction of epithelial cells. PKC appears to be intimately involved in regulation of cytoskeletal function.

GENOMIC STRUCTURE AND REGULATION OF PKC EXPRESSION

PKC α and β expression in rat brain starts at fetus, whereas PKC γ in rat brain parallels synapse formation between granule and Purkinje cells and hence may have linked developmental regulatory mechanisms (F. L. Huang et al., 1990). In light of the fact that PKC can phosphorylate transcription factors and regulate gene expression (Abate et al., 1991; Yamamoto et al., 1988; Gonzalez et al., 1989; Sakurai et al., 1991; Mahoney et al., 1992), the earlier expression of PKC α and β relative to PKC γ suggests that PKC α and/or β may possibly be necessary to turn on the PKC γ gene. F. L. Huang and colleagues (1990) have reported differential expression of PKC α, β, and γ in developing rat cerebellar cell types.

PKC γ gene expression is tissue specific, occurring in brain and the spinal cord.

The 5′ flanking region of rat PKC γ gene contains cis elements for the common transcription factors SP1, AP2, adenovirus major late promoter, and progesterone receptor (Chen et al., 1990, and unpublished results). The transcription initiation site for PKC γ lies 243 base pair (bp) upstream from the translation initiation site and genomic fragments corresponding to −1612 to +243 and −163 to +243 bp has similar full promoter activity in transfected nonneuronal 293 cells (Chen et al., 1990). The PKC γ promoter, similar to housekeeping genes, lacks TATA and CAAT boxes in the normal positions. The promoters of other brain-specific genes aldolase C, thy-1, and γ-enolase, have no cis elements in common with the PKC γ promoter (Chen et al., 1990). A cis element (M-box) approximately 670 bp upstream from the transcription initiation start site appears to be a negative regulatory element, since newborn rat brain nuclear extracts contain a protein that binds this element, yet adult rat brain lacks this binding activity (Chen et al., unpublished results).

The 5′ flanking region of human PKC β contains cis elements for OBP, SP1, AP1, AP2, and E boxes (Obeid et al., 1992). The transcriptional initiation site is 197 bp upstream from the translational initiation site and the promoter lacks TATA and CAAT boxes at their normal positions (Obeid et al., 1992). Deletion analysis indicated that a genomic fragment containing −111 to +43 bp could confer maximal promoter activity (Obeid et al., 1992). The promoter was stimulated 8- to 20-fold by phorbol ester, which is consistent with PKC β mRNA up-regulation by phorbol ester in HL60 and PKC β-transfected K562 cells (Mc-Swine-Kennick et al., 1991; Obeid et al., 1992). Phorbol ester responsiveness of the PKC β promoter is conferred by the −111 to +43 bp genomic fragment, which has no AP1 sites (Obeid et al., 1992). The 5′ flanking region of human PKC β has also been characterized by Niino and co-workers (1992). In contrast, they found a CAAT box at −110 bp but no TATA box, and the transcription initiation site was 484 bp upstream from the translation initiation site (Niino et al., 1992). Deletion analysis of genomic PKC β revealed that fragment −234 to +179 bp provided nearly maximal promoter activity in transfected GH_4C_1 cells, whereas mutants containing a 5′ flanking region > 1.9 kilobase (kb) had similar levels of activity in transfected cells as expression of endogenous PKC β (Niino et al., 1992). Positive (P1, P2, and PN) and negative (N1, and PN) cis elements, which had differential responses in transfected GH_4C_1, P19, and 3Y1 cells, were found in the 1.9-kb region upstream from the transcription initiation site (Niino et al., 1992).

Mouse neuroblastoma 2a cells induced to differentiate by various agents have large decreases in PKC α and ε mRNA and a lesser change in PKC ζ mRNA levels (Wada et al., 1989). HL60 promyelocytic cells can be induced to differentiate into a neutrophil or monocyte phenotype by retinoic acid or vitamin D_3, respectively (Devalia et al., 1992). HL60 cells induced to differentiate into a neutrophil phenotype with retinoic acid display decreased PKC α mRNA levels (Devalia et al., 1992). Isolated neutrophils have no detectable PKC α mRNA or protein. In contrast, HL60 cells induced to differentiate into a monocyte phenotype by vitamin D_3 had no decrease in PKC α mRNA, but had an increase in PKC β mRNA (Obeid et al., 1990; Devalia et al., 1992). DMSO or retinoic acid induction of differentiation of HL60 cells to a neutrophil phenotype resulted in more than twofold increases in all three PKC α, β, and γ proteins as detected by immunoblots (Makowske et al., 1988). T cell line CCRF-CEM when treated with iron-transferrin undergoes more than sixfold enhancement in PKC activity, which is correlated with a great enhancement in PKC β mRNA levels (Alcantara et al., 1991). No PKC α or γ mRNA was detected in these cells. Little is known about the mechanism of physiological regulation of PKC gene expression.

ENZYMATIC PROPERTIES OF PKC ISOZYMES

Members of the PKC family are structurally similar; however, they are sufficiently different to allow chromatographic separation. The various PKC isoforms have been isolated from different tissues and from COS or insect cells transfected with the various PKC cDNAs. Brain PKC I (γ), PKC II (βI + βII), and PKC III (α) purified by hydroxyapatite column chromatography exhibit distinctive differences in their sensitivities to stimulation by activators, to proteolytic degradation, and to inactivation by acidic phospholipids (K.-P. Huang et al., 1986b; F. L. Huang et al., 1989a; F. L. Huang and Huang, 1991). In addition, these enzymes also display minor differences in substrate specificity. With use of a phospholipid/detergent mixed micelles assay, these enzymes appear to have similar requirement for Ca^{2+} either in the assay for kinase activity or for phorbol ester binding. The phospholipid requirements of these PKC isozymes are distinguishable: PKC γ is more sensitive to stimulation by cardiolipin in the absence of DAG than are PKC α and β (K.-P. Huang et al., 1988). In the presence of cardiolipin, the concentrations of DAG and phorbol 12,13-dibutyrate (PDBu) required for half-maximal activation of PKC γ are nearly an order of magnitude lower than those for PKC α and β. In the presence of PS, binding of PDBu to PKC γ evokes a corresponding stimulation of the kinase activity, whereas binding of this phorbol ester to PKC α or β produces a lesser degree of stimulation in kinase activity. These findings indicate that binding of phorbol ester to the various PKC isozymes may evoke different degrees of activation of these enzymes. The various PKC isozymes also respond differently to stimulation by arachidonic acid (AA) (Naor et al., 1988) and its metabolites (Shearman et al., 1989a). PKC γ is activated by micromolar concentrations of AA either in the presence or absence of Ca^{2+}; at a higher concentration, this fatty acid becomes inhibitory. PKC β is slightly activated by low concentrations of AA only in the presence of Ca^{2+}. PKC α is similar to PKC β in its requirement for Ca^{2+}; however, a maximal stimulation of PKC α requires higher concentrations ($>$100 μM) of AA. Lipoxin A, a metabolite of AA, selectively activates PKC γ at micromolar concentrations (Shearman et al., 1989a). In addition to their different responses to the various activators, the three PKC isozymes also have different K_m values for the various protein substrates (K.-P. Huang et al., 1988; Marais and Parker, 1989). It is likely that these enzymes may be differentially stimulated by a variety of stimuli at different cellular locations where they may phosphorylate their preferred substrates.

PKC δ, ϵ, ζ, η, and θ expressed in COS cells and insect cells have been characterized as Ca^{2+}-independent PKCs (Ohno et al., 1988; Ono et al., 1988; Osada et al., 1990; Schaap and Parker, 1990; Bacher et al., 1991; Liyanage et al., 1992). PKC δ purified from porcine spleen (Leibersperger et al., 1990) and rat brain (Ogita et al., 1992) behaves similarly to the PKC δ expressed in COS cells; this kinase mostly associates with the particulate fractions and can be extracted by nonionic detergent. PKC ε from both rat (Koide et al., 1992) and rabbit (Saido et al., 1992a) brains has been purified from the soluble fraction of these homogenates. Both PKC δ and ε have similar substrate specificity (Olivier and Parker, 1991) and can be phosphorylated by other protein kinase and undergo autophosphorylation, and the phosphorylated enzymes exhibit retarded mobility upon sodium dodecyl sulfate-polyacrylamide gel electrophoresis (SDS-PAGE). It is unknown whether phosphorylation plays any role in the catalytic function. Although PKC ε, as well as other group B PKCs, has been characterized as Ca^{2+}-independent enzymes using histone as a substrate, Ca^{2+} exerts both positive and negative effects on rabbit brain PKC ε depending on the substrate and phospholipids used in the assay (Saido et al., 1992a). With peptide ε, homologous to the pseudosubstrate region of PKC ε with a substitution of Ala by Ser as a substrate and either PS or cardiolipin as a supporting phospholipid, PKC ε is stimulated by phorbol ester, PMA, and the PMA-stimulated activity is inhibited by Ca^{2+}. In comparison, with MBP_{4-14} peptide as a substrate and cardiolipin as a supporting phospholipid, PKC ε is not stimulated by PMA or Ca^{2+}. With this same peptide as a substrate and PS as a supporting phospholipid, Ca^{2+} stimulates the kinase activity either in the presence or absence of PMA. In addition, PKC ε appears to have a higher K_d for phorbol ester than that of the group A PKCs (Akita et al., 1990a) and, unlike the rest of the PKCs, phosphorylates histone H1 poorly compared with its phosphorylation of myelin basic protein, but the histone kinase activity can be increased by limited proteolysis (Schaap et al., 1990). PKC ζ expressed in COS cells is stimulated by PS but does not seem to bind phorbol ester or to be activated by DAG (Ono et al., 1989b); the native enzyme purified from bovine kidney behaves similarly to that expressed in COS cells (Nakanishi and Exton, 1992). PKC ζ is specifically stimulated by PIP_3, which is elevated on stimulation by growth factors (Nakanishi et al., 1993). PKC η, θ, and λ have not yet been purified from any tissue. The different activation requirements and substrate specificities of the group B PKCs also point to distinct regulatory functions for these enzymes.

PKCs autophosphorylate intramolecularly in both the regulatory and catalytic domains (K.-P. Huang et al., 1986a; Mochly-Rosen and Koshland, 1987). Autophosphorylation of group A PKCs stimulates enzymatic activity by lowering the K_m for histone twofold, by lowering the Ca^{2+} requirement for activation fourfold when assayed with PS without DAG, and by increasing the affinity for phorbol ester binding (K.-P. Huang et al., 1986; Mochly-Rosen and Koshland, 1987). Autophosphorylation of PKC has a ten-fold lower K_m for ATP (1.5 μM) relative to exogenous substrate phosphorylation (K.-P. Huang et al., 1986a). PKM, the 50-kDa catalytic fragment of PKC, cannot autophosphorylate (Mochly-Rosen and Koshland, 1987) and has a five-fold higher K_m for histone H1 relative to PKC. Autophosphorylation of PKC causes dissociation of PKC from the membrane

(Wolf et al., 1985b). PKC α and γ primarily autophosphorylate at serine residues, whereas PKC β autophosphorylation occurs at both serine and threonine residues (K.-P. Huang et al., 1986a). Peptide mapping of autophosphorylated PKC α, β, and γ provided evidence that the autophosphorylation sites are located in the variable sequence regions (K.-P. Huang et al., 1986a). Flint and colleagues (1990) have mapped the autophosphorylation sites in PKC βII at three different domains; the N-terminus, the hinge region, and the C-terminus. A serine and threonine in the N-terminus, two threonines in the hinge region, and two threonines in the C-terminus were delineated as the autophosphorylation sites, indicating great flexibility in the PKC polypeptide (Flint et al., 1990). Protamine can induce autophosphorylation of PKC (Chauhan and Chauhan, 1992), as can low pH (pH 4.6) (McFadden et al., 1989).

MECHANISM OF ACTIVATION OF PKC

Both biochemical and biophysical approaches have been used to define the specificity of phospholipid and PKC interaction and the mechanism of activation of the enzyme. It should be emphasized that PKC and phospholipid interaction is required but does not necessarily evoke kinase activity. PKC interacts with acidic phospholipids in either the presence or absence of divalent metal ion. In the absence of divalent metal ion, both the regulatory and catalytic domains of the kinase bind PS and polyphosphoinositides, resulting in an inactivation of the kinase activity (K.-P. Huang and Huang, 1990; F. L. Huang and Huang, 1991). The divalent metal ion-independent binding of PS to the regulatory domain causes no ill effect on PDBu binding; however, binding of PIP_2 to this domain reduces the binding affinity for PDBu. Under these conditions PKC probably interacts electrostatically with acidic phospholipids having high surface charge density, thus causing conformational changes and inactivation of the kinase. The phospholipid-induced inactivation of PKC can be reduced by diluting the phospholipid surface charge density with neutral phospholipids or by divalent metal ion. In the presence of Ca^{2+}, the acidic phospholipids interact preferentially with the regulatory domain of PKC with no strong preference for a particular head group (Bazzi and Nelsestuen, 1987), and DAG or phorbol ester does not appear to influence the stability of the reversible PKC/phospholipid complex. Phospholipids in vesicles, as monolayer (Bazzi and Nelsestuen, 1988; Souvignet et al., 1991), or as phospholipid/Triton X-100 mixed micelles (Hannun et al., 1986a,b) can all interact with PKC, suggesting that the bilayer structure of the membrane is not required for the reversible association. When phospholipid/Triton X-100 mixed micelles are used, PKC preferentially interacts with PS over PG and PI (Hannun et al., 1986a). Analysis of PKC interaction with PS/PC (1:4) monolayer suggests that in the presence of Ca^{2+}, PKC can penetrate into the lipid core and express DAG- or PMA-stimulated kinase activity (Souvignet et al., 1991). Penetration of PKC into the phospholipid bilayer also has been suggested on the basis of labeling studies with a lipid-soluble probe (Snoek et al., 1986).

Although several acidic phospholipids promote the interaction of PKC with the membrane, PS is the most effective one to stimulate PKC activity. PS activates PKC in a highly cooperative and specific manner when the PS/Triton X-100 mixed micelles assay is used (Newton and Koshland, 1989). Micelles containing 12 or more molecules of PS activate PKC maximally with a Hill coefficient of 8. It is apparent that sufficient surface charge density is essential to attract PKC to the micelle surface, and the interaction of PKC with multiple PS molecules is required to elicit maximal catalytic potential of the enzyme. Several functional groups within the phospho-L-serine polar head groups are essential for the binding and activation of PKC (Lee and Bell, 1989). Both the carboxyl and amino moieties are important for activation; modification of either functional group results in an inactive phospholipid. The distance between the phosphate moiety and the carboxyl and amino functional groups as well as the stereospecificity for L-serine are important structural features of PS to support PKC activity. The stereochemistry within the glycerol backbone is not crucial for activation—both 1,2-rac-phosphatidyl-L-serine and 1,3-phosphatidyl-L-serine are fully active. The specificity of phospholipid in supporting PDBu binding is less stringent than that for the activation of the kinase activity. On the basis of specific interaction of PKC with PS, Bell and co-workers concluded that each PKC molecule makes three or more points of contact with free carboxyl, amino, and phosphate groups in a stereospecific manner. Other anionic phospholipids, such as phosphatidic acid, phosphatidylglycerol, cardiolipin, and phosphatidylinositol, are less effective in supporting PDBu binding and kinase activation. Thus, these phospholipids, lacking the essential free carboxyl and amino moieties of PS, may interact with and activate PKC by different mechanisms.

Activation of PKC by DAG is stereospecific since sn-1,2-DAG is active but neither sn-2,3-DAG nor sn-1,3-DAG is (Rando, 1988). These latter two inactive isomers do not even inhibit the sn-1,2-DAG-stimulated kinase activity. The stereospecificity for the diglyceride backbone is rather stringent; however, little specificity is directed toward the fatty acyl side chains. As long as the diacyl chains are suitably hydrophobic to permit insertion into the bilayer membrane, an active DAG will result. The active endogenous DAGs are believed to contain saturated fatty acids at the 1 position and unsaturated fatty acids at the 2 position. Under *in vitro* assay conditions the DAGs containing diC8:0 is equally active as diC18:1. In contrast to the relative lack of specificity with respect to the alkyl side chains, the chain length of the glycerol backbone is highly specific. Any modification leads to either a strong decrease in activity or no activity at all (Ganong et al., 1986). Both carboxyl moieties of the esters and the 3-hydroxyl group of DAG are essential for maximal activation. It is envisioned that PKC forms at least three contact points at the hydroxyl and the two carboxyl esters of DAG, with the hydrophobic domain inserted into the membrane.

Several structurally diverse tumor promoters also activate PKC in a fashion similar to that of DAG. The question arises as to how the DAG-binding site in PKC can accommodate structurally diverse tumor promoters such as phorbol esters, aplysiatoxins, teleocidins, bryostatins, and ingenols. Structural analysis has yet to pinpoint a consensus homology between these tumor promoters and the functional groups of DAG (Jeffrey and Liskamp, 1986; Wender et al., 1986).

Recently, Kong and colleagues (1990) demonstrated that PKC activation is also stereospecific with respect to the second chiral center of 3-methyl-1,2-diacyl-sn-glycerols. The absolute configuration at C2 and C3 of the active 3-methylated diacylglycerols is the same as the absolute configuration of C29 and C30 of the naturally occurring tumor promoter debromoaplysiatoxins (Nakamura et al., 1989; Kong et al., 1991), suggesting a structural link between tumor promoters and diacylglycerols for PKC activation. At least three hydrophilic moieties, with defined stereospecificity, within these tumor promoters appear to make contact with PKC, and the spatially corresponding hydrophobic moieties are inserted into the membrane.

Members of the PKC enzyme family do not contain a typical Ca^{2+}-binding site exemplified by the EF-hand structure. Only PKC α contains a sequence segment resembling the F helix of the EF hand, which is not conserved in PKC β and γ. However, PKCs do interact with Ca^{2+} without added phospholipid, as evidenced by the Ca^{2+}-induced quenching of PKC intrinsic fluorescence (K.-P. Huang, 1989) and enhancement of hydrophobicity to interact with hydrophobic matrix (Walsh et al., 1984). This interaction appears to be low affinity and low stoichiometry (Bazzi and Nelsestuen, 1990). PKC also interacts with a frequently used trivalent lanthanide luminescent probe, terbium (Tb^{3+}), for Ca^{2+}-binding site in proteins; the sites of interaction of Tb^{3+} with PKC appear to be different from those of Ca^{2+}, and this metal ion behaves similarly to several divalent heavy metal ions in causing inhibition of PKC kinase and phorbol ester-binding activity (Walters and Johnson, 1990). In the presence of phospholipid vesicles, PKC binds Ca^{2+} with high affinity and stoichiometry; at least eight Ca^{2+} ions per PKC molecule were detected by equilibrium dialysis and gel filtration chromatography (Bazzi and Nelsestuen, 1990). Phorbol ester (PDBu) has little effect on Ca^{2+} binding under these experimental conditions, in contrast to the phorbol ester- or DAG-induced reduction of Ca^{2+} requirement for the activation of the kinase activity. Half-maximal binding of Ca^{2+} to PKC, in the presence of PS/PC (1:3) occurs at 40 μM Ca^{2+}, whereas the concentration of Ca^{2+} required for half-maximal kinase activation is in the submicromolar concentration. These quantitative measurements under equilibrium conditions may involve additional Ca^{2+} binding sites, which are not essential to evoke kinase activation. Under equilibrium, PDBu does enhance PKC-membrane interaction; the resulting complexes are not readily dissociated by Ca^{2+} chelator (Bazzi and Nelsestuen, 1990). Another interesting feature of Ca^{2+} binding to PKC is the dependency on the compositions of membrane phospholipids; an increase in the PS content in the membrane reduces the Ca^{2+} requirement. It has been proposed that Ca^{2+} functions as a bridge to hold PKC and phospholipid complexes together (Bell, 1986). The Ca^{2+} binding sites may be generated by alignment of free carboxyl groups of PKC with acidic phospholipids. This class of phospholipid-dependent Ca^{2+}-binding proteins has been identified in several members of the lipocortin family (Klee, 1988).

Arachidonic acid (AA) (McPhail et al., 1984), cis-unsaturated fatty acids (Murakami et al., 1986), and lipoxin A (Hansson et al., 1986) activate PKC independent of PS. The effects of Ca^{2+} and DAG on the fatty acid-mediated activation are variable. AA-mediated stimulation of PKC γ is independent of

Ca^{2+}, and stimulation of PKC α and β is dependent on Ca^{2+} (Naor et al., 1988). Oleic acid activation of PKC has been shown to be independent of Ca^{2+} (Murakami et al., 1986) or totally dependent on Ca^{2+} (Sekiguchi et al., 1987; Seifert et al., 1988), and this fatty acid only activates PKC in the soluble form but not those associated with membrane (el Touny et al., 1990). In addition, activation of PKC by oleate is more resistant to inhibition by sphingosine than that by PS/DAG, and oleate fails to stimulate PKC autophosphorylation or inhibit PDBu binding to PKC. Synergistic activation of PKC α, β, and γ occurs in the presence of cis-unsaturated fatty acid and DAG (Shinomura et al., 1991). In the presence of PS and DAG, cis-unsaturated fatty acids further increase the apparent affinity of PKC to Ca^{2+} and render the enzyme fully active at the basal level of Ca^{2+} concentration. Under these conditions, effective concentrations of cis-unsaturated fatty acid are in the range of 20 to 50 μM. It is evident that the mechanism of activation of PKC by free fatty acids is different from that by PS and DAG. The physiological significance of free fatty acid-mediated activation of PKC is not yet clear.

PIP_2 has been shown to stimulate PKC in a PS- and Ca^{2+}-dependent manner (Chauhan and Brockerhoff, 1988), resembling that by DAG but to a lesser extent. Mechanistically, PIP_2 functions as a lipid activator and as a cofactor to replace some PS molecules (Lee and Bell, 1991). PKC binds PIP_2 and PS at distinct regions and each causes characteristic conformational changes of PKC, as evidenced by distinctive changes in the intrinsic fluorescence of the enzyme (F. L. Huang and Huang, 1991). The PIP_2 binding site does not seem to overlap with the DAG or PDBu binding site, as PIP_2 does not compete with PDBu for binding to PKC. This polyanionic phospholipid may interact with the basic region of PKC molecule, such as the pseudosubstrate segment, to unmask the active site of PKC. It has been proposed that PIP_2 serves as a membrane-anchoring site for PKC, which readily interacts with this phospholipid in the presence of Mg^{2+} under basal physiological condition when Ca^{2+} is low and Mg^{2+} is at millimolar concentration (F. L. Huang and Huang, 1991). Association of PKC with membrane rich in polyphosphoinositides may allow the kinase to be activated proximal to DAG generated by the agonist-induced hydrolysis of these phospholipids.

It is envisaged that an increase in Ca^{2+} influx or Ca^{2+} release from the internal stores triggered by IP_3 sets the stage for the activation of PKC (Figure 2.4). A rise in $[Ca^{2+}]_i$ promotes the interaction of cytosolic or the loosely membrane-bound PKC α, β, and γ with PS to form a preactivated form of the kinase. At this stage, those PKC molecules associated simultaneously with both PS and PIP_2 become partially activated. An increase in DAG near the site of PKC/membrane attachment induces a higher degree of activation of the enzyme. It is believed that the group B PKCs are also activated when DAG is generated by the hydrolysis of membrane phospholipids by PLCs. Depending on the duration and nature of the stimulatory signal, DAGs generated from transient activation of phospho-inositide-specific PLCs or from a more sustained activation of phosphatidylcholine-specific PLC are all active in the stimulation of PKC. Other effective DAGs are derived from the hydrolysis of phospholipids by phospholipase D and a phosphatidate phosphatase, hydrolysis of PI-glycan by a specific phospholipase C, as well as that formed during the biosynthesis of phospholipids. In addition, acti-

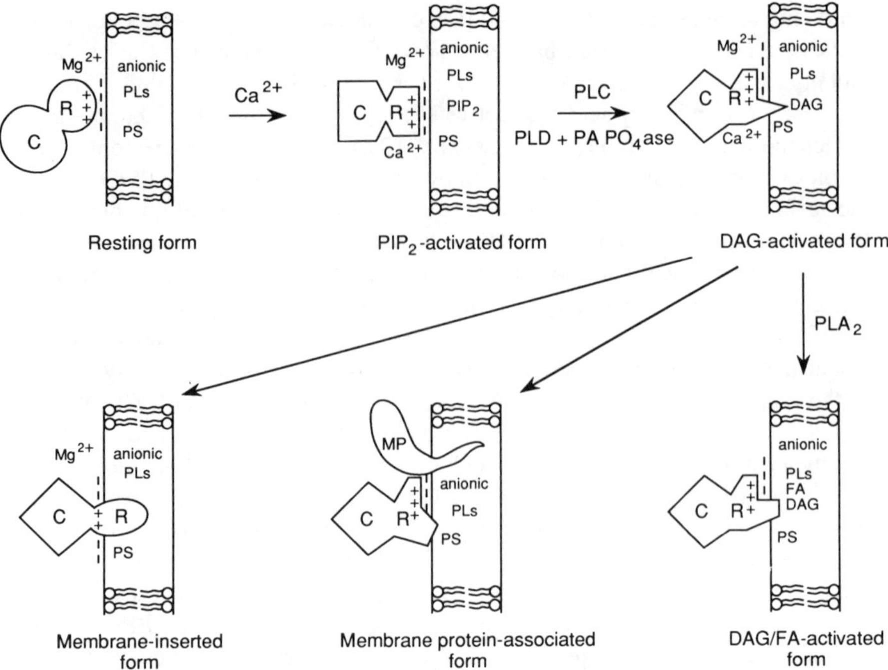

Figure 2.4 Mechanism of activation of PKC. Under the basal physiological conditions with millimolar concentration of Mg^{2+} and submicromolar concentration of Ca^{2+}, PKCs exist in the resting state as cytosolic or anionic phospholipid-bound form. Increase in $[Ca^{2+}]_i$, as a result of Ca^{2+} influx or release from the internal stores, promotes the interaction of Ca^{2+}-dependent PKCs with PS to form a preactivated form of the enzyme. Those enzymes interacting simultaneously with both PS and PIP_2 are partially activated. An increase in DAG resulting from hydrolysis of membrane phospholipids by PLCs or PLD and phosphatidate phosphatase causes activation of both Ca^{2+}-dependent and -independent PKCs. The kinase activity can be further increased by unsaturated fatty acids generated by the activation of PLA_2. Following sustained stimulation of the cells, fraction of PKC may be converted into membrane-inserted form or membrane protein-associated form. The membrane-inserted form may be responsible for a prolonged activation of the kinase; however, the activation state of the membrane protein (MP)-associated form is unclear. ''R'' represents the regulatory domain and ''C'' represents the catalytic domain of PKC.

vation of phospholipase A_2 that generates AA and cis-unsaturated fatty acids could synergize with DAG to further activate PKC even at basal level of $[Ca^{2+}]_i$. Sustained stimulation of the cells may cause the conversion of PKC into membrane-inserted form or membrane protein-associated form. The former enzyme form may be responsible for a prolonged activation of the kinase; however, the state of activation of the latter is unknown.

The Ca^{2+} signal that triggers the activation of PKC varies in its frequency and magnitude. $[Ca^{2+}]$ oscillations that make use of intracellular stores and/or extracellular sources following receptor occupation could convey information by varying the frequency of the Ca^{2+} spike while maintaining low levels of $[Ca^{2+}]_i$. The

Ca^{2+} spike can act as a signal to activate Ca^{2+}-dependent PLC and PLA_2 that generate DAG and AA, respectively, to activate PKC. The activated PKC in turn dampens the Ca^{2+} signal by affecting agonist-receptor interaction, G-protein coupling, IP_3 metabolism, IP_3 receptor Ca^{2+} channel, Ca^{2+}/CaM-dependent Ca^{2+} pump, and the PI-specific PLC. Within the duration of receptor occupancy, repetitive fluctuation in $[Ca^{2+}]_i$ may regulate protein phosphorylation/dephosphorylation intermittently and thus an oscillation of cellular responses. During this repetitive cycle, a fraction of PKC may also be converted into effector-independent kinase for a sustained activation.

INHIBITORS OF PKC

The regulatory and catalytic domains can be targeted, separately, by inhibitors such as sphingosine (Hannun et al., 1986c), calphostin C (Kobayashi et al., 1989), AMG (1-O-alkyl-2-O-methylglycerol) (Kramer et al., 1989), and NPC 15437 (2,6,diamino-N-([1-(oxotridecyl)-2-peperidinyl]methyl)-hexanamide) (Sullivan et al., 1991) for the regulatory domain and H-7 (Hidaka et al., 1984), staurosporine (Tamaoki et al., 1986), and bisindolylmaleimides (Toullec et al., 1991; Nixon et al., 1992) for the catalytic domain. Many of the inhibitors act by interfering with ATP binding or substrate recognition and hence are not specific, since most protein kinases (and many enzymes) utilize ATP and have similar substrate specificities. Several inhibitors appear to be specific for PKC. Calphostin C, which interferes with phorbol ester binding, inhibits group A PKCs (IC_{50} = 50 nM). Calphostin C has IC_{50} values for PKA and tyrosine protein kinase that are more than 1000-fold higher than that for PKC (Kobayashi et al., 1989). Calphostin C inhibition of PKC is light dependent (fluorescent) and the mechanism appears to be via irreversible covalent modification of PKC (Bruns et al., 1991), which of course may limit its usefulness *in vivo*. Whether calphostin C can inhibit Ca^{2+} and unsaturated free fatty acid stimulation of PKC in the absence of DAG or phorbol ester is unknown. It is expected that calphostin C will inhibit group B but not group C PKCs. Hypericin, a red perlenequinone similar in structure to calphostin C, also inhibits group A PKC in a light-dependent manner (IC_{50} = 1.5 µg/ml) and has antiproliferative effects on Balb 3T3/H-ras cells (IC_{50} = 15 µg/ml) (Takahashi et al., 1989). Bisindolylmaleimides, structural analogs of staurosporine, are potent selective inhibitors of PKC (IC_{50} = 5–70 nM) (Toullec et al., 1991). Like staurosporine, they are competitive with respect to ATP, and they inhibit PKC α, β, and γ similarly (IC_{50} = 16–20 nM) (Toullec et al., 1991). The EGF-, PDGF-, and insulin-receptor tyrosine kinases are inhibited by bisindolylmaleimide GF 109203X with IC_{50} values greater than 2000-fold higher than that for PKC (Toullec et al., 1991). Bisindolymaleimide Ro 31-8425 potently inhibits group A PKC (IC_{50} = 8 nM) but has IC_{50} values for PKA and Ca^{2+}/calmodulin protein kinase more than 160-fold higher than that for PKC (Nixon et al., 1992). Synthetic NPC 15,437 competitively inhibits group A PKC (IC_{50} = 19 µM) with respect to phorbol ester and PS (Sullivan et al., 1992). NPC 15,437 has IC_{50} values more than 15-fold higher for PKA and MLCK than for PKC, and it inhibited PDBu-stimulated phosphorylation of the 47-kDa protein in intact platelets at similar inhibitory doses (Sullivan et al., 1992). Chelerythrine inhibits group A PKC (IC_{50} = 0.7 μM) competitively with respect to histone III-S. Its IC_{50} values for PKA, Ca^{2+}/calmodulin-dependent protein kinase, and tyrosine protein kinase are more than 100-fold higher than for PKC (Herbert et al., 1990). Staurosporine, although it has been widely used as a potent PKC inhibitor (IC_{50} = 3 nM), is nonspecific. Staurosporine inhibits PKA, p60[v-src] tyrosine kinase, and neutrophil tyrosine kinase activity with IC_{50} = 8, 6, and <10 nM, respectively (Tamaoki et al., 1986; Nakano et al., 1987; Badwey et al., 1991).

PKC subtypes α, β, and γ are differentially inactivated by PS and PIP_2, in the absence of Ca^{2+}, with sensitivity of inactivation being PKC $\gamma > \beta > \alpha$ (IC_{50} = 5, 45, 120 μM PS, respectively) (K.-P. Huang and Huang, 1990; F. L. Huang and Huang, 1991). PIP_2, in the absence of Ca^{2+}, inactivates PKC γ, β, and α with IC_{50} = 2, 4, and 11 μM, respectively, in an irreversible manner (F. L. Huang and Huang, 1991). Suramin, an anti-HIV reverse transciptase agent, preferentially inhibits PKC γ (K_i = 17 μM) relative to PKC α and β (K_i = 31 and 27 μM, respectively) (Mahoney et al., 1990). Suramin, a competitive inhibitor with respect to ATP, can also activate group A PKCs (200–400%) in the absence of lipid, and in the presence of Ca^{2+}, by acting as a negatively charged phospholipid analog (Mahoney et al., 1990). A need remains for selective PKC subtype inhibitors for physiological studies.

Several endogenous protein inhibitors have been found to inhibit PKC; these inhibitors may counteract the stimulatory effects of Ca^{2+} and DAG *in vivo*. Previously, calmodulin (CaM) was identified as an inhibitor of PKC (Albert et al., 1984); the inhibition is partly due to CaM-mediated blockade of the PKC phosphorylation sites of certain substrates. PKC inhibitor-1 (M_r 17,000) from bovine brain, a 125-amino acid Zn^{2+}-binding protein, has been shown to inhibit PKC specifically, and the inhibition cannot be prevented by elevated levels of Ca^{2+}, DAG, or phospholipid (McDonald and Walsh, 1985). The inhibitory mechanism of this protein is unknown; it has been speculated that this protein may modulate PKC binding to membrane by sequestering cytosolic free Zn^{2+} or by associating with PKC in a Zn^{2+}-mediated fashion (Pearson et al., 1990). In addition, this protein is also rich in lysine and thus may exert its inhibitory effect analogous to that of polylysine (House and Kemp, 1987), which exhibits substrate-dependent inhibition of PKC. Microinjection of the purified inhibitor into isolated chick dorsal root ganglion neuronal cells has been shown to block the attenuating effect of norepinephrine on L-type dihydropyridine-sensitive voltage-dependent Ca^{2+} channels in these cells (Rane et al., 1989). Another group of putative endogenous inhibitors of PKC has been identified from sheep brain (Toker et al., 1990). The amino acid sequences of these M_r = 29 to 33 kDa proteins are similar to a neuron-specific protein termed 14-3-3 (Ichimura et al., 1988) and the carboxyl terminus of the lipocortin family of phospholipid-dependent Ca^{2+}-binding proteins (Klee, 1988). The inhibitory activities of these proteins are heat labile and cannot be reversed by increasing the substrate, cofactor, or ATP concentration. Furthermore, these inhibitors do not inhibit PDBu binding to PKC. Recently, annexin V, a phospholipid-dependent Ca^{2+}-binding protein located on the cytosolic face of the plasma membrane, has been identified as a potential inhibitor of PKC (Schlaepfer et al., 1992). This protein inhibits PKC most likely by direct interaction with the kinase rather than by competitive binding to phospholipid. On the basis of estimates of PKC inhibitor-1 level in brain (McDonald et al., 1987) and annexin V in many cell types (Schlaepfer et al., 1992), there are sufficient inhibitors present to inactivate all the PKCs. Thus, these inhibitor proteins must be regulated by a separate mechanism, such as that for the attenuation of the activity of cAMP-dependent protein kinase inhibitor by tyrosine phosphorylation (Van Patten et al., 1987).

ASSOCIATION OF PKC WITH OTHER CELLULAR PROTEINS

PKC has been located in a variety of intracellular compartments including Golgi apparatus (Saito et al., 1989), the nucleus (Masmoudi et al., 1989), the perinucleus (Halsey et al., 1987; Fields et al., 1989; Leach et al., 1989), and cytoskeletal elements (Jaken et al., 1989; Papadopoulos and Hall, 1989), in addition to plasma membrane. Treatment of cells with phorbol ester or ligand that induces a rise in DAG further enhances the association of PKC with the membrane fraction, a phenomenon known as translocation or stimulator-induced redistribution of PKC. At least two types of membrane-associated PKC are distinguishable, one that can be extracted by metal ion chelators and the other that is stable to chelators and high salt but can be dissociated with detergents. The detergent-extractable PKC behaves like an integral membrane protein that may result from membrane insertion or high-affinity interaction with other membrane proteins. The PMA-induced membrane binding of PKC has been shown to be sensitive to prior treatment of the isolated membrane with protease or phospholipase (Gopalakrishna et al., 1986). Two proteins (M_r = 30 and 33 kDa) from the detergent-insoluble membrane fractions have been identified as putative intracellular receptors for activated PKC (RACKS) (Mochly-Rosen et al., 1991a). Binding of PKC to these proteins requires the presence of PS and Ca^{2+} and can be further increased with the addition of DAG or PMA. In addition, binding of PKC to these proteins is concentration dependent and saturable. Several members of the annexin family, such as annexin I, and to a lesser extent annexins II and VII, can also serve as anchoring sites for PKC. Mochly-Rosen proposed that activation of PKC by Ca^{2+}/PS/DAG or PMA results in an exposure of a binding site that is not overlapping with the activators and substrate-binding sites of the kinase. A synthetic peptide corresponding to the carboxy-terminal end of annexin I can bind PKC in a fashion similar to that of the intact protein, and inhibits PKC binding to the receptor protein in a dose-dependent manner (Mochly-Rosen et al., 1991b). This PKC-binding sequence motif is also present in the sheep brain PKC inhibitor described by Aitken and colleagues (1990). In addition to Ca^{2+}, PS, and DAG, Zn^{2+}ion has been shown to enhance the redistribution of PKC to plasma membrane (Csermely et al., 1988) and membrane cytoskeleton, possibly actin (Zalewski et al., 1990). PKC-binding proteins (100 and 115 kDa) from human white blood cells bind PS in a divalent cation-independent manner, yet PKC binding to these proteins is both Ca^{2+} and PS dependent (Wolf and Baggiolini, 1990). PKC-binding proteins (110 and 115 kDa) from erythrocytes are highly enriched in the cytoskeletal fraction, whereas the 115-kDa PKC-binding protein in brain is in the cytosol, postsynaptic densities, and nuclei (Wolf and Sayhoun, 1986). Phosphorylation of these proteins markedly inhibited Ca^{2+}- and PS-dependent PKC binding (Wolf and Sayhoun, 1986). Interaction of the various PKC isozymes with specific membrane protein or cytoskeletal protein may confer the unique localization of each of these enzymes.

PROTEOLYTIC DEGRADATION OF PKC

Phorbol ester-mediated activation of PKC in intact cells frequently results in limited proteolysis followed by complete degradation, or down-regulation, of the enzyme. *In vitro,* PKC can be degraded by trypsin or calpain into two fragments containing the catalytic and regulatory domains. The protease-sensitive sites are located within the hinge region between these two domains. The various PKC isoforms have different susceptibility to proteolysis *in vitro* as well as *in vivo*. In the absence of phospholipid, the rate of degradation is PKC γ > PKC β > PKC α (F. L. Huang et al., 1989a; Kishimoto et al., 1989). In the presence of Ca^{2+} and phospholipid, the rate of proteolysis is greatly increased and DAG or phorbol ester reduces the concentration of Ca^{2+} required for proteolysis. The effects of DAG and phorbol ester in promoting PKC degradation in the presence of Ca^{2+} are different with respect to the specificity for phospholipids (F. L. Huang et al., 1989a). In the presence of DAG, only those phospholipids such as PS, cardiolipin, and PA, which are active in supporting kinase activation, are active in promoting proteolysis. In contrast, the PMA-stimulated PKC degradation has a broader specificity for phospholipids; it is active with most acidic phospholipids. The neutral phospholipids, such as PC and PE, which are ineffective in the activation of PKC, are also ineffective in supporting PKC degradation either in the presence of DAG or PMA. Activators of PKC, when they bind to the regulatory domain, dislodge the pseudosubstrate region from the catalytic domain, which results in exposure of the protease-sensitive sites. The sensitivity of the various PKC isozymes to proteolysis in the PMA-treated cells, unlike that seen *in vitro,* appears to be variable depending on the cell type under study. PKC β from rat basophilic leukemia cells (F. L. Huang et al., 1989a) and human T cell (Ase et al., 1988) is more susceptible to proteolysis than PKC α, and both of these enzymes in rat synaptosomal preparations are degraded at faster rates than PKC γ (Oda et al., 1991). *In vivo,* Ca^{2+}-dependent proteolysis of PKCs is thought to be catalyzed by calpain (Melloni et al., 1985; Girard et al., 1986; Adachi et al., 1990). Two mammalian forms of calpain, μ-and m-calpain, are activated by micromolar and millimolar $[Ca^{2+}]$, respectively. Recently, Saido and colleagues have identified polyphosphoinositides as activators of μ-calpain. They function by lowering the Ca^{2+} requirement for proteolysis to the physiological nanomolar range (Saido et al., 1992b). Although DAG *in vitro* stimulates proteolytic degradation of PKC (Ase et al., 1988), DAG (OAG or diC8:0-DG) did not down-regulate PKC α in

Swiss-3T3 cells (Issandou et al., 1988) or PKC in breast cancer cell line MCF-7 (Issandou and Rozengurt, 1989).

Human T cell leukemic cells (Jurkat), when chronically treated with phorbol ester, have reduced PKC α protein levels, but levels of PKC β and γ protein remain unchanged (Isakov et al., 1990). Similar mRNA levels for PKC α, β, and γ were found in phorbol ester treated and untreated Jurkat cells, indicating that PKC α is selectively down-regulated in these cells (Isakov et al., 1990). In GH_4C_1 cells, both thyrotropin-releasing hormone (TRH) and phorbol ester specifically down-regulate PKC ϵ, yet PKC α and β are down-regulated only by phorbol ester and not by TRH (Akita et al., 1990b). PKC α is relatively resistant, whereas PKC β is susceptible to phorbol ester down-regulation in GH_4C_1 cells (Akita et al., 1990b), similar to KM3 (Ase et al., 1988) and RBL-2H3 cells (F. L. Huang et al., 1989a). In RBL-2H3 cells, PMA causes membrane association of PKC α, β, δ, and ϵ but not PKC ζ, all of which coexist in this cell (Ozawa et al., 1993a). PKC δ and ϵ also respond to antigen stimulation of RBL-2H3 cells for translocation either in the presence or absence of Ca^{2+} without degradation. In U937 and MDCK-D1 cells, PKC α is susceptible, but PKC β is resistant to down-regulation (Strulovici et al., 1989; Godson, et al., 1990). The induction of differentiation in SH-SY5Y neuroblastoma cells by PMA has been linked to down-regulation, but not to activation, of PKC (Heikkila et al., 1989). Inhibition of PKC by H7-enhanced PMA-induced differentiation, and PMA-induced down-regulation of PKC, was associated with differentiation in these cells. PMA-induced down-regulation of PKC in breast carcinoma cells is associated with growth inhibition and the appearance of membrane-associated 77- and 80-kDa PKC-related proteins, which are immunoreactive with antibody against group A PKC. These proteins are kinase inactive and are not able to bind phorbol ester (Borner et al., 1988), and likely represent certain modified forms of the kinase. Induction of differentiation in promyelocytic leukemic cells by PMA results in an increase in a PKC unable to either translocate or undergo down-regulation (Homma et al., 1986; Solanski et al., 1981), suggesting the possible presence of another modified form of PKC. PKC η, which is exclusively in the nucleus (Greif et al., 1992), is PMA down-regulated in mouse epidermis, and PKC ζ, which is PMA insensitive, does not translocate or down-regulate (Gschwendt et al., 1992).

The catalytic fragment (PKM) generated by limited proteolysis is constitutively active without any activator and also exhibits a broader substrate specificity than the holoenzyme. PKM can phosphorylate myosin light chain at both sets of sites normally recognized separately by PKC and myosin light-chain kinase (Nakabayashi et al., 1991). Proteolytic degradation of PKC ϵ also increases its activity toward histone, which is not a preferred substrate for this PKC (Schaap et al., 1990). More strikingly, PKM has been shown to phosphorylate phosphatidylinositol-4-phosphate using either ATP or GTP, whereas neither PKC nor PKM utilizes GTP to phosphorylate histone (Tusupov et al., 1991). Since proteolysis of PKC is facilitated by its association with the membrane, it is believed that activation of PKC in intact cells will enhance its degradation. Point mutation of PKC α at the putative ATP-binding site eliminates the kinase activity as well as PMA-mediated down-regulation, even though phorbol ester-binding activity and the association of PKC with membrane fractions remain intact (Ohno et al., 1990). It was proposed that autophosphorylation of PKC α is a prerequisite for proteolytic cleavage that leads to down-regulation. However, a separate study using PKC γ kinase-deficient mutant revealed that autophosphorylation of this kinase or expression of kinase activity was not a prerequisite for down-regulation (Freiseminkel et al., 1991). In Swiss-3T3 fibroblasts, treatments of cells with PMA in the presence of K252a (potent staurosporine analog), in spite of the inhibition of PKC autophosphorylation and substrate phosphorylation by the inhibitor, led to a normal degree of down-regulation (Lindner et al., 1991). Although down-regulation of PKC has been observed in numerous PMA-treated cells, it is not yet clear whether PKM play a role in the amplification of the signal leading to cellular responses.

PKC SUBSTRATES

Numerous cell surface receptors, cytoskeletal proteins, ion channels, and cytosolic and nuclear proteins have been identified as potential PKC substrates (Nishizuka, 1986). Phosphorylation of these proteins by PKC manifests the external signal locally at the site of activation or globally by amplification of other signaling pathways. Available evidences indicate that PKC is involved in the control of movement of certain ions across membrane through ion pump, exchanger, and channel and thus modulates the intracellular ion concentration, membrane potential, and electrical signal (for review see Shearman et al., 1989b). Other signaling pathways that make use of cAMP, cGMP, Ca^{2+}, and AA and its metabolites have been shown to be affected by PKC either in a positive or negative fashion. Recently, a group of CaM-binding proteins, such as MARCKS (Stumpo et al., 1989), neuromodulin (GAP-43, B-50, F-1) (Basi et al., 1987; Cimler et al., 1987; Karns et al., 1987; Snipes et al., 1987; Nielander et al., 1987; Meiri et al., 1988), and neurogranin (RC3) (Watson et al., 1990; Baudier et al., 1991) have been identified as the most prominent substrates of PKC α, β, and γ in neural tissue. MARCKS protein binds CaM in the presence of Ca^{2+}, whereas neuromodulin and neurogranin bind CaM in the absence of Ca^{2+}. Synthetic peptides corresponding to the sites of phosphorylation of numerous PKC substrates, with consensus sequence motif of S/TXK/R,K/RXXS/T,K/RXXS/TXK/R,K/RXS/T, and K/RXS/TXK/R, are convenient substrates for measuring cofactor-stimulated PKC activity. Synthetic peptides corresponding to the site of phosphorylation of myelin basic protein, MBP_{4-14}, and that of neurogranin (RC3), $RC3_{29-48}$, are specific substrates of PKCs (unpublished results). The substrate preferences for PKC δ is $MBP_{4-14} \sim$ EGFR peptide $\gg$ ϵ peptide, in the presence of cardiolipin, whereas in the presence of PS it is ϵ peptide $\gg MBP_{4-14} \sim$ EGFR peptide (Mizuno et al., 1991). Parker's group reported substrate preference for PKC δ to be δ peptide $\gg$ α peptide $>$ protamine sulphate $\gg$ histone III-S $\sim$ myelin basic protein (MBP) using PS/Triton X-100 mixed micelles (Olivier and Parker, 1991). PKC ϵ, like PKC δ, has a similar preference for pseudosubstrate peptides: ϵ peptide $>$ α peptide $\gg$ MBP $\gg$ histone III-S in the PS/Triton X-100 micelle assay (Schaap and Parker, 1990). Koide and co-workers (1992) found a similar substrate preference for PKC ϵ: ϵ peptide $\sim$ δ peptide $\sim$ ζ peptide $\sim$ α peptide $\gg$ MBP$_{4-14}$ > MBP $\gg$ histone H1, in the presence of PS and DAG. PKC ζ, independent of PMA and Ca^{2+}, has a substrate preference like that of group B PKCs, ϵ peptide $\gg$ α peptide > protamine sulphate > MBP $\gg$ histone, in the presence of PS (Nakanishi and Exton, 1992). Physiological substrates for group B and C PKCs remain largely unknown. A constructed fusion protein containing the N-terminal regulatory domain of PKC ϵ fused to the C-terminal catalytic domain of PKC γ results in a protein with substrate preference similar to that of PKC ϵ (Pears et al., 1991). PKM generated from PKC ϵ phosphorylates histone III-S, unlike its parent PKC ϵ (Pears et al., 1991). Hence, it appears that the N-terminal extended regulatory domain of PKC ϵ determines its restricted substrate specificity.

Phosphorylation of MARCKS (myristoylated alanine-rich C-kinase substrate) has been used widely as a marker or index of PKC activation in intact cells (Rozengurt et al., 1983; Blackshear et al., 1986). MARCKS is an elongated acidic protein of M_r 28 to 32 kDa, which on SDS-PAGE exhibits an apparent molecular mass of 60 to 87 kDa (Albert et al., 1987; Patel and Kligman, 1987; Morris and Rozengurt, 1988; Graff et al., 1989a). In CNS, the distribution of the MARCKS protein does not match exactly with distribution of PKC subspecies (F. L. Huang et al., 1988; Kitano et al., 1987; Mochly-Rosen et al., 1987; Saito et al., 1988). Molecular cloning of MARCKS cDNA from bovine (Stumpo et al., 1989), rat (Erusalimsky et al., 1991), human (Harlan et al., 1991), mouse (Seykora et al., 1991), and chicken (Graff et al., 1989a) reveals extensive sequence identity within the N-terminal half of the molecule containing myristoylation, CaM binding, and PKC-phosphorylation domains. Three serine residues located within a 25-amino acid stretch of highly basic region have been identified as potential PKC phosphorylation sites. This basic region also serves as Ca^{2+}-dependent CaM-binding sites. These conserved regions may contribute to the physiological function of this protein. The C-terminal half of the MARCKS protein appears to diverge greatly. Phosphorylation of MARCKS by PKC leads to its translocation from membrane to cytosol (Wang et al., 1989) and reduces its binding affinity to CaM (Graff et al., 1989b; McIlroy et al., 1989). MARCKS protein also appears to have a role in reversibly linking the actin cytoskeleton to the plasma membrane (Rosen et al., 1990).

Neuromodulin (GAP-43, B-50, F-1) is a M_r 24 kDa CNS-specific phosphoprotein that appears to play a role in neuronal growth, synaptic plasticity, neurotransmitter release, and phosphoinositide metabolism (for review see Coggins and Zwiers, 1991). In adult rat brain, high levels of neuromodulin are present in the CA1 field of hippocampus, in layer I of the cortex, and in several subcortical structures including caudate-putamen, olfactory tubercle, and amygdala (Benowitz et al., 1988). After synthesis this protein is relocalized by rapid axoplasmic transport to the neuronal growth cone membrane (Pfenninger et al., 1983; Skene et al., 1986). Molecular cloning of neuromodulin cDNA from mouse (Cimler et al., 1987), rat (Karns et al., 1987), and human (Ng et al., 1988) reveals extensive sequence identity. The N-terminal domain includes proposed sites (Cys^3 and

Cys[4]) of fatty acylation and membrane binding (Skene and Virág, 1989) and for binding of CaM (Alexander et al., 1988). The CaM-binding sequence segment forms a distinct region of positive charge and reduced hydrophilicity. The PKC phosphorylation site (Ser[41]) is located adjacent to the CaM-binding region (Apel et al., 1990). Stoichiometric phosphorylation of neuromodulin by PKC weakens its binding to CaM. The N-terminal first 24-amino acid peptide of neuromodulin stimulates GTP-γ-S binding to G_o, which is co-localized in the growth cone, to the same level as the holoprotein (Strittmatter et al. 1990). It is unknown whether palmitylation and phosphorylation have any effect on such stimulation. The C-terminal domain sequence is less conserved among the various species examined; this domain is strikingly deficient in hydrophobic residues and predicts a highly extended negatively charged rod structure (Masure et al., 1986). It is possible that this region of the molecule may interact with cytoskeletal or other cytoplasmic element (Allsopp and Moss, 1989; Meiri and Gordon-Weeks, 1990; Moss et al., 1990). Liu and Storm (1990) proposed that neuromodulin binds and concentrates CaM on the growth cone membrane and that stimulation of PKC releases high concentrations of CaM locally in the growth cone. Interactions between the released CaM and cytoskeleton proteins may affect polymerization, crosslinking, and membrane attachment of cytoskeleton polymer and thus trigger the initial events in filopodia formation and neurite extension. Recently, neuromodulin was shown to interact with acidic phospholipids by electrostatic interaction with the cluster of basic amino acids at the CaM-binding domain (Houbre et al., 1991). In the presence of CaM, binding of neuromodulin to PS is inhibited, resulting in a total inhibition of phosphorylation by PKC. Thus, an increase in $[Ca^{2+}]_i$ can cause release of neuromodulin-bound CaM for the Ca^{2+}/CaM-dependent enzymes; released neuromodulin is also free to interact with phospholipid and to be phosphorylated by PKC. It is not yet clear whether the phosphorylation of neuromodulin or the release of CaM or both mediate the subsequent cellular responses.

Neuromodulin can also be phosphorylated by casein kinase II predominantly at two serine residues and less extensively at three threonine residues located at the C-terminal half of the molecule (Apel et al., 1991). Phosphorylation by casein kinase II does not affect the ability of neuromodulin to bind CaM. However, binding of CaM inhibits the phosphorylation of neuromodulin by casein kinase II in a fashion similar to that by PKC. It has been suggested that phosphorylation of the C-terminal domain by casein kinase II may regulate the interaction of neuromodulin with the components of the membrane skeleton. Phosphorylation of neuromodulin by PKC prevents CaM from binding to this protein and thus indirectly enhances the phosphorylation of neuromodulin by casein kinase II.

Neurogranin (RC3) is a recently discovered CNS-specific PKC substrate (Watson et al., 1990; Baudier et al., 1991). This protein is expressed in limited areas of the brain but with high concentration in cerebral cortex and hippocampus and is nearly undetectable in cerebellum, thalamus, and brain stem. Neurogranin has a molecular mass of 7500 to 7800, yet on SDS-PAGE the protein monomer migrates abnormally with an apparent M_r of 15 to 19 kDa. This protein has been shown to be phosphorylated in hippocampal slices on stimulation by phorbol ester

(Baudier et al., 1991), suggesting that it is likely an *in vivo* substrate of PKC. *In vitro,* neurogranin is phosphorylated by PKC and not by other protein kinases. A serine residue adjacent to the CaM-binding domain, which exhibits extensive sequence homology to the same domain in neuromodulin, has been identified as the sole phosphorylation site. The amino acid sequence surrounding the site of phosphorylation (KIQASFRGH) is conserved between neurogranin and neuro-modulin, suggesting a functional relationship between these two proteins. The deduced amino acid sequences of rat and mouse neurogranin cDNAs are identical and the bovine sequence is identical to the rodent protein in at least 73 of 78 residues. The N-terminal domain of neurogranin is rich in dicarboxylic amino acids and cysteine and the C-terminal domain is rich in glycine. The cysteine-rich domain is similar to domains of snake venom neurotoxins (Tu, 1973) and the glycine-rich hydrophobic domain resembles collagen (Fietzek et al., 1973). The expression of neurogranin in developing rat brain parallels that of PKC γ, and both of these two proteins are concentrated in the cell bodies and dendrites of numerous cortical and hippocampal neurons (F. L. Huang et al., 1988; Represa et al., 1990; Watson et al., 1990). Since the cerebral cortex is the most recently evolved structure in mammalian brain, neurogranin, which is expressed in high level in this region, may be involved in higher order brain function.

Protein kinases phosphorylate a number of transcription factors including c-jun, c-fos, CREB, SRF, SP1, Oct2, Myb, Max, myogenin, HSF, Gal-4, ADR1, C/EBP α and β, and the vitamin D receptor (for reviews see Bohman, 1990; Hsieh et al., 1991; Hunter and Karin, 1992; Mahoney et al., 1992; Wegner et al., 1992). Since various PKC subtypes have been found in and can translocate to the nucleus depending on the cell type (see above), and because activation of PKC is associated with altered gene expression, it can be expected that PKC phospho-rylates some transcription factors and thereby regulates gene expression. PKC phosphorylates CREB, the cAMP-responsive element-binding protein, at the same and additional sites as cAMP-dependent protein kinase (PKA), and as a result CRE binding is dramatically enhanced (Yamamoto et al., 1988). C/EBP α is phosphorylated by PKC α, β, and γ at Ser^{248}, Ser^{277}, and Ser^{299}, which results in greatly attenuated DNA binding. With use of mutant-truncated proteins, the attenuation of DNA binding by PKC phosphorylation of C/EBP α was mapped to Ser^{299}, which lies in the basic DNA binding domain (Mahoney et al., 1992). Truncated C/EBP α containing the C-terminal basic DNA-binding domain and the downstream leucine zipper/dimerization domain, when bound to its cognitive DNA probe, is only weakly phosphorylated by PKC, indicating that the phos-phorylation site Ser^{299} is blocked in the bound state (Mahoney et al., 1992). On the basis of amino acid sequence homology of other transcription factors relative to the Ser^{299} region of C/EBP α, it was predicted that GCN4, CPC-1, HBP-1, TGA-1, Opague 2, and v-jun should be PKC substrates, but not c-jun (Mahoney et al., 1992). In contrast, C/EBP β is a substrate for Ca^{2+}/calmodulin kinase II, but not C/EBP α (Wegner et al., 1992). The Ca^{2+}/calmodulin kinase II phos-phorylation site on C/EBP β was delineated as being Ser^{276}, which lies in the C-terminal leucine zipper/dimerization domain. Co-transfection of both Ca^{2+}/cal-modulin kinase II and C/EBP β constructs into G/C cells resulted in a dramatic stimulation of transcription from the linked reporter gene, whereas replacement of C/EBP β with the C/EBP α construct showed very weak stimulation of transcription (Wegner et al., 1992). The Ca^{2+}/calmodulin kinase II phosphorylation site, Ser^{276}, in C/EBP β is not conserved in C/EBP α or δ and C/EBP β was not phosphorylated by PKA (Wegner et al., 1992). NF-κB, in the resting cell, is predominantly cytosolic as a complex with I-κB. On activation of PKC by PMA, I-κB becomes phosphorylated, the complex dissociates, and NF-κB translocates into the nucleus as an active transcription factor (Baeuerle and Baltimore, 1988; Ghosh and Baltimore, 1990). c-jun, a PMA-responsive transcription factor, is also regulated by PKC. Epithelial and fibroblastic cells, when stimulated by PMA, undergo dephosphorylation of c-jun at sites that negatively regulate its DNA binding activity (Boyle et al., 1991). GSK-3 phosphorylation of c-jun occurs at Ser^{239}, Ser^{243}, and Ser^{249} just upstream from the basic DNA binding domain and results in attenuated binding to its PMA-responsive element (Boyle et al., 1991). c-jun phosphorylated by GSK-3, in the presence of PKC and PMA, undergoes dephosphorylation, which results in dramatically enhanced binding to its cognitive DNA element (Boyle et al., 1991). It is postulated that in the resting cell, c-jun is in a phosphorylated inactive state and that on activation of PKC, dephosphorylation occurs at one or several of the GSK-3 sites, thereby activating DNA binding and gene expression. The vitamin D_3 receptor (VDR), a member of the steroid/thyroid hormone receptor superfamily, binds vitamin D_3, associates with its cognitive DNA element, and thereby alters transcription of genes such as the osteocalcin gene (Hsieh et al., 1991). PKC β phosphorylates VDR at Ser^{51} and mutation of Ser^{51} markedly inhibits transcriptional activation by vitamin D_3 (Hsieh et al., 1991). PMA treatment of Swiss-3T3 cells resulted in a dramatic decrease in both VDR mRNA and protein levels, while substantially stimulating proliferation (Krishnan and Feldman, 1991).

LOWER EUKARYOTIC PKC

In addition to mammals, PKC or PKC-related proteins have been found in lower eukaryotes such as yeast (Levin et al., 1990; Ogita et al., 1990; Simon et al., 1991; Iwai et al., 1992). *Drosophila* (Rosenthal et al., 1987; Schaeffer et al., 1989; Choi et al., 1991), *Caenorhabditis elegans* (Tabuse et al., 1989), *Dictyostelium* (Jimenez et al., 1989; Luderus et al., 1989), sea urchin eggs (Shen and Ricke, 1989), and *Xenopus laevis* oocytes (Chen et al., 1988; Otte et al., 1988, 1991; Dominquez et al., 1992; Otte and Moon, 1992; Sahara et al., 1992) (Figure 2.5). *Saccharomyces cerevisiae* appears to contain three distinct group A-like PKCs, as well as unique subtypes. Three enzymatically active fractions, obtained from hydroxyapaptite chromatography, were activated by Ca^{2+}, PS, and DAG or PMA, and utilized histone III-S as a substrate (Simon et al., 1991). Western blotting using anti-rat brain group A PKC antibody revealed a strongly cross-reacting 85-kDa protein in one of the peaks from hydroxyapaptite fractionation (Simon et al., 1991). In contrast, Ogita and co-workers (1990) isolated a 90-kDa PKC from *S. cerevisiae*, which was activated by Ca^{2+}, PS, and DAG but not by PMA. This

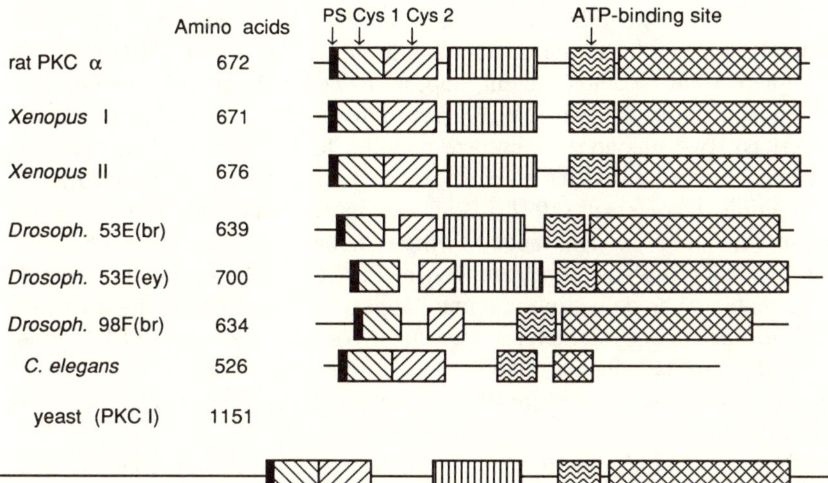

Figure 2.5 Schematic structure of lower eukaryotic PKC members. All lower eukaryotic PKCs contain two Cys-rich Zn^{2+} fingers; C2, the putative Ca^{2+} binding domain (with the exception of *C. elegans* and *Drosophila* 98F(br)); a pseudosubstrate domain; and C3/C4 catalytic domains. Yeast PKC1, considerably larger than all other PKCs, contains extended variable V1 and V5 regions. Rat PKC α is shown for comparative purposes. See Figure 2.1 for symbol presentation.

yeast PKC preferentially phosphorylated Thr rather than Ser residues, in contrast to mammalian group A PKCs, and had a substrate preference: MBP ≫ histone H1 > protamine sulphate ≫ MBP_{4-14} ~ S6 peptide ~ EGF receptor peptide (Ogita et al., 1990). Moreover, this yeast PKC phosphorylated MBP at amino acid residues (Thr^{19}, Thr^{34}, Thr^{65}) different from those phosphorylated by the rat brain group A PKC (Ser^{8}, Ser^{46}, Ser^{55}, Ser^{110}, Ser^{132}, Ser^{151}, Ser^{161}) (Iwai et al., 1992). MBP_{4-14}, which is a good substrate for group A and B (under select conditions), is a poor substrate for this yeast PKC, whereas MBP_{12-23} is a good substrate for the latter, due to the different phosphorylation sites for the yeast PKC compared with mammalian PKC (Iwai et al., 1992). A cDNA for PKC1, cloned from *S. cerevisiae*, predicted a 132-kDa protein containing two Cys-rich Zn^{2+} fingers, a C2 region, and the catalytic domain (Levin et al., 1990) (Figure 2.5). No enzymatic characterization of the predicted 132-kDa PKC was reported, but such a PKC would be expected to have characteristics similar to those of mammalian group A PKC, although it is considerably larger than the latter. Deletion of PKC1 in *S. cerevisiae* resulted in recessive lethality (Levin et al., 1990). Cells depleted of PKC1 gene product had a phenotype similar to that of cell cycle-dependent (cdc) mutants in which cell division was arrested subsequent to DNA replication but prior to mitosis (Levin et al., 1990).

Three PKC genes—53E(br), 98F(br), and 53E(ey)—have been identified from *Drosophila*. The first two are most abundantly transcribed in brain, and the last exclusively in photoreceptor cells (Rosenthal et al., 1987; Schaeffer et al., 1989). dPKC 53E(br) gene spans about 20 kb and contains 14 exons (Rosenthal et al.,

1987). dPKC 53E(br) and 53E(ey), coding for 639 and 700 amino acids protein, respectively, are similar to the mammalian group A PKCs and dPKC 98F (br), coding for 634 amino acids protein, appears to lack C2 and is homologous to PKC δ (Schaeffer et al., 1989). PKCs from *Drosophila* have not yet been well characterized; two enzymes, one corresponding to Ca^{2+}/PS- and the other to PDBu/PS-stimulated form, have been identified from the extracted head membrane fraction (Choi et al., 1991). Both enzyme activities are reduced in the *Drosophila* learning-deficient mutant *turnip* (Choi et al., 1991), although an 84-kDa protein that cross-reacts with anti-bovine PKC antibody is present in the wild type and mutant. The defect in the mutant appears to be related to the activation of PKC; *turnip*[+] is not a structural gene for PKC since *Drosophila* PKC genes map elsewhere in the genome relative to the former. Nevertheless, the wild-type *turnip* gene product is required for PKC activation and is correlated with learning in *Drosophila*.

A cDNA for PKC, cloned from the soil nematode *C. elegans*, predicted a gene product of 526 amino acids containing two Cys-rich Zn^{2+} fingers and a catalytic domain, but not C2 (Tabuse et al., 1989). PMA causes disturbances in the behavior and growth of *C. elegans*, and PMA-resistant mutants show little PMA-induced phenotypic change (Tabuse et al., 1989). PKC has been purified to near homogeneity from the lower eukaryotic slime mold *Dictyostelium discoideum* (Jimenez et al., 1989). The enzyme, a 140-kDa phosphoprotein, is stimulated by PS but not by DAG or Ca^{2+}, and utilizes histone H1 as phosphoacceptor. In contrast, Luderus and colleagues (1989) partially purified PKC from *D. discoideum;* activity in the presence of PS and PMA was inhibited by Ca^{2+}, but in the presence of either PS or DAG was stimulated by Ca^{2+}, using EGF receptor peptide as a substrate. PKC, which has been implicated in the respiratory burst associated with fertilization (Heinecke and Shapiro, 1992) has been partially purified from sea urchin eggs, a commonly used fertilization model system (Shen and Ricke, 1989). The enzyme is activated by Ca^{2+}, PS, and DAG or PMA and utilized histone H1 as a substate. Two cDNAs for PKC have been cloned from *Xenopus laevis* oocytes (Chen et al., 1988). These two cDNAs predict gene products containing the conserved C1–C4 and variable V1–V5 domains, like group A PKC, and consist of 671 and 676 amino acids, respectively. In agreement with this, Sahara and co-workers (1992) have partially purified and biochemically characterized two distinct PKCs, xPKC I and II, from *Xenopus laevis* oocytes. Both xPKCs cross-reated with anti-rat group A PKC-catalytic, $-$C1, and $-$C2 domain antibodies, and revealed a single 80-kDa band on Western blots. Both xPKCs had similar phospholipid and phorbol ester preferences and degrees of activation relative to rat PKC α. xPKC I and II are synergistically activated to a greater extent in the presence of PS, DAG, and 50-μM arachidonic acid at low $[Ca^{2+}]$ (1 μM), compared with rat PKC α (Sahara et al., 1992). xPKC II is significantly activated by arachidonic acid alone independent of $[Ca^{2+}]$ (Sahara et al., 1992). xPKC II was nearly totally down-regulated in PMA-treated (300-nM) oocytes within 1 hour, whereas xPKC I activity and protein levels as determined by Western blots were essentially unchanged up to 2 hours under the same conditions (Sahara et al., 1992). Induction of neural differentiation in *Xenopus laevis* oocyte ectoderm by mesoderm or PMA results in translocation of PKC from the cytosol to particulate fraction (Otte and Moon, 1992). PKC α and γ are preferentially abundant in dorsal and ventral ectoderm, respectively, whereas PKC β is uniformly distributed in *Xenopus laevis* oocytes (Otte et al., 1988, and Otte and Moon, 1992). PMA can induce neural differentiation in *Xenopus* oocyte dorsal ectoderm, but not in ventral ectoderm (Otte et al., 1991). In contrast, ventral ectoderm of *Xenopus* oocyte can be induced to differentiate into a neural phenotype by mesoderm (Otte and Moon, 1992). Overexpression of PKC α elevated the neural competence of ventral ectoderm to that of dorsal ectoderm. Hence, it appears that localized PKC subtypes play an intimate role in regulating competence and neural differentiation in early development of *X. laevis*. In addition to group A PKCs, PKC ζ, but not PKC δ or ϵ, is detected in *X. laevis* oocytes, using Western blotting (Otte and Moon, 1992). Insulin, p21[ras], and PC-PLC action, which can induce mitogenesis and maturation in oocytes independently of group A PKCs, was associated with PKC ζ activity in *X. laevis* oocytes (Dominquez et al., 1992). In contrast, progesterone-induced mitogenesis/maturation in this system was not associated with PKC ζ activity (Dominquez et al., 1992).

CONCLUSION

Molecular cloning and mutagenesis analysis of the PKC gene family members have yielded a wealth of information on the structure/function relationship of these enzymes. Structural features important for Zn^{2+}, phorbol ester, or DAG binding have been elucidated; however, those for the binding of PS, Ca^{2+}, PIP_2, lysophospholipids, and unsaturated fatty acids remain to be defined. Activation of PKC by various stimulators has been illustrated *in vitro* under specific sets of assay conditions; the effects of these putative activators generated *in situ* by the various PLCs, PLD, and PLA_2 have largely yet to be demonstrated. Similarly, the mechanism of activation of the Ca^{2+}-independent subgroup of PKCs by the various lipid activators also requires further investigation. The ultimate proof for the activation of PKC requires a direct demonstration of the ligand-induced effect in intact cells. Recently, monitoring of the activation of cAMP-dependent protein kinase by fluorescence ratio imaging has been developed (Adams et al., 1991); a similar approach may be applicable for PKC. *In situ* determination of PKC activation at selective cellular locations in response to a local rise in Ca^{2+}, such as that seen in the dendritic spine of the hippocampal pyramidal cells (Müller and Connor, 1991), should be correlated with the phosphorylation of a specific substrate.

Activation of PKC under a variety of conditions likely involves multiple isozymes at distinct cellular locations. It remains a challenge to define the role of each isoform in the various physiological responses. Selective translocation and/or down-regulation of each isozyme can be followed by using antibodies specific for each kinase. Supplementation of selective PKC isozymes to the washed permeabilized cells for the analysis of secretory response and phospholipid turnover has been proven to be useful for defining the roles of each PKC isozymes (Naor et al., 1989; Ozawa et al., 1993b). It would be most useful if selective activators, inhibitors, or substrates for each PKC isozyme can be identified; however, progress has yet to be made in this respect. Likewise, the functional roles of many PKC substrates are still poorly defined. Identification of prominent PKC substrates, MARCKS, neuromodulin (GAP-43, B-50, F-1), and neurogranin (RC3), as CaM-binding proteins, is an important advance in understanding the amplification of PKC-mediated signal transduction. These proteins may also possess CaM-regulated enzyme activity, which could be regulated by PKC phosphorylation.

Many of the long-term events regulated by PKC involve the regulation of gene transcription, as in the case of genes containing phorbol ester-responsive elements. The mechanism of signal transduction from cell surface receptors for growth factors, mitogens, and neurotransmitters to the nucleus remains largely unknown. Nuclear membrane is rich in phosphoinositides and contains functional IP_3 receptors (Malviya et al., 1990) and an ATP-stimulated Ca^{2+} uptake system. In addition, nuclear extracts contain PKC and DAG kinase, a key enzyme in attenuating PKC activity. Thus, nuclear PKCs may be stimulated by Ca^{2+}, PS, and DAG in a fashion similar to cytosolic enzymes. Several transcriptional factors, histone, and nonhistone chromatin proteins are likely physiological substrates of nuclear PKC. The detailed mechanisms related to the generation of DAG and transport of PKC into the nucleus have yet to be defined. Identification of PKC subtypes homologous to the mammalian PKCs in the lower eukaryotes will facilitate the elucidation of the PKC-mediated nuclear events by genetic analysis.

REFERENCES

Abate, C., D. R. Marshak, and T. Curran, 1991. Fos is phosphorylated by p34[cdc2], cAMP-dependent protein kinase, and protein kinase C at multiple sites clustered within regulatory regions. *Oncogene* 6:2179–2185.

Adachi, Y., M. Maki, K. Ishii, M. Hatanaka, and T. Murachi. 1990. Possible involvement of calpain in down-regulation of protein kinase C. *Adv. Sec. Mess. Phosphoprot. Res.* 24:478–484.

Adams, S. R., A. T. Harootunian, Y. J. Buechler, S. S. Taylor, and R. Y. Tsien. 1991. Fluorescence ratio imaging of cyclic AMP in single cells. *Nature* 349:694–697.

Ahmed, S., R. Kozma, J. Lee, C. Monfries, N. Harden, and L. Lim. 1991. The cysteine-rich domain of human proteins, neuronal chimaerin, protein kinase C, and diacylglycerol kinase binds zinc. *Biochem. J.* 280:233–241.

Aitken, A., C. A. Ellis, A. Harris, L. A. Sellers, and A. Toker. 1990. Kinases and neurotransmitters. *Nature* 344:594.

Akita, Y., S. Ohno, Y. Konno, A. Yano, and K. Suzuki. 1990a. Expression and properties of two distinct classes of the phorbol ester receptor family, four conventional protein kinase C types and a novel protein kinase C. *J. Biol. Chem.* 265:354–362.

Akita, Y., S. Ohno, Y. Yajima, and K. Suzuki. 1990b. Possible role of Ca^{2+}-independent protein kinase C isozyme, nPKCε, in thyrotropin-releasing hormone-stimulated signal transduction; differential down-regulation of nPKCε in GH₄C₁ cells. *Biochem. Biophys. Res. Commun.* 172:184–189.

Albert, K. A., W. C.-S. Wu, A. C. Nairn, and P. Greengard. 1984. Inhibition by calmodulin of calcium/phospholipid-dependent protein phosphorylation. *Proc. Natl. Acad. Sci. U.S.A.* 81:3622–3625.

Albert, K. A., A. C. Nairn, and P. Greengard. 1987. The 87kDa protein, a major specific substrate for protein kinase C: Purification from bovine brain and characterization. *Proc. Natl. Acad. Sci. U.S.A.* 84:7046–7050.

Alcantara, O., M. Javors, and D. H. Boldt. 1991. Induction of protein kinase C mRNA in cultured lymphoblastoid T cells by iron-transferrin but not by soluble iron. *Blood* 77:1290–1297.

Alessenko, A., W. A. Khan, W. C. Westel, and Y. A. Hannun. 1992. Selective changes in protein kinase C isoenzymes in rat liver nuclei during liver regeneration. *Biochem. Biophys. Res. Commun.* 182:1333–1339.

Alexander, K. A., B. T. Wakim, G. S. Doyle, K. A. Walsh, and D. R. Storm. 1988. Identification and characterization of the calmodulin-binding domain of neuromodulin, a neurospecific calmodulin-binding protein. *J. Biol. Chem.* 263:7544–7549.

Allsopp, T. E., and D. J. Moss. 1989. A developmentally regulated chicken neuronal protein associated with the cortical cytoskeleton. *J. Neurosci.* 9:13–24.

Apel, E. D., M. F. Byford, D. Au, K. A. Walsh, and D. R. Storm. 1990. Identification of the protein kinase C phosphorylation site in neuromodulin. *Biochemistry* 29:2330–2335.

Apel, E. D., D. W. Litchfield, R. H. Clark, E. G. Krebs, and D. K. Storm. 1991. Phosphorylation of neuromodulin (GAP-43) by casein kinase II. Identification of phosphorylation sites and regulation by calmodulin. *J. Biol. Chem.* 266:10544–10551.

Ase, K., N. Berry, U. Kikkawa, A. Kishimoto, and Y. Nishizuka. 1988. Differential down-regulation of protein kinase C subspecies in KM3 cells. *FEBS Lett.* 236:396–400.

Ashendel, C. L., J. M. Staller, and R. K. Boutwell. 1983. Identification of a calcium- and phospholipid-dependent phorbol ester binding activity in the soluble fraction of mouse tissues. *Biochem. Biophys. Res. Commun.* 111:340–345.

Bacher, N., T. Zisman, E. Berent, and E. Livneh. 1991. Isolation and characterization of PKC-L, a new member of the protein kinase C-related gene family specifically expressed in lung, skin, and heart. *Molec. Cell Biol.* 11:126–133.

Badwey, J. A., R. W. Erickson, and J. T. Curnutte. 1991. Staurosporine inhibits the soluble and membrane-bound protein tyrosine kinases of human neutrophils. *Biochem. Biophys. Res. Commun.* 178:423–429.

Baeuerle, P. A., and D. Baltimore. 1988. IκB: A specific inhibitor of NF-κB transcription factor. *Science* 242:540–546.

Basi, G. S., R. D. Jacobson, I. Virag, J. Schilling, and J. H. P. Skene. 1987. Primary structure and transcriptional regulation of GAP-43, a protein associated with nerve growth. *Cell* 49:785–791.

Baudier, J., J. C. Deloulme, A. Van Dorsselaer, D. Black, and H. W. D. Matthes. 1991. Purification and characterization of a brain-specific protein kinase C substrate, neurogranin (p 17). *J. Biol. Chem.* 266:229–237.

Bazzi, M. D., and G. L. Nelsestuen. 1987. Role of substrate in imparting calcium and phospholipid requirements to protein kinase C activation. *Biochemistry* 26:1974–1982.

Bazzi, M. D., and G. L. Nelsestuen. 1988. Properties of membrane-inserted protein kinase C. *Biochemistry* 27:7589–7593.

Bazzi, M. D., and G. L. Nelsestuen. 1990. Protein kinase C interaction with Ca^{2+}: A phospholipid-dependent process. *Biochemistry* 29:7624–7630.

Bell, R. M. 1986. Protein kinase C activation by diacylglycerol second messengers. *Cell* 45:631–632.

Bell, R. M., and D. J. Burns. 1991. Lipid activation of protein kinase C. *J. Biol. Chem.* 266:4661–4664.

Benowitz, L. I., P. J. Apostolides, N. Perrone-Bizzozero, S. P. Finklestein, and H. Zwiers. 1988. Anatomical distribution of the growth-associated protein GAP-43/B-50 in the adult brain. *J. Neurosci.* 8:339–352.

Berg, J. M. 1990. Zinc finger domains: Hypotheses and current knowledge. *Annu. Rev. Biophys. Biophys. Chem.* 19:405–421.

Blackshear, P. J., L. Wen, B. P. Glynn, and L. A. Witters. 1986. Protein kinase C-stimulated phosphorylation *in vitro* of a M$_r$ 80,000 protein phosphorylated in response to phorbol esters and growth factors in intact fibroblasts. *J. Biol. Chem.* 261:1459–1469.

Bohman, D. 1990. Transcription factor phosphorylation: A link between signal transduction and the regulation of gene expression. *Cancer Cells* 2:337–344.

Bonner, T. I., H. Oppermann, P. Seeburg, S. B. Kerby, M. A. Gunnel, A. C. Young, and U. R. Rapp. 1986. The complete coding sequence of the human *raf* oncogene and the corresponding structure of the C-raf-1 gene. *Nucleic Acids Res.* 14:1009–1015.

Borner, C., U. Eppenberger, R. Wyss, and D. Fabbro. 1988. Continuous synthesis of two PKC-related proteins after down-regulation by phorbol esters. *Proc. Natl. Acad. Sci. U.S.A.* 85:2110–2114.

Boyle, W. J., T. Smeal, L. H. K. Defize, P. Angel, J. R. Woodgett, M. Karin, and T. Hunter. 1991. Activation of protein kinase C decreases phosphorylation of c-Jun at sites that negatively regulate its DNA-binding activity. *Cell* 64:573–584.

Bruns, R. F., F. D. Miller, R. L. Merriman, J. J. Howbert, W. F. Heath, E. Kobayashi, I. Takahashi, T. Tamaoki, and H. Nakano. 1991. Inhibition of protein kinase C by calphostin C is light-dependent. *Biochem. Biophys. Res. Commun.* 176:288–293.

Buchner, K., H. Otto, R. Hilbert, C. Lindschau, H. Haller, and E. Hucho. 1992. Properties of protein kinase C associated with nuclear membranes. *Biochem. J.* 286:369–375.

Burns, D. J., and R. M. Bell. 1991. Protein kinase C contains two phorbol ester binding domains. *J. Biol. Chem.* 266:18330–18338.

Cambier, J. C., M. L. Newell, L. B. Justement, J. C. McGuire, K. L. Leach, and Z. Z. Chen. 1987. Ia binding ligands and cAMP stimulate nuclear translocation of protein kinase C in B lymphocytes. *Nature* 327:629–632.

Chauhan, V. P. S., and H. Brockerhoff. 1988. Phosphatidylinositol 4,5-bisphosphate may antecede diacylglycerol as activator of protein kinase C. *Biochem. Biophys. Res. Commun.* 155:18–23.

Chauhan, V. P. S., and A. Chauhan. 1992. Protamine induces autophosphorylation of protein kinase C: Stimulation of protein kinase C-mediated protamine phosphorylation by histone. *Life Sci.* 51:537–544.

Chen, K.-H., Z-G. Peng, S. Lavu, and H-F. Kung. 1988. Molecular cloning and sequence analysis of two distinct types of *Xenopus laevis* protein kinase C. *Sec. Mess. Phosphoprot. Res.* 12:251–260.

Chen, K.-H., S. G. Widen, S. H. Wilson, and K-P. Huang. 1990. Characterization of the 5′-flanking region of the rat protein kinase C γ gene. *J. Biol. Chem.* 265:19961–19965.

Choi, K.-W., R. F. Smith, R. M. Buratowski, and W. G. Quinn. 1991. Deficient protein kinase C activity in *turnip,* a *Drosophila* learning mutant. *J. Biol. Chem.* 266: 15999–16006.

Cimler, B. M., D. H. Giebelhaus, B. T. Wakim, D. R. Storm, and R. T. Moon. 1987. Characterization of murine cDNAs encoding P-57, a neural-specific calmodulin binding protein. *J. Biol. Chem.* 262:12158–12163.

Clark, J. D., L.-L. Lin, R. W. Kriz, C. S. Ramesha, L. A. Sultzman, A. Y. Lin, N. Milona, and J. L. Knopf. 1991. A novel arachidonic acid-selective cytosolic PLA_2 contains a Ca^{2+}-dependent translocation domain with homology to protein kinase C and GAP. *Cell* 65:1043–1051.

Coggins, P. J., and H. Zwiers. 1991. B-50 (GAP-43): Biochemistry and functional neurochemistry of a neuron-specific phosphoprotein. *J. Neurochem.* 56:1095–1106.

Csermely, P., M. Szamel, K. Resch, and J. Somogyi. 1988. Zinc can increase the activity of protein kinase C and contributes to its binding to plasma membranes in T lymphocytes. *J. Biol. Chem.* 263:6487–6490.

Devalia, D., N. S. Thomas, P. J. Roberts, H. M. Jones, and D. C. Linch. 1992. Downregulation of human protein kinase C α is associated with terminal neutrophil differentiation. *Blood* 80:68–76.

Divecha, N., H. Banfic, and R. F. Irvine. 1991. The polyphosphoinositide cycle exists in the nuclei of Swiss 3T3 cells under control of a receptor (for IGF-1) in the plasma membrane, and stimulation of the cycle increases nuclear diacylglycerol and apparently induces translocation of protein kinase C to the nucleus. *EMBO J.* 10:3207–3214.

Dominquez, I., M. T. Diaz-Meco, M. M. Municio, E. Berra, A. Garcia de Herreros, M. E. Cornet, L. Sanz, and J. Moscat. 1992. Evidence for a role of protein kinase C ζ subspecies in maturation of *Xenopus laevis* oocytes. *Molec. Cell. Biol.* 12:3776–3783.

el Touny, S., W. Kahn, and Y. Hannun. 1990. Regulation of platelet protein kinase C by oleic acid. Kinetic analysis of allosteric regulation and effects on autophosphorylation, phorbol ester binding, and susceptibility to inhibition. *J. Biol. Chem.* 265: 16437–16443.

Erusalimsky, J. D., S. F. Brooks, T. Herget, C. Morris, and E. Rozengurt. 1991. Molecular cloning and characterization of the acidic 80-kda protein kinase C substrate from rat brain. Identification as a glycoprotein. *J. Biol. Chem.* 266:7073–7080.

Fields, A. P., S. M. Pincus, A. S. Kraft, and W. S. May. 1989. Interleukin-3 and bryostatin 1 mediate rapid nuclear envelope protein phosphorylation in growth factor-dependent FDC-P1 hematopoietic cells. A possible role for nuclear protein kinase C. *J. Biol. Chem.* 264:21896–21901.

Fietzek, P. P, F. Rexrodt, K. Hopper, and K. Kuhn. 1973. The covalent structure of collagen. *Eur. J. Biochem.* 28:396–400.

Flint, A. J., R. D. Paladini, and D. E. Koshland Jr. 1990. Autophosphorylation of protein kinase C at three separated regions of its primary sequence. *Science* 249:408–411.

Freiseminkel, I., D. Riethmacher, and S. Stabel. 1991. Downregulation of protein kinase C-γ is independent of a functional kinase domain. *FEBS Lett.* 280:262–266.

Ganong, B. R., C. R. Loomis, Y. A. Hannun, and R. M. Bell. 1986. Specificity and mechanism of protein kinase C activation by sn-1,2-diacylglycerols. *Proc. Natl. Acad. Sci. U.S.A.* 83:1184–1188.

Ghosh, S., and D. Baltimore. 1990. Activation *in vitro* of Nf-κB by phosphorylation of its inhibitor IκB. *Nature* 344:678–682.

Girard, P. R., G. J. Mazzei, and J. F. Kuo. 1986. Immunological quantitation of phospholipid/Ca^{2+}-dependent protein kinase and its fragments. Tissue levels, subcellular distribution, and ontogenetic changes in brain and heart. *J. Biol. Chem.* 261:370–375.

Godson, C., B. A. Weiss, and P. A. Insel. 1990. Differential activation of protein kinase C α is associated with arachidonate release in Madin-Darby canine kidney cells. *J. Biol. Chem.* 265:8369–8372.

Gonzalez, G. A., K. K. Yamamoto, W. H. Biggs, and M. R. Montminy. 1989. A cluster of phosphorylation sites on the cyclic AMP regulated nuclear factor CREB predicted by its sequence. *Nature* 337:749–752.

Gopalakrishna, R., S. H. Barsky, T. P. Thomas, and W. B. Anderson. 1986. Factors influencing chelator-stable, detergent-extractable, phorbol diester-induced membrane association of protein kinase C. *J. Biol. Chem.* 261:16438–16445.

Graff, J. M., D. J. Stumpo, and P. J. Blackshear. 1989a. Molecular cloning, sequence, and expression of a cDNA encoding the chicken myristoylated alanine-rich C kinase substrate (MARCKS). *Molec. Endocrinol.* 3:1903–1906.

Graff, J. M., T. M. Young, J. D. Johnson, and P. J. Blackshear. 1989b. Phosphorylation-regulated calmodulin binding to a prominent cellular substrate for protein kinase C. *J. Biol. Chem.* 264:21818–21823.

Greif, H., J. Ben-Chaim, T. Shimon, E. Bechor, H. Eldar, and E. Livneh. 1992. The protein kinase C-related PKC-*L* (η) gene product is localized in the cell nucleus. *Molec. Cell. Biol.* 12:1304–1311.

Gschwendt, M., H. Leibersperger, W. Kittstein, and F. Marks. 1992. Protein kinase C ζ and η in murine epidermis: TPA induces down-regulation of PKC η but not PKC ζ. FEBS *Lett* 307:151–155.

Hall, C., C. Monfries, P. Smith, H. H. Lim, R. Kozma, S. Ahmed, V. Vannasingham, T. Leung, and L. Lim. 1990. Novel human brain cDNA encoding a 34,000 M_r protein n-chimaerin, related to both the regulatory domain of protein kinase C and BCR, the product of the breakpoint cluster region gene. *J. Molec. Biol.* 211:11–16.

Halsey, D. L., P. R. Girard, J. F. Kuo, and P. J. Blackshear. 1987. Protein kinase C in fibroblasts. Characteristics of its intracellular location during growth and after exposure to phorbol esters and other mitogens. *J. Biol. Chem.* 262:2234–2243.

Hannun, Y. A., C. R. Loomis, and R. M. Bell. 1986a. Phorbol ester binding and activation of protein kinase C on triton X-100 mixed micelles containing phosphatidylserine. *J. Biol. Chem.* 261:9341–9347.

Hannun, Y. A., C. R. Loomis, and R. M. Bell. 1986b. Protein kinase C activation in mixed micelles. Mechanistic implications of phospholipid, diacylglycerol, and calcium interdependencies. *J. Biol. Chem.* 261:7184–7190.

Hannun, Y. A., C. R. Loomis, A. H. Merill Jr., and R. M. Bell. 1986c. Sphingosine inhibition of protein kinase C activity and of phorbol dibutyrate binding *in vitro* and in human platelets. *J. Biol. Chem.* 261:12604–12609.

Hansson, A., C. N. Serham, J. Haeggstrom, M. Ingelman-Sundberg, and B. Samuelsson. 1986. Activation of protein kinase C by lipoxin A and other eicosanoids. Intracellular action of oxygenation products of arachidonic acid. *Biochem. Biophys. Res. Commun.* 134:1215–1222.

Hansson, A., G. Skoglund, I. Lassing, U. Lindberg, and M. Ingelman-Sundberg. 1988. Protein kinase C-dependent phosphorylation of profilin is specifically stimulated by phosphatidylinositol bisphosphate (PIP_2). *Biochem. Biophys. Res. Commun.* 150:526–531.

Harlan, D. M., J. M. Graff, D. J. Stumpo, R. L. Eddy Jr., T. B. Shows, J. M. Boyle, and P. J. Blackshear, 1991 The human myristoylated alanine-rich C kinase substrate (MARCKS) gene (MACS). *J. Biol. Chem.* 266, 14399–14405.

Hartwig, J. H., M. Thelen, A. Rosen, P. A. Janmez, A. C. Nairn, and A. Aderem. 1992. MARKS is an actin filament crosslinking protein regulated by protein kinase C and calcium-calmodulin. *Nature* 356:618–622.

Heikkila, J. E., G. Akerlind, and K. E. O. Akerman. 1989. Protein kinase C activation and down-regulation in relation to phorbol ester-induced differentiation of SH-SY5Y human neuroblastoma cells. *J. Cell Physiol.* 140:593–600.

Heinecke, J. W., and B. M. Shapiro. 1992. The respiratory burst oxidase of fertilization. *J. Biol. Chem.* 267:7959–7962.

Herbert, J. M., J. M. Augereau, J. Gleye, and J. P. Maffrand. 1990. Chelerythrine is a potent and specific inhibitor of protein kinase C. *Biochem. Biophys. Res. Commun.* 172:993–999.

Hidaka, H., M. Inagaki, S. Kawamoto, and Y. Sasaki. 1984. Isoquinolinesulfonamides, novel and potent inhibitors of cyclic nucleotide dependent protein kinase and protein kinase C. *Biochemistry* 23:5036–5041.

Hocevar, B. A., and A. P. Fields. 1991. Selective translocation of βII protein kinase C to the nucleus of human promyelocytic (HL60) leukemia cells. *J. Biol. Chem.* 266:28–33.

Homma, Y., C. B. Henning-Chub, and E. Huberman. 1986. Translocation of protein kinase C in human leukemia cells susceptible or resistant to differentiation induced by phorbol 12-myristate 13-acetate. *Proc. Natl. Acad. Sci. U.S.A.* 83:7316–7319.

Hosoda, K., N. Saito, A. Kose, A. Ito, T. Tsujino, K. Ogita, U. Kikkawa, Y. Ono, K. Igarashi, Y. Nishizuka, and C. Tanaka. 1989. Immunocytochemical localization of the β1 subspecies of protein kinase C in rat brain. *Proc. Natl. Acad. Sci. U.S.A.* 86:1393–1397.

Houbre, D., G. Duportail, J. C. Deloulme, and J. Baudier. 1991. The interactions of the brain-specific calmodulin-binding protein kinase C substrate, neuromodulin (GAP-43), with membrane phospholipids. *J. Biol. Chem.* 266:7121–7131.

House, C., and B. E. Kemp, 1987. Protein kinase C contains a pseudosubstrate prototope in its regulatory domain. *Science* 238:1726–1728.

Hsieh, J.-C., P. W. Jurutka, M. A. Galligan, C. M. Terpening, C. A. Haussler, C. S. Samuels, Y. Shimizu, N. Shimizu, and M. R. Haussler. 1991. Human vitamin D receptor is selectively phosphorylated by protein kinase C on serine 51, a residue crucial to its trans-activation function. *Proc. Natl. Acad. Sci. U.S.A.* 88:9315–9319.

Huang, F. L., and K.-P. Huang. 1991. Interaction of protein kinase C isozymes with phosphatidylinositol 4,5-bisphosphate. *J. Biol. Chem.* 266:8727–8733.

Huang, F. L., Y. Yoshida, H. Nakabayashi, and K.-P. Huang. 1987. Differential distribution of protein kinase C isozymes in the various regions of brain. *J. Biol. Chem.* 262:15714–15720.

Huang, F. L., Y. Yoshida, H. Nakabayashi, W. S. Young III, and K.-P. Huang. 1988. Immunocytochemical localization of protein kinase C isozymes in rat brain. *J. Neurosci.* 8:4734–4744.

Huang, F. L., Y. Yoshida, J. R. Cunha-Melo, M. A. Beaven, and K.-P. Huang. 1989a. Differential down-regulation of protein kinase C isozymes. *J. Biol. Chem.* 264:4238–4243.

Huang, F. L., Y. Yoshida, H. Nakabayashi, D. P. Friedman, I. G. Ungerleider, W. S. Young III and K.-P. Huang. 1989b. Type I protein kinase C isozyme in the visual information-processing pathway of monkey brain. *J. Cell Biochem.* 39:401–410.

Huang, F. L., W. S. Young III, Y. Yoshida, and K.-P. Huang. 1990. Developmental expression of protein kinase C isozymes in rat cerebellum. *Dev. Brain Res.* 52:121–130.

Huang, K.-P. 1989. The mechanism of protein kinase C activation. *Trends Neurosci.* 12: 425–432.

Huang, K.-P. 1990. Role of protein kinase C in cellular regulation. *BioFactors* 2:171–178.

Huang, K.-P., and F. L. Huang. 1990. Differential sensitivity of protein kinase C isozymes to phospholipid-induced inactivation. *J. Biol. Chem.* 265:738–744.

Huang, K.-P., K.-F. J. Chan, T. J. Singh, H. Nakabayashi, and F. L. Huang. 1986a. Autophosphorylation of rat brain Ca^{2+}-activated and phospholipid-dependent protein kinase. *J. Biol. Chem.* 261:12134–12140.

Huang, K.-P., H. Nakabayashi, and F. L. Huang. 1986b. Isozymic forms of rat brain Ca^{2+}-activated and phospholipid-dependent protein kinase. *Proc. Natl. Acad. Sci. U.S.A.* 83:8535–8539.

Huang, K.-P., F. L. Huang, H. Nakabayashi, and Y. Yoshida. 1988. Biochemical characterization of rat brain protein kinase C isozymes. *J. Biol. Chem.* 263:14839–14845.

Huang, K.-P., F. L. Huang, C. W. Mahoney, and K.-H. Chen. 1991. Protein kinase C subtypes and their respective roles. *Prog. Brain Res.* 89:143–155.

Hubbard, S. R., W. R. Bishop, P. Kirschmeier, S. J. George, S. P. Cramer, and W. A. Hendrickson. 1991. Identification and characterization of zinc binding sites in protein kinase C. *Science* 254:1776–1779.

Hunter, T., and M. Karin. 1992. The regulation of transcription by phosphorylation. *Cell* 70:375–387.

Ichimura, T., T. Isobe, T. Okuyama, N. Takahashi, K. Araki, R. Kuwano, and Y. Takahashi. 1988. Molecular cloning of cDNA coding for brain-specific 14-3-3 protein, a protein kinase-dependent activator of tyrosine and tryptophan hydroxylases. *Proc. Natl. Acad. Sci. U.S.A.* 85:7084–7088.

Inagaki, M., N. Gonda, M. Matsuyama, K. Nishizawa, Y. Nishi, and C. Sata. 1988. Intermediate filament reconstitution *in vitro. J. Biol. Chem.* 263:5970–5978.

Isakov, N., P. McMahon, and A. Altman. 1990. Selective post-transcriptional down-regulation of protein kinase C isozymes in leukemic T cells chronically treated with phorbol ester. *J. Biol. Chem.* 265:2091–2097.

Ishikawa, F., F. Takaku, M. Nagao, and T. Sugimura. 1986. Cysteine-rich regions conserved in amino-terminal halves of *raf* gene family products and protein kinase C. *GANN* 77:1183–1187.

Issandou, M., F. Bayard, and J. M. Darbon. 1988. Inhibition of MCF-7 cell growth by 12-0-tetradecanoylphorbol-13-acetate and 1,2 dioctanoyl-sn-glycerol: Distinct effects on protein kinase C activity. *Canc. Res.* 48:6943–6950.

Issandou, M., and E. Rozengurt. 1989. Diacylglycerols, unlike phorbol esters, do not induce homologous desensitization or down-regulation of protein kinase C in Swiss 3T3 cells. *Biochem. Biophys. Res. Commun.* 163:201–206.

Iwai, T., N. Fujisawa, K. Ogita, and U. Kikkawa. 1992. Catalytic properties of yeast protein kinase C: Difference between the yeast and mammalian enzymes. *J. Biochem.* 112: 7–10.

Jaken, S., K. Leach, and T. Klauck. 1989. Association of type 3 protein kinase C with focal contacts in rat embryo fibroblasts. *J. Cell. Biol.* 109:697–704.

James, G., and E. Olson. 1992. Deletion of the regulatory domain of protein kinase C α exposes regions in the hinge and catalytic domains that mediate nuclear targeting. *J. Cell. Biol.* 116:863–874.

Jeffrey, A. M., and R. J. M. Liskamp, 1986. Computer-assisted molecular modeling of tumor promoters: Rationale for the activity of phorbol esters, teleocidin B, and aplysiatoxin. *Proc. Natl. Acad. Sci. U.S.A.* 83:241–245.

Jimenez, B., A. Pestana, and M. Fernandez-Renart. 1989. A phosholipid-stimulated protein kinase from *Dictyostelium discoideum. Biochem. J.* 260:557–561.

Kaibuchi, K., T. Fukumoto, N. Oku, Y. Takai, K. Arai, and M. Muromatsu. 1989. Molecular genetic analysis of the regulatory and catalytic domains of protein kinase C. *J. Biol. Chem.* 264:13489–13496.

Karns, L. R., S.-C. Ng, J. A. Freeman, and M. C. Fishman. 1987. Cloning of complementary DNA for GAP-43, a neuronal growth-related protein. *Science* 236:597–600.

Kawamoto, S., and H. Hidaka. 1984. Ca^{2+}-activated phospholipid-dependent protein kinase catalyzes the phosphorylation of actin-binding proteins. *Biochem. Biophys. Res. Commun.* 118:736–742.

Kiley, S. C., P. J. Parker, D. Fabbro, and S. Jaken. 1992. Hormone-and phorbol ester-activated protein kinase C isozymes mediate a reorganization of the actin cytoskeleton associated with prolactin secretion in GH_4C_1 cells. *Molec. Endocrinol.* 6:120–131.

Kishimoto, A., M. Mikawa, K. Hashimoto, I. Yasuda, S. Tanaka, M. Tominaga, T. Kuroda, and Y. Nishizuka. 1989. Limited proteolysis of protein kinase C by calcium-dependent neutral protease (calpain). *J. Biol. Chem.* 264:4088–4092.

Kitano, T., T. Hashimoto, U. Kikkawa, K. Ase, N. Saito, C. Tanaka, Y. Ichimori, K. Tsukamoto, and Y. Nishizuka. 1987. Monoclonal antibodies against rat brain protein kinase C and their application to immunocytochemistry in nervous tissues. *J. Neurosci.* 7:1520–1525.

Klee, C. B. 1988. Ca^{2+}-dependent phospholipid-(and membrane-) binding proteins. *Biochemistry* 27:6645–6653.

Kobayashi, E., H. Nakano, M. Morimoto, and T. Tamaoki. 1989. Calphostin C (UCN-1028C), a novel microbial compound, is a highly potent and specific inhibitor of protein kinase C. *Biochem. Biophys. Res. Commun.* 159:548–553.

Koide, H., K. Ogita, U. Kikkawa, and Y. Nishizuka. 1992. Isolation and characterization of the ε subspecies of protein kinase C from rat brain. *Proc. Natl. Acad. Sci. U.S.A.* 89:1149–1153.

Kong, F. H., T. Kishi, D. Perez-Sala, and R. R. Rando. 1990. The stereochemical requirement for protein kinase C activation by 3-methyldiglycerides matches that found in naturally occurring tumor promoters aplysiatoxins. *FEBS Lett.* 274:203–206.

Kong, F. H., T. Kishi, D. Perez-Sala, and R. R. Rando. 1991. The pharmacophore of debromoaplysiatoxin responsible for protein kinase C activation. *Proc. Natl. Acad. Sci. U.S.A.* 88:1973–1976.

Kose, A., N. Saito, H. Ito, U. Kikkawa, Y. Nishizuka, and C. Tanaka. 1988. Electron microscopic localization of type I protein kinase C in rat purkinje cells. *J. Neurosci.* 8:4262–4268.

Kraft, A. S., and W. B. Anderson. 1983. Phorbol esters increase the amount of Ca^{2+}, phospholipid-dependent protein kinase associated with plasma membrane. *Nature* 301:621–623.

Kraft, A. S., W. B. Anderson, H. L. Cooper, and J. J. Sando. 1982. Decrease in cytosolic calcium/phospholipid-dependent protein kinase activity following phorbol ester treatment of EL4 thyoma cells. *J. Biol. Chem.* 257:13193–13196.

Kraft, A., J. Reeves, and C. Ashendel. 1988. Differing modulation of protein kinase C by bryostatin 1 and phorbol ester in JB6 mouse epidermal cells. *J. Biol. Chem.* 263:8437–8442.

Kramer, I. M., R. L. van der Bend, A. T. J. Tool, W. J. van Blitterswijk, D. Roos, and A. J. Verhoeven. 1989. 1-O-Hexadecyl-2-O-methylglycerol, a novel inhibitor of protein kinase C, inhibits the respiratory burst in human neutrophils. *J. Biol. Chem.* 264:5876–5884.

Krishnan, A. V., and D. Feldman. 1991. Activation of protein kinase C inhibits vitamin D receptor gene expression. *Molec. Endocrinol.* 5:605–612.

Leach, K. L., E. A. Powers, V. A. Ruff, S. Jaken, and S. Kaufmann. 1989. Type 3 protein kinase C localization to the nuclear envelope of phorbol ester-treated NIH3T3 cells. *J. Cell Biol.* 109:685–695.

Leach, K. L., V. A. Ruff, M. B. Jarpe, L. D. Adams, D. Fabbro, and D. M. Raben. 1992. α-Thrombin stimulates nuclear diacylglycerol levels and differential nuclear localization of protein kinase C isozymes in IIC9 cells. *J. Biol. Chem.* 267:21816–21822.

Lee, M. H., and R. M. Bell. 1989. Phospholipid functional groups involved in protein kinase C activation, phorbol ester binding, and binding to mixed micelles. *J. Biol. Chem.* 264:14797–14805.

Lee, M. H., and R. M. Bell. 1991. Mechanism of protein kinase C activation by phosphatidylinositol 4,5-bisphosphate. *Biochemistry* 30:1041–1049.

Lee, S. H., J. W. Karaszkiewicz, and W. B. Anderson. 1992. Elevated levels of nuclear protein kinase C in multidrug-resistant MCF-7 human breast carcinoma cells. *Canc Res.* 52:3750–3759.

Leibersperger, H., M. Gschwendt, and F. Marks. 1990. Purification and characterization of a calcium-unresponsive, phorbol ester/phospholipid-activated protein kinase from porcine spleen. *J. Biol. Chem.* 265:16108–16115.

Levin, D. E., F. O. Fields, R. Kunisawa, J. M. Bishop, and J. Thorner. 1990. A candidate protein kinase C gene, *PKC1,* is required for the *S. cerevisiae* cell cycle. *Cell* 62:213–224.

Lindner, D., M. Gschwendt, and F. Marks. 1991. Down-regulation of protein kinase C in Swiss 3T3 fibroblasts is independent of its phosphorylating activity. *Biochem. Biophys. Res. Commun.* 176:1227–1231.

Liu, J. D., J. G. Wood, R. L. Raynor, Y. C. Wang, T. A. Jr., Norland, A. A. Ansari, and J. F. Kuo. 1989. Subcellular distribution and immunocytochemical localization of protein kinase C in myocardium and phosphorylation of troponin in isolated myocytes stimulated by isoproterenol or phorbol ester. *Biochem. Biophys. Res. Commun.* 162:1105–1110.

Liu, Y., and D. R. Storm. 1990. Regulation of free calmodulin by neuromodulin: Neuron growth and regeneration. *Trends Pharmacol. Sci.* 11:107–111.

Liyanage, M., D. Frith, E. Livneh, and S. Stabel. 1992. Protein kinase C group B members, PKC—δ, -ε, -ζ, and PKC-L (η): Comparison of properties of recombinant proteins *in vitro* and *in vivo. Biochem. J.* 283:781–787.

Luderus, M. E., R. G. Van der Most, A. P. Otte, and R. Van Driel. 1989. A protein kinase C-related enzyme activity in *Dictyostelium discoidium. FEBS Lett.* 253:71–75.

Luisi, B. F., W. X. Xu, Z. Otwinoski, L. P. Freedman, K. R. Yamamoto, and P. B. Sigler. 1991. Crystallographic analysis of the interaction of the glucocorticoid receptor with DNA. *Nature* 352:497–505.

Mahoney, C. W., A. Azzi, and K.-P. Huang. 1990. Effects of suramin, an anti-human immunodeficiency virus reverse transcriptase agent, on protein kinase C: Differential activation and inhibition of protein kinase C isozymes. *J. Biol. Chem.* 265:5424–5428.

Mahoney, C. W., J. Shuman, S. L. McKnight, H.-C. Chen, and K.-P. Huang. 1992. Phosphorylation of CCAAT-enhancer binding protein by protein kinase C attenuates site-selective DNA binding. *J. Biol. Chem.* 267:19396–19403.

Makowske, M., R. Ballester, Y. Cayre, and O. M. Rosen. 1988. Immunochemical evidence that three protein kinase C isozymes increase in abundance during HL-60 differentiation induced by DMSO and retinoic acid. *J. Biol. Chem.* 263:3402–3410.

Malviya, A. N., P. Rouge, and G. Vincendon. 1990. Stereospecific inositol 1,4,5-[^{32}P]trisphosphate binding to isolated rat liver nuclei: Evidence for inositol trisphosphate receptor-mediated calcium release from the nucleus. *Proc. Natl. Acad. Sci. U.S.A.* 87:9270–9274.

Maraganore, J. 1987. Structural elements for protein-phospholipid interactions may be shared in protein kinase C and phospholipase A$_2$. *Trends Biochem. Sci.* 12:176–177.

Marais, R. M., and P. J. Parker. 1989. Purification and characterization of bovine brain protein kinase C isotypes α, β, and γ. *Eur. J. Biochem.* 182:129–137.

Masliah, E., K. Yoshida, S. Shimohama, F. H. Gage, and T. Saitoh. 1991. Differential expression of protein kinase C isozymes in rat glial cell cultures. *Brain Res.* 549: 106–111.

Masmoudi, A., G. Labourdete, M. Mersel, F. L. Huang, K.-P. Huang, G. Vincendon, and A. N. Malviya. 1989. Protein kinase C located in rat liver nuclei. Partial purification and biochemical and immunochemical characterization. *J. Biol. Chem.* 262:2234–2243.

Masure, H. R., K. A. Alexander, B. T. Wakim, and D. R. Storm. 1986. Physicochemical and hydrodynamic characterization of P-57, a neurospecific calmodulin binding protein. *Biochemistry* 25:7553–7560.

McDonald, J. R., and M. Walsh. 1985. Ca^{2+}-binding proteins from bovine brain including a potent inhibitor of protein kinase C. *Biochem. J.* 232:559–567.

McDonald, J. R., U. Gröschel-Stewart, and M. P. Walsh. 1987. Properties and distribution of the protein inhibitor (M$_r$ 17000) of protein kinase C. *Biochem. J.* 242:695–705.

McFadden, P. N., A. Mandpe, and D. E. Koshland Jr. 1989. Calcium- and lipid-independent protein kinase C activation at low pH. *J. Biol. Chem.* 264:12765–12771.

McIlroy, B. K., J. D. Walters, P. J. Blackshear, and J. D. Johnson. 1989. Phosphorylation-dependent binding of a synthetic MARCKS peptide to calmodulin. *J. Biol. Chem.* 264:21818–21823.

McPhail, L. C., C. C. Clayton, and R. Snyderman. 1984. A potential second messenger role for unsaturated fatty acids: Activation of Ca^{2+}-dependent protein kinase. *Science* 224:622–625.

McSwine-Kennick, R. L., E. M. McKeegan, M. D. Johnson, and M. J. Morin. 1991. Phorbol diester-induced alterations in the expression of protein kinase C isozymes and their mRNAs: Analysis in wild type and phorbol diester-resistant HL60 cell clones. *J. Biol. Chem.* 266:15135–15143.

Meiri, K. F., and P. R. Gordon-Weeks. 1990. GAP-43 in growth cones is associated with areas of membrane that are tightly bound to substrate and is a component of a membrane skeleton subcellular fraction. *J. Neurosci.* 10:256–266.

Meiri, K. F., W. Willard, and M. I. Johnson. 1988. Distribution and phosphorylation of the growth-associated protein GAP-43 in regenerating sympathetic neurons in culture. *J. Neurosci.* 8:2571–2581.

Melloni E., S. Pontremolli, M. Michetti, O. Sacco, B. Sparatore, F. Salamino, and B. L. Horrecker. 1985. Binding of protein kinase C to neutrophil membranes in the presence of Ca^{2+} and its activation by a Ca^{2+}-requiring proteinase. *Proc. Natl. Acad. Sci. U.S.A.* 82:6435–6439.

Mizuno, K., K. Kubo, T. C. Saido, Y. Akita, S.-I. Osada, T. Kuroki, S. Ohno, and K.

Suzuki. 1991. Structure and properties of a ubiquitously expressed protein kinase C, nPKCδ. *Eur. J. Biochem.* 202:931–940.

Mochly-Rosen, D., and D. E. Koshland Jr. 1987. Domain structure and phosphorylation of protein kinase C. *J. Biol. Chem.* 262:2291–2297.

Mochly-Rosen, D., A. I. Basbaum, and D. E. Koshland Jr. 1987. Distinct cellular and regional localization of immunoreactive protein kinase C in rat brain. *Proc. Natl. Acad. Sci. U.S.A* 84:4660–4664.

Mochly-Rosen, D., C. J. Henrich, L. Cheever, H. Khaner, and P. C. Simpson. 1990. A protein kinase C isozyme is translocated to cytoskeletal elements on activation. *Cell Regul.* 1:693–706

Mochly-Rosen, D., H. Khaner, and J. Lopez. 1991a. Identification of intracellular receptor proteins for activated protein kinase C. *Proc. Natl. Acad. Sci. U.S.A.* 88:3997–4000.

Mochly-Rosen, D., H. Khaner, J. Lopez, and B. L. Smith. 1991b. Intracellular receptors for activated protein kinase C. *J. Biol. Chem.* 266:14866–14868.

Morris, C., and E. Rozengurt. 1988. Purification of a phosphoprotein from rat brain closely related to the 80kDa substrate of protein kinase C identified in Swiss 3T3 fibroblasts. *FEBS Lett.* 231:311–316.

Moss, D. J., P. Fernyhough, K. Chapman, L. Baizer, D. Bray, and T. Allsopp. 1990. Chicken growth-associated protein GAP-43 is tightly bound to the actin-rich neuronal membrane skeleton. *J. Neurochem.* 54:729–736.

Müller, W., and J. A. Connor. 1991. Dendritic spine as individual neuronal compartments for synaptic Ca^{2+} responses. *Nature* 354:73–76.

Murakami, K., S. Y. Chan, and A. Routtenberg. 1986. Protein kinase C activation by cis-fatty acid in the absence of Ca^{2+} and phospholipids. *J. Biol. Chem.* 261:15424–15429.

Nakabayashi, H., J. R. Sellers, and K.-P. Huang. 1991. Catalytic fragment of protein kinase C exhibits altered substrate specificity toward smooth muscle myosin light chain. *FEBS Lett.* 294:144–148.

Nakamura, H., Y. Kishi, M. A., Pajares, and R. R. Rando, 1989. Structural basis of protein kinase C activation by tumor promoters. *Proc. Natl. Acad. Sci. U.S.A.* 86:9672–9276.

Nakanishi, H., and J. H. Exton. 1992. Purification and characterization of the ζ isoform of protein kinase C from bovine kidney. *J. Biol. Chem.* 267:16347–16354.

Nakanishi, H., K. A. Brewer, and J. H. Exton. 1993. Activation of the ζ isoform of protein kinase C by phosphatidylinositol 3,4,5 trisphosphate. *J. Biol. Chem.* 268:13–16.

Nakano, H., Y. Kobayashi, Y. Takahashi, T. Tamaoki, Y. Kuzuu, and Y. Iba. 1987. Staurosporine inhibits tyrosine-specific protein kinase activity of Rous sarcoma virus transforming protein p60. *J. Antibiotics* 40:706–709.

Naor, Z., M. S. Shearman, A, Kishimoto, and Y. Nishizuka. 1988. Calcium-independent activation of hypothalamic type I protein kinase C by unsaturated fatty acids. *Molec. Endocrinol.* 2:1043–1048.

Naor, Z., H. Dan-Cohen, J. Hermon, and R. Limor. 1989. Induction of exocytosis in permeabilized pituitary cells by α- and β- type protein kinase C. *Proc. Natl. Acad. Sci. U.S.A.* 86:4501–4504.

Newton, A. C., and D. E. Koshland Jr. 1989. High cooperativity, specificity, and multiplicity in the protein kinase C-lipid interaction. *J. Biol. Chem.* 264:14909–14915.

Ng, S.-C., S. M. De La Monte, G. L. Conboy, L. R. Karns, and M. C. Fishman. 1988. Cloning of human GAP-43: Growth associated and ischemic resurgence. *Neuron* 1:133–139.

Nielander, H. B., L. H. Schrama, A. J. Van Rozen, M. Kasperaitis, A. B. Oestreicher, P. N. E. De Graan, W. H. Gispen, and P. Schotman. 1987. Primary structure of the neuron-specific phosphoprotein B-50 is identical to growth-associated protein GAP-43. *Neurosci. Res. Commun.* 1:163–172.

Niino, Y. S., S. Ohno, and K. Suzuki. 1992. Positive and negative regulation of the transcription of the human protein kinase C β gene. *J. Biol. Chem.* 267:6158–6163.

Nishizuka, Y. 1986. Studies and perspectives of protein kinase C. *Science* 233:305–312.

Nishizuka, Y. 1988. The molecular heterogeneity of protein kinase C and its implications for cellular regulation. *Nature* 334:661–665.

Nishizuka, Y. 1992. Intracellular signaling by hydrolysis of phospholipids and activation of protein kinase C. *Science* 258:607–614.

Nixon, J. S., J. Bishop, D. Bradshaw, P. D. Davis, C. H. Hill, L. H. Elliot, H. Kumar, G. Lawton, E. J. Lewis, M. Mulqueen, D. Westmacott, J. Wadsworth, S. E. Wilkinson. 1992. The design and biological properties of potent and selective inhibitors and protein kinase C. *Biochem. Soc. Trans.* 20:419–425.

Obeid, L. M., T. Okazaki, L. A. Karolak, and Y. A. Hannun. 1990. Transcriptional regulation of protein kinase C by 1,25 dihydroxyvitamin D_3 in HL-60 cells. *J. Biol. Chem.* 265:2370–2374.

Obeid, L. M., G. C. Blobe, L. A. Karolak, and Y. A. Hannun. 1992. Cloning and characterization of the major promoter of the human protein kinase C β gene. *J. Biol. Chem.* 267:20804–20810.

Oda, T., M. S. Shearman, and Y. Nishizuka. 1991. Synaptosomal protein kinase C subspecies. B. Down-regulation promoted by phorbol ester and its effect on evoked norepinephrine release. *J. Neurochem.* 56:1263–1269.

Ogita, K., S.-I. Miyamoto, H. Koide, T. Iwai, M. Oka, K. Ando, A. Kishimoto, K. Ikeda, Y. Fukami, and Y. Nishizuka. 1990. Protein kinase C in *Saccharomyces cerevisiae:* Comparison with the mammalian enzyme. *Proc. Natl. Acad. Sci. U.S.A.* 87:5011–5015.

Ogita, K., S. Miyamoto, K. Yamaguchi, H. Koide, N. Fujisawa, U. Kikkawa, S. Sahara, Y. Fukami, and Y. Nishizuka. 1992. Isolation and characterization of δ-subspecies of protein kinase C from rat brain. *Proc. Natl. Acad. Sci. U.S.A.* 89:1592–1596.

Ohno, S., Y. Akita, Y. Konno, S. Imajoh, and K. Suzuki. 1988. A novel phorbol ester receptor/protein kinase, nPKC, distantly related to the protein kinase C family. *Cell* 53:731–741.

Ohno, S., T. Konno, Y. Akita, A. Yano, and K. Suzuki. 1990. A point mutation at the putative ATP-binding site of protein kinase C α abolishes the kinase activity and renders it down-regulation-insensitive. *J. Biol. Chem.* 265:6296–6300.

Olivier, A. R., and P. J. Parker. 1991. Expression and characterization of protein kinase C δ. *Eur. J. Biochem.* 200:805–810.

Omary, M. B., G. T. Baxter, C.-F. Chou, C. L. Riopel, W. Y. Lin, and B. Strulovici. 1992. Protein kinase C-ε related kinase associates with and phosphorylates cytokeratin 8 and 18. *J. Cell. Biol.* 117:583–593.

Ono, Y., U. Kikkawa, K. Ogita, T. Fujii, T. Kurokawa, Y. Asaoka, K. Sekiguchi, K. Ase, K. Igarashi, and Y. Nishizuka. 1987. Expression and properties of two types of protein kinase C: alternative splicing from a single gene. *Science* 236:1116–1120.

Ono, Y., T. Fujii, K. Ogita, U. Kikkawa, K. Igarashi, and Y. Nishizuka. 1988. The structure, expression, and properties of additional members of the protein kinase C family. *J. Biol. Chem.* 263:6927–6932.

Ono, Y., T. Fujii, K. Igarashi, T. Kuno, C. Tanaka, U. Kikkawa, and Y. Nishizuka. 1989a.

Phorbol ester binding to protein kinase C requires a cysteine-rich zinc-finger-like sequence. *Proc. Natl. Acad. Sci. U.S.A.* 86:4868–4871.

Ono, Y., T. Fujii, K. Ogita, U. Kikkawa, K. Igarashi, and Y. Nishizuka. 1989b. Protein kinase C subspecies ζ from rat brain: its structure, expression, and properties. *Proc. Natl. Acad. Sci. U.S.A.* 86:3099–3103.

Osada, S., K. Mizuno, T. C. Saido, Y. Akita, K. Suzuki, T. Kuroki, and S. Ohno. 1990 A phorbol ester receptor/protein kinase, nPKCη, a new member of the protein kinase C family predominantly expressed in lung and skin. *J. Biol. Chem.* 265:22434–22440.

Osada, S-I., K. Mizuno, T. C. Saido, K. Suzuki, T. Kuroki, and S. Ohno. 1992. A new member of the protein kinase C family, nPKCθ, predominantly expressed in skeletal muscle. *Mol Cell. Biol.* 12:3930–3938.

Otte, A. P., and R. T. Moon. 1992. Protein kinase C isozymes have distinct roles in neural induction and competence in *Xenopus. Cell* 68:1021–1029.

Otte, A. P., C. H. Koster, C. T. Snoek, and A. J. Durston. 1988. Protein kinase C mediates induction in *Xenopus laevis. Nature* 334:618–620.

Otte, A. P., I. M. Kramer, and A. J. Durston. 1991. Protein kinase C and regulation of the local competence of *Xenopus* ectoderm. *Science* 251:570–573.

Ozawa, K., Z. Szallasi, M. G. Kazanietz, P. M. Blumberg, H. Mishak, J. F. Mushinski, and M. A. Beaven. 1993a. Ca^{2+}-dependent and Ca^{2+}-independent isozymes of protein kinase C mediate exocytosis in antigen-stimulated rat basophilic RBL-2H3 cells. *J. Biol. Chem.* 268:1749–1756.

Ozawa, K., K. Yamada, M. G. Kazanietz, P. M. Blumberg, and M. A. Beaven. 1993b. Different isozymes of protein kinase C mediate feedback inhibition of phospholipase C and stimulatory signals for exocytosis in rat RBL-2H3 cells. *J. Biol. Chem.* 268:2280–2283.

Pan, T. and J. E. Coleman, 1989. The structure and function of the Zn(II) binding site within the DNA binding domain of the Ga14 transcription factor. *Proc. Natl. Acad. Sci. U.S.A.* 86:3145–3149.

Papadopoulas, V., and P. F. Hall. 1989. Isolation and characterization of protein kinase C from Y-1 adrenal cell cytoskeleton. *J. Cell Biol.* 108:553–567.

Patel, J., and D. Kligman. 1987. Purification and characterization of an Mr 87,000 protein kinase C substrate from rat brain. *J. Biol. Chem.* 262:16686–16691.

Pavletich, N. P. and C. O. Pabo. 1991. Zinc finger-DNA recognition: crystal structure of a Zif268-DNA complex at 2.1 A°. *Science* 252:809–817.

Pears, C., D. Schaap, and P. J. Parker. 1991. The regulatory domain of PKC ε restricts the catalytic domain-specificity. *Biochem. J.* 276:257–260.

Pearson, J. D., D. B. De Wold, W. R. Mathews, N. M. Mozier, H. A. Zürcher-Neely, R. L. Heinrikson, M. A. Morris, W. D. McCubbin, J. R. McDonald, E. D. Fraser, H. J. Vogel, C. M. Kay, and M. P. Walsh. 1990. Amino acid sequence and characterization of a protein inhibitor of protein kinase C. *J. Biol. Chem.* 265:4583–4591.

Perin, M. S., V. A. Fried, G. A. Migney, R. John, and T. C. Südhof. 1990. Phospholipid binding by a synaptic vesicle protein homologous to the regulatory region of protein kinase C. *Nature* 345:260–263.

Perletti, G., A. Ghessi, E. Raffaldoni, and F. Piccinini. 1991. The activity of a β subtype of protein kinase C purified from nuclei of human neutrophils is enhanced by treatment with phorbol 12-myristate 13-acetate. *Biochem. Biophys. Res. Commun.* 181:348–352.

Pfenninger, K. H., L. Ellis, M. P. Johnson, L. B. Friedman, and S. Somlo. 1983. Nerve growth cones isolated from fetal brain: subcellular fractionation and characterization. *Cell* 35:573–584.

Quest, A. F. G., J. Blumenthal, E. S. G. Bardes, and R. M. Bell. 1992. The regulatory domain of protein kinase C coordinates four atoms of zinc. *J. Biol. Chem.* 267: 10193–10197.

Rando, R. R. 1988. Regulation of protein kinase C activity by lipids. *FASEB J.* 2:2348–2355.

Rane, S. G., M. P. Walsh, J. R. McDonald, and K. Dunlap. 1989. Specific inhibitors of protein kinase C block transmitter-induced modulation of sensory neuron calcium current. *Neuron* 3:239–245.

Represa, A., J. C. Deloulme, M. Sensenbrenner, Y. Ben-Ari, and J. Baudier. 1990. Neurogranin: immunocytochemical localization of a brain-specific protein kinase C substrate. *J. Neurosci.* 10:3782–3792.

Rhodes, D., and A. Klug. 1993. Zinc fingers. *Scientific Am.* 268:56–65.

Rogue, P., G. Labourdette, A. Masmoudi, Y. Yoshida, F. L. Huang, K-P. Huang, J. Zwiller, G. Vincendon, and A. N. Malviya. 1990. Rat liver nuclei protein kinase C is isozyme type II. *J. Biol. Chem.* 265:4161–4165.

Rosen, A., K. F. Keenan, M. Thelen, A. C. Nairn, and A. Aderem. 1990. Activation of protein kinase C results in the displacement of its myristoylated alanine-rich substrate from punctuate structures in macrophage filopodia. *J. Exp. Med.* 172:1211–1215.

Rosenthal, A., L. Rhee, L. Yadegari, R. Paro, A. Ullrich, and D. V. Goeddel. 1987. The structure and nucleotide sequence of a *Drosophila melanogaster* protein kinase C gene. *EMBO J.* 6:433–441.

Rozengurt, E., M. Rodriguez-Pena, and K. A. Smith. 1983. Phorbol esters, phospholipase C, and growth factors rapidly stimulate the phosphorylation of a M_r 80,000 protein in intact quiescent 3T3 cells. *Proc. Natl. Acad. Sci. U.S.A.* 80:7244–7248.

Sahara, S., K-I. Sato, M. Aoto, T. Ohnishi, H. Kaise, H. Koide, K. Ogita, and Y. Fukami. 1992. Characterization of protein kinase C in *Xenopus* oocytes. *Biochem. Biophys. Res. Commun.* 182:105–114.

Saido, T. C., K. Mizuno, Y. Konno, S. Osada, S. Ohno, and K. Suzuki. 1992a. Purification and characterization of protein kinase C ε from rabbit brain. *Biochemistry* 31:482–490.

Saido, T. C., M. Shibata, T. Takenawa, H. Murofushi, and K. Suzuki. 1992b. Positive regulation of μ-calpain action by polyphosphoinositides. *J. Biol. Chem.* 267: 24585–24590.

Saito, N., U. Kikkawa, Y. Nishizuka, and C. Tanaka. 1988. Distribution of protein kinase C-like immunoreactive neurons in rat brain. *J. Neurosci.* 8:369–382.

Saito, N., A. Kose, A. Ito, K. Hosoda, M. Mori, M. Hirata, K. Ogita, U. Kikkawa, Y. Ono, K. Igarashi, Y. Nishizuka, and C. Tanaka. 1989 Immunocytochemical localization of βII subspecies of protein kinase C in rat brain. *Proc. Natl. Acad. Sci. U.S.A.* 86: 3409–3413.

Sakane, F., K. Yamada, H. Kanoh, C. Yokoyama, and C. Tanabe. 1990. Porcine diacylglycerol kinase sequence has zinc finger and E-F hand motif. *Nature* 344:345–348.

Sakurai, A., T. Maekawa, T. Sudo, S. Ishii, and A. Kishimoto. 1991. Phosphorylation of cAMP response element-binding protein, CREBP1, by cAMP-dependent protein kinase and protein kinase C. *Biochem. Biophys. Res. Commun.* 181:629–635.

Schaap, D., and P. J. Parker, 1990. Expression, purification, and characterization of protein kinase C-ε. *J. Biol. Chem.* 265:7301–7307.

Schaap, D., J. Hsuan, N. Totty, and P. J. Parker. 1990. Proteolytic activation of protein kinase C-ε. *Eur. J. Biochem.* 191:431–435.

Schaeffer, E., D. Smith, G. Mardon, W. Quinn, and C. Zucker. 1989. Isolation and characterization of two new *Drosophila* protein kinase C genes, including one specifically expressed in photoreceptor cells. *Cell* 57:403–412.

Schlaepfer, D. D., J. Jones, and H. T. Haigler. 1992. Inhibition of protein kinase C by annexin V. *Biochemistry* 31:1886–1891.

Schwabe, J. W. R., and D. Rhodes. 1991. Beyond zinc fingers: steroid hormone receptors have a novel structural motif for DNA recognition. *Trends Biochem. Sci.* 16:291–296.

Seifert, R., C. Schächtele, W. Rosenthal, and G. Shultz. 1988. Activation of protein kinase C by cis-and trans-fatty acids and its potentiation by diacylglycerol. *Biochem. Biophys. Res. Commun.* 154:20–26.

Sekiguchi, K., M. Tsukuda, K. Ogita, U. Kikkawa, and Y. Nishizuka. 1987. Three distinct forms of rat brain protein kinase C: differential response to unsaturated fatty acids. *Biochem. Biophys. Res. Commun.* 145:797–802.

Seykora, J. T., J. V. Ravetch, and A. Aderem. 1991. Cloning and molecular characterization of the murine macrophage ''68-kDa'' protein kinase C substrate and its regulation by bacterial lipopolysaccharide. *Proc. Natl. Acad. Sci. U.S.A.* 88:2505–2509.

Shearman, M. S., Z. Naor, K. Sekiguchi, A. Kishimoto, and Y. Nishizuka. 1989a. Selective activation of the γ-subspecies of protein kinase C from bovine cerebellum by arachidonic acid and its lipoxygenase metabolites. *FEBS Lett.* 243:177–182.

Shearman, M.S., K. Sekiguchi, and Y. Nishizuka, 1989b. Modulation of ion channel activity: a key function of the protein kinase C enzyme family. *Pharmacol. Rev.* 41:211–237.

Shen, S. S. and L. A. Ricke. 1989. Protein kinase C from sea urchin eggs. *Comp. Biochem. Physiol.* 92B:251–254.

Shinomura, T., Y. Asaoka, M. Oka, K. Yoshida, and Y. Nishizuka. 1991. Synergistic action of diacylglycerol and unsaturated fatty acid for protein kinase C activation: its possible implications. *Proc. Natl. Acad. Sci. U.S.A.* 88:5149–5153.

Simon, A. J., Y. Milner, S. P. Saville, A. Dvir, D. Mochly-Rosen, and E. Orr. 1991. The identification and purification of a mammalian-like protein kinase C in the yeast *Saccharomyces cerevisiae. Proc. R. Soc. London (Biol.)* 243:165–171.

Skene, J. H. P., R. D. Jacobson, G. J. Snipes, C. B. McGuire, J. J. Norden, and J. A. Freeman. 1986. A protein induced during growth (GAP-43) is a major component of growth-cone membranes. *Science* 233:783–786.

Skene, J. H. P., and I. Virág, 1989. Post-translational membrane attachment and dynamic fatty acylation of a neuronal growth cone protein, GAP-43. *J. Cell Biol.* 108:613–624.

Snipes, G. L., S. Y. Chan, C. B. McGuire, B. R. Costello, J. J. Norden, J. A. Freeman, and A. Routtenberg. 1987. Evidence for the coidentification of GAP-43, a growth associated protein, and F1, a plasticity-associated protein. *J. Neurosci.* 7:4066–4075.

Snoek, G. T., I. Rosenberg, S. W. de Laat, and C. Gitler. 1986. The interaction of protein kinase C and other specific cytoplasmic proteins with phospholipid bilayers. *Biochim. Biophys. Acta.* 860:336–344.

Solanski, V., T. J. Slaga, M. Callaham, and E. Huberman. 1981. Down regulation of specific binding of [20-^{3}H] phorbol 12,13 dibutyrate and phorbol ester-induced differentiation of human promyelocytic leukemia cells. *Proc. Natl. Acad. Sci. U.S.A.* 78:1722–1725.

Souvignet, C., J.-M. Pelosin, S. Daniel, E. M. Chambaz, S. Ransac, and R. Verger. 1991. Activation of protein kinase C in lipid monolayers. *J. Biol. Chem.* 266:40–44.

Stahl, M. L., C. R. Ferenz, K. L. Kelleher, R. W. Kriz, and J. L. Knopf. 1988. Sequence similarity of phospholipase C with the non-catalytic region of src. *Nature* 332:269–272.

Strittmatter, S. M., D. Valenzuela, T. E. Kennedy, E. J. Neer, and M. C. Fishman. 1990. G_o is a major growth cone protein subject to regulation by GAP-43. *Nature* 344:836–841.

Strulovici, B., S. Daniel-Issakani, E. Otto, J. Nestor Jr., H. Chan, and A-P. Tsou. 1989. Activation of distinct protein kinase C isozymes by phorbol esters: correlation with induction of interleukin 1β gene expression. *Biochemistry* 28:3569–3576.

Stumpo, D. J., J. M. Graff, K. A. Albert, P. Greengard, and P. Blackshear. 1989. Molecular cloning, characterization, and expression of a cDNA encoding the ''80- to 87-kDa'' myristoylated alanine-rich C kinase substrate: a major cellular substrate for protein kinase C. *Proc. Natl. Acad. Sci. U.S.A.* 86:4012–4016.

Sullivan, J. P., J. R. Connor, B. G. Shearer, and R. M. Burch. 1992. 2,6-diamino-N-([1-(1-oxotridecyl)-2-piperidinyl]methyl) hexanamide (NPC 15437): a novel inhibitor of protein kinase C interacting at the regulatory domain. *Mol. Pharmacol.* 41:38–44.

Sullivan, J. P., J. R. Connor, C. Tiffany, B. G. Shearer, and R. M. Burch. 1991. NPC 15437 interacts with the C1 domain of protein kinase C. An analysis using mutant PKC constructs. *FEBS Lett.* 285:120–123.

Tabuse, Y., K. Nishiwaki, J. Miwa. 1989. Mutations in a protein kinase C homolog confer phorbol ester resistance on *Caenorhabditis elegans*. *Science* 243:1713–1716.

Takahashi, I., S. Nakanishi, E. Kobayashi, H. Nakano, K. Suzuki, and T. Tamaoki. 1989. Hypericin and pseudohypericin specifically inhibit protein kinase C: possible relation to their antiretroviral activity. *Biochem. Biophys. Res. Commun.* 165:1207–1212.

Tamaoki, T., H. Nomoto, I. Takahashi, Y. Kato, M. Morimoto, and F. Tomita. 1986. Staurosporine, a potent inhibitor of phospholipid/Ca^{2+} dependent protein kinase. *Biochem. Biophys. Res. Commun.* 135:397–402.

Thomas, T. P., H. S. Talwar, and W. B. Anderson. 1988. Phorbol ester-mediated association of protein kinase C to the nuclear fraction in NIH 3T3 cells. *Canc. Res.* 48:1910–1919.

Toker, A., C. A. Ellis, L. A. Sellers, and A. Aitken. 1990. Protein kinase C inhibitor proteins. Purification from sheep brain and sequence similarity to lipocortins and 14-3-3 protein. *Eur. J. Biochem.* 191:421–429.

Toullec, D., P. Pianetti, H. Coste, P. Bellevergue, T. Grand-Perret, M. Ajakane, V. Baudet, P. Boissin, E. Boursier, F. Coriolle, L. Duhanel, D. Charon, and J. Kirilovsky. 1991. The bisindolylmaleimide GF 109203X is a potent and selective inhibitor of protein kinase C. *J. Biol. Chem.* 266:15771–15781.

Tu, A. T. 1973. Neurotoxins of animal venoms: Snake. *Annu. Rev. Biochem.* 42:235–258.

Tusupov, O. K., S. E. Severin, and V. I. Shvets. 1991. Proteolytic fragment of protein kinase C (kinase M) phosphorylates in vitro phosphatidylinositol-4-phosphate. *Biochem. Biophys. Res. Commun.* 176:1007–1013.

Van Patten, S. M., G. J. Heisermann, H.-C. Cheng, and D. A. Walsh. 1987. Tyrosine kinase catalyzed phosphorylation of the inhibitor protein of the cAMP-dependent protein kinase. *J. Biol. Chem.* 262:3398–3403.

Venema, R. C., and J. F. Kuo. 1993. Protein kinase C-mediated phosphorylation of troponin-I and C-protein in isolated myocardial cells is associated with inhibition of myofibrillar MgATPase. *J. Biol. Chem.* 268:2705–2711.

Vogel, U. S., R. A. F. Dixon, M. D. Schaber, R. E. Diehl, M. S. Marshall, E. M. Scholnick, I. S. Sigal, and J. B. Gibbs. 1988. Cloning of bovine GAP and its interaction with oncogenic ras p21. *Nature* 335:90–93.

Wada, H., S. Ohno, K. Kubo, C. Taya, S. Tsuji, S. Yonehara, and K. Suzuki. 1989. Cell type-specific expression of the genes for the protein kinase C family: down-regulation of mRNAs for PKC α and nPKCε upon *in vitro* differentiation of a mouse neuroblastoma cell line Neuro 2a. *Biochem. Biophys. Res. Commun.* 165:533–538.

Walsh, M. P., K. A. Valentine, P. K. Ngai, C. A. Carruthers, and M. D. Hollenberg. 1984. Ca^{2+}-dependent hydrophobic-interaction chromatography. Isolation of a novel Ca^{2+}-binding protein and protein kinase C from bovine brain. *Biochem. J.* 224:117–124.

Walters, J. D., and J. D. Johnson, 1990. Terbium as a luminescent probe of metal-binding sites in protein kinase C. *J. Biol. Chem.* 265:4223–4226.

Wang, J. K. T., S. I. Walaas, T. Sihra, A. Aderem, and P. Greengard. 1989. Phosphorylation and associated translocation of the 87-kDa protein, a major protein kinase C substrate, in isolated nerve terminals. *Proc. Natl. Acad. Sci. U.S.A.* 86:2253–2256.

Watson, J. B., E. F. Battenberg, K. K. Wong, F. E. Bloom, and J. G. Sutcliffe. 1990. Subtractive cDNA cloning of RC3, a rodent cortex-enriched mRNA encoding a novel 78 residue protein. *J. Neurosci. Res.* 26:397–408.

Ways, D. K., P. P. Cook, C. Webster, and P. J. Parker. 1992. Effect of phorbol esters on protein kinase C-ζ. *J. Biol. Chem.* 267:4799–4805.

Wegner, M., Z. Cao, and M. C. Rosenfeld. 1992. Calcium-regulated phosphorylation within the leucine zipper of C/EBP β. *Science* 256:370–373.

Wender, P. A., K. F. Koehler, N. A. Sharkey, M. L. Dell'Aquila, and P. M. Blumberg. 1986. Analysis of the phorbol ester pharmacophore on protein kinase C as a guide to the rational design of new classes of analogs. *Proc. Natl. Acad. Sci. U.S.A.* 83:4214–4218.

Werth, D. K., and I. Pastan. 1984. Vinculin phosphorylation in response to calcium and phorbol esters in intact cells. *J. Biol. Chem.* 259:5264–5270.

Wolf, M., and M. Baggiolini. 1990. Identification of phosphatidylserine-binding proteins in human white blood cells. *Biochem. J.* 269:723–728.

Wolf, M., and N. Sayhoun. 1986. Protein kinase C and phosphatidylserine bind to Mr 110,000/115,000 polypeptides enriched in cytoskeletal and postsynaptic density preparations. *J. Biol. Chem.* 261:13227–13232.

Wolf, M., P. Cuatrecasas, and N. Sayhoun. 1985a. Interaction of protein kinase C with membranes is regulated by Ca^{2+}, phorbol esters, and ATP. *J. Biol. Chem.* 260:15718–15722.

Wolf, M., H. Levine, W. S. May, P. Cuatrecasas, and N. Sayhoun. 1985b. A model for intracellular translocation of protein kinase C involving synergism between Ca^{2+} and phorbol esters. *Nature* 317:546–549.

Yamamoto, K. K., G. A. Gonzalez, W. H. Biggs, III, and M. R. Montminy. 1988. Phosphorylation-induced binding and transcriptional efficacy of nuclear factor CREB. *Nature* 334:494–498.

Yoshida, Y., F. L. Huang, H. Nakabayashi, and K.-P. Huang. 1988. Tissue distribution and developmental expression of protein kinase C isozymes. *J. Biol. Chem.* 263:9869–9873.

Yoshihara, C., N. Saito, K. Taniyama, and C. Tanaka. 1991. Differential localization of four subspecies of protein kinase C in the rat striatum and substania nigra. *J. Neurosci.* 11:690–700.

Young, W. S. III. 1988. Expression of 3 (and a putative 4) protein kinase C genes in brains of rat and rabbit. *J. Chem. Neuroanat.* 1:177–194.

Zalewski, P. D., I. J. Forbes, C. Giannakis, P. A. Cowled, and W. H. Betts. 1990. Synergy between zinc and phorbol ester in translocation of protein kinase C to cytoskeleton. *FEBS Lett.* 273:131–134.

3

The Molecular Mechanism of Protein Kinase C Regulation by Lipids

ANDREW F. G. QUEST
ROBERT M. BELL

The cell membrane is an essential barrier required to maintain physical and chemical gradients between cell plasma and the external environment. In addition, a typical eukaryotic cell utilizes extensive internal membrane systems to create functionally distinct organelles. To coordinate intracellular functions between compartments and to enable cells to respond to changes in the extracellular environment, information must cross membranes. Membranes contain phospholipid bilayers that comprise the permeability barrier. Proteins in the membrane facilitate the transmission of signals across this barrier into the cell.

While some cellular stimuli like steroid and thyroid hormones are sufficiently hydrophobic to pass through the plasma membrane, the majority of chemical stimuli impinging on the cell surface are not. The discovery of cyclic AMP (cAMP) in the early 1960s and the subsequent development of the second-messenger model of cell activation by Sutherland (1972) represented a breakthrough in the conceptualization of intercellular as well as intracellular communication following two decades of research aimed at demonstrating that extracellular signaling molecules were cofactors for intracellular enzymes (Hechter, 1955). The adenylate cyclase signal-transducing system became the first of many transmembrane signaling systems to be studied. At the same time, the role of calcium as the factor coupling excitation to neurosecretion (Douglas and Rubin, 1961; Katz, 1966) or skeletal muscle contraction (Ebashi and Endo, 1968) was discovered. Much of our understanding of signal transduction events stems from attempts to

The encouraging and active support of lab members and in particular also of Lisette Leyton is gratefully acknowledged. This work was supported by National Institutes of Health grant GM 38737.

elucidate the respective roles of cAMP and calcium as intracellular second messengers.

As a consequence of this research it was discovered that regulation of the phosphorylation state of proteins is an important means of controlling cell function and a means by which extracellular signals influence intracellular events (Krebs, 1972; Cohen, 1988). The central role of phosphorylation in cellular regulation is underscored by the discovery of an ever increasing number of protein kinases (Hunter, 1987) and their counterparts in determining levels of protein phosphorylation, the protein phosphatases (Cohen, 1989). Among the multitude of existing kinases, one family of serine/threonine kinases, referred to as protein kinase C's (PKCs), remains unique by virtue of its specific regulation by lipids (Nishizuka, 1984,1986; Bell, 1986) and activation by a variety of tumor promoters (Castagna et al., 1982; Rando, 1988)

Protein kinase C, initially described as a cyclic nucleotide-independent kinase from brain tissue requiring proteolytic cleavage to be catalytically active (Inoue et al., 1977; Takai et al., 1977), is now known to be a ubiquitous kinase in tissues and organs (Kuo et al., 1980; Minakuchi et al., 1981). Later, the reversible activation of PKC by a lipid-soluble, membrane-associated factor in the presence of calcium was demonstrated (Takai et al., 1979a,c; Kaibuchi et al., 1981). While phospholipids such as phosphatidylserine (PS), phosphatidylinositol (PI), phosphatidic acid (PA), or phosphatidylethanolamine (PE) could replace the factor and sustain kinase activity, other lipids such as phosphatidylcholine (PC) or sphingomyelin were less effective in this respect under similar conditions (Kaibuchi et al., 1981; Wise et al., 1982; Schatzman et al., 1983). In addition to phospholipids, a small amount of the neutral lipid sn-1,2-diacylglycerol, particularly with an unsaturated fatty acid chain, was found to greatly enhance the reaction velocity with a concomitant decrease in the calcium concentration required for full activation (Takai et al., 1979a,b; Kishimoto et al., 1980). The presence of diacylglycerol increases the affinity of PKC for calcium to submicromolar concentrations, but this effect is essentially observed only in the presence of phosphatidylserine and not the other phospholipids (Kaibuchi et al., 1981; Wise et al., 1982; Schatzman et al., 1983).

Concomitant with these discoveries it was suggested that the hormone-stimulated turnover of inositol phospholipids, initially observed by Hokin and Hokin (1953), might precede and lead to intracellular calcium mobilization (Michell, 1975). Water-soluble hydrolysis products of phosphatidylinositol turnover, inositol-1,4,5-trisphosphate or inositol-1,4-biphosphate, were suggested to function as second messengers controlling calcium release (Berridge, 1983). Meanwhile, this hypothesis has received strong experimental support (Berridge, 1984; Hirasawa and Nishizuka, 1985; Hokin, 1985). Thus, hydrolysis of phosphoinositides leads to the formation of two physically distinct classes of second-messenger molecules, hydrophilic inositol phosphates (and subsequently calcium) and hydrophobic, membrane-bound diacylglycerol. Until this point diacylglycerol was considered an ordinary intermediate of glycerolipid synthesis and breakdown (Bell and Coleman, 1980). A role in signal transduction had been unthinkable. The ensuing studies on the regulation of PKC by lipids have transformed our view of the role of both glycerolipids and sphingolipids in cells (Bishop and Bell, 1986; Hannun and Bell, 1987, 1989; Nishizuka, 1988, 1992). Given the number of distinct lipids in eukaryotic membranes—a conservative estimate suggests the existence of roughly a thousand molecular species (Raetz, 1986)—one may only anticipate the potential diversity of lipid-derived second messengers. This potential, at least partially substantiated, complexity (Hannun and Bell, 1987, 1989; Michell, 1992; Nishizuka, 1992; Menniti et al., 1993), dwarfs that of any other known second-messenger system.

This review summarizes our current understanding of how lipids interact with protein kinase C and lead to its activation. Knowledge derived from such studies influences our mechanistic perception of signal transduction events at membrane surfaces and protein-lipid interactions in general. These insights should provide a basis for understanding the regulation of other signaling proteins that may possibly be controlled by lipids. Given the diversity of lipid second messengers, the complexity of PKC-lipid interactions, and the evolutionary tendency to conserve important mechanisms, it appears likely that a considerable group of such proteins will be identified.

PROTEIN KINASE C FAMILY MEMBERS

PKC is implicated in a wide variety of biological processes, such as development (Otte et al., 1991) and memory (Alkon and Rasmussen, 1988; Alkon, 1989). The discovery of PKC as a high-affinity receptor for phorbol esters, a class of compounds that affect proliferation, differentiation, and tumor promotion in animal cells and tissues (Boutwell, 1974; Blumberg, 1980; Castagna et al., 1982; Sharkey et al., 1984), further suggests a role of PKC in carcinogenesis (Ashendel, 1985). Given the diversity and complexity of PKC-related cellular functions, the existence of PKC as a family of multiple, closely related enzymes does not appear surprising (Nishizuka, 1988, 1989a,b; Parker et al., 1989; K. P. Huang, 1989; Bell and Burns, 1991).

To date ten different PKC subspecies have been identified in mammalian tissues (Table 3.1). With the exception of three of these isoforms (η, λ, θ), all were initially identified as cDNAs obtained from brain cDNA libraries, which also reflects the abundance of PKC in this tissue. Biochemical and genetic studies have distinguished functionally two domains within PKC molecules (Lee and Bell, 1986; Ono et al., 1989b). The amino terminal domain, which contains all the known lipid interaction sites essential for lipid-dependent regulation of the kinase, is referred to as the regulatory domain. In contrast, the carboxy terminal catalytic domain is more similar to regions of other serine/threonine kinases in that it contains consensus sequences for ATP- and substrate-binding sites (Hanks et al., 1988; Bell and Burns, 1991). This domain exhibits constitutive kinase activity in the absence of the regulatory domain and has been referred to as M-kinase in the literature (Woodgett et al., 1987).

Within individual PKC isoforms exist conserved regions (C) of extended-sequence homology between family members, as well as variable (V) regions.

On the basis of the alignment of constant and variable regions within the regulatory domain, PKC isoforms are divided into three groups. Group A contains the four classical or conventional PKCs (cPKC): α, βI, βII, and γ. Each of these isoforms has four conserved and five variable regions (Nishizuka, 1988, 1989a,b, 1992; Bell and Burns, 1991). Because activation of the cPKCs by phosphatidylserine, diacylglycerol, or phorbol esters is enhanced in the presence of calcium, they are also known as calcium-dependent PKCs. A second group of novel, noncalcium-dependent PKC isoforms (nPKC), group B, also contains four members: δ, ϵ, $\eta(L)$, and θ (Nishizuka, 1988, 1989a,b, 1992; Leibersperger et al., 1990; Osada et al., 1990; Schaap and Parker, 1990; Bacher et al., 1991; Bell and Burns, 1991; Koide et al., 1992; Liyanage et al., 1992; Ogita et al., 1992; Saido et al., 1992). A third group, group C, consists of two atypical PKC isoforms (aPKC): ζ and λ (Nishizuka, 1988, 1992; Ono et al., 1989a; Bell and Burns, 1991; Nakanishi and Exton, 1992; Ways et al., 1992; Baier et al., 1993; Nakanishi et al., 1993). The contribution of different regions of the regulatory domain in conferring the lipid-dependent activation properties characteristic of PKCs will be described subsequently. However, it should be mentioned that since cPKCs were the first isoforms to be characterized in detail with respect to their activation by lipids, models describing these events rely heavily on data acquired with this group of PKCs. Detailed enzymological characterizations of η (L), λ, and θ are not yet available.

PROTEIN KINASE C ACTIVATION IN VIVO BY MEMBRANE TRANSLOCATION

PKC is readily recovered as a soluble protein from resting cells homogenized in the presence of divalent cation chelators (Kikkawa et al., 1983a). On cell activation by agonists leading to the stimulation of phosphatidylinositol-specific phospholipase C activity, a redistribution of PKC to the particulate fraction is observed (Nishizuka, 1989a; Jaken, 1990). Under these circumstances, cellular levels of two PKC cofactors, calcium and diacylglycerol, are increased transiently (Nishizuka, 1992). These results established a temporal correlation between the cellular distribution of PKC and its putative activation state. This conclusion is further substantiated by experiments with phorbol esters. PKC was identified as the high-affinity intracellular receptor of phorbol esters, a class of tumor promoters with profound effects in animal cells (Ashendel, 1985). Phorbol esters activate PKC *in vitro* and can substitute for diacylglycerol in platelet activation (Castagna et al., 1982). Diacylglycerol inhibits competitively phorbol ester binding to PKC, indicating that interaction sites with these activators in a calcium-lipid-PKC complex are similar if not identical (Sharkey et al., 1984; Sharkey and Blumberg, 1985). Stoichiometric levels of phorbol ester binding to PKC require the presence of lipids (Castagna et al., 1982; Hannun and Bell, 1986), although interactions among phorbol esters, calcium, and PKC in solution may actually "prime" the enzyme for the subsequent translocation to a lipid surface (Boscá and Morán, 1993). Because phorbol esters have more than 1000-fold higher affinity than diacylglycerol for PKC in the presence of lipids (Rando, 1988) and activate PKC for a longer period of time than diacylglycerol, they have been used extensively as probes for PKC activation both *in vitro* and *in vivo*.

Treatment of parietal yolk sack cells with phorbol esters elicits a rapid decrease in cytosolic PKC that is accompanied by a significant increase in level of plasma membrane-associated enzyme (Kraft and Anderson, 1983). Results from experiments employing red blood cell membranes in a reconstitution system indicate that calcium and phorbol esters act synergistically to elicit PKC translocation to the membrane and subsequent activation (May et al., 1985; Wolf et al., 1985). Taken together, these observations suggest that translocation to the plasma membrane and interaction there with either physiological activators, such as diacylglycerol, or pharmacological reagents, like phorbol esters, represent the sequence of events responsible for PKC activation in cells. On activation, both *in vivo* and *in vitro,* PKC phosphorylates a large number of proteins (Nishizuka, 1986), as well as catalyzes intrapeptide autophosphorylation at multiple sites (Newton and Koshland, 1987; Flint et al., 1990).

LIPID-INDUCED CONFORMATIONAL CHANGES LEAD TO PKC ACTIVATION

For a number of protein kinases, including cAMP-dependent protein kinase, calcium/calmodulin-dependent protein kinase II as well as PKC specific sequences, within either the intermolecular regulatory subunit or the intramolecular regulatory domain have been identified that, by interacting with the catalytic domain, provide a means of controlling activity (Hardie, 1988; Soderling, 1990). These sequences have been termed pseudodubstrate sites because they are similar to consensus sequences for phosphorylation, except that a nonphosphorylatable amino acid replaces the hydroxyl-containing residue. Such a pseudosubstrate was first identified in PKC by showing that a synthetic peptide mimicking the residues 19–31 of PKC α (Table 3.1, V1 region) was a potent inhibitor of activity (House and Kemp, 1987, 1990). This initial observation was subsequently supported by a variety of different experimental approaches. Antibodies specific for this region of PKCα were able to activate the enzyme in the absence of lipid cofactors, presumably by sequestering this autoinhibitory region (Makowsky and Rosen, 1989). Replacement of alanine 25 with glutamic acid by site-directed mutagenesis of PKC α leads to elevated effector-independent (constitutive) activity. Deletion of the pseudosubstrate region produces constitutively active PKC α (Pears et al., 1990). Both results are consistent with the idea that electrostatic interactions between positive charges from basic residues within the pseudosubstrate domain and negatively charged amino acids within the substrate-binding site are crucial for autoinhibition (House et al., 1989).

Activation of PKC by release of autoinhibitory restraints imposed by intramolecular binding between regulatory and catalytic domain would be expected to lead to detectable conformational changes in the PKC molecule. Evidence for such changes upon membrane binding is provided by studies monitoring intrinsic tryptophan fluorescence as well as in circular dichroism measurements of PKC (Brumfeld and Lester, 1990; Shah and Shipley, 1992; Boscá and Morán, 1993).

In addition, proteases have been employed to monitor such conformational changes. Upon binding to membranes, the proteolytic sensitivity of PKC is enhanced within the hinge region (Table 3.1, V3 region) (Kishimoto et al., 1983; F. L. Huang et al., 1989; Newton and Koshland, 1989; Nishizuka, 1989a), a region of PKC anticipated to be exquisitely flexible (Flint et al., 1990). Conformational changes specific for activation by phosphatidylserine and diacylglycerol have also been reported. In the presence of phosphatidylserine, an essential cofactor for PKC activation *in vitro* (see discussion later), the pseudosubstrate domain of PKC βII is cleaved after the first residue, arginine 19, by the endoprotease Arg-C. Exposure of this residue, and susceptibility to proteolytic cleavage, is enhanced in the presence of diacylglycerol. In contrast, association of PKC with nonactivating lipids does not lead to exposure of the autoinhibitory pseudosubstrate domain (Orr et al., 1992).

Taken together, these results suggest that interactions between specific lipids and the regulatory domain of PKC lead to conformational changes that release autoinhibitory restraints imposed by the pseudosubstrate domain and induce PKC activation (Figure 3.1). The rest of the discussion will focus primarily on lipid activators and regions of PKC involved in lipid-protein interactions in the regulatory domain.

LIPID ACTIVATORS OF PKC

Diacylglycerol (DAG)

With the exception of one known PKC isoform, PKC ζ (Ono et al., 1989a; Nakanishi and Exton, 1992; Ways et al., 1992; Nakanishi et al., 1993), all are activated by diacylglycerol and phorbol esters. By kinetic inference the latter are thought to interact with the same site in PKC as diacylglycerol (Sharkey et al., 1984). Studies both with phosphatidylserine vesicles (König et al., 1985) and with detergent lipid mixed micelles (Hannun et al., 1985) suggest that one molecule of diacylglycerol is sufficient to activate PKC. Likewise, similar stoichiometries are observed for binding to and activation of PKC by phorbol esters (Kikkawa et al., 1983b; Hannun and Bell, 1986).

Activation of PKC by diacylglycerol is stereospecific; only *sn*-1,2-diacylglycerols (but neither 1,3 nor 2,3-diacylglycerols) were effective in both assays employing lipid vesicles or detergent lipid mixed micelles (Boni and Rando, 1985; Hannun et al., 1985, 1986; Ganong et al., 1986; Bonser et al., 1988). Analog studies have indicated that both the oxygen esters (1,2-positions) and the primary hydroxyl group (3-position) are essential for diacylglycerol function (Ganong et al., 1986), while the acyl chain composition does not appear to be critical, although a preference for unsaturated side chains has been reported (Takai et al., 1979b; Kishimoto et al., 1980; Mori et al., 1982). The major requirement regarding acyl chains appears to be their length. *In vitro,* diacylglycerols with both acyl

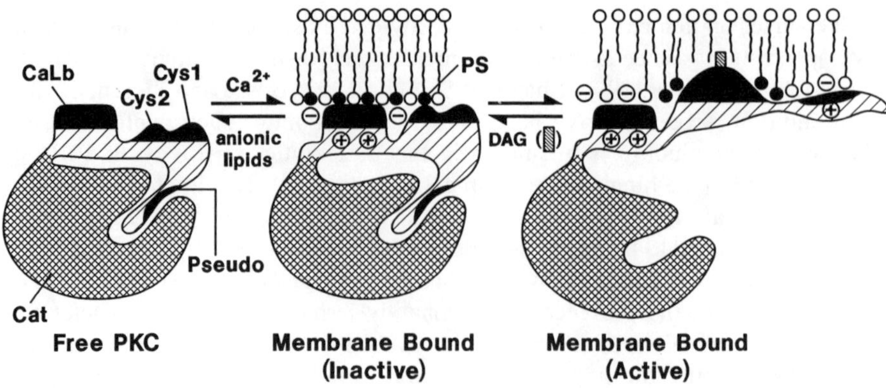

Figure 3.1 Mechanism of cPKC activation by membrane-bound lipids. Lipid interaction sites in the regulatory domain of PKC are indicated as dark shaded areas in the respective PKC regions: pseudo, pseudosubstrate motif in V1 region; Cys1, Cys2, cysteine-rich motifs in C1 region; CaLB, calcium-dependent lipid-binding site in C2 region. No structural organization within the catalytic domain (cat) is indicated. Figure shows three steps in the sequence of events leading to PKC activation. (1) PKC resides in the cytosol in an inactive state with the pseudosubstrate motif buried within the catalytic domain (left). Initial PKC association with the membrane mediated through the CaLB motif is shown in the presence of calcium and acidic phospholipids, but in the absence of diacylglycerol (DAG). This interaction is most likely driven by electrostatic forces, although for phosphatidylserine in the presence of calcium additional forces are implicated. (2) Upon association, in particular with phosphatidylserine (PS, black circles), a conformational change exposes the hinge region (between CaLB and cat) and could bring the C1 region closer to the membrane to allow subsequent interactions with DAG. In this state the pseudosubstrate domain has not yet been exposed and the kinase is not active (middle). (3) DAG binding within the cysteine-rich regions promotes cooperative binding to PS by increasing the affinity for PS (black circles) but not for other anionic lipids (white circles). These additional interactions within the C1 region promote insertion of the protein into the hydrophobic core of the membrane, which releases the autoinhibitory pseudosubstrate motif from the catalytic domain and activates the kinase. Despite the presence of two potential interaction sites for DAG in the C1 region, only one molecule appears necessary for activation. Binding of one DAG molecule within the C1 region is indicated with no preference given to either of the cysteine-rich motifs. Subsequent binding of the pseudosubstrate motif to the membrane through electrostatic forces could provide additional stabilization and/or prevent this region from further interference with catalytic activity (right). The minimum of four PS molecules required in conjunction with DAG for maximum PKC activity is shown. Other anionic lipids may replace PS at additional, less crucial, PKC-lipid interaction sites.

chains containing six or more carbon atoms induce maximal kinase activity, suggesting that diacylglycerols must be sufficiently hydrophobic to favor partitioning into the micelle or the membrane (Davis et al., 1985; Lapentina et al., 1985; Hannun et al., 1986). Water-soluble molecules like 1-oleoyl-2-acetyl-*sn*-glycerol or diacylglycerols with two short-chain saturated fatty acids of six to ten carbons in length are efficiently delivered to cells and have been used to study PKC function *in vivo* (Castagna et al., 1982; Davis et al., 1985; Lapentina et al., 1985).

The high specificity of these interactions is consistent with the role of diacyl-glycerols as intracellular second messengers and activators of PKC. Activation of PKC by diacylglycerol is believed to result from two effects elicited by dia-cylglycerols: (1) The affinity of PKC for phosphatidylserine increases and strengthens PKC association with the membrane (Orr and Newton, 1992a,b; Mosier and Epand, 1993). 2) Diacylglycerol together with phosphatidylserine promotes the conformational change that exposes the pseudosubstrate domain (Orr et al., 1992).

Stereospecific requirements similar to those documented for diacylglycerols have also been noted for phorbol esters (Castagna et al., 1982; Rando, 1988). Phorbol esters that enhance PKC association with lipids both *in vivo* and *in vitro* (Kraft and Anderson, 1983; Hannun and Bell, 1986) also induce conformational changes in PKC (Boscá and Morán, 1993) and enhance proteolytic sensitivity in the hinge region (Kishimoto et al., 1983; Tapley and Murray, 1985; Melloni et al., 1986). Since phorbol esters induce substrate phosphorylation (Castagna et al., 1982), it must be assumed that, as with diacylglycerol, the pseudosubstrate domain becomes exposed. However, this has not been shown explicitly.

Phosphatidyl-L-Serine (PS)

Although several negatively charged lipid effectors may partially activate PKC, phosphatidyl-L-serine is the most effective phospholipid in its ability to activate PKC (Hannun et al., 1986; Burns et al., 1990) and is thought to represent the principal physiological phospholipid cofactor. Studies with phosphatidylserine analogs have revealed a high degree of specificity in their ability to support PKC activity, [^{3}H]phorbol dibutyrate (PDBu) binding, and binding to mixed micelles (Lee and Bell, 1989). Stereospecific interactions between PKC and the amino, carboxyl, and phosphate moieties of the phospho-L-serine head group in phos-phatidylserine appear critical, whereas no such stereospecificity is observed for the glycerol backbone. Thus, 1,2-*sn*-phosphatidylserine, 1,3-*sn*-phosphatidylser-ine, and 1,2-*rac*-phosphatidylserine are equally effective in supporting PKC activ-ity. Surprisingly, however, lyso-phosphatidylserine and 1-oleoyl-2-acetyl phos-phatidylserine are not, indicating that both acyl chains are critical even though positioning is not (Lee and Bell, 1989). This may reflect a requirement for an appropriate interfacial lipid conformation.

In contrast to diacylglycerol-dependent activation of PKC, activation by phos-phatidylserine is highly sigmoidal. Hill plot analysis of the phosphatidylserine-dependent activation of PKC at saturating calcium and diacylglycerol concentra-tions in mixed micellar assays yields Hill coefficients ranging from 4 to 11 (Hannun et al., 1985, 1986; Newton and Koshland, 1989; Burns et al., 1990). Slightly lower Hill coefficients are obtained in phosphatidylserine-dependent PDBu-binding experiments (Hannun and Bell, 1986). From these results two inferences have been made. First, a minimal number of about four phosphatidyl-serine molecules is required to observe PKC activation; when 10 to 12 phosphatidylserine-dependent activation is highly cooperative in the presence of calcium and diacylglycerol (Newton and Koshland, 1989). However, this second conclusion is controversial.

Since aggregation of phosphatidylserine-containing bilayers by calcium is well documented in the literature (Mclaughlin et al., 1981; Ekerdt and Papahadjopoulos, 1982; Feigenson, 1986), it has been proposed that cooperativity of PKC activation may arise due to the formation of calcium (phosphatidylserine)$_n$ complexes. Several lines of evidence suggest that this is not the case: different cPKCs display varying phosphatidylserine dependencies (K.-P. Huang et al., 1988; Burns et al., 1990); phosphatidylserines with different affinities for calcium bind PKC with the same calcium dependence (Keranen et al., 1992); despite the absence of any known calcium-binding motif (Klee et al., 1980), direct high-affinity binding of calcium to PKCβ in the absence of lipids has been demonstrated (Maurer et al., 1992), although binding of multiple calcium ions requires the presence of lipids (Bazzi and Nelsestuen, 1990); and calcium-independent nPKCs are activated cooperatively in the absence of calcium (Schaap and Parker, 1990).

Hill coefficients often serve as an indication of both subunit interactions and the number of interaction sites. A well-known example of a protein exhibiting cooperative interactions is hemoglobin with four subunits and a Hill coefficient of 2.8 for oxygen binding. For protein subunits, Hill coefficients of 5 are among the highest reported (Tallant and Cheung, 1984). Thus, the values obtained for phosphatidylserine-dependent activation of PKC are exceptionally high and may, in part, reflect the change in dimensionality of diffusion once PKC is bound to a lipid surface. For instance, cooperativity observed in the binding of basic peptides to acidic lipids is due to the electrostatic potential created by the acidic lipids as well as to the reduction in dimensionality that results after binding of the first basic peptide residue to the membrane (Mosier and McLaughlin, 1991, 1992a,b). Two results in studies with PKC indicate that at least the degree of cooperativity observed cannot be explained solely on the basis of electrostatic interactions and a reduction in dimensionality. First, diacylglycerol is required in conjunction with phosphatidylserine to yield highly cooperative interactions between PKC and phosphatidylserine. Second, surface-associated PKC continues to be cooperatively activated by phosphatidylserine, although to a decreased extent (Orr and Newton, 1992a). This, together with the observation that the Hill coefficient for phosphatidylserine-dependent PKC binding to membranes increases as the total lipid concentration decreases (Mosier and Epand, 1993), is consistent with apparent cooperativity being partially due to a reduction in dimensionality.

In summary, therefore, multiple highly specific interactions occur between phosphatidylserine and PKC. The phosphatidylserine-dependent PKC activation, especially in the presence of diacylglycerol, is highly cooperative, although probably not to the extent reflected in the Hill coefficients. This property allows PKC to sense and respond to very small changes in the membrane lipid composition. Phosphatidylserine concentrations of 15 mol% of total lipid have been reported in the plasma membrane inner leaflet for most cells (Verkleij et al., 1973; White, 1973; Zwaal and Bevers, 1983; Hannun and Bell, 1986) and are more than sufficient to totally activate PKC in the presence of saturating amounts of calcium and diacylglycerol. The concentrations of both calcium and DAG fluctuate depending on the activation state of the cell, and the concentration of phosphatidylserine available in the membrane is probably considerably lower due to sequestration by binding to membrane proteins (Boggs et al., 1977; Ong, 1984), cytoskeletal components (Sato and Ohnishi, 1983) and other calcium phospholipid-binding proteins (Geisow and Walker, 1986). Thus, cooperative activation of PKC by phosphatidylserine in conjunction with calcium and diacylglycerol at the plasma membrane inner surface could represent a physiologically relevant mechanism.

Other Phospholipids

Among the various phospholipids tested, phosphatidylserine, phosphatidylinositol, phosphatidylethanolamine, phosphatidic acid, phosphatidylglycerol, and cardiolipin are capable of supporting PKC activity, albeit to variable extents and at relatively high concentrations of calcium (10^{-4}–10^{-3} M). However, at low calcium concentrations (10^{-6} M), only phosphatidylserine in conjunction with diacylglycerol leads to high levels of kinase activity. Some differences in the degree of activity supported by various phospholipids are seen in different assay systems. Compare, for instance, results with phosphatidylethanolamine, phosphatidylinositol, and phosphatidic acid in liposomal versus mixed micellar assays (Kaibuchi et al., 1981; Hannun et al., 1986).

Despite these differences, it is apparent from results obtained using both assays that preferentially negatively charged, acidic phospholipids support activity. In contrast to the highly specific interaction between PKC and phosphatidylserine in the presence of diacylglycerol, these interactions appear to reflect the capacity of acidic lipids in general to bind PKC (Bazzi and Nelsestuen, 1987a; Lee and Bell, 1989; Orr and Newton, 1992b) and are probably due to unspecific, electrostatic effects. Three aspects of PKC-phospholipid interactions support this view. First, PKC binding to phosphatidylserine in lipid-detergent mixed micelles, compared with that of other phospholipids, is relatively insensitive to increasing ionic strength, indicating some degree of specificity beyond electrostatic interactions (Orr and Newton, 1992b). PKC binding to vesicles containing 30 percent phosphatidylserine is also unaffected up to 300 mM NaCl (Bazzi and Nelsestuen, 1987b). Similarly, NaCl does not affect the binding of phorbol esters to PKC (Hannun and Bell, 1990). Second, neither diacylglycerol nor calcium alter the binding of PKC to phospholipids other than phosphatidylserine (Orr and Newton, 1992b). Third, only binding to phosphatidylserine in the presence of diacylglycerol leads to conformational changes in PKC that are associated with activation (Orr et al., 1992).

In striking contrast to results discussed so far, both short-chain phosphatidylserine and phosphatidylcholine in micellar form have been shown to fully support kinase activity (Walker and Sando, 1988; Walker et al., 1990). Alternatively, others have shown that binding of PKC to neutral membranes containing either mixtures of naturally occurring phosphatidylserine and phosphatidylcholine or phosphatidylcholine alone in the presence of diacylglycerol is possible, but only at millimolar concentrations of calcium, and only mixtures of phosphatidyletha-nolamine/phosphatidylcholine support activity (Bazzi et al., 1992). To interpret the results with short-chain phosphatidylcholines, it was suggested that a loosely packed micellar structure may permit direct access of PKC to the interfacial and hydrophobic region (see section "Model for PKC Activation," in this chapter). Activation would occur where micelles form and would perturb the structure of PKC.

Interestingly, a precursor of diacylglycerol, phosphatidylinositol-4,5-bisphos-phate, has also been shown to activate PKC directly and effectively (O'Brian et al., 1987; Chauhan and Brockerhoff, 1988; Chauhan et al., 1990). Inhibition of [^{3}H]phorbol dibutyrate binding to PKC in the presence of phosphatidylinositol-4,5-bisphosphate indicated that phosphatidylinositol-4,5-bisphosphate interacted with PKC at the diacylglycerol/phorbol ester binding site (Chauhan et al., 1989). Others, however, have reported that while phosphatidylinositol-4,5-bisphosphate can activate PKC, it is not able to displace phorbol esters, indicating that phos-phatidylinositol-4,5-bisphosphate interacts with PKC at site(s) distinct from the phorbol ester-binding site (Lee and Bell, 1991; F. L. Huang and Huang, 1991). Additionally, phosphatidylinositol-4,5-bisphosphate, phosphatidylinositol-4,5-monophosphate, or phosphatidylinositol may function to spare, in part, the phos-phatidylserine phospholipid cofactor requirement of PKC (Lee and Bell, 1991). This possible role of phosphatidylinositol lipids in PKC activation *in vivo* is more consistent with estimated concentrations of these lipids in the plasma membrane inner leaflet. For phosphatidylinositol-4,5-bisphosphate this estimate lies in a sub-optimal concentration range for PKC activation *in vitro* of 0.05 to 0.25 mol% (Lee and Bell, 1991). It would be interesting to extend the above studies to phos-phatidylinositol-3,4,5-trisphosphate, which was recently implicated in the acti-vation of PKCζ (Nakanishi et al., 1993). Such studies may help shed light on how interactions of PKC with phosphatidylinositol lipids lead to activation and the domain within PKC where these interactions occur.

In summary, these studies have revealed that interactions between PKC and a variety of phospholipids are possible. PKC binds almost exclusively to acidic lipids, which points toward electrostatic interactions as the driving force for these associations. PKC activation by such interactions occurs predominantly in the presence of high calcium concentrations. By contrast, binding to phosphatidyl-serine is stereospecific, of higher affinity, enhanced in the presence of diacyl-glycerol, and highly cooperative. Several phosphatidylserine interaction sites appear to exist on PKC. For maximum PKC activity, only some, but not all, may be replaced by other anionic phospholipids. Activation occurs at the lipid mem-brane surface, and binding to this surface represents a prerequisite for activation.

Fatty Acids

In the absence of phosphatidylserine and calcium, cis-unsaturated fatty acids (FFA) such as arachidonic, linoleic, and oleic acid can activate PKC although to various degrees depending on the isoform (McPhail et al., 1984; Murakami and

Routtenberg, 1985; Hansson et al., 1986; Murakami et al., 1986; Sekiguchi et al., 1987, 1988; Naor et al., 1988; Nishikawa et al., 1988; Buday and Farago, 1990; Burns et al., 1990). Saturated or trans-unsaturated fatty acids are inactive. Furthermore, fatty acids have been shown to activate PKC synergistically with diacylglycerol in the presence of phosphatidylserine and to yield nearly fully active PKC at basal cytosolic calcium concentrations (El Touny et al., 1990; Lester et al., 1991; Shearman et al., 1991; Shimomura et al., 1991). While the majority of the results favor a physiological role for cis-unsaturated fatty acids in the regulation of PKC activity, others, in contrast, propose that fatty acids may exert their effect by infiltrating the PKC regulatory domain and destabilizing the autoinhibitory association between regulatory and catalytic domain (Chauhan et al., 1990).

The mechanism of PKC activation by fatty acids is distinct from that by diacylglycerol. While arachidonic acid can inhibit phorbol ester binding by a mechanism distinct from that of diacylglycerol (Sharkey and Blumberg, 1985), oleic acid is completely unable to do so (El Touny et al., 1990; Lester, 1990), indicating that neither interact with PKC at the phorbol ester/diacylglycerol-binding site. Fatty acids have also been proposed to activate PKC by modifying the biophysical properties of the membrane and promoting diacylglycerol availability (Lester, 1990; Lester et al., 1991). However, recent studies strongly suggest that only free, non-membrane-bound fatty acids are able to activate soluble PKC, indicating that activation of PKC *in vivo* may occur both in the cytosol by fatty acids and membrane-associated by diacylglycerol (Khan et al., 1992). In platelets where several PKC isoforms coexist (α, βI, βII, and δ), sodium oleate causes membrane translocation of predominantly PKC α, βII, and δ. Calcium-independent PKC δ is preferentially activated by oleate (EC_{50} 5 μM versus EC_{50} 50 μM for PKC α and β, respectively) and leads to calcium-independent phosphorylation of a 40-kDa substrate in platelet cytosol fractions (Khan et al., 1993).

Thus, PKC activation by fatty acids appears physiologically relevant but the mechanism of activation is distinct from that of diacylglycerol. Cis-unsaturated fatty acids may act synergistically with diacylglycerol at basal intracellular concentrations to activate both calcium-dependent and calcium-independent PKC isoforms while, in the absence of diacylglycerol, only calcium-independent isoforms would be activated. Diacylglycerol can be produced by phospholipase C-catalyzed hydrolysis of phosphatidylinositol and phosphatidylcholine or via prior formation of phosphatidic acid through phospholipase D-catalyzed hydrolysis (Besterman et al., 1986; Pelech and Vance, 1989; Exton, 1990). Fatty acids and lysophosphatidylcholine are produced by phospholipase A_2. Lysophosphatidylcholine may also elicit some of its effects on cells by direct potentiation of diacylglycerol-dependent activation, but, unlike cis-unsaturated fatty acids, it does not increase calcium sensitivity of PKC activation (Asaoka et al., 1992; Yoshida et al., 1992). In summary, PKC activation is a point of convergence for signal-induced, membrane phospholipid hydrolysis products formed by the activation of phospholipases A_2, C, and D (Nishizuka, 1992) and may also be regulated by sphingolipid metabolites (Hannun and Bell, 1989). Lipid activators of PKC are summarized in Table 3.1.

Table 3.1 Properties of Known Mamallian Protein Kinase C Isozymes[a]

Group/subspecies		Molecular size, (Daltons)	Lipid activators[a]	Zinc stoichiometry (mol Zn/mol PKC)	Tissues	Domain structure[b]
cPKC	α	76799	PS, Ca^{2+}, DAG, FFA	4.2	Universal	
	β$_1$	76790	PS, Ca^{2+}, DAG, FFA	?	Some tissues	
	β$_{11}$	76933	PS, Ca^{2+}, DAG, FFA	3.4	Many tissues	
	γ	78366	PS, Ca^{2+}, DAG, FFA	4.0	Brain	
nPKC	δ	77517	PS, DAG	?	Universal	
	ε	83478	PS, DAG, FFA	?	Brain and others	
	η(L)	77972	?	?	Lung, skin, heart	
	θ	81571	?	?	Skeletal muscle	
aPKC	ζ	67740	PS, FFA, PIP$_3$	?	Universal	
	λ	67200	?	?	Ovary, testis	

[a]PS, phosphatidylserine; DAG, diacylglycerol; FFA, cis-unsaturated fatty acids; PIP$_3$, phosphatidylinositol-3,4,5-triphosphate.

[b]Variable regions are solid (V1–V5); conserved regions either blank (ATP-binding domain, C3) or shaded (C1, C2, and C4).

Sources: V. Nishizuka. 1992. Intracellular signalling by hydrolysis of phospholipids and activation of protein kinase C. *Science* 258:607–614. Values for zinc stoichiometry from A. F. G. Quest, J. Bloomenthal, E. S. G. Bardes, and R. M. Bell, 1992. The regulatory domain of protein kinase C coordinates four atoms of zinc. *J. Biol. Chem.* 267:10193 10197.

REGIONS OF SIGNIFICANT PKC–LIPID INTERACTIONS

Some evidence for lipid interaction sites in the catalytic domain exists. For instance, phospholipids modulate substrate phosphorylation by a catalytic fragment of PKC (Nakadate et al., 1987) and acidic phospholipids in the absence of calcium (nonactivating conditions) differentially inactivate different PKC isoforms, as assessed by subsequent lipid-independent phosphorylation of protamine (K.-P. Huang and Huang, 1990). All regions of PKC identified so far that interact with lipids to produce the conformational changes required for PKC activation are located within the regulatory domain. In the following section the discussion will focus on interactions with lipids in three regions of mammalian PKC: C1, C2, and V1 (see Table 3.1).

The C1 Domain

The C1 region of PKC harbors two cysteine-rich motifs, referred to as Cys1 and Cys2,[1] with striking homology to zinc-coordinating motifs of transcription factors involved in the regulation of transcriptional activity (Nishizuka, 1988; Bell and Burns, 1991). Although some PKC is found in the nucleus (Leach et al., 1989), and PKC fragments including sequence elements close to the hinge region (V3) translocate to the nucleus (James and Olson, 1992) there is no direct evidence that PKC or its cysteine-rich regions are directly involved in transcriptional regulation even though PKC binding to DNA has been reported (Testori et al., 1988). The cysteine-rich regions do, however, confer phospholipid-dependent phorbol dibutyrate binding, and either Cys1 or Cys2 is sufficient for binding (Kaibuchi et al., 1989; Ono et al., 1989b; Cazaubon et al., 1990; Burns and Bell, 1991). Mutations of conserved cysteine residues in each cysteine-rich motif destroy binding activity (Ono et al., 1989b). Therefore, the cysteine-rich motifs contain sequence elements crucial for phorbol ester binding and, by inference, also for the interaction with diacylglycerol (Sharkey et al., 1984; König et al., 1985). The majority of mammalian PKC isoforms contain two such cysteine-rich motifs (cPKCs and nPKCs; Table 3.1), whereas members of the aPKC group (PKC λ and ζ; Table 3.1) contain only one such motif. PKC ζ neither binds phorbol esters nor is activated by diacylglycerols (Ono et al., 1989a; Ways et al., 1992; Nakanishi and Exton, 1993). However, recent work suggests a role for phosphatidylinositol-3,4,5-trisphosphate in PKC ζ activation (Nakanishi et al., 1993). No equivalent data on the enzymological properties of PKC λ are currently available. The results with PKC ι indicate that one cysteine-rich motif in the context of the entire protein may not suffice to bind phorbol esters or interact with diacylglycerol and that both motifs in tandem may be required for this function. Alternatively, sequence

---

[1] Cys1, first PKC$_\gamma$ cysteine-rich region (amino acids 25–93); Cys2, second PKC$_\gamma$ cysteine-rich region (amino acids 92–173, numbered as in J. L. Knopf, M.-H. Lee, L. A. Sultzman, R. W. Kris, C. R. Loomis, R. M. Hewick, and R. M. Bell. 1986. Cloning and expression of multiple protein kinase C cDNAs. *Cell* 46: 492 502.).

elements essential for these interactions may simply be missing. However, none of the six cysteines and two histidines that are conserved between cysteine-rich motifs present in all known PKC isoforms, as well as in other proteins, are missing (Ahmed et al., 1991; Hubbard et al., 1991).

Data available on the stoichiometry of interactions between diacylglycerol and rat brain PKC (predominantly PKC α, β, and γ) and on phorbol ester binding indicate that only one of the two potential high-affinity binding sites is utilized in cPKCs (Kikkawa et al., 1983b; Hannun et al., 1985; Hannun and Bell, 1986). Recent binding studies with phorbol esters and related molecules provide indirect evidence for the presence of a high- and a low-affinity binding site. Calcium-dependent activation was proposed to occur through ligand binding to the former site, whereas subsequent membrane insertion would result from additional binding to the latter site (Kazanietz et al., 1992). With two inequivalent binding sites for lipophilic agonists, PKC would resemble its kinase counterparts that are activated by the hydrophilic ligands cAMP and cGMP. Both cAMP- and cGMP-dependent kinases contain nonequivalent cyclic nucleotide binding sites within their respective regulatory subunits or domains (Weber et al., 1989; Taylor et al., 1990, 1993).

Initial direct qualitative evidence for the presence of zinc in PKC and other proteins was provided by the capacity of cysteine-rich regions immobilized on nitrocellulose to bind radioactive zinc (Ahmed et al., 1991). Zinc quantitation by electrothermal atomic absorption spectrometry of three cPKCs—PKC α, PKC βII, and PKC γ—as well as of the regulatory domain of PKC α, provided direct evidence for the coordination of four atoms of zinc within the regulatory domain of PKC (Quest et al., 1992). On the basis of the similarity to zinc-coordinating domains essential for DNA binding (Coleman, 1992), possible structures for the PKC cysteine-rich motifs have been inferred (Bell and Burns, 1991). Indirect structural evidence obtained by extended x-ray absorption fine structure (EXAFS) analysis favors a model in which two zinc atoms are tetracoordinated between six conserved cysteines (C) and either zero, one, or two conserved histidines (H) in the 50-amino-acid C_6H_2 consensus present in cysteine-rich motifs of all known PKCs (Hubbard et al., 1991).

Peptides as small as 86 amino acids from the Cys2 region of PKC γ are capable of binding phorbol dibutyrate, albeit with significantly lower affinity (30–60 nM) than that of intact PKC γ (2–5 nM) (Burns and Bell, 1991). With use of PKC γ regions expressed as fusion proteins with glutathione-S-transferase, definitive evidence for the coordination of two atoms of zinc within the 50-amino-acid C_6H_2 consensus of PKC γ Cys2 has been provided (Quest et al., 1993, 1994a,b,c). By deletion analysis, a minimal protein sequence of 45 amino acids has been defined for phorbol dibutyrate binding, which closely resembles but is not identical to the C_6H_2 consensus sequence (Quest et al., 1994b). Furthermore, the fusion protein glutathione-S-transferase/Cys2 with only 82 amino acids of PKC γ (amino acids 92–173, numbered as in Knopf et al., 1986) was shown to bind stereospecifically 4β-hydroxy phorbol dibutyrate with high affinity (K_d 23 nM) in a cooperative manner akin to that of PKC γ with respect to phosphatidylserine, and direct evidence for stereospecific interactions between diacylglycerol and Cys2 was provided. Finally, association of glutathione-S-transferase/Cys2 with phosphatidyl-serine-/phosphatidylcholine liposomes was found to increase in the presence of either phorbol dibutyrate or diacylglycerol in a stereospecific manner (Quest et al., 1994b). The implications of these studies are that the features of lipid-dependent regulation considered to be unique to PKC are conferred by a remarkably small section of the regulatory domain. Thus, any proteins with cysteine-rich motifs, like those in diacylglycerol kinase, n-chimaerin, β-chimaerin, *unc*-13, *raf/ mil,* and *vav* oncogene products (Ishikawa et al., 1986; Katzau et al., 1990; Ahmed et al., 1990; Hall et al., 1990; Schaap et al., 1990; Ahmed et al., 1991; Hubbard et al., 1991; Maruyama and Brenner, 1991; Leung et al., 1993) may be physiologically regulated by diacylglycerol in a manner similar to that of PKC. These observations also suggest that caution is required when interpreting effects of phorbol esters and cell-permeable diacylglycerols *in vivo* as being solely due to interactions with PKC.

The C2 Domain

A number of PKC family members (nPKCs, aPKCs) lack the C2 region, which is present only in cPKCs. PKCδ and PKCε do not require calcium for effector-dependent phosphorylation (Ohno et al., 1988; Ono et al., 1988; Akita et al., 1990; Schaap and Parker, 1990; Olivier and Parker, 1991). In contrast to PKC α and β, neither PKC δ nor PKC ε bind to membranes in a calcium-dependent manner (Kiley et al., 1990; Olivier and Parker, 1991). From these observations the C2 domain has been implicated in conferring calcium dependence of phorbol ester binding and activation of cPKCs. Molecular analysis of PKC γ has shown that phorbol dibutyrate binding to deletion proteins lacking the C2 domain are substantially less calcium dependent, again implicating the C2 region in mediating calcium dependence of PKC-lipid interactions (Kaibuchi et al., 1989; Ono et al., 1989b).

Several other proteins also contain sequence elements homologous to the C2 region of cPKCs: Phospholipase C-γ1, an enzyme that hydrolyzes phosphatidyl-inositol-4,5-bisphosphate to produce the second messengers diacylglycerol and inositol trisphosphate (Stahl et al., 1988); the GTPase-activating protein (GAP), which binds specifically acidic phospholipids (Vogel et al., 1988; Tsai et al., 1991); a synaptic vesicle protein, called synaptotagmin, which also binds acidic phospholipids, in particular phosphatidylserine, and may function as a calcium sensor on the synaptic vesicle surface (Perin et al., 1990; Geppert et al., 1991; Brose et al., 1992); a cytosolic phospholipase A_2 ($cPLA_2$) (Clark et al., 1991), and a novel phorbol ester-binding protein (*unc*-13) that was cloned from *Caenorhabditis elegans* (Maruyama and Brenner, 1991).

An amino terminal fragment of $cPLA_2$ was shown to translocate to natural membrane vesicles in a calcium-dependent fashion (Clark et al., 1991) and this behavior was attributed to the presence of the C2-like domain, in particular a 30- to 40-amino-acid stretch with several basic residues called the CaLB (*Ca*lcium-dependent *L*ipid *B*inding) domain. Similarly, expression in *Escherichia coli* of the regions of *unc*-13 homologous to C1 and C2 domains of cPKCs (Maruyama and Brenner, 1991) yielded a similar calcium dependence of phorbol ester binding as the corresponding elements from PKC γ (Ono et al., 1989b). Taken together, these results strongly implicate the C2 domain in mediating calcium-dependent interactions with anionic lipid surfaces. Although phosphatidylserine-dependent calcium-binding activity was recently demonstrated for the regulatory domain of PKC βI (Luo et al., 1993), conclusive evidence for a direct role of the CaLB motif in mediating calcium-dependent lipid interactions or lipid-dependent calcium binding is still missing.

Experiments determining phorbol ester binding to PKC deletions lacking the C2 region show a markedly reduced calcium dependence (Ono et al., 1989b), but a slight dependence is still apparent, indicating that other regions may contribute partially to the calcium dependence of phorbol ester binding observed for cPKCs. Indeed, phorbol dibutyrate binding to glutathione-S-transferase fusion proteins with either Cys1 or Cys2, or with Cys1Cys2 together, in the absence of the C2 domain, were strongly calcium dependent, albeit to a lesser extent than PKC γ itself, indicating that the cysteine-rich regions themselves partially contribute to the calcium dependence observed for cPKCs (Quest et al., 1994c).

In summary, the results implicate the C2 domain in mediating calcium-dependent lipid interactions, but do not exclude possible contributions from other lipid interaction sites to the overall calcium dependence observed in intact cPKCs. From a mechanistic point of view, the C1 and C2 domains differ markedly in the types of lipid interactions they mediate. Thus, in models for the activation of cPKCs, it is proposed that the C2 domain mediates an initial, calcium-dependent translocation to the membrane and the C1 domain is considered responsible for subsequent tighter interactions that lead to the conformational changes affiliated with PKC activation (see below).

The V1 Region

Additional lipid interaction sites in the pseudosubstrate site of PKC have been predicted on the basis of the observation that synthetic peptides mimicking the pseudosubstrate region are able to bind to acidic lipids in membranes (Mosier and McLaughlin, 1991, 1992a,b). This interaction could provide about 6 kcal mol^{-1} of stabilization energy. Because the pseudosubstrate domain does not seem to be exposed prior to activation (Orr et al., 1992), and the loss in entropy resulting from translocation of a 100-kDa protein (PKCs are roughly 80 kDa; see Table 3.1) to the membrane could be as much as 6 kcal mol^{-1} larger than that for a 1-kDa peptide (Janin and Chothia, 1978; Dwyer and Bloomfield, 1981), it appears unlikely that the pseudosubstrate domain is involved in the initial membrane translocation event. However, after PKC activation, this region could associate with the acidic phospholipids and provide additional stabilization at the membrane surface. This interaction could also help to prevent the pseudosubstrate region from folding back into the kinase domain and inhibiting kinase-substrate interactions after activation. Indirect evidence supporting such a role for the pseudosubstrate domain stems from measurements of the phorbol dibutyrate binding affinity for glutathione-S-transferase fusion proteins with cysteine-rich motifs in the presence or absence of the pseudosubstrate, which indicates that this region increases binding affinity significantly (Quest et al., 1994c).

MODEL FOR PKC ACTIVATION

Recently, several papers describing molecular mechanisms of PKC activation have been published (Stabel and Parker, 1991; Burns and Bell, 1992; Orr and Newton, 1992b; Zidovetski and Lester, 1992). All originated in a proposal by Hannun and Bell (Bell, 1986; Hannun and Bell, 1986). A model is proposed here that resembles previous suggestions with some subtle differences concerning the contributions of different parts within the regulatory domain (Figure 3.1).

As pointed out before, the majority of data available are from work with cPKCs, and thus the model focuses on providing a mechanistic basis for understanding the activation of this group *in vivo* (Table 3.1). In an inactive state PKC resides within the cytosol. On elevation of intracellular calcium levels, cPKCs translocate to the membrane in an initial step mediated through the C2 domain. In this state the enzyme remains inactive, but a conformational change has occurred (Nelsestuen and Bazzi, 1991; Boscá and Morán, 1993). This initial association event is not sufficient for activation and could be mediated both by nonactivating lipids and by phosphatidylserine.

Interactions with diacylglycerols or phorbol esters at the membrane surface would then induce additional conformational changes that release the pseudosubstrate domain from its site of attachment within the catalytic domain. Diacylglycerols from either phosphatidylinositol (Leach et al., 1991) or phosphatidylcholine hydrolysis (Diaz-Laviada et al., 1990) would lead to activation. These activators may release the inhibitory pseudosubstrate motif from the substrate-binding site in the catalytic domain by promoting a transition within the regulatory domain from a membrane-associated to a membrane-inserted state (Bazzi and Nelsestuen, 1988, 1989; Brumfeld and Lester, 1990; Lester et al., 1990). PKC activity is known to be sensitive to the physical state of membranes (Epand et al., 1988; Epand and Lester, 1990; Bolen and Sando, 1992), and although diacylglycerol and phorbol esters have been suggested to activate membrane-bound PKC through local changes in the lipid environment, which would allow PKC to penetrate deeper into the membrane (Souvignet et al., 1991), the stereospecificity of activation favors a direct interaction between PKC and the activating molecules. The cysteine-rich domains of PKC that bind phorbol esters and contain interaction sites with diacylglycerol have been shown to insert into membrane surfaces (Lester et al., 1990). Stoichiometric analysis of PKC interactions with either phorbol esters or diacylglycerols suggest that only one molecule of these agonists binds to PKC and is sufficient for activation despite the presence of two potential binding sites. Which cysteine-rich motif actually binds these agonists, how the other one is prevented from doing so, and how phorbol esters may lead to irreversible PKC–membrane association even in the absence of calcium (Bazzi and Nelsestuen, 1989) are still unresolved questions. It is, of course, possible that the present preparations of PKC contain enzyme incapable of binding phorbol esters and thus prevent stoichiometries approaching 2 from being measured.

Finally, liberation of the pseudosubstrate domain from inhibitory interactions in the catalytic domain could allow it to associate subsequently with the lipid surface, perhaps providing additional stabilization energy and preventing further interactions with the catalytic domain.

CONCLUSIONS

The model proposed in Figure 3.1 is consistent with a majority of the data available for cPKCs. Since nPKCs lack the C2 domain and activity is calcium independent, the initial step in this model must be modified to account for activation of these isoforms. nPKCs have an extended V1 region compared with cPKCs (Table 3.1), which in the case of PKC δ and PKC ε modulates the substrate specificity (Olivier and Parker, 1991; Pears et al., 1991). This may result from involvement of the additional parts of the extended V1 region besides the pseudosubstrate site in PKC-membrane and/or -cytoskeleton interactions (Akita et al., 1990; Leibersperger et al., 1990). As PKC δ and PKC ε are largely recovered in the particulate fraction of cell homogenates, these isoforms may already be located very close to the membrane, making translocation unnecessary, or the extended V1 region could mediate such a translocation in calcium-independent fashion.

The model cannot explain how, for instance, different isoforms of the same PKC group may be independently regulated by diacylglycerols unless, perhaps, pools of lipid precursors for diacylglycerol exist with differential accessibility. In addition to the experiments with PKC δ and PKC ε that suggested a role for the regulatory domain in modulating substrate specificity, the association of different isoforms with distinct intracellular, in particular cytoskeletal, elements (Jaken et al., 1989; Kiley and Jaken, 1990; Kiley et al., 1990; Mochley-Rosen et al., 1990; Kiley et al., 1991; Mochley-Rosen et al., 1991a,b; Spudich et al., 1992) may help define the substrate targets of PKC isozymes.

Finally, the model as such cannot explain the possible role for phosphorylation in the activation process. Protein kinase C undergoes autophosphorylation at several sites (Flint et al., 1990) and appears also to be additionally trans-phosphorylated *in vivo*. This latter process has been suggested to represent a "priming" event that is required for autophosphorylation and substrate phosphorylation (Pears et al., 1992). Interestingly, all autophosphorylation sites of PKC βII lie in or near regions critical for the events described in the model: two sites lie in V1 (near pseudosubstrate), two lie in V3 (hinge region), and two lie in C4 (roughly 100 amino acids from substrate-binding site). One might envisage how autophosphorylation at all sites could "lock" PKC into an open, active conformation, unable to fold back around the hinge region into an inactive state. This conformation may correlate with irreversibly membrane-associated PKC induced in the presence of phorbol esters (Bazzi and Nelsestuen, 1989). Such a "rigid" open conformation might yield PKC more susceptible to proteolytic degradation and provide an explanation for PKC down-regulation, a common response on phorbol ester treatment of cells. This possibility is supported by the following observations. First, PKC α with a point mutation in the catalytic domain that abolishes kinase activity is not down-regulated in cells on phorbol ester treatment although it does translocate from cytosol to the particular fraction (Ohno et al., 1990). An alternative interpretation of this experiment is that PKC-dependent substrate phosphorylation is required to activate a down-regulation pathway, for instance through calpains I and II (Kishimoto et al., 1983, 1989). Second, down-regulation of PKC activity in human breast cancer cells by phorbol ester treatment correlates with the disappearance of phosphorylated PKC protein bands of 77- and 80-kDa and the simultaneous accumulation of a 74-kDa precursor as judged by sodium dodecyl sulfate polyacrylamide electrophoresis (Borner et al., 1989). This latter report also implicates posttranslational PKC phosphorylation in functional maturation.

In summary, therefore, many questions concerning PKC activation remain to be answered before we can fully appreciate the complexity and intricacy of its regulation by lipids. However, work in this area has already yielded insights into mechanisms of lipid association and possible regulation of other proteins besides PKC involved in signal transduction. Furthermore, our current level of understanding of PKC activation and the apparent problems indicate that future analysis will need to focus on the role of individual PKC subdomains in mediating protein-lipid interactions, as well as their potential to mediate protein-protein interactions. An approach with considerable promise in this respect is the expression and characterization of PKC subdomains.

REFERENCES

Ahmed, S., R. Kozma, C. Monfries, C. Hall, H. H. Lim, and P. Smith. 1990. Human brain n-chimaerin cDNA encodes a novel phorbol ester receptor. *Biochem. J.* 272:767–773.

Ahmed, S., R. Kozma, J. Lee, C. Monfries, N. Harden, and L. Lim. 1991. The cysteine-rich domain of human proteins, neuronal chimaerin, protein kinase C and diacylglycerol kinase binds zinc. *Biochem. J.* 280:233–241.

Akita, Y., S. Ohno, Y. Konno, A. Yano, and K. Suzuki. 1990. Expression and properties of two distinct classes of the phorbol ester receptor family, four conventional protein kinase C types, and a novel protein kinase C. *J. Biol. Chem.* 265:354–362.

Alkon, D. L. 1989. Memory storage and neural systems. *Sci. Am.* 261:42–50.

Alkon, D. L., and H. Rasmussen. 1988. A spatial-temporal model of cell activation. *Science* 239:998–1005.

Asaoka, Y., M. Oka, K. Yoshida, Y. Sasaki, and Y. Nishizuka. 1992. Role of lysophosphatidylcholine in T-lymphocyte activation: Involvement of phospholipase A_2 in signal transduction through protein kinase C. *Proc. Natl. Acad. Sci. U.S.A.* 89:6447–6451.

Ashendel, C. L. 1985. The phorbol ester receptor: A phospholipid-regulated kinase. *Biochem. Biophys. Acta* 822:219–242.

Bacher, N., Y. Zisman, E. Berent, and E. Livneh. 1991. Isolation and characterization of protein kinase C-L, a new member of the protein kinase C-related gene family specifically expressed in lung, skin and heart. *Molec. Cell. Biol.* 11:126–133.

Baier, G., D. Telford, L. Giampa, K. M. Coggeshall, G. Baier-Bitterlich, N. Isakov, and A. Altman. 1993. Molecular cloning and characterization of protein kinase C θ, a novel member of the protein kinase C (PKC) gene family expressed predominantly in hematopoietic cells. *J. Biol. Chem.* 268:4997–5004.

Bazzi, M. D., and G. L. Nelsestuen. 1987a. Association of protein kinase C with phospholipid vesicles. *Biochemistry* 26:115–122.

Bazzi, M. D., and G. L. Nelsestuen. 1987b. Role of substrate in imparting calcium and phospholipid requirements to protein kinase C activation. *Biochemistry* 26:1974–1982.

Bazzi, M. D., and G. L. Nelsestuen. 1988. Association of protein kinase C with phospholipid monolayers: Two-stage irreversible binding. *Biochemistry* 27:6776–6783.

Bazzi, M. D., and G. L. Nelsestuen. 1989. Differences in the effects of phorbol esters and diacylglycerols on protein kinase C. *Biochemistry* 28:9317–9323.

Bazzi, M. D., and G. L. Nelsestuen. 1990. Protein kinase C interaction with calcium: A phospholipid-dependent process. *Biochemistry* 29:7624–7630.

Bazzi, M. D., A Yanakim, and G. L. Nelsestuen. 1992. Importance of phosphatidylethanolamine for association of protein kinase C and other cytoplasmic proteins with membranes. *Biochemistry* 31:1125–1134.

Bell, R. M. 1986. Protein kinase C activation by diacylglycerol second messengers. *Cell* 45:631–632.

Bell, R. M., and D. J. Burns. 1991. Lipid activation of protein kinase C. *J. Biol. Chem.* 266:4661–4664.

Bell, R. M., and R. A. Coleman. 1980. Enzymes of glycerolipid synthesis in eukaryotes. *Annu. Rev. Biochem.* 49:459–487.

Berridge, M. J. 1983. Rapid accumulation of inositol triphosphate reveals that agonists hydrolyze polyphosphoinositides instead of phosphatidylinositol. *Biochem. J.* 212:849–858.

Berridge, M. J. 1984. Inositol triphosphate and diacylglycerol as second messengers. *Biochem. J.* 220:345–360.

Besterman, J. M., V. Duronio, and P. Cuatrecasas. 1986. Rapid formation of diacylglycerol from phosphatidylcholine: A pathway for generation of a second messenger. *Proc. Natl. Acad. Sci. U.S.A.* 83:6785–6789.

Bishop, W. R., and R. M. Bell. 1988. Functions of diacylglycerol in glycerolipid metabolism, signal transduction and cellular transformation. *Oncogene Res.* 2:205–208.

Blumberg, P. M. 1980. *In vitro* studies on the mode of action of the phorbol esters, potent tumor promoters. *CRC Crit. Rev. Toxicol.* 8:153–197.

Boggs, J. M., D. D. Wood, M. A. Moscarello, and D. Papahadjopoulos. 1977. Lipid phase separation induced by a hydrophobic protein in phosphatidylserine-phosphatidylcholine vesicles. *Biochemistry* 16:2325–2329.

Bolen, E. J., and J. J. Sando. 1992. Effect of phospholipid unsaturation on protein kinase C activation. *Biochemistry* 31:5945–5951.

Boni, L. T., and R. R. Rando. 1985. The nature of protein kinase C activation by physically defined phospholipid vesicles and diacylglycerols. *J. Biol. Chem.* 260:10819–10825.

Bonser, R. W., N. T. Thompson, H. F. Hodson, R. M. Beams, and L. G. Garland. 1988. Evidence that a second stereochemical centre in diacylglycerols defines interaction at the recognition site on protein kinase C. *FEBS Lett.* 234:341–344.

Borner, C., I. Filipuzzi, M. Wartmann, U. Eppenberger, and D. Fabbro. 1989. Biosynthesis and post-translational modifications of protein kinase C in human breast cancer cells. *J. Biol. Chem.* 264:13902–13909.

Boscá, L., and F. Morán. 1993. Circular dichroism analysis of ligand-induced conformational changes in protein kinase C. *Biochem. J.* 290:827–832.

Boutwell, R. K. 1974. The function and mechanism of promotors of carcinogenesis. *CRC Crit. Rev. Toxicol.* 2:419–443.

Brose, N., A. G. Petrenko, T. C. Südhof, and R. Jahn. 1992. Synaptotagmin: A calcium sensor on the synaptic vesicle surface. *Science* 256:1021–1025.

Brumfeld, V., and D. S. Lester. 1990. Protein kinase C penetration into lipid bilayers. *Arch. Biochem. Biophys.* 277:318–323.

Buday, L., and A. Farago. 1990. Dual effect of arachidonic acid on protein kinase C isozyme isolated from rabbit thymus cells. *FEBS Lett.* 276:223–226.

Burns, D. J., and R. M. Bell. 1991. Protein kinase C contains two phorbol ester binding domains. *J. Biol. Chem.* 266:18330–18338.

Burns, D. J., and R. M. Bell. 1992. Lipid regulation of protein kinase C. In: *Protein Kinase C: Current Concepts and Future Perspectives,* edited by D. Lester and R. M. Epand, pp. 25–40. Horwood, Chichester, England.

Burns, D. J., J. Bloomenthal, M.-H. Lee, and R. M. Bell. 1990. Expression of the α, βII and γ protein kinase C isozymes in the Baculovirus-insect cell expression system. *J. Biol. Chem.* 265:12044–12051.

Castagna, M., Y. Takai, K. Kaibuchi, K. Sano, U. Kikkawa, and Y. Nishizuka. 1982. Direct activation of calcium-activated, phospholipid-dependent protein kinase by tumor-promoting phorbol esters. *J. Biol. Chem.* 257:7847–7851.

Cazaubon, S., C. Webster, L. Camoin, A. D. Strosberg, and P. J. Parker. 1990. Effector dependent conformational changes in protein kinase Cγ through epitope mapping with inhibitory monoclonal antibodies. *Eur. J. Biochem.* 194:799–804.

Chauhan, V. P. S., and H. Brockerhoff. 1988. Phosphatidylinositol-4,5-bisphosphate may antecede diacylglycerol as activator of protein kinase C. *Biochem. Biophys. Res. Comm.* 155:18–23.

Chauhan, A., V. P. S. Chauhan, D. S. Deshmukh, and H. Brockerhoff. 1989. Phosphatidylinositol-4,5-biphosphate competitively inhibits phorbol ester binding to protein kinase C. *Biochemistry* 28:4952–4956.

Chauhan, V. P. S., A. Chauhan, D. S. Deshmukh, and H. Brockerhoff. 1990. Lipid activators of protein kinase C. *Life Sci.* 47:981–986.

Clark, J. D., L. L. Lim, R. W. Kriz, C. S. Ramesha, L. A. Sultzman, A. Y. Lin, N. Milona, and J. L. Knopf. 1991. A novel arachidonic acid selective cytosolic PLA$_2$ contains a Ca^{2+}-dependent translocation domain with homology to PKC and GAP. *Cell* 65: 1043–1051.

Cohen, P. 1988. Protein phosphorylation and hormone action. *Proc. Roy. Soc. Lond. B* 234:115–144.

Cohen, P. 1989. The structure and function of protein phosphatases. *Annu. Rev. Biochem.* 58:453–508.

Coleman, J. E. 1992. Zinc proteins: Enzymes, storage proteins, transcription factors, and replication proteins. *Annu. Rev. Biochem.* 61:897–946.

Davis, R. J., B. R. Ganong, R. M. Bell, and M. P. Czech. 1985. Structural requirements for diacylglycerols to mimic tumor-promoting phorbol diester action on the epidermal growth factor receptor. *J. Biol. Chem.* 260:5315–5322.

Diaz-Laviada, I., P. Larrodera, M. T. Diaz-Meco, M. E. Cornet, P. H. Guddal, T. Johansen, and J. Moscat. 1990. Evidence for a role of phosphatidylcholine hydrolysing phospholipase C in the regulation of protein kinase C by ras and src oncogenes. *EMBO J.* 9:3907–3912.

Douglas, W. W., and R. P. Rubin. 1961. The role of calcium in the secretory response of the adrenal medulla to acetylcholine. *J. Physiol.* 159:40–57.

Dwyer, J. B., and V. A. Bloomfield. 1981. Binding of multivalent ligands to mobile receptors in membranes. *Biopolymers* 20:2323–2336.

Ebashi, S., and M. Endo. 1968. Calcium and muscle contraction. *Progr. Biophys. Molec. Biol.* 18:123–183.

Ekerdt, R., and D. Papahadjopoulos. 1982. Intermembrane contacts after calcium binding to phospholipid vesicles. *Proc. Natl. Acad. Sci. U.S.A.* 79:2273–2277.

El Touny, S., W. Khan, and Y. Hannun. 1990. Regulation of platelet protein kinase C by oleic acid. *J. Biol. Chem.* 265:16437–16443.

Epand, R. M., and D. S. Lester. 1990. The role of membrane biophysical properties in the regulation of protein kinase C. *Trends Pharmacol. Sci.* 11:317–320.

Epand, R. M., A. R. Stafford, J. J. Cheetam, R. Bottega, and E. H. Ball. 1988. The relationship between the bilayer to hexagonal phase transition temperature in membranes and protein kinase C activity. *Biosci. Rep.* 8:49–54.

Exton, J. H. 1990. Signaling through phosphatidylcholine breakdown. *J. Biol. Chem.* 265:1–4.

Feigenson, G. W. 1986. On the nature of calcium ion binding between phosphatidylserine lamellae. *Biochemistry* 25:5819–5825.

Flint, A. J., R. D. Palachini, and D. E. Koshland Jr. 1990. Autophosphorylation of protein kinase C at three separate regions of its primary sequence. *Science* 249:408–411.

Ganong, B. R., C. R. Loomis, Y. A. Hannun, and R. M. Bell. 1986. Specificity and mechanism of protein kinase C activation by *sn*-1,2-diacylglycerol. *Proc. Natl. Acad. Sci. U.S.A.* 83:1184–1188.

Geisow, M. J., and J. H. Walker. 1986. New proteins involved in cell regulation by Ca^{2+} phospholipids. *Trends Biochem. Sci.* 11:420–423.

Geppert, M., B. T. Archer, and T. C. Südhof. 1991. Synaptotagmin II: A novel differentially distributed form of synaptotagmin. *J. Biol. Chem.* 266:13548–13552.

Hall, C., C. Monfries, P. Smith, H. H. Lim, R. Kozma, S. Ahmed, V. Vanniasingham, T. Leung, and L. Lim. 1990. Novel human brain cDNA encoding a 34 000 M_r protein n-chimaerin, related to both the regulatory domain of protein kinase C and BCR, the product of the breakpoint cluster region gene. *J. Molec. Biol.* 211:11–16.

Hanks, S. K., A. M. Quinn, and T. Hunter. 1988. The protein kinase C family: Conserved features and deduced phylogeny of the catalytic domains. *Science* 241:45–52.

Hannun, Y. A., and R. M. Bell. 1986. Phorbol ester binding and activation of protein kinase C on Triton X-100 mixed micelles containing phosphatidylserine. *J. Biol. Chem.* 261:9341–9347.

Hannun, Y. A., and R. M. Bell. 1987. Lysosphingolipids inhibit protein kinase C: Implications for the sphingoliposes. *Science* 235:670–674.

Hannun, Y. A., and R. M. Bell. 1989. Functions of sphingolipids and sphingolipid breakdown products in cellular regulation. *Science* 243:500–507.

Hannun, Y. A., and R. M. Bell. 1990. Rat brain protein kinase C: Kinetic analysis of substrate-dependence, allosteric regulation, and autophosphorylation. *J. Biol. Chem.* 265:2962–2972.

Hannun, Y. A., C. R. Loomis, and R. M. Bell. 1985. Activation of protein kinase C by Triton X-100 mixed micelles containing diacylglycerol and phosphatidylserine. *J. Biol. Chem.* 260:10039–10043.

Hannun, Y. A., C. R. Loomis, and R. M. Bell. 1986. Protein kinase C activation in mixed micelles. *J. Biol. Chem.* 261:7184–7190.

Hansson, A., C. N. Serhan, J. Haeggstrom, M. Ingelman-Sandberg, and B. Samuelsson. 1986. Activation of protein kinase C by lipoxin A and other eicosanoids. Intracellular action of oxygenation products of arachidonic acid. *Biochem. Biophys. Res. Comm.* 134:1215–1222.

Hardie, G. 1988. Pseudosubstrates turn off protein kinases. *Nature* 335:592–593.

Hechter, O. 1955. Concerning possible mechanism of hormone action. *Vitamin Horm.* 13: 293–346.

Hirasawa, K., and Y. Nishizuka. 1985. Phosphatidylinositol turnover in receptor mechanism and signal transduction. *Annu. Rev. Pharmacol. Toxicol.* 25:147–170.

Hokin, L. E. 1985. Receptors and phosphoinositide-generated second messengers. *Annu. Rev. Biochem.* 54:205–235.

Hokin, M. R., and L. E. Hokin. 1953. Enzyme secretion and the incorporation of ^{32}P into phospholipids of pancreas slices. *J. Biol. Chem.* 203:967–977.

House, C., and B. E. Kemp. 1987. Protein kinase C contains a pseudosubstrate prototype in its regulatory domain. *Science* 238:1726–1728.

House, C., and B. E. Kemp. 1990. Protein kinase C pseudosubstrate prototype: Structure-function relationships. *Cell. Signal.* 2:187–190.

House, C., P. J. Robinson, and B. E. Kemp. 1989. A synthetic peptide of the putative substrate-binding motif activates protein kinase C. *FEBS Lett.* 249:243–247.

Huang, F. L., and K. P. Huang. 1991. Interactions of protein kinase C isozymes with phosphatidylinositol-4,5-bisphosphate. *J. Biol. Chem.* 266:8727–8733.

Huang, F. L., Y. Yoshida, J. R. Cunha-Melo, M. A. Beaven, and K.-P. Huang. 1989. Differential down-regulation of protein kinase C isozymes. *J. Biol. Chem.* 264: 4238–4243.

Huang, K.-P. 1989. The mechanism of protein kinase C activation. *Trends Neurosci.* 12: 425–432.

Huang, K.-P., and F. L. Huang. 1990. Differential sensitivity of protein kinase C isozymes to phospholipid-induced inactivation. *J. Biol. Chem.* 265:738–744.

Huang, K.-P., F. L. Huang, H. Nakabayashi, and Y. Yoshida. 1988. Biochemical characterization of rat brain protein kinase C isozymes. *J. Biol. Chem.* 263:14839–14845.

Hubbard, S. R., W. R. Bishop, P. Kirschmeier, S. J. George, S. P. Cramer, and W. A. Hendrickson. 1991. Identification and characterization of zinc binding sites in protein kinase C. *Science* 254:1776–1779.

Hunter, T. 1987. A thousand and one protein kinases. *Cell* 50:823–829.

Inoue, M., A. Kishimoto, Y. Takai, and Y. Nishizuka. 1977. Studies on a cyclic nucleotide-independent protein kinase and its proenzyme in mammalian tissue II. *J. Biol. Chem.* 252:7610–7616.

Ishikawa, F., F. Takaka, M. Nagao, and T. Sugimura. 1986. Cysteine-rich regions conserved in amino-terminal halves of raf gene family products and protein kinase C. *Jpn. J. Can. Res.* 77:1183–1187.

Jaken, S. 1990. Protein kinase C and tumor promoters. *Curr. Opin. Cell Biol.* 2:192–197.

Jaken, S., K. Leach, and T. Klauck. 1989. Association of type 3 protein kinase C with focal contacts in rat embryo fibroblasts. *J. Cell Biol.* 109:697–704.

James, G., and E. Olson. 1992. Deletion of the regulatory domain of protein kinase Cα exposes regions in the hinge and catalytic domains that mediate nuclear targeting. *J. Cell Biol.* 116:863–874.

Janin, J., and C. Chothia. 1978. Role of hydophobicity in true binding of coenzymes. *Biochemistry* 17:2943–2948.

Kaibuchi, K., Y. Takai, and Y. Nishizuka. 1981. Cooperative roles of various membrane phospholipids on the activation of calcium-activated, phospholipid-dependent protein kinase. *J. Biol. Chem.* 256:7146–7149.

Kaibuchi, K., Y. Fukumoto, N. Oku, Y. Takai, K.-I. Arai, and M. Muramatsu. 1989. Molecular genetic analysis of the regulatory and catalytic domain of protein kinase C. *J. Biol. Chem.* 264:13489–13496.

Katz, B. 1966. *Nerve, Muscle and Synapse.* McGraw-Hill, New York.

Katzau, S., D. Martin-Zanca, and M. Barbacid. 1990. VAV. *EMBO J.* 8:2283–2290.

Kazanietz, M. G., K. W. Krausz, and P. M. Blumberg. 1992. Differential irreversible insertion of protein kinase C into phospholipid vesicles by phorbol esters and related activators. *J. Biol. Chem.* 267:20878–20886.

Keranen, L. M., J. W. Orr, and A. C. Newton. 1992. Role of Ca^{2+} in the interaction of protein kinase C with phosphatidylserine. *Biophys. J.* 61:A89.

Khan, W. A., G. C. Blobe, and Y. A. Hannun. 1992. Activation of protein kinase C by oleic acid. *J. Biol. Chem.* 267:3605–3612.

Khan, W. A., G. C. Blobe, A. Halpern, W. Taylor, W. C. Wetsel, D. J. Burns, C. R. Loomis, and Y. A. Hannun. 1993. Selective regulation of protein kinase C isozymes by oleic acid in human platelets. *J. Biol. Chem.* 268:5063–5068.

Kikkawa, U., R. Minakuchi, Y. Takai, and Y. Nishizuka. 1983a. Calcium-activated, phospholipid-dependent protein kinase (protein kinase C) from rat brain. *Meth. Enzymol.* 99:288–298.

Kikkawa, U., Y. Takai, Y. Tanaka, R. Miyake, and Y. Nishizuka. 1983b. Protein kinase C as a possible receptor protein of tumor-promoting phorbol esters. *J. Biol. Chem.* 258:11442–11445.

Kiley, S. C., and S. Jaken. 1990. Activation of a protein kinase C leads to association with detergent-insoluble components of GH_4C_1 cells. *Molec. Endocrinol.* 4:59–67.

Kiley, S., D. Schaap, P. Parker, L.-L. Hsieh, and S. Jaken. 1990 Protein kinase C heterogeneity in GH_4C_1 rat pituitary cells. *J. Biol. Chem.* 265:15704–15712.

Kiley, S., P. Parker, D. Fabbro, and S. Jaken. 1991. Differential regulation of protein kinase C isozymes by thyrotropin-releasing hormone in GH_4C_1 cells. *J. Biol. Chem.* 266: 23761–23768.

Kishimoto, A., Y. Takai, T. Mori, U. Kikkawa, and Y. Nishizuka. 1980. Activation of calcium and phospholipid-dependent protein kinase by diacylglycerol, its possible relation to phosphatidylinositol turnover. *J. Biol. Chem.* 255:2273–2276.

Kishimoto, A., N. Kajikawa, M. Shiota, and Y. Nishizuka. 1983. Proteolytic activation of calcium-activated, phospholipid-dependent protein kinase C by calcium-dependent neutral protease. *J. Biol. Chem.* 258:1156–1164.

Kishimoto, A., K. Mikawa, K. Hashimoto, I. Yasuda, S. Tanaka, M. Tominaga, T. Kuroda, and Y. Nishizuka. 1989. Limited proteolysis of protein kinase C subspecies by calcium-dependent neutral protease (calpain). *J. Biol. Chem.* 264:4088–4092.

Klee, C. B. 1988. Calcium-dependent phospholipid- (and membrane-) binding proteins. *Biochemistry* 27:6645–6654.

Klee, C. B., T. H. Crouch, and P. G. Richmond. 1980. Calmodulin. *Annu. Rev. Biochem.* 49:489–515.

Knopf, J. L., M.-H. Lee, L. A. Sultzman, R. W. Kris, C. R. Loomis, R. M. Hewick, and R. M. Bell. 1986. Cloning and expression of multiple protein kinase C cDNAs. *Cell* 46:491–502.

Koide, H., H. Ogita, U. Kikkawa, and Y. Nishizuka. 1992. Isolation and characterization of the ε subspecies of protein kinase C from rat brain. *Proc. Natl. Acad. Sci. U.S.A.* 89:1149–1153.

König, B., P. A. Di Nitto, and P. M. Blumberg. 1985. Stoichiometric binding of diacylglycerol to the phorbol ester receptor. *J. Cell. Biochem.* 29:37–44.

Kraft, A. S., and W. B. Anderson. 1983. Phorbol esters increase the amount of Ca^{2+}, phospholipid-dependent protein kinase associated with plasma membrane. *Nature* 301:621–623.

Krebs, E. G. 1972. Protein kinases. *Curr. Topics Cell. Reg.* 5:99–133.

Kuo, J. F., R. G. G. Andersson, B. C. Wise, L. Mackerlova, I. Salomonsson, N. L. Brackett, N. Katoh, M. Shoji, and R. W. Wrenn. 1980. Calcium-dependent protein kinase: Widespread occurrence in various tissues and phyla of the animal kingdom and comparison of effects of phospholipid, calmodulin, trifluoperazine. *Proc. Natl Acad. Sci. U.S.A.* 77:7039–7043.

Lapentina, E. G., B. Reep, B. R. Ganong, and R. M. Bell. 1985. Exogenous *sn*-1,2-diacylglcerols containing saturated fatty acids function as bioregulators of protein kinase C in human platelets. *J. Biol. Chem.* 260:1358–1361.

Leach, K. L., E. A. Powers, V. A. Ruff, S. Jaken, and S. Kaufman. 1989. Type 3 protein kinase C localization to the nuclear envelope of phorbol ester treated NIH 3T3 cells. *J. Cell Biol.* 109:685–695.

Leach, K. L., V. A. Ruff, T. M. Wright, M. S. Pessin, and D. M. Raben. 1991. Dissociation of protein kinase C activation and *sn*-1,2-diacylglycerol formation. *J. Biol. Chem.* 266:3215–3221.

Lee, M.-Y., and R. M. Bell. 1986. The lipid-binding, regulatory domain of protein kinase C. *J. Biol. Chem.* 261:14867–14870.

Lee, M.-H., and R. M. Bell. 1989. Phospholipid functional groups involved in protein kinase C activation, phorbol ester binding and binding to mixed micelles. *J. Biol. Chem.* 264:14797–14805.

Lee, M.-H., and R. M. Bell. 1991. Mechanism of protein kinase C activation by phosphatidylinositol-4,5-bisphosphate. *Biochemistry* 30:1041–1049.

Leibersperger, H., M. Gschwendt, and F. Marks. 1990. Purification and characterization of a calcium-unresponsive, phorbol-ester/phospholipid-activated protein kinase from porcine spleen. *J. Biol. Chem.* 265:16108–16115.

Lester, D. S. 1990. *In vitro* linoleic acid activation of protein kinase C. *Biochim. Biophys. Acta* 1054:297–303.

Lester, D. S., C. Collin, R. Etcheberrigaray, and D. L. Alkon. 1991. Arachidonic acid and diacylglycerol act synergistically to activate protein kinase C *in vitro* and *in vivo*. *Biochem. Biophys. Res. Comm.* 179:1522–1528.

Lester, D. S., L. Doll, V. Brumfeld, and I. R. Miller. 1990. Lipid-dependence of surface conformations of protein kinase C. *Biochim. Biophys. Acta* 1039:33–41.

Leung, T., B.-E. How, E. Manser, and L. Lim. 1993. Germ cell β-chimaerin, a new GTPase-activating protein for p21rac, is specifically expressed during the acrosomal assembly stage in rat testis. *J. Biol. Chem.* 268:3813–3816.

Liyanage, M., D. Frith, E. Livneh, and S. Stabel. 1992. Protein kinase C group B members PKC-δ,-ε,-ζ and PKC-L(η). *Biochem. J.* 283:781–787.

Luo, J.-H., S. Kahn, K. O'Driscoll, and B. Weinstein. 1993. The regulatory domain of protein kinase C $β_1$ contains phosphatidylserine- and phorbol ester-dependent calcium binding activity. *J. Biol. Chem.* 268:3715–3719.

Makowsky, M., and O. M. Rosen. 1989. Complete activation of protein kinase C by antipeptide antibody directed against the pseudosubstrate prototype. *J. Biol. Chem.* 264:16155–16159.

Maruyama, I. N., and S. Brenner. 1991. A phorbol ester/diacylglycerol-binding protein encoded by the unc-13 gene of *Caenorhabditis elegans*. *Proc. Natl. Acad. Sci. U.S.A.* 88:5729–5733.

Maurer, M. C., J. J. Sando, and C. M. Grisham. 1992. High-affinity Ca^{2+}- and substrate binding sites on·protein kinase Cα as determined by nuclear magnetic resonance spectroscopy. *Biochemistry* 31:7714–7721.

May Jr., W. S., N. Sahyoun, M. Wolf, and P. Cuatrecasas. 1985. Role of intracellular calcium mobilization in the regulation of protein kinase C-mediated membrane processes. *Nature* 317:549–551.

McLaughlin, S., N. Mulrine, T. Gresalfi, G. Vaio, and A. McLaughlin. 1981. Adsorption of divalent cations to bilayer membranes containing phosphatidylserine. *J. Gen. Physiol.* 77:445–473.

McPhail, L. C., C. C. Clayton, and R. Snyderman. 1984. A potential second messenger role for unsaturated fatty acids: Activation of Ca^{2+}-dependent protein kinase C. *Science* 224:622–625.

Melloni, E., S. Petremoli, M. Michetti, O. Sacco, B. Sparatore, and B. L. Horecker. 1986. The involvement of calpain in the activation of protein kinase C in neutrophils stimulated by phorbol myristic acid. *J. Biol. Chem.* 261:4101–4105.

Menniti, F. S., K. G. Oliver, J. W. Putney Jr., and S. B. Shears. 1993. Inositol phosphates and cell signalling: New views of InsP$_5$ and InsP$_6$. *Trends Biochem. Sci.* 18:53–56.

Michell, R. H. 1975. Inositol phospholipids and cell surface receptor function. *Biochem. Biophys. Acta* 415:81–147.

Michell, R. H. 1992. Inositol lipids in cellular signalling mechanisms. *Trends Biochem. Sci.* 17:274–276.

Minakuchi, R., Y. Takai, and Y. Nishizuka. 1981. Widespread occurrence of calcium-activated, phospholipid-dependent protein kinase in mammalian tissues. *J. Biochem. (Tokyo)* 89:1651–1654.

Mochley-Rosen, D., C. J. Heinrich, L. Cheever, H. Khaner, and P. C. Simpson. 1990. A protein kinase C isozyme is translocated to cytoskeletal elements in activation. *Cell Reg.* 1:689–706.

Mochley-Rosen, D., H. Khaner, and J. Lopez. 1991a. Identification of intracellular receptor proteins for activated protein kinase C. *Proc. Natl. Acad. Sci. U.S.A.* 88:3997–4000.

Mochley-Rosen, D., H. Khaner, J. Lopez, and B. L. Smith. 1991b. Intracellular receptors for activated protein kinase C. *J. Biol. Chem.* 266:14866–14868.

Mori, T., Y. Takai, B. Yu, J. Takahashi, Y. Nishizuka, and T. Fujikura. 1982. Specificity of fatty acyl moieties of diacylglycerol for the activation of calcium-activated, phospholipid-dependent protein kinase. *J. Biochem. (Tokyo)* 91:427–431.

Mosier, M., and R. M. Epand. 1993. Mechanism of activation of protein kinase C: Roles of diolein and phosphatidylserine. *Biochemistry* 32:66–75.

Mosier, M., and S. McLaughlin. 1991. Peptides that mimic the pseudosubstrate region of protein kinase C bind to acidic lipids in membranes. *Biophys. J.* 60:149–159.

Mosier, M., and S. McLaughlin. 1992a. Binding of basic peptides to acidic lipids in membranes: Effects of inserting alanine(s) between the basis residues. *Biochemistry* 31:1767–1773.

Mosier, M., and S. McLaughlin. 1992b. Electrostatics and dimensionality can produce apparent cooperativity when protein kinase C and its substrates bind to acidic lipids in membranes. In: *Protein Kinase C: Current Concepts and Future Perspectives,* edited by D. Lester and R. M. Epand, pp. 157–180. Horwood, Chichester, England.

Murakami, K., and A. Routtenberg. 1985. Direct activation of purified protein kinase C by unsaturated fatty acids (oleate and arachidonate) in the absence of phospholipids and Ca^{2+}. *FEBS Lett.* 192:189–193.

Murakami, K., S. Y. Chan, and A. Routtenberg. 1986. Protein kinase activation by cis-fatty acid in the absence of calcium and phospholipids. *J. Biol. Chem.* 261:15424–15429.

Nakadate, T., A. Y. Jeng, and P. M. Blumberg. 1987. Effects of phospholipids on substrate phosphorylation by a catalytic fraction of protein kinase C. *J. Biol. Chem.* 262:11507–11513.

Nakanishi, H., and J. H. Exton. 1992. Purification and characterization of the ζ isoform of protein kinase C from bovine kidney. *J. Biol. Chem.* 267:16347–16354.

Nakanishi, H., K. A. Brewer, and J. H. Exton. 1993. Activation of the ζ isozyme of protein kinase C by phosphatidylinositol 3,4,5-trisphosphate. *J. Biol. Chem.* 268:13–16.

Naor, Z., M. S. Shearman, A. Kishimoto, and Y. Nishizuka. 1988. Calcium-independent activation of hypothalamic type I protein kinase C by unsaturated fatty acids. *Molec. Endocrinol.* 2:1043–1048.

Nelsestuen, G. L., and M. D. Bazzi. 1991. Activation and regulation of protein kinase C enzymes. *J. Bioenerget. Biomembr.* 23:43–61.

Newton, A. C., and D. E. Koshland Jr. 1987. Protein kinase C autophosphorylates by an intrapeptide reaction. *J. Biol. Chem.* 262:10185–10188.

Newton, A. C., and D. E. Koshland Jr. 1989. High cooperativity, specificity, and multiplicity in the protein kinase C-lipid interaction. *J. Biol. Chem.* 264:14909–14915.

Nishikawa, M., H. Hidaka, and S. Shirakawa. 1988. Possible involvement of direct stimulation of protein kinase C by unsaturated fatty acids in platelet activation. *Biochem. Pharmacol.* 37:3079–3089.

Nishizuka, Y. 1984. The role of protein kinase C in cell surface signal transduction and tumor promotion. *Nature* 308:693–695.

Nishizuka, Y. 1986. Studies and perspectives of protein kinase C. *Science* 233:305–312.

Nishizuka, Y. 1988. The molecular heterogeneity of protein kinase C and its implications for cellular regulations. *Nature* 334:661–665.

Nishizuka, Y. 1989a. Studies and perspectives of the protein kinase C family for cellular regulation. *Cancer* 63:1892–1903.

Nishizuka, Y. 1989b. The family of protein kinase C for signal transduction. *JAMA* 262:1826–1833.

Nishizuka, Y. 1992. Intracellular signalling by hydrolysis of phospholipids and activation of protein kinase C. *Science* 258:607–614.

Ogita, K., S. Miyamoto, K. Yamagachi, H. Koide, N. Fujisawa, U. Kikkawa, S. Sahara, Y. Fukami, and Y. Nishizuki. 1992. Isolation and characterization of δ-subspecies of protein kinase C from rat brain. *Proc. Natl. Acad. Sci. U.S.A.* 89:1592–1596.

Ohno, S., Y. Akita, Y. Konno, Imajoh, S., and K. Suzuki. 1988. A novel phorbol ester receptor/protein kinase, nPKC, distantly related to the protein kinase C family. *Cell* 53:731–741.

Ohno, S., Y. Konno, Y. Akita, A. Yano, and K. Suzuki. 1990. A point mutation at the putative ATP-binding site of protein kinase Cα abolishes the kinase activity and renders it downregulation-insensitive. *J. Biol. Chem.* 265:6296–6230.

Olivier, A. R., and P. J. Parker. 1991. Expression and characterization of PKC-δ. *Eur. J. Biochem.* 200:805–810.

Ong, R. L. 1984. ^{31}P and ^{19}F NMR studies of glycophorin-reconstituted membranes: Preferential interaction of glycophorin with phosphatidylserine. *J. Membr. Biol.* 78:1–7.

Ono, Y., T. Fujii, K. Ogita, U. Kikkawa, K. Igarashi, and Y. Nishizuka. 1988. The structure, expression, and properties of additional members of the protein kinase C family. *J. Biol. Chem.* 263:6927–6932.

Ono, Y., T. Fujii, K. Ogita, U. Kikkawa, K. Igarashi, and Y. Nishizuka. 1989a. Protein kinase C ζ from rat brain: Its structure, expression and properties. *Proc. Natl. Acad. Sci. U.S.A.* 86:3099–3103.

Ono, Y., T. Fujii, K. Igarashi, T. Kuno, C. Tanaka, U. Kikkawa, and Y. Nishizuka. 1989b. Phorbol ester binding to protein kinase C requires a cysteine-rich zinc-finger-like sequence. *Proc. Natl. Acad. Sci. U.S.A.* 86:4868–4871.

Orr, J. W., and A. C. Newton. 1992a. Interaction of protein kinase C with phosphatidyl-serine. 1. Cooperativity in lipid binding. *Biochemistry* 31:4661–4667.

Orr, J. W., and A. C. Newton. 1992b. Interaction of protein kinase C with phosphatidyl-serine. 2. Specificity and regulation. *Biochemistry* 31:4667–4673.

Orr, J. W., L. M. Keranen, and A. C. Newton. 1992. Reversible exposure of the pseudo-substrate domain of protein kinase C by phosphatidylserine and diacylglycerol. *J. Biol. Chem.* 267:15263–15266.

Osada, S., K. Mizuro, T. C. Saido, Y. Akita, K. Suzuki, T. Kuroki, and S. Ohno. 1990. A phorbol ester receptor/protein kinase C, nPKCη, a new member of the protein kinase C family predominantly expressed in lung and skin. *J. Biol. Chem.* 265: 22434–22440.

Otte, A. P., I. M. Kramer, and A. J. Durston. 1991. Protein kinase C and the regulation of local competence of Xenopus ectoderm. *Science* 251:570–573.

Parker, P. J., G. Kour, R. M. Marais, F. Mitchell, C. J. Pears, D. Schaap, S. Stabel, and C. Liebster. 1989. Protein kinase C—a family affair. *Molec. Cell. Endocrinol.* 65: 1–11.

Pears, C. J., G. Kour, C. House, B. E. Kemp, and P. J. Parker. 1990. Mutagenesis of the pseudosubstrate site of protein kinase C leads to activation. *Eur. J. Biochem.* 194: 89–94.

Pears, C. J., D. Schaap, and P. J. Parker. 1991. The regulatory domain of protein kinase C-ε restricts the catalytic-domain specificity. *Biochem. J.* 276:257–260.

Pears, C. J., S. Stabel, S. Cazaubon, and P. J. Parker. 1992. Studies on the phosphorylation of protein kinase C-α. *Biochem. J.* 283:515–518.

Pelech, S. L., and D. E. Vance. 1989. Signal transduction via phosphatidylcholine cycles. *Trends Biochem. Sci.* 14:28–30.

Perin, M. S., V. A. Fried, G. A. Mignery, R. Jahn, and T. C. Südhof. 1990. Phospholipid binding by a synaptic vesicle protein homologous to the regulatory region of pro-tein kinase C. *Nature* 345:250–263.

Quest, A. F. G., J. Bloomenthal, E. S. G. Bardes, and R. M. Bell. 1992. The regulatory domain of protein kinase C coordinates four atoms of zinc. *J. Biol. Chem.* 267: 10193–10197.

Quest, A. F. G., E. S. G. Bardes, J. Bloomenthal, R. A. Borchardt, and R. M. Bell. 1993. Protein kinase C is a zinc metalloprotein: Quantitation of zinc by atomic absorption spectrometry. *Meth. Neurosci.* 8:138–153.

Quest, A. F. G., E. S. G. Bardes, and R. M. Bell. 1994a. A phorbol ester binding domain of protein kinase C γ: High affinity binding to a glutathione-S-transferase/ Cys2 fusion protein. *J. Biol. Chem.* 269, in press.

Quest, A. F. G., E. S. G. Bardes, and R. M. Bell. 1994b. A phorbol ester binding domain of protein kinase C γ: Deletion analysis of the Cys2 domain defines a minimal 45 amino acid peptide. *J. Biol. Chem.* 269, in press.

Quest, A. F. G., R. A. Borchardt, E. S. G. Bardes, and R. M. Bell. 1994c. Functional characterization of lipid interaction sites in the regulatory domain of protein kinase Cγ. Submitted.

Raetz, C. R. H. 1986. Molecular genetics of membrane phospholipid synthesis. *Am. Rev. Genet.* 20:253–295.

Rando, R. R. 1988. Regulation of protein kinase C activity by lipids. *FASEB J.* 2:2348–2355.

Saido, T. C., K. Mizuno, Y. Konno, S. Osada, S. Ohno, and K. Suzuki. 1992. Purification and characterization of protein kinase C ε from rabbit brain. *Biochemistry* 31:482–490.

Sato, S. B., and S. Ohnishi. 1993. Interaction of a peripheral protein of the erythrocyte membrane, band 4.1, with phosphatidylserine-containing liposomes and erythrocyte inside-out vesicles. *Eur. J. Biochem.* 130:19–25.

Schaap, D., and P. Parker. 1990. Expression, purification and characterization of protein kinase C- ε. *J. Biol. Chem.* 265:7301–7307.

Schaap, D., J. de Widt, J. van der Wal, J. Vanderkerckhove, J. van Damme, D. Gussow, H. L. Ploegh, W. I. van Blitterswijk, and R. L. van der Bend. 1990. Purification, cDNA-cloning and expression of human diacylglycerol kinase. *FEBS Lett.* 275: 151–158.

Schatzman, R. C., R. L. Raynor, R. B. Fritz, and J. F. Kuo. 1983. Purification to homogeneity, characterization and monoclonal antibodies of phospholipid-sensitive Ca^{2+}-dependent protein kinase from spleen. *Biochem. J.* 209:435–443.

Sekiguchi, K., M. Tsukuda, K. Ogita, U. Kikkawa, and Y. Nishizuka. 1987. Three different isoforms of rat brain protein kinase C: Differential response to unsaturated fatty acids. *Biochem. Biophys. Res. Comm.* 145:797–802.

Sekiguchi, K., M. Tsukuda, K. Ase, U. Kikkawa, and Y. Nishizuka. 1988. Mode of activation and kinetic properties of three distinct forms of protein kinase C from rat brain. *J. Biochem. (Tokyo)* 103:759–765.

Shah, J., and G. G. Shipley. 1992. Circular dichroic studies of protein kinase C and its interaction with calcium and lipid vesicles. *Biochim. Biophys. Acta* 1119:19–26.

Sharkey, N. A., and P. M. Blumberg. 1985. Kinetic evidence that 1,2-diolein inhibits phorbol ester binding via a competitive mechanism. *Biochem. Biophys. Res. Comm.* 133:1051–1056.

Sharkey, N. A., K. L. Leach, and P. M. Blumberg. 1984. Competitive inhibition by diacylglycerol of specific phorbol ester binding. *Proc. Natl. Acad. Sci. U.S.A.* 81:607–610.

Shearman, M. S., T. Shinomura, T. Oda, and Y. Nishizuka. 1991. Protein kinase C subspecies in adult rat hippocampal synaptosomes: Activation by diacylglycerol and arachidonic acid. *FEBS Lett.* 279:261–264.

Shimomura, T., Y. Asaoka, M. Oka, K. Yoshida, and Y. Nishizuka. 1991. Synergistic action of diacylglycerol and unsaturated fatty acid for protein kinase C activation; its possible implications. *Proc. Natl. Acad. Sci. U.S.A.* 88:5149–5153.

Soderling, T. R. 1990. Protein kinases. Regulation by autoinhibitory domains. *J. Biol. Chem.* 265:1823–1826.

Souvignet, C., J. M. Pelosin, S. Daniel, E. M. Chambaz, S. Ransac, and R. Verger. 1991. Activation of protein kinase C in lipid monolayers. *J. Biol. Chem.* 266:40–44.

Spudich, A., T. Meyer, and L. Stryer. 1992. Association of the β isoform of protein kinase C with vimentin filaments. *Cell Motil. Cytoskel.* 22:250–256.

Stabel, S., and P. J. Parker. 1991. Protein kinase C. *Pharm. Therap.* 51:71–95.

Stahl, M. L., C. R. Ferenz, K. L. Kelleher, R. W. Kriz, and J. L. Knopf. 1988. Sequence similarity of phospholipase C with the non-catalytic region of *src*. *Nature* 332: 269–272.

Sutherland, E. W. 1972. Studies on the mechanism of hormone action. *Science* 177:401–408.

Takai, Y., A. Kishimoto, M. Inoue, and Y. Nishizuka. 1977. Studies on a cyclic nucleotide-independent protein kinase and its proenzyme in mammalian tissue. I. Purification and characterization of an active enzyme from bovine cerebellum. *J. Biol. Chem.* 252:7603–7609.

Takai, Y., A. Kishimoto, Y. Iwasa, Y. Kawahara, T. Mori, and Y. Nishizuka. 1979a. Calcium-dependent activation of a multifunctional protein kinase by membrane phospholipids. *J. Biol. Chem.* 254:3692–3695.

Takai, Y., A. Kishimoto, U. Kikkawa, T. Mori, and Y. Nishizuka. 1979b. Unsaturated diacylglycerol as a possible messenger for the activation of a calcium-activated, phospholipid-independent protein kinase system. *Biochem. Biophys. Res. Commun.* 91:1218–1224.

Takai, Y., A. Kishimoto, Y. Iwasa, Y. Kawahara, T. Mori, and Y. Nishizuka. 1979c. A role of membranes in the activation of a new multifunctional protein kinase system. *J. Biochem. (Tokyo)* 86:575–578.

Tallant, E. A., and W. Y. Cheung. 1984. Characterization of bovine brain calmodulin-dependent protein phosphatase. *Arch. Biochem. Biophys.* 232:269–279.

Tapley, P. M., and A. W. Murray. 1985. Evidence that treatment of platelets with phorbol ester causes proteolytic activation of Ca^{2+}-activated, phospholipid-dependent protein kinase. *Eur. J. Biochem.* 151:419–423.

Taylor, S. S., J. A. Buechler, and W. Yonemoto 1990. cAMP-dependent protein kinase: Framework for a diverse family of regulatory enzymes. *Annu. Rev. Biochem.* 59:971–1005.

Taylor, S. S., D. R. Knighton, J. Zheng, J. M. Sowadski, C. S. Gibbs, and M. T. Zoller. 1993. A template for the protein kinase family. *Trends Biochem. Sci.* 18:84–89.

Testori, A., C. S. T. Hii, A. Fournier, L. A. Burgoyne, and A. W. Murray. 1988. DNA binding proteins in protein kinase C preparations. *Biochem. Biophys. Res. Comm.* 156:222–227.

Tsai, M.-H., M. Roudebush, S. Dobrovolski, C.-L. Lu, J. B. Gibbs, and D. W. Stacey. 1991. *Ras* GTPase-activity protein physically associates with mitogenically active phospholipids. *Molec. Cell. Biol.* 11:2785–2793.

Verkleij, A. J., R. F. A. Zwaal, B. Rodofsen, P. Comfurius, D. Kastelijn, and L. L. M. van Deenen. 1973. The asymmetric distribution of phospholipids in the human red cell membrane: A combined study using phospholipases and freeze-etch electron microscopy. *Biochim. Biophys. Acta* 323:178–193.

Vogel, U. S., R. A. F. Dixon, M. D. Schaber, R. D. Diehl, M. S. Marshall, E. M. Scolnick, I. S. Sigal, and J. B. Gibbs. 1988. Clonong of bovine GAP and its interaction with oncogenic *ras* p21. *Nature* 335:90–93.

Walker, J. M., and J. J. Sando. 1988. Activation of protein kinase C by short chain phosphatidylcholines. *J. Biol. Chem.* 263:4537–4540.

Walker, J. M., E. C. Honan, and J. J. Sando. 1990. Differential activation of protein kinase C isozymes by short chain phosphatidylserines and phosphatidylcholines. *J. Biol. Chem.* 265:8016–8021.

Ways, D. K., P. P. Cook, C. Webster, and P. J. Parker. 1992. Effect of phorbol ester on protein kinase C-ζ. *J. Biol. Chem.* 267:4799–4805.

Weber, I. T., J. B. Shabb, and J. D. Corbin. 1989. Predicted structures of the cGMP binding domains of the cGMP-dependent protein kinase: A key alanine/threonine difference in evolutionary divergence of cAMP and cGMP binding sites. *Biochemistry* 28:6122–6127.

White, D. A. 1973. In: *Form and Function of Phospholipids,* edited by G. B. Ansell, J. N. Hawthorne, and R. M. C. Dawson, pp. 441–482. Elsevier Scientific Publishing Co., Amsterdam.

Wise, B. C., R. L. Raynor, and J. F. Kuo. 1982. Phospholipid-sensitive calcium-dependent protein kinase from heart: Purification and general properties. *J. Biol. Chem.* 257: 8481–8488.

Wolf, M., H. Le Vinett, W. S. May Jr., P. Cuatrecasas, and N. Sahyoun. 1985. A model for intracellular translocation of protein kinase C involving synergism between Ca^{2+} and phorbol esters. *Nature* 317:546–549.

Woodgett, J. R., T. Hunter, and K. L. Gould. 1987. Protein kinase C and its role in cell growth. In: *Cell Membranes: Methods and Reviews,* edited by E. Elson, W. Frazier, and L. Glaser, pp. 215–340. Plenum Press, New York.

Yoshida, K., Y. Asaoka, and Y. Nishizuka. 1992. Platelet activation by simultaneous actions of diacylglycerol and unsaturated fatty acids. *Proc. Natl. Acad. Sci. U.S.A.* 89:6443–6446.

Zidovetski, R., and D. S. Lester. 1992. The mechanism of activation of protein kinase C: A biophysical perspective. *Biochim. Biophys. Acta* 1134:261–272.

Zwaal, R. F. A., and E. M. Bevers. 1983. Platelet phospholipid asymmetry and its significance in hemostasis. *Subcell. Biochem.* 9:299–334.

4

Protein Kinase C Inhibitors

CATHERINE A. O'BRIAN
J. F. KUO

Since the discovery of protein kinase C (PKC) in 1977, intensive efforts have been dedicated to the search for potent and selective inhibitors of this enzyme (Bottega and Epand, 1992; Huang, 1989). More recently these efforts have been redirected to a search for isozyme-specific inhibitors, as a result of the discovery that PKC is a family of at least ten closely related isozymes (Huang, 1989; Nishizuka, 1992; Chapter 2). It is anticipated that studies now directed at an understanding of the structural and functional differences among PKC isozymes will provide a basis for the development of isozyme-specific PKC inhibitors.

PKC isozymes play critical roles in multifarious processes that include cell growth and differentiation, muscle contraction, neurotransmission, platelet activation, and tumor promotion (Huang, 1989; Kikkawa et al., 1989; Kuo et al., 1989; O'Brian and Ward, 1989a; Weinstein, 1990). Early observations that phorbol-ester tumor promoters can down-regulate PKC activity (Kikkawa et al., 1989) led to the use of phorbol ester-induced PKC down-regulation to measure specific cellular responses in cells thought to lack PKC, in order to assess whether PKC was required for the cellular response examined. However, it is now clear that complete elimination of PKC generally cannot be achieved by prolonged exposure to phorbol-ester tumor promoters (Strulovici et al., 1991; Borner et al., 1992; Ways et al., 1992). In particular, the isozyme nPKC-ζ appears to be completely recalcitrant to phorbol-ester-induced down-regulation (Ways et al., 1992). Isozyme-specific PKC inhibitors therefore appear to be necessary to determine whether PKC isozymes are essential to cellular responses of interest. In fact, even

Supported by National Institutes of Health grants CA 36777 and HL 15696 (J.F.K.) and CA 52460 (C.A.O.) and by Robert A. Welch Foundation grant #G1141 (C.A.O.).

We are grateful to Nancy E. Ward and Karen R. Gravitt for their contributions to our studies on PKC. We thank Patherine Greenwood for her expert preparation of the manuscript.

the development of a selective PKC inhibitor that cannot distinguish among the isozymes could greatly advance the current understanding of the role of PKC in specific cellular responses. For example, multi-drug-resistant (MDR) tumor cells often express high levels of P glycoprotein, which is an integral membrane protein that extrudes structurally diverse anticancer drugs from the cells. A number of PKC inhibitors antagonize the MDR phenotype, but, due to the nonselective nature of the PKC inhibitors, it is still not clear whether PKC activity is essential to the maintenance of the MDR phenotype (O'Brian et al., in press). In principle, the contribution of PKC to MDR in a given tumor cell line could be directly and unambiguously measured through the use of selective PKC inhibitors.

Although currently available PKC inhibitors lack specificity, they have provided highly important information about the active site chemistry of PKC and the regulation of PKC activity. The understanding of the regulation of PKC activity by the interactions of allosteric cofactors (Ca^{2+}, phosphatidylserine, 1,2-*sn*-diacylglycerol, phorbol-ester tumor promoters) with the regulatory domain of the enzyme (Kikkawa et al., 1989) has been advanced through studies of PKC inhibitors that antagonize activation of the enzyme. These inhibitors include agents with demonstrated anticancer activity, such as tamoxifen (O'Brian et al., 1985; Su et al., 1985) and cytotoxic alkyl lysophospholipids (Helfman et al., 1983). Thus, studies directed at an understanding of the importance of PKC inhibition to the cellular responses elicited by these agents are clearly warranted and may lead to improved anticancer therapies. Inhibitory peptide substrate analogs have shed light on the structure of the active site of PKC, the substrate specificity of the enzyme, the kinetics of the catalytic mechanism of PKC (House and Kemp, 1987; for other references see below), and the effects of occupation of the protein-substrate binding site of PKC on the bond-breaking step of its protein kinase reaction (ATPase reaction) (O'Brian and Ward, 1990a). Inhibitors that compete with the substrate ATP, such as the isoquinolinesulfonamides, have been used to characterize the kinetics of PKC catalysis, to probe the topology of the nucleotide-substrate binding site of PKC, and to assess the relatedness between the active sites of PKC and other protein kinases (Hidaka and Hagiwara, 1987).

In recent years, a tremendous amount of work has been accomplished in the area of PKC inhibitor research. For the sake of clarity, this review will focus on several areas of this research that appear to be especially promising for the development of selective PKC inhibitors, for the improvement of state-of-the art therapeutics that act in part by PKC inhibition, and for the elucidation of the structure and function of PKC isozymes.

PHOSPHOLIPID METABOLITES AND ANALOGUES

Because phosphatidylserine (PS), whether in the form of vesicles, micelles, or biomembranes, functions as a cofactor for PKC, it is not unexpected that a number of phospholipid metabolites have been shown to regulate PKC activity. Oishi and colleagues (1988a) reported that lysophosphatidylcholine (lyso-PC), a degradation product of membrane phosphatidylcholine that is formed by phospholipase

A_2 (PLA$_2$) catalysis, acts synergistically with diacylglycerol to activate PKC at low concentrations (about 5 μM) and inhibits PKC at high concentrations ($>$ 30 μM). Other hydrolytic products of PLA$_2$ catalysis include the fatty acids arachidonic and oleic acid, both of which are PKC activators (McPhail et al., 1984). In addition to phospholipid hydrolysis, degradation of membrane sphingolipids appears to play a role in transmembrane signaling, because sphingosine inhibits PKC (Hannun et al., 1986). The effects of lyso-PC and sphingosine on biological membranes, however, are not limited to alterations in PKC activity; these agents also inhibit synaptosomal Na,K-ATPase or Na pump in HL60 cells with potencies similar to those observed in their inhibition of PKC (Oishi et al., 1990). This suggests that Na pump is also a potential target for the putative second messengers. (The mechanism of PKC inhibition by sphingosine is discussed in the next section.) Palmitoylcarnitine, a phospholipid metabolite shown to accumulate in the hypoxic myocardium, inhibits PKC (Katoh et al., 1981; Wise and Kuo, 1983), suggesting involvement of PKC in ischemic heart diseases.

A number of alkyl lyso-PC analogues have been synthesized and tested for their biological activities. The ether lipid 1-O-octadecyl-2-O-methyl-*rac*-glycero-3-phosphocholine (ET-18-OCH$_3$) and the thioether lipid 1-S-hexadecyl-2-methoxymethyl-*rac*-glycero-3-phosphocholine (BM 41.440) are the two prototypes of this class of compounds. They are experimental anticancer agents currently undergoing clinical trials, largely on the basis of their abilities to inhibit cancer cell growth (Andreesen et al., 1978; Modolell et al., 1979; Berdel et al., 1980, 1983; Tidwell et al., 1981; Berger et al., 1984; Fromm et al., 1987). Hexadecylphosphocholine (HePC, D18506, miltefosine) is a new antineoplastic agent that is novel insofar as it is a simple ''phospholipid'' lacking the glycerol backbone (Eibl et al., 1992). Its structure is compared with lyso-PC and its analogues as follows:

CH$_2$-R$_1$
|
CH$_2$-R$_2$ O$^-$
| |
| CH$_3$-(CH$_2$)$_{15}$-O-P-O-(CH$_2$)$_2$-N$^+$(CH$_3$)$_3$
| O$^-$ ||
| | O
CH$_2$-O-P-O-(CH$_2$)$_2$-N$^+$(CH$_3$)$_3$
      || HePC
      O

Lyso-PC: R$_1$, O(CH$_2$)$_{17}$-CH$_3$;   R$_2$, OH
ET-18-OCH$_3$:   R$_1$, O(CH$_2$)$_{17}$-CH$_3$;   R$_2$, O-CH$_3$
BM 41.440:   R$_1$, S-(CH$_2$)$_{15}$-CH$_3$;   R$_2$, CH$_2$-O-CH$_3$

HePC has just been approved by the German Health Organization for topical treatment of skin metastasis in breast cancer patients. Although the molecular mechanisms of action of ET-18-OCH$_3$, BM 41.440, and HePC are unclear, inhibition of PKC (Helfman et al., 1983; Zheng et al., 1990; Shoji et al., 1991) and Na, K-ATPase or Na pump (Shoji et al., 1988; Oishi et al., 1988b; Zheng et al., 1990), and modifications of phospholipid metabolism (Modollel et al., 1979), might partially account for their antineoplastic effects. The notion that they might act via PKC inhibition is further supported by the findings that they counteract the effects of the highly specific PKC activator 12-O-tetradecanoylphorbol-13-acetate (TPA) in promoting PKC translocation and down-regulation in HL60 and other cells and in inducing terminal differentiation of HL60 cells (Shoji et al., 1988, 1991).

In addition to alkyl lyso-PC analogues, analogues of PC itself can modulate PKC activity. Charp and colleagues (1988b) reported that distearoyl PC, a poor inhibitor of PKC, becomes a PKC inhibitor as potent as ET-18-OCH$_3$ when the 8-position of the fatty acid residue at *sn*-2 is methylated, whereas it becomes an activator as potent as diacylglycerol (i.e., sn-1,2-dioctanoylglycerol) when the same 8-position of the fatty acid in both *sn*-1 and *sn*-2 are butylated. These findings indicate that the PC analogues having different branched-chain fatty alkyl chains exhibit opposing regulatory effects on PKC. This presumably reflects, at least in part, some unique physicochemical properties of the analogues that render them capable of critically modifying the activation process of PKC involving PKC/PS/Ca^{2+}/diacylglycerol (TPA) interactions.

The inhibitory or stimulatory effects of all of the PKC-modulating lipid substances discussed above were characterized in studies that used either a mixture of PKC isozymes (unfractionated) or the individual isozymes (α, β, and γ) of the conventional group of PKC purified from pig or rat brain extracts. Little or no selectivity in the inhibition of the PKC isozymes has been noted with these agents, with the peptide/protein class of PKC inhibitors, or with cationic amphiphilic PKC inhibitors in general (to be discussed later). It is unclear at present whether the lipid, peptide/protein, or other classes of PKC inhibitors can specifically inhibit certain subtypes or species of the novel and atypical classes of PKC isozymes (δ, ϵ, η/L, θ, ζ, and λ), although substrate specificities of PKC isozymes suggest that certain inhibitory peptide/protein substrate analogs may be isozyme-selective.

CATIONIC AMPHIPHILIC PKC INHIBITORS AND INHIBITION OF THE PHOSPHOLIPID-DEPENDENT ACTIVATION OF THE ENZYME

A wide variety of cationic amphiphiles have been shown to inhibit PKC activity (Bottega and Epand, 1992). A number of these inhibitors are naturally occurring amphiphilic peptides and proteins, for example, the microbial antibacterial antibiotic polymyxin B (Mazzei et al., 1982), human neutrophil antibiotic defensins (Charp et al., 1988a), cobra venom cardiotoxin and marine worm cytotoxin (Kuo et al., 1983), bee venom toxin melittin (Katoh et al., 1982), and wasp venom toxin mastoparan (Raynor et al., 1991). In addition, a number of synthetic compounds, such as the phenothiazine chlorpromazine (Mori et al., 1980; Aftab et

Table 4.1   Cationic Amphiphilic PKC Inhibitors that Exhibit Reduced Inhibitory
Potencies in the Presence of Elevated Concentrations of the Phospholipid Cofactor

Cationic amphiphile	References
4-Hydroxytamoxifen	O'Brian et al., 1986
Acridine orange	Hannun and Bell, 1988
Cardiotoxin	Kuo et al., 1983
Chlorpromazine	Mori et al., 1980
Clomiphene	O'Brian et al., 1986
Dequalinium	Rotenberg et al., 1990
Dibucaine	Mori et al., 1980
Imipramine	Mori et al., 1980
Melittin	Katoh et al., 1982
N-desmethyltamoxifen	O'Brian et al., 1986
N-myristoyl-KRTLR	O'Brian et al., 1990a
N-myristoyl-RKRTLRRL	O'Brian et al., 1991
Nafoxidine	Su et al., 1985
Phentolamine	Mori et al., 1980
Polymyxin B	Mazzei et al., 1982
Rhodamine 6G	O'Brian and Weinstein, 1987
Sphingosine	Senisterra and Epand, 1992
Tamoxifen	Su et al., 1985; O'Brian et al., 1985
Tetracaine	Mori et al., 1980
Verapamil	Mori et al., 1980
W7	Tanaka et al., 1982

al., 1991), the anticarcinoma agent dequalinium (Rotenberg et al., 1990), and the
antiestrogen tamoxifen (O'Brian et al., 1985; Su et al., 1985) are cationic amphiphilic PKC inhibitors. Initial reports concerning the inhibition of Ca^{2+}- and phospholipid-dependent PKC activity by cationic amphiphiles demonstrated an
inverse correlation between the inhibitory potency of the cationic amphiphile and
the phospholipid cofactor concentration present in the assay mixture (Mori et al.,
1980; Katoh et al., 1982; Su et al., 1985; O'Brian et al., 1985, 1986), and in
recent years this finding has been extended to numerous cationic amphiphilic PKC
inhibitors (Table 4.1). It has been inferred from the observed correlation that the
inhibitory mechanism shared by the cationic amphiphiles involves interference
with the phospholipid cofactor in the activation of PKC (Mori et al., 1980; Katoh
et al., 1982; Su et al., 1985; O'Brian et al., 1985, 1986). In fact, it has been
demonstrated that the lethal effect of cobra cardiotoxin subcutaneously admin-
istered to mice is greatly diminished when the toxin is premixed with PS (Kuo
et al., 1983). Binding studies provide independent support to the proposed role
for the phospholipid cofactor in the mechanism of PKC inhibition by cationic
amphiphiles. Direct binding interactions between the phospholipid cofactor phosphatidylserine and the cationic amphiphilic PKC inhibitor N-(6-aminohexyl)-5-
chloro-1-naphthalenesulfonamide (W7) have been demonstrated by gel filtration
chromatography (Tanaka et al., 1982). More recently, polymyxin B has been
shown to antagonize binding interactions between PKC and a fluorescent phosphatidylserine analog (Rodriguez-Paris et al., 1989). However, it remains unclear

whether inhibition of Ca^{2+}- and phospholipid-dependent PKC activity by cationic amphiphiles generally involves binding of the cationic amphiphile to the phospholipid binding region in the regulatory domain of PKC, to the phospholipid cofactor itself, or to both the lipid cofactor and the lipid cofactor binding site.

Studies of PKC inhibition are often complicated by the presence of multiple allosteric cofactors (Ca^{2+}, phosphatidylserine, phorbol-ester tumor promoter, *sn*-1,2-diacylglycerol) in the assay mixtures. Unsaturated fatty acids and phosphatidylserine activate PKC by related but distinct mechanisms (Shinomura et al., 1991). Unlike phosphatidylserine, arachidonic acid and other unsaturated fatty acids can activate PKC in the absence of other allosteric cofactors. Mechanistic studies of the inhibition of PKC by the cationic amphiphile rhodamine 6G showed that rhodamine 6G could potently inhibit the lipid-dependent activity of PKC when the only allosteric cofactor present was arachidonic acid, but the inhibitor was without effect on the basal enzyme activity observed in the absence of lipid cofactor (O'Brian and Weinstein, 1987). Thus, the allosteric cofactors Ca^{2+} and diacylglycerol are not required for inhibition of PKC by rhodamine 6G. Other studies have shown that the protein substrate is not essential to the mechanism of PKC inhibition by cationic amphiphiles. PKC catalyzes a Ca^{2+}- and phospholipid-dependent ATPase reaction that represents the bond-breaking step of its protein kinase reaction (O'Brian and Ward, 1990a; Ward and O'Brian, 1992a). In the absence of protein and peptide substrates, the Ca^{2+}- and phospholipid-dependent ATPase reaction of PKC is inhibited by melittin, polymyxin B, and N-desmethyltamoxifen, and the inhibitory potencies of the cationic amphiphiles against the ATPase reaction are similar to their potencies against the Ca^{2+}- and phospholipid-dependent histone kinase reaction of PKC (O'Brian and Ward, 1990a). Thus, histone does not dramatically influence the inhibitory potencies of the cationic amphiphiles against PKC. In contrast, the nature of the lipid cofactor profoundly affects the inhibitory potencies of the cationic amphiphiles. This has been demonstrated in studies that showed that the inhibitory potencies of rhodamine 6G, chlorpromazine, and polymyxin B are altered 3- to 16-fold when the lipid cofactor arachidonic acid is used in place of phosphatidylserine in assays of the Ca^{2+}- and lipid-dependent protein kinase activity of PKC (O'Brian and Weinstein, 1987; Robinson, 1992). Substitution of the lipid cofactor PS with arachidonic acid increases the inhibitory potency of rhodamine 6G (O'Brian and Weinstein, 1987) and decreases the inhibitory potencies of chlorpromazine and polymyxin B (Robinson, 1992).

Studies with mastoparan and cardiotoxin have shed light on their mechanisms of inhibition of phospholipid-dependent PKC activation through detailed characterizations of the kinetics of inhibition and the structure–activity relationships among these inhibitors and their isoforms. Kinetic analysis of the mechanisms of action of mastoparan (Raynor et al., 1991) and cardiotoxin (Chiou et al., 1993) indicates that the toxins inhibit PKC competitively with respect to PS, in a mixed manner with respect to Ca^{2+} or diacylglycerol (diolein), noncompetitively with respect to protein substrate (e.g., histone H1), and uncompetitively with respect to ATP. Hydrophobic interactions of the agents with PKC/PS/Ca^{2+} complex are critical for their inhibitory activity. The study with the cardiotoxin isoforms and neurotoxin from cobra venom clearly supports this hypothesis (Chiou et al., 1993); a close relationship between the hydrophobicity and PKC inhibitory activity of the isoforms exists and their relative potency, in decreasing order, is cardiotoxin-1 ~ cardiotoxin-3 > cardiotoxin-4 >>> neurotoxin (inactive). The hydrophobicity-activity relationship, however, is not evident for mastoparan analogues (Raynor et al., 1992). One possible reason might be that mastoparan, a small tetradecapeptide (molecular weight of about 1500), compared with cardiotoxin having 60 amino acid residues (molecular weight of about 6700), lacks certain high-order structures essential for well-defined functional interactions with the hydrophobic regulatory domain of PKC.

Because of its amphiphilicity, actions of cardiotoxin could be attributed to its potential generalized, nonspecific interactions with PS or membranes. This possibility is largely eliminated because (1) cardiotoxin potently inhibits PKC activity assayed using synaptosomal membrane as phospholipid cofactor, with an IC_{50} (concentration causing 50% inhibition) of 1 μM, without affecting Na,K-ATPase activity in the same membrane preparation, (2) the isoforms of cardiotoxin inhibit proliferation of various cancer cell lines with an order of potency similar to their inhibition of PKC activity and TPA binding to the enzyme, and (3) cardiotoxin inhibits proliferation and TPA-induced differentiation of HL60 cells at concentrations that do not cause cell lysis. It has been noted that the IC_{50} of cardiotoxin for the above cellular effects (0.1 μM) is lower than that for PKC inhibition *in vitro* (1 μM). It is unclear at present whether cardiotoxin is highly membrane bound or even permeable to cells, or exerts it cellular effects in part through PKC inhibition. It is entirely possible that cardiotoxin exerts its cytotoxicity (hence its alternative name cytotoxin) and other biological effects via additional mechanisms that are independent of PKC. Biochemical mechanisms of cardiotoxin action other than PKC inhibition, however, are yet to be identified. It should be noted that cardiotoxin (like defensins, polymyxin B, and ET-18-OCH$_3$) inhibits PKC much more potently (IC_{50} of 1μM) than Ca^{2+}/calmodulin-dependent protein kinase II or myosin light-chain kinase (IC_{50} of about 100 μM), whereas mastoparan (like melittin) inhibits all of the enzymes potently and nonselectively (IC_{50} of 1–8 μM) (for a summary, see Raynor et al., 1992; see also Table 4.2).

While it is clear from the studies reviewed above that the ability of cationic amphiphiles to inhibit PKC is strongly influenced by the lipid cofactor of the enzyme, biophysical studies indicate that the inhibitory action of the cationic amphiphiles is not simply a function of their abilities to alter membrane bilayer stability (Bottega and Epand, 1992; Epand and Lester, 1990). The ability of zwitterionic and uncharged lipids, such as *sn*-1,2-diacylglycerol, to activate PKC correlates very well with their bilayer-destabilizing activity. Furthermore, uncharged and zwitterionic lipid-interacting inhibitors of PKC tend to be membrane stabilizers. (It is important to note that the opposing effects of neutral bilayer stabilizers and destabilizers on PKC activity presumably reflect their differential effects on the bulk properties of the lipid environment and do not necessarily reflect altered bilayer stability itself.) In contrast, cationic amphiphiles that inhibit PKC include both bilayer stabilizers and bilayer destabilizers (Epand and Lester, 1990; Lester and Baumann, 1991; Bottega and Epand, 1992). Nonetheless, bilayer-stabilizing

Table 4.2 Targets of Cationic Amphiphilic PKC Inhibitors

Target	Cationic amphiphile	Action	Reference
Calmodulin	Polymyxin B	Inhibition	Mazzei et al., 1982
	W7	Inhibition	Tanaka et al., 1982
	Calmidazolium	Inhibition	Mazzei et al., 1984
	Melittin	Inhibition	Katoh et al., 1982
	cis- and *trans*-tamoxifen	Inhibition	O'Brian et al., 1990b; Lam, 1984
Calpain	Melittin	Inhibition	Brumley and Wallace, 1989
	Calmidazolium	Inhibition	Brumley and Wallace, 1989
	W7	Inhibition	Brumley and Wallace, 1989
	Trifluoperazine	Inhibition	Brumley and Wallace, 1989
Phosphatidylinositol kinase	Chlorpromazine	Biphasic	Husebye and Flatmark, 1988
	Calmidazolium	Biphasic	Husebye and Flatmark, 1988
	Trifluoperazine	Biphasic	Husebye and Flatmark, 1988
Na, K-ATPase	Melittin	Inhibition	Raynor et al., 1991
	Mastoparan	Inhibition	Raynor et al., 1991
	Sphingosine	Inhibition	Oishi et al., 1990

activity may contribute to the inhibitory potency of a cationic amphiphile against PKC. On the basis of structure-activity studies, bilayer-stabilizing cationic amphiphiles appear to be more potent than bilayer-destabilizing cationic amphiphiles in inhibiting PKC (Bottega and Epand, 1992). Moreover, a highly anionic lipid environment converts the bilayer-destabilizing cationic amphiphilic PKC inhibitor sphingosine into a weak PKC activator, but does not abolish the inhibitory activity of a structurally related bilayer-stabilizing cationic amphiphilic PKC inhibitor (Senisterra and Epand, 1992).

CATIONIC AMPHIPHILIC PKC INHIBITORS: INHIBITION OF THE LIPID COFACTOR-INDEPENDENT ACTIVITY OF THE ENZYME

It is now clear that inhibition of PKC by cationic amphiphiles often entails not only inhibition of lipid cofactor-mediated activation of the enzyme, but also inhibition of the basal activity of the enzyme (Table 4-3). On the basis of inhibitory properties of cationic amphiphiles studies to date, it appears that the inhibitory potency of a cationic amphiphile against the basal activity of PKC is generally weaker than its potency against the lipid-dependent activity of the enzyme (Table 4-3). Due to this discrepancy in inhibitory potencies, early studies of cationic amphiphilic PKC inhibitors did not reveal inhibitory activity against basal PKC activity.

Inhibition of the lipid cofactor-independent activity of PKC by cationic amphiphiles appears to be a consequence of direct interactions between the inhibitors and the catalytic domain of PKC. Direct and reversible binding interactions between PKC and inhibitory cationic amphiphiles have been demonstrated in the absence of PKC cofactors and substrates by the quantitative binding of PKC to the inhibitor-coupled resins W7-agarose, 2-chloro-10-(3-aminopropyl) phenothiazine (CAPP)-agarose, *N*-didesmethyltamoxifen-agarose, and melittin-agarose,

Table 4.3 Cationic Amphiphiles Shown to Inhibit Both the Phospholipid-Dependent Activation (PDA) and the Lipid Cofactor-Independent Activity (LCIA) of PKC

Cationic amphiphile	IC$_{50}$[a] (PDA)	IC$_{50}$ (LCIA)	Reference
Melittin	3 μM	25 μM	O'Brian and Ward, 1989c
N-didesmethyltamoxifen	40 μM	110 μM	O'Brian et al., 1988
Sphingosine	30 μM	100 μM	Nakadate et al., 1988
W7	140 μM	340 μM	Inagaki et al., 1986
Acridine yellow G	40 μM	170 μM	Hannun and Bell, 1988

[a]IC$_{50}$ = concentration causing 50 percent inhibition.

and by the elution of PKC from the inhibitor-coupled resins with Triton-X-100 (O'Brian et al., 1987, 1988; O'Brian and Ward, 1989c). A fully active, Ca^{2+}- and phospholipid-independent catalytic fragment of PKC that contains only the catalytic domain of the enzyme can be produced by limited proteolysis of PKC (Kikkawa et al., 1989). The histone kinase activity of the catalytic fragment is inhibited by various cationic amphiphiles, including W7 (Inagaki et al., 1986), N-desmethyltamoxifen (O'Brian et al., 1988), and melittin (O'Brian and Ward, 1989c) in the absence of allosteric cofactors of PKC, such as phospholipid. The catalytic fragment of PKC also binds reversibly to W7-agarose (O'Brian and Ward, 1989b), N-didesmethyltamoxifen-agarose (O'Brian et al., 1988), and melittin-agarose (O'Brian and Ward, 1989c) in the absence of substrates and cofactors. Taken together, these results provide strong evidence that the cationic amphiphiles inhibit the catalytic fragment of PKC by binding it directly.

The chromatographic properties of the catalytic fragment of PKC on the inhibitor-coupled resins have been shown to be indistinguishable from those of PKC itself, providing evidence that the direct binding interactions between PKC and the cationic amphiphiles primarily involve the catalytic domain of the enzyme (O'Brian et al., 1988; O'Brian and Ward, 1989b, c). Elution of PKC from melittin-agarose (Figure 4.1), N-didesmethyltamoxifen-agarose, and W7-agarose has also been achieved with MgATP (O'Brian et al., 1988; O'Brian and Ward, 1989b,c). The specificity of the MgATP-sensitive binding interactions between PKC and melittin is indicated by the dramatic purification of rat brain PKC effected by chromatography on melittin-agarose according to silver-stained polyacrylamide gel analysis (O'Brian and Ward, 1989c). Because the regulatory domain of PKC lacks nucleotide-binding sites (Kikkawa et al., 1989), the elution of PKC from the inhibitor-coupled resins with MgATP provides another line of evidence that the enzyme binds the inhibitor-coupled resins through its catalytic domain. In addition, the elution of PKC with MgATP suggests that the binding of PKC to the inhibitor-coupled resins may be through the nucleotide-substrate binding site of the enzyme. However, a consensus sequence for a potential second nucleotide-binding site is conserved in the catalytic domains of several PKC isozymes, and that putative nucleotide-binding site could be responsible for the binding of PKC to the inhibitor-coupled resins (O'Brian and Ward, 1989c). The consensus sequence for the putative second nucleotide-binding site (Gly-X-Gly-X-X-Gly ... Lys) is found in all of the cPKC isozymes but is substantially altered in

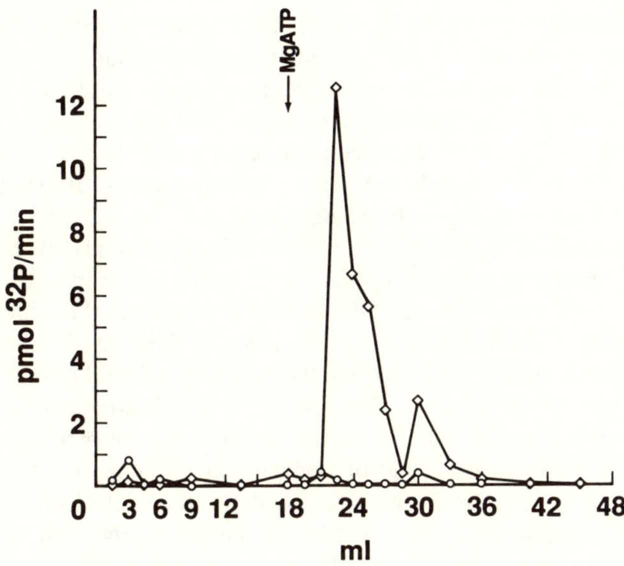

Figure 4.1 Partially purified rat brain PKC (approximately 10% pure) was loaded onto a 1.5-ml melittin-agarose column equilibrated in 20 mM Tris-HCl, 4 mM EDTA, 4 mM EGTA, 0.2 M KCl, 15 mM 2-mercaptoethanol pH 8.3. The column was washed with 20 ml equilibration buffer, and PKC activity was eluted with 10 mM MgCl$_2$ plus 1 mM ATP in equilibration buffer. Open diamonds = histone kinase activity observed in PKC assay mixtures containing Ca^{2+} and PS (pmol ^{32}P/min); open circles = histone kinase activity observed in the absence of the activating cofactors (pmol ^{32}P/min). Arrow indicates application of MgATP-containing buffer to the column. Pmol ^{32}P/min is the pmols ^{32}P transferred from [γ ^{32}P]ATP to histone III-S per minute under the above-described conditions.

nPKCε and nPKC ζ (Ono et al., 1988; O'Brian and Ward, 1992). Recently we found that the rat brain isozymes cPKC α, cPKC β, cPKC γ, nPKC ζ, and nPKC ε bind to melittin-agarose and can be eluted from the resin with MgATP. The chromatographic properties of nPKC ε and nPKC ζ were indistinguishable from those of cPKC α, cPKC β, and cPKC γ despite profound structural differences among the isozymes at the putative second nucleotide-binding site (Ward et al., manuscript in preparation). These results rule out an important role for the putative second nucleotide binding site in the binding of PKC to melittin-agarose, and they provide strong evidence that melittin binds to the nucleotide-substrate-binding site of PKC isozymes. However, the alternative possibility, that melittin and the substrate MgATP bind distinct sites of PKC and induce mutually exclusive conformations of the enzyme (O'Brian and Ward, 1989c), cannot be ruled out.

The mechanisms of inhibition of the lipid cofactor-independent activity of PKC by cationic amphiphiles have been defined through kinetic analyses of the inhibition of the catalytic fragment of PKC. These studies have shown that cationic amphiphiles inhibit the lipid cofactor-independent activity of PKC by diverse mechanisms, which presumably reflect the structural diversity within this broad class of PKC inhibitors. In the first report to show inhibition of the lipid cofactor-independent activity of PKC by a cationic amphiphile, Inagaki and co-workers (1986) demonstrated that W7 and four other naphthalenesulfonamides inhibited histone phosphorylation catalyzed by the catalytic fragment of PKC through competition with ATP with K_is that ranged from 47 to 340 μM. Other cationic amphiphiles shown to inhibit the lipid-independent activity of PKC by competition with ATP include the triphenylethylene N-desmethyltamoxifen (O'Brian et al., 1988), a hydroxylated triphenylacrylonitrile derivative (Bignon et al., 1990), the acridine derivatives acridine orange and acridine yellow G (Hannun and Bell, 1988), and the anthracycline-metal complex adriamycin-iron (III) (Hannun et al., 1989). The competition of W7 and N-desmethyltamoxifen with ATP in the inhibition of the catalytic fragment of PKC (Inagaki et al., 1986; O'Brian et al., 1988) complements the observation that PKC binds to the immobilized inhibitors in a MgATP-sensitive manner (O'Brian et al., 1988; O'Brian and Ward, 1989b). Thus, W7 and N-desmethyltamoxifen bind to the nucleotide-substrate-binding site of PKC according to two independent lines of evidence. In addition, the observation that *cis*- and *trans*-tamoxifen both inhibit the catalytic fragment of PKC, albeit with somewhat different potencies (50% inhibition is achieved by 130-μM *cis*-tamoxifen and 230-μM *trans*-tamoxifen) (O'Brian et al., 1990b), indicates that triphenylethylenes bind to the active site of PKC in a nonstereospecific manner. In contrast, melittin and N-myristoylated cationic peptides inhibit histone phosphorylation catalyzed by the catalytic fragment of PKC noncompetitively with respect to ATP (O'Brian and Ward, 1989c; O'Brian et al., 1990a, 1991). (Inhibition of PKC by the N-myristoylated peptides is discussed in detail in the next section.)

INHIBITORY PEPTIDE SUBSTRATE ANALOGS OF PKC

The division of labor among protein kinases is based on differences in compartmentalization, mode of regulation, and substrate specificity. Several synthetic analogues of substrate peptides have been found to be selective PKC inhibitors. Examples are the histone H1 peptide analogue Arg-Arg-Lys-Ala-Ala-Gly-Pro-Pro-Val (O'Brian et al., 1984), in which the phosphoacceptor serine is replaced by alanine, and the bovine myelin basic protein (MBP) peptide analogues [Ala[107]]MBP(104–118) and [Ala[113]]MBP(104–118) (Turner et al., 1985; Su et al., 1986), in which the arginine recognition sites at positions 107 and 113, respectively, are replaced by alanine. Inhibitory sequences that occur in the regulatory domains of protein kinases and resemble phosphorylation sites of their protein substrates are called pseudosubstrate prototopes. Several protein kinases have been shown to be selectively inhibited by synthetic peptides that correspond to their pseudosubstrate prototopes (House and Kemp, 1987; Soderling, 1990). The selectivity of the pseudosubstrate synthetic-peptide inhibitors indicates the importance of substrate specificity in the division of labor among protein kinases and the feasibility of designing selective protein kinase inhibitors based on substrate specificities. In particular, studies with pseudosubstrate synthetic peptides targeted against PKC suggest that it may be possible to design highly selective PKC inhibitors based on substrate specificity. A synthetic peptide corresponding to the pseudosubstrate sequence of PKC-α (residues 19–31) inhibits PKC-catalyzed peptide phosphorylation with an IC_{50} of 92 $\pm$ 5 nM by competition with the peptide substrate (House and Kemp, 1987). The PKC α (19–31) pseudosubstrate peptide appears to be a selective PKC inhibitor, because it is without significant inhibitory effects against cAMP-dependent protein kinase and myosin light-chain kinase (House and Kemp, 1987). Unfortunately, the PKC α (19–31) pseudosubstrate peptide is too bulky and hydrophilic to enter cells, and studies of its pharmacologic action have been limited to permeabilized cell systems (Alexander et al., 1989; Eichholtz et al., 1990).

Because small PKC peptide substrates are typically positively charged (O'Brian and Ward, 1989a), certain *N*-acylated derivatives of the peptide substrates may be cationic amphiphiles that readily enter cells and thereby overcome the limitation posed by the impermeability of mammalian cells to the pseudosubstrate synthetic-peptide PKC inhibitor PKC α (19–31). In principle, *N*-acylation of small PKC peptide substrates and nonphosphorylatable substrate analogs should allow the identification of cell-permeable PKC inhibitors that selectively recognize the active site of PKC and compete with endogenous substrates of the enzyme. *N*-acylated peptide substrate analogs may achieve selective inhibition of PKC by exploiting not only the substrate specificity of PKC (O'Brian et al., 1990a, 1991), but also the compartmentalization of the enzyme, because activated PKC is generally membrane associated and amphiphilic peptides tend to concentrate at cell membranes (Epand, 1992). In fact, two *N*-myristoylated PKC peptide substrates have been shown to potently inhibit the histone kinase activity of purified PKC and to antagonize PKC-mediated events in intact cells (O'Brian et al., 1990a, 1991; Ioannides et al., 1990). These *N*-acylated, cationic peptide-substrate analogs of PKC form a new class of cationic amphiphilic PKC inhibitors that may be useful model compounds for the development of selective PKC inhibitors (O'Brian et al., 1990a, 1991; O'Brian and Ward, 1992). Presumably, optimal recognition of the peptide sequence of an *N*-acylated peptide by the active site of PKC should allow the design of moderately amphiphilic *N*-acylated peptides that potently inhibit PKC but have negligible inhibitory effects against other common targets of cationic amphiphiles (Table 4.2). In principle, the *N*-acylated cationic peptides could even provide a starting point for the development of isozyme-specific PKC inhibitors, since fundamental differences have been observed among the substrate specificities of several PKC isozymes (Huang et al., 1988; Marais and Parker, 1989; Burns et al., 1990; Koide et al., 1992; Liyanage et al., 1992).

PKC catalyzes the phosphorylation of the epidermal growth factor receptor at Thr 654, which is located in the sequence -Arg-Lys-Arg-*Thr*-Leu-Arg-Arg-Leu (-RKRTLRRL-) (Hunter et al., 1984). PKC also phosphorylates the synthetic octapeptide RKRTLRRL ($K_{m\,app}$ = 20 μM) (O'Brian et al., 1991) and the related pentapeptide KRTLR ($K_{m\,app}$ = 300 μM) (DeBont et al., 1989). *N*-myristoylation endows each of these synthetic peptide substrates of PKC with potent inhibitory activity against PKC-catalyzed histone phosphorylation, without improving their abilities to serve as PKC substrates (O'Brian et al., 1990a, 1991). Like a number of other cationic amphiphilic PKC inhibitors, *N*-myristoyl-KRTLR and *N*-myristoyl-RKRTLRRL each inhibits PKC activity by dual mechanisms, and both lipid-dependent and lipid-independent components of PKC-catalyzed histone kinase activity are inhibited by the N-acylated peptides (O'Brian et al., 1990a, 1991). N-myristoyl-KRTLR and N-myristoyl-RKRTLRRL inhibit the lipid-dependent activation of PKC with IC_{50}s of 75 μM and 5 μM, respectively, when the lipid cofactor is PS. Under these conditions, the nonmyristoylated parent peptides KRTLR and RKRTLRRL are without effect on the lipid-dependent activation of PKC at concentrations up to and including 600 μM and 50 μM, respectively (O'Brian et al., 1990a, 1991). The histone kinase activity of the catalytic fragment of PKC is inhibited by N-myristoyl-KRTLR and N-myristoyl-RKRTLRRL with IC_{50}s of 200 μM and 80 μM, respectively; the respective non-myristoylated parent peptides are without effect on this activity at concentrations up to and including 600 μM and 200 μM (O'Brian et al., 1990a, 1991). Similarly, myristic acid lacks inhibitory activity against the catalytic fragment (O'Brian et al., 1990a, 1991). Because the N-myristoylated peptides are not superior to the nonmyristoylated parent peptides as PKC substrates, it is clear that the superior inhibitory potencies of the N-myristoylated peptides against both the lipid-independent and the lipid-dependent histone kinase activities of PKC do not merely reflect the relative potencies of the N-myristoylated and nonmyristoylated peptides as PKC substrates (O'Brian et al., 1990a, 1991).

The inhibition of Ca^{2+}- and PS-dependent PKC activity by N-myristoyl-KRTLR and N-myristoyl-RKRTLRRL can be overcome by increasing the concentration of the cofactor PS, providing evidence that the N-acylated peptides interfere with the lipid cofactor in the activation of PKC (O'Brian et al., 1990a, 1991). Thus, the N-myristoylated peptides behave as classic cationic amphiphiles in the inhibition of the lipid-dependent activity of PKC (Table 4.1).

At concentrations where the N-myristoylated peptides inhibit the lipid-independent activity of the catalytic fragment of PKC, the peptides also serve as PKC substrates. It is therefore clear that the N-myristoylated peptides bind the peptide substrate-binding site of PKC when present at inhibitory concentrations (O'Brian et al., 1990a, 1991). The kinetic mechanism of PKC-catalyzed histone phosphorylation is steady-state preferred ordered with ATP binding first (Leventhal and Bertics, 1991). Therefore, a PKC inhibitor that mimics the behavior of the substrate histone would be expected to yield uncompetitive kinetics with respect to ATP, because the inhibitor would bind the complex PKC-ATP, but not PKC_{free} (Leventhal and Bertics, 1991). In fact, these kinetics are observed with the dead-end inhibitors poly-L-lysine (Leventhal and Bertics, 1991) and PKC α (19–36) (House and Kemp, 1987). In contrast, the apparent kinetics of inhibition of the histone kinase activity of the catalytic fragment of PKC by N-myristoyl-KRTLR and by N-myristoyl-RKRTLRRL are noncompetitive with respect to ATP (O'Brian et al., 1990a, 1991). The noncompetitive kinetics with respect to ATP observed with the N-myristoylated peptides suggest that these peptides may differ from the substrate histone in that they may bind to the peptide substrate-binding site of both PKC-ATP and PKC_{free}, although other mixed inhibitory mechanisms cannot be excluded. While the studies with N-myristoyl-KRTLR and N-myristoyl-RKRTLRRL described above lend support to a mechanism of inhibition of the catalytic fragment that involves binding of the inhibitory peptides at the peptide substrate-binding site of PKC (O'Brian et al., 1990a, 1991), studies on the kinetics of inhibition of the histone kinase activity of the catalytic fragment by nonphosphorylatable analogs of these *N*-acylated peptides, (i.e., dead-end inhibitors), are still needed to test this model of PKC inhibition by *N*-acylated peptides. Nonphosphorylatable analogues of *N*-myristoyl-RKRTLRRL in which the residue Thr is replaced by Ala or Tyr are just as potent as the Thr-containing peptide in inhibiting the histone kinase activity of the catalytic fragment of PKC (N. E. Ward and C. A. O'Brian, unpublished observations), and studies on the kinetics of inhibition of the catalytic fragment by these peptides are in progress. Regardless of the kinetic mechanism of inhibition, it is clear that the inhibitory potencies of the *N*-myristoylated peptides compare favorably with other inhibitory synthetic peptide substrate analogs. For example, the pseudosubstrate peptide PKC α (19–31) is without effect on the histone kinase reaction of PKC (House and Kemp, 1987), whereas the *N*-myristoylated peptides readily inhibit the phosphorylation of this highly potent PKC substrate (O'Brian et al., 1990a, 1991).

Importantly, *N*-myristoyl-KRTLR and *N*-myristoyl-RKRTLRRL inhibit PKC-mediated events in intact mammalian cells. PKC activation is implicated in the induction of the interleukin-2 (IL-2) receptor and IL-2 production in the T lymphoblastoid cell line Jurkat, and both of these events have been shown to be suppressed in the cultured cell system by *N*-myristoyl-KRTLR at peptide concentrations that lack nonspecific cytotoxic effects (Ioannides et al., 1990). PKC activation is also implicated in the multi-drug-resistance phenotype of the murine fibrosarcoma cell line UV-2237M-ADR[R]. *N*-myristoyl-RKRTLRRL has been shown to partially reverse the resistance of this cell line to Adriamycin, without causing nonspecific cytotoxicity itself (O'Brian et al., 1991). Furthermore, in an important extension of these studies, Eichholtz and colleagues (1993) have demonstrated that an *N*-myristoylated nonapeptide containing a sequence that occurs in the pseudosubstrate prototope of cPKC α inhibits TPA-induced phosphorylation of the PKC substrate MARCKS and bradykinin-induced phospholipase D activation (which is a strictly PKC-dependent event) in HF cells. Because the investigation of the *N*-myristoylated pseudosubstrate peptide was restricted to the effects of the peptide in an intact cell system (Eichholtz et al., 1993), further studies are needed to evaluate the authors' claim that the acylated peptide is a specific PKC inhibitor, particularly in view of its cationic amphiphilic structure. In addition, studies are still needed to determine whether the *N*-acylated cationic peptides actually enter cells or merely associate with the plasma membrane. However, their abilities to inhibit PKC-requiring pathways in intact cells suggest the potential usefulness of this class of cationic-amphiphilic PKC inhibitors as antagonists of PKC activation *in vivo*.

ISOQUINOLINESULFONAMIDE PROTEIN KINASE INHIBITORS

Isoquinolinesulfonamide protein kinase inhibitors were developed from naphthalenesulfonamide calmodulin inhibitors such as W7 (Tanaka et al., 1982; Inagaki et al., 1986) by replacing the naphthalene structure with an isoquinoline ring (Hidaka et al., 1984). Unlike naphthalenesulfonamide calmodulin inhibitors, the isoquinolinesulfonamides do not bind calmodulin and do not interact with phospholipids. Thus, the isoquinolinesulfonamides lack inhibitory activity against calmodulin and do not interfere with PS in the activation of PKC (Hidaka et al., 1984; Hidaka and Hagiwara, 1987). However, the isoquinolinesulfonamides have in common with certain naphthalenesulfonamides that bear short alkyl chains (Inagaki et al., 1986) the ability to potently inhibit protein kinases by competition with MgATP (Hidaka et al., 1984). The K_i values observed with isoquinolinesulfonamides in the inhibition of protein kinases are often as low as 0.5 to 10 μM (Hidaka et al, 1984; Hidaka and Kobayashi, 1992).

Kinetic analysis indicates that the isoquinolinesulfonamides inhibit cAMP-dependent protein kinase, cGMP-dependent protein kinase, myosin light-chain kinase, and PKC competitively with respect to MgATP and noncompetitively with respect to protein substrate (Hidaka et al., 1984). This kinetic behavior would be expected for dead-end inhibitors that resemble the nucleotide substrate in steady-state ordered protein kinase reaction mechanisms with nucleotide substrate binding first, and evidence has been presented that PKC (Leventhal and Bertics, 1991) and cAMP-dependent protein kinase (Whitehouse et al., 1983) each can operate by this type of mechanism. Consistent with the ordered mechanisms proposed for PKC and cAMP-dependent protein kinase, dead-end nucleotide inhibitors of PKC and cAMP-dependent protein kinase such as AMPPNP inhibit the enzymes competitively with respect to MgATP and noncompetitively with respect to protein substrate (Whitehouse et al., 1983; Leventhal and Bertics, 1991). The similarity of the inhibitory kinetics observed with the isoquinolinesulfonamides and with dead-end nucleotide inhibitors provides strong evidence that, despite profound structural differences between the isoquinolinesulfonamides and ATP, the isoquinolinesulfonamides function as dead-end inhibitors that mimic the nucleotide substrate. In other words, at least in the case of PKC and cAMP-dependent protein kinase, the isoquinolinesulfonamides appear to bind to the same enzyme form and at the same site as the substrate MgATP, and the isoquinolinesulfonamides also resemble MgATP in that their binding to the enzyme appears to be independent of protein substrate.

Further kinetic support for this inhibitory mechanism is provided in a study of the ATPase reaction catalyzed by PKC. The ATPase reaction of PKC represents the bond-breaking step of the protein kinase reaction of the enzyme. This partial reaction allows a simplified analysis of the utilization of the nucleotide substrate by PKC in the absence of protein substrate (O'Brian and Ward, 1991; Ward and O'Brian, 1992a). Consistent with the inhibitory mechanism proposed for isoquinolinesulfonamides from the kinetics of protein kinase inhibition, the isoquinolinesulfonamide H7 inhibits the ATPase reaction of PKC with competitive kinetics (Ward and O'Brian, 1992a). Also in accordance with the proposed inhibitory mechanism, direct binding interactions between protein kinases and the isoquinolinesulfonamide H9 have been demonstrated in a study that shows the binding of PKC, cAMP-dependent protein kinase, and cGMP-dependent protein kinase to H9-coupled sepharose (Inagaki et al., 1985). In addition, the observation that

H7 protects PKC from inactivation by the nucleotide-binding site label 5'-p-fluorosulfonylbenzoyladenosine (FSBA) provides evidence that H7 binds PKC at the nucleotide substrate binding site (Ohta et al., 1988). Taken together, the straightforward kinetics observed in the inhibition of protein kinases by isoquinolinesulfonamides along with the inhibitor-binding data provide convincing evidence that the isoquinolinesulfonamides generally inhibit protein kinases by a single mode of action (i.e., competition with nucleotide substrate) rather than by a mixed mechanism.

Studies of protein kinase inhibition by isoquinolinesulfonamides have provided important topological information regarding the nucleotide-substrate binding sites of protein kinases. At micromolar concentrations, the isoquinolinesulfonamides generally recognize protein kinase nucleotide-substrate binding sites, but fail to recognize nucleotide substrate binding sites in other types of enzymes. For example, micromolar concentrations of H7 are virtually without effect on the activities of actomyosin ATPase, $(Ca^{2+}-Mg^{2+})$-ATPase, and adenylate cyclase, but do inhibit diverse protein kinases to various extents (Hidaka et al., 1984; Hagiwara et al., 1988). The ability of isoquinolinesulfonamides to distinguish protein kinase nucleotide-substrate binding sites from nucleotide binding sites that occur in other classes of enzymes provides evidence that the nucleotide-substrate binding sites of protein kinases share a fundamentally unique structure. This conclusion finds strong support in the recent elucidation of the crystal structure of the catalytic subunit of the cAMP-dependent protein kinase, which is the first solved crystal structure of a mammalian protein kinase catalytic domain. According to the crystal structure, the structure of the nucleotide-substrate binding site of the protein kinase is, in fact, novel (Knighton et al., 1991).

In addition to revealing shared topological features of protein kinase nucleotide-substrate binding sites, studies with isoquinolinesulfonamides have shown that profound structural differences exist among the nucleotide-substrate binding sites of different protein kinases (Hidaka et al., 1984; Hidaka and Kobayashi, 1992). A number of isoquinolinesulfonamide protein kinase inhibitors have been shown to successfully discriminate between the nucleotide-substrate binding sites of even very closely related protein kinases (Hidaka et al., 1984; Hidaka and Kobayashi, 1992). For example, the catalytic domains of PKC, cGMP-dependent protein kinase, and cAMP-dependent protein kinase are very closely related, according to a comparison of the primary structures of 65 eukaryotic protein kinases (Hanks et al., 1988), but the isoquinolinesulfonamide H89 inhibits cAMP-dependent protein kinase (K_i = 0.05 μM) much more effectively than either cGMP-dependent protein kinase (K = 0.5 μM) or PKC(K_i = 30 μM) (Hidaka and Kobayashi, 1992). Furthermore, H7 can be used in conjunction with HA1004 to assess the effects of inhibition of cellular PKC (Ioannides et al., 1990). H7 inhibits PKC and cyclic nucleotide-dependent protein kinases with similar potencies, but causes only negligible inhibition of myosin light-chain kinase, casein kinase I, and casein kinase II. HA1004 inhibits cyclic nucleotide-dependent protein kinases with about the same potency as H7, but is much weaker than H7 in its inhibition of PKC (Hidaka et al., 1984; Hidaka and Kobayashi, 1992). On the basis of the inhibitory selectivities of H7 and HA1004, it appears likely that biological effects observed with H7 but not with HA1004 reflect PKC inhibition. However, it should be noted that the protein kinase family is very large (Hanks et al., 1988), so that the possibility that inhibition of protein kinases other than PKC may be responsible for biological effects observed with H7 but not HA1004 cannot be entirely ruled out. In any case, it is clear that the isoquinolinesulfon-amides can serve as probes of the topological relatedness of protein kinase nucle-otide-substrate binding sites, and thereby complement available primary structure information. Moreover, the striking novelty of the nucleotide-substrate binding site of nPKC ζ (O'Brian and Ward, 1992) suggests that exploitation of unique features of that site by an isoquinolinesulfonamide might even allow potent, specific inhibition of nPKC ζ.

CONCLUSIONS

In recent years, a vast body of work has accrued in the field of PKC inhibitor research. For the sake of clarity, this review provides a detailed treatment of several important classes of PKC inhibitors rather than a comprehensive review of the field. A number of areas of PKC inhibitor research that fall outside of the scope of this review are, however, clearly deserving of close attention.

Notably, the indole carbazole staurosporine has provoked wide interest due to its remarkable potency in the inhibition of PKC (IC_{50} = 3 nM) (Tamaoki et al., 1986). Staurosporine appears to inhibit PKC by binding near the nucleotide-substrate binding site, because the binding of radiolabeled staurosporine to PKC is inhibited by H7 (Gross et al., 1990), and the kinetics of inhibition of the catalytic fragment of PKC by staurosporine are mixed competitive/noncompetitive with respect to MgATP (Ward and O'Brian, 1992b). Furthermore, the inhibition of PKC by staurosporine is completely independent of the regulatory domain of the enzyme (Nakadate et al., 1988). However, the usefulness of staurosporine as a probe of PKC function is limited by its highly nonselective nature (O'Brian and Ward, 1990b). In addition to PKC, staurosporine potently inhibits widely divergent members of the protein kinase family, including protein-tyrosine kinases (O'Brian and Ward, 1990b). The biological effects of staurosporine indicate the nonselective nature of its inhibitory activity against PKC in cellular systems. Some of the biological responses to staurosporine oppose effects elicited by phorbol ester PKC activators, but other responses to staurosporine parallel the effects achieved with phorbol ester PKC activators (Sako et al., 1988; Dlugosz and Yuspa, 1991). Thus, staurosporine is best described as a general protein kinase inhibitor with potent and unique biological properties, which clearly warrants further investigation.

The aminoalkyl bisindolylmaleimide GF109203X (Gö 6850) is a staurosporine analog with selective inhibitory activity against PKC, according to a comparison of the inhibitory potencies of GF109203X against isolated PKC (IC_{50} = 10 nM), phosphorylase kinase (IC_{50} = 700 nM), cAMP-dependent protein kinase (IC_{50} = 2 μM), and three receptor protein-tyrosine kinases (IC_{50} values > 50 μM) (Toullec et al., 1991). Importantly, studies of diverse effects of GF109203X in intact cells provide evidence for selective recognition of cellular PKC by the inhibitor and therefore suggest the usefulness of GF109203X as a probe of PKC function in cellular systems (Toullec et al., 1991). It has been reported recently that the indole carbazole Gö 6976, another staurosporine analog, is a potent PKC inhibitor (Martiny-Baron et al., 1993). Gö 6976, unlike staurosporine and GF109203X, selectively inhibits the Ca^{2+}-dependent PKC isozymes α and β_I (IC_{50} = 2–6 nM) without affecting the Ca^{2+}-independent isozymes ϵ, δ, and ζ at concentrations up to 3 μM. It appears that Gö 6976 can be used to study, for the first time, functional roles of the two groups of PKC isozymes in intact cells.

The benzophenanthridine alkaloid chelerythrine is another potent PKC inhibitor (IC_{50} = 0.7 μM). Chelerythrine appears to inhibit PKC selectively, based on its lack of significant inhibitory activity against cAMP-dependent protein kinase, calcium/calmodulin-dependent protein kinase, and a lymphoma-derived protein-tyrosine kinase (Herbert et al., 1990). Whereas GF109203X inhibits PKC by competition with MgATP (Toullec et al., 1991), chelerythrine inhibits the enzyme by competition with the protein substrate (Herbert et al., 1990). The inhibitory potency of chelerythrine appears to be independent of the protein substrate, because nearly identical inhibitory potencies have been observed with five different protein substrates of PKC. Thus, inhibition appears to result from direct interactions between PKC and chelerythrine (Herbert et al., 1990). Although further studies will be necessary to evaluate the selectivity of chelerythrine in the inhibition of PKC, the novel structure and inhibitory mechanism of chelerythrine indicate its potential value as an analytical tool for dissecting the mechanism of PKC catalysis. Finally, it should be noted that calphostin C, which was originally described as a highly potent and specific PKC inhibitor (Kobayashi et al., 1989), was recently shown to inhibit PKC by a complicated, light-dependent mechanism (Bruns et al., 1991), casting into doubt its usefulness as a probe of PKC function (Rando and Kishi, 1992).

REFERENCES

Aftab, D. T., L. M. Ballas, C. R. Loomis, and W. N. Hait. 1991. Structure-activity relationships of phenothiazines and related drugs for inhibition of protein kinase C. *Molec. Pharmacol.* 40:798–805.

Alexander, D. R., J. M. Hexham, S. C. Lucas, J. D. Graves, D. A. Cantrell, and M. J. Crumpton. 1989. A protein kinase C pseudosubstrate peptide inhibits phosphorylation of the CD3 antigen in streptolysin-0-permeabilized human T lymphocytes. *Biochem. J.* 260:893–901.

Andreesen, R., M. Modolell, H. U. Weltzien, H. Eibl, H. H. Common, G. W. Loehr, and P. G. Munder. 1978. Selective destruction of human leukemic cells by alkyl-lysophospholipids. *Cancer Res.* 38:3894–3899.

Berdel, W. E., W. R. Bausert, H. U. Weltzien, M. Modolell, K. H. Widmann, and P. G. Munder. 1980. The influence of alkyl-lysophospholipid-activated macrophages on the development of metastasis of 3-Lewis lung carcinoma. *Eur. J. Cancer* 16:1199–1204.

Berdel, W. E., M. Fromm, U. Fink, W. Pahlke, U. Bicker, A. Reichert, and J. Rastetter. 1983. Cytotoxicity of thioether lysophospholipids in leukemias and tumors of human origin. *Cancer Res.* 43:5538–5543.

Berger, M. E., P. G. Munder, D. Schmahl, and O. Westphal. 1984. Influence of the alkyl-lysophospholipid ET-18-OCH$_3$ on methylnitrosourea-induced rat mammary carcinomas. *Oncology* 41:109–113.

Bignon, E., M. Pons, J. Gilbert, and Y. Nishizuka. 1990. Multiple mechanisms of protein kinase C inhibition by triphenylacrylonitrile antiestrogens. *FEBS Lett.* 271:54–58.

Borner, C., S. N. Guadagno, D. Fabbro, and I. B. Weinstein. 1992. Expression of four protein kinase C isoforms in rat fibroblasts. *J. Biol. Chem.* 267:12892–12899.

Bottega, R., and R. M. Epand. 1992. Inhibition of protein kinase C by cationic amphiphiles. *Biochemistry* 31:9025–9030.

Brumley, L. M. and R. W. Wallace. 1989. Calmodulin and protein kinase C antagonists also inhibit the Ca^{2+}-dependent protein protease, calpain I. *Biochem. Biophys. Res. Commun.* 159:1297–1303.

Bruns, R. F., D. Miller, R. L. Merriman, J. J. Howbert, W. F. Heath, E. Kobayashi, I. Takahashi, T. Tamaoki, and H. Nakano. 1991. Inhibition of protein kinase C by calphostin C is light-dependent. *Biochem. Biophys. Res. Commun.* 176:288–293.

Burns, D. J., J. Bloomenthal, M. H. Lee, and R. M. Bell. 1990. Expression of the α, β_{II}, and γ protein kinase C isozymes in the baculovirus-insect cell expression system. *J. Biol. Chem.* 265:12044–12051.

Charp, P. A., W. G. Rice, R. L. Raynor, E. Reimund, J. M. Kinkade, Jr., T. Ganz, M. E. Selsted, R. I. Lehrer, and J. F. Kuo. 1988a. Inhibition of protein kinase C by defensins, antibiotic peptides from human neutrophils. *Biochem. Pharmacol.* 37: 951–956.

Charp, P. A., Q. Zhou, M. G. Wood, Jr., R. L. Raynor, F. M. Menger, and J. F. Kuo. 1988b. Synthetic branched-chain analogues of distearoylphosphatidylcholine: Structure-activity relationship in inhibiting and activating protein kinase C. *Biochemistry* 27:4607–4612.

Chiou, S.-H., R. L. Raynor, B. Zheng, T. C. Chambers, and J. F. Kuo. 1993. Cobra venom cardiotoxin (cytotoxin) isoforms and neurotoxin: Comparative potency of protein kinase C inhibition and cancer cell cytotoxicity, and mode of enzyme inhibition. *Biochemistry* 32:2062–2067.

DeBont, H. B. A., R. M. J. Liskamp, C. A. O'Brian, C. Erkelens, G. H. Veeneman, and J. H. van Boom. 1989. Synthesis of a substrate of protein kinase C and its corresponding phosphopeptide. *Intl. J. Peptide Protein Res.* 33:115–123.

Dlugosz, A. A. and S. H. Yuspa. 1991. Staurosporine induces protein kinase C agonist effects and maturation of normal and neoplastic mouse keratinocytes *in vitro*. *Cancer Res.* 51:4677–4684.

Eibl, H., P. Hilgard, and C. Unger. 1992. *Alkylphosphocholines: New Drugs in Cancer Therapy*. Karger AG, Basel.

Eichholtz, T., J. Alblas, M. van Overveld, W. Moolenaar, and H. Ploegh. 1990. A pseudosubstrate peptide inhibits protein kinase C-mediated phosphorylation in permeabilized rat-1 cells. *FEBS Lett.* 261:147–150.

Eichholtz, T., D. B. DeBont, J. de Widt, R. M. J. Liskamp, and H. L. Ploegh. 1993. A myristoylated pseudosubstrate peptide, a novel protein kinase C inhibitor. *J. Biol. Chem.* 268:1982–1986.

Epand, R. M. 1992. Effect of PKC modulators on membrane properties. In: *Protein Kinase C: Current Concepts and Future Perspectives,* edited by D. S. Lester and R. M. Epand, pp. 135–156. Ellis Horwood, New York.

Epand, R. M., and D. S. Lester. 1990. The role of membrane biophysical properties in the regulation of protein kinase C activity. *Trends Pharm. Sci.* 11:317–320.

Fromm, M., W. E. Berdel, H. D. Schick, U. Fink, W. Pahlke, U. Bicher, A. Richert, and J. Rastetter. 1987. Antineoplastic activity of the thioether lysophospholipid derivative BM 41.440 *in vitro. Lipids* 22:916–918.

Gross, J. L., W. F. Herblin, U. H. Do, J. S. Pounds, L. J. Buenaga, and L. E. Stephens. 1990. Characterization of specific [3H]dimethylstaurosporine binding to protein kinase C. *Biochem. Pharmacol.* 40:343–350.

Hagiwara, M., M. Inagaki, J. Takahashi, T. Yoshida, and H. Hidaka. 1988. A fluorescent protein kinase C inhibitor: 1-(1-hydroxy-5-isoquinolinylsulfonyl)piperazine. *Pharmacology* 36:365–370.

Hanks, S. K., A. M. Quinn, and T. Hunter. 1988. The protein kinase family: Conserved features and deduced phylogeny of the catalytic domains. *Science* 241:42–52.

Hannun, Y. A., and R. M. Bell. 1988. Aminoacridines, potent inhibitors of protein kinase C. *J. Biol. Chem.* 263:5124–5131.

Hannun, Y. A., C. R. Loomis, A. H. Merrill, and R. M. Bell. 1986. Sphingosine inhibition of protein kinase C activity and of phorbol dibutyrate binding in vitro and in human platelets. *J. Biol. Chem.* 261:12604–12609.

Hannun, Y. A., R. J. Foglesong, and R. M. Bell. 1989. The Adriamycin-iron (III) complex is a potent inhibitor of protein kinase C. *J. Biol. Chem.* 264:9960–9966.

Helfman, D. M., K. C. Barnes, J. M. Kinkade, Jr., W. R. Vogler, M. Shoji, and J. F. Kuo. 1983. Phospholipid sensitive Ca^{2+}-dependent protein phosphorylation system in various types of leukemic cells from human patients and in human leukemic cell lines HL60 and K562, and its inhibition by alkyl-lysophospholipid. *Cancer Res.* 43:2955–2961.

Herbert, J. M., J. M. Augereau, J. Gleye, and J. P. Maffrand. 1990. Chelerythrine is a potent and specific inhibitor of protein kinase C. *Biochem. Biophys. Res. Commun.* 172:993–999.

Hidaka, H. and M. Hagiwara. 1987. Pharmacology of the isoquinoline sulfonamide protein kinase C inhibitors. *Trends Pharmacol. Sci.* 8:162–164.

Hidaka, H. and R. Kobayashi. 1992. Pharmacology of protein kinase inhibitors. *Annu. Rev. Pharmacol. Toxicol.* 32:377–397.

Hidaka, H., M. Inagaki, S. Kawamoto, and Y. Sasaki. 1984. Isoquinolinesulfonamides, novel and potent inhibitors of cyclic nucleotide-dependent protein kinase and protein kinase C. *Biochemistry* 23:5036–5041.

House, C. and B. E. Kemp. 1987. Protein kinase C contains a pseudosubstrate prototope in its regulatory domain. *Science* 238:1726–1728.

Huang, K.-P. 1989. The mechanism of protein kinase C activation. *Trends Neurosci.* 12: 425–432.

Huang, K. P., F. L. Huang, H. Nakabayashi, and Y. Yoshida. 1988. Biochemical characterization of rat brain protein kinase C isozymes. *J. Biol. Chem.* 263:14839–14845.

Hunter, T., N. Ling, and J. A. Cooper. 1984. Protein kinase C phosphorylation of the EGF receptor at a threonine residue close to the cytoplasmic face of the plasma membrane. *Nature* 311:480–485.

Husebye, E. S., and T. Flatmark. 1988. Phosphatidylinositol kinase of bovine adrenal chromaffin granules. *Biochem. Pharmacol.* 37:4149–4156.

Inagaki, M., M. Watanabe, and H. Hidaka. 1985. *N*-(2-aminoethyl)-5-isoquinolinesulfonamide, a newly synthesized protein kinase inhibitor, functions as a ligand in affinity chromatography. *J. Biol. Chem.* 260:2922–2925.

Inagaki, M., S. Kawamoto, H. Itoh, M. Saitoh, M. Hagiwara, J. Takahashi, and H. Hidaka.

1986. Naphthalenesulfonamides as calmodulin antagonists and protein kinase inhibitors. *Molec. Pharmacol.* 29:577–581.

Ioannides, C. G., R. S. Freedman, R. M. Liskamp, N. E. Ward, and C. A. O'Brian. 1990. Inhibition of IL-2 receptor induction and IL-2 production in the human leukemic cell line Jurkat by a novel peptide inhibitor of protein kinase C. *Cell. Immunol.* 131:242–252.

Katoh, N., R. W. Wrenn, B. C. Wise, M. Shoji, and J. F. Kuo. 1981. Substrate proteins for calmodulin-sensitive and phospholipid-sensitive Ca^{2+}-dependent protein kinases in heart, and inhibition of their phosphorylation by palmitoylcarnitine. *Proc. Natl. Acad. Sci. U.S.A.* 78:4813–4817.

Katoh, N., R. L. Raynor, B. C. Wise, R. C. Schatzman, R. S. Turner, D. M. Helfman, J. N. Fain, and J. F. Kuo. 1982. Inhibition by melittin of phospholipid-sensitive and calmodulin-sensitive Ca^{2+}-dependent protein kinases. *Biochem. J.* 202:217–224.

Kikkawa, U., A. Kishimoto, and Y. Nishizuka. 1989. The protein kinase C family: Heterogeneity and its implications. *Annu. Rev. Biochem.* 58:31–44.

Knighton, D. R., J. Zheng, L. F. T. Eyck, V. A. Ashford, N. H. Xuong, S. S. Taylor, and J. Sowadski. 1991. Crystal structure of the catalytic subunit of cyclic adenosine monophosphate-dependent protein kinase. *Science* 253:407–414.

Kobayashi, E., H. Nakano, M. Morimoto, and T. Tamaoki. 1989. Calphostin C (UCN-1028C), a novel microbial compound, is a highly potent and specific inhibitor of protein kinase C. *Biochem. Biophys. Res. Commun.* 159:548–553.

Koide, H., K. Ogita, U. Kikkawa, and Y. Nishizuka. 1992. Isolation and characterization of the ε subspecies of protein kinase C from rat brain. *Proc. Natl. Acad. Sci.* U.S.A. 89:1149–1153.

Kuo, J. F., R. L. Raynor, G. J. Mazzei, R. C. Schatzman, R. S. Turner, and W. R. Kem. 1983. Cobra polypeptide cytotoxin I and marine worm polypeptide cytotoxin A-IV are potent and selective inhibitors of phospholipid-sensitive Ca^{2+}-dependent protein kinase. *FEBS Lett.* 153:183–186.

Kuo, J. F., M. Shoji, Z. Kiss, P. R. Girard, E. Deli, K. Oishi, and W. R. Vogler. 1989. Protein kinase C in cell growth and differentiation. *Adv. Exptal. Medic. Biol.* 255: 8–20.

Lam, H. Y. P. 1984. Tamoxifen is a calmodulin antagonist in the activation of cAMP phosphodiesterase. *Biochem. Biophys. Res. Commun.* 118:27–32.

Lester, D. S., and D. Baumann. 1991. Action of organic solvents on protein kinase C. *Eur. J. Pharmacol.* 206:301–308.

Leventhal, P. S., and P. J. Bertics. 1991. Kinetic analysis of protein kinase C: Product and dead-end inhibition studies using ADP, poly(L-Lysine), nonhydrolyzable ATP analogues, and diadenosine oligophosphates. *Biochemistry* 30:1385–1390.

Liyanage, M., D. Frith, E. Livneh, and S. Stabel. 1992. Protein kinase C group B members PKC-δ, ε, ζ, and PKC-L (η). *Biochem. J.* 283:781–787.

Marais, R. M. and P. J. Parker. 1989. Purification and characterization of bovine brain protein kinase C isotypes α, β, and γ. *Eur. J. Biochem.* 182:129–137.

Martiny-Baron, G., M. G., Mazanietz, H. Mischak, P. M. Blumberg, G. Kochs, H. Hug, D. Marmé, and C. Schächtele, 1993. Selective inhibition of protein kinase C isozymes by the indolocarbazole Gö 6976. *J. Biol. Chem.* 268:9194–9197.

Mazzei, G. J., N. Katoh, and J. F. Kuo. 1982. Polymyxin B is a more selective inhibitor for phospholipid-sensitive Ca^{2+}-dependent protein kinase than for calmodulin-sensitive Ca^{2+}-dependent protein kinase. *Biochem. Biophys. Res. Commun.* 109:1129–1133.

Mazzei, G. J., R. C. Schatzman, R. S. Turner, W. R. Vogler, and J. F. Kuo. 1984. Phospholipid-sensitive Ca^{2+}-dependent protein kinase inhibition by R-24571, a calmodulin antagonist. *Biochem. Pharmacol.* 33:125–130.

McPhail, L. C., C. C. Clayton, and R. A. Snyderman. 1984. Potential second messenger role for unsaturated fatty acids: Activation of Ca^{2+}-dependent protein kinase. *Science* 224:622–625.

Modolell, M., R. Andreesen, W. Pahlke, U. Brugger, and P. G. Munder. 1979. Disturbance of phospholipid metabolism during the selective destruction of tumor cells induced by alkyl-lysophospholipids. *Cancer Res.* 39:4681–4686.

Mori, T., Y. Takai., R. Minakuchi, B. Yu, and Y. Nishizuka. 1980. Inhibitory action of chlorpromazine, dibucaine, and other phospholipid-interacting drugs on calcium-activated, phospholipid-dependent protein kinase. *J. Biol. Chem.* 255:8378–8380.

Nakadate, T., A. Y. Jeng, and P. M. Blumberg. 1988. Comparison of protein kinase C functional assays to clarify mechanisms of inhibitor action. *Biochem. Pharmacol.* 37:1541–1545.

Nishizuka, Y. 1992. Intracellular signaling by hydrolysis of phospholipids and activation of protein kinase C. *Science* 258:607–614.

O'Brian, C. A., and N. E. Ward. 1989a. Biology of the protein kinase C family. *Cancer Met. Revs.* 8:199–214.

O'Brian, C. A., and N. E. Ward. 1989b. Binding of protein kinase C to *N*-(6-aminohexyl)-5-chloro-1-naphthalenesulfonamide through its ATP binding site. *Biochem. Pharmacol.* 38:1737–1742.

O'Brian, C. A., and N. E. Ward. 1989c. ATP-sensitive binding of melittin to the catalytic domain of protein kinase C. *Molec. Pharmacol.* 36:355–359.

O'Brian, C. A., and N. E. Ward. 1990a. Characterization of a Ca^{2+}- and phospholipid-dependent ATPase reaction catalyzed by rat brain protein kinase C. *Biochemistry* 29:4278–4282.

O'Brian, C. A., and N. E. Ward. 1990b. Staurosporine: A prototype of a novel class of inhibitors of tumor cell invasion? *J. Natl. Cancer Inst.* 82:1734–1735.

O'Brian, C. A., and N. E. Ward. 1991. Stimulation of the ATPase activity of rat brain protein kinase C by phosphoacceptor substrates of the enzyme. *Biochemistry* 30: 2549–2554.

O'Brian, C. A., and N. E. Ward. 1992. The phosphotransferase reaction of protein kinase C: Mechanistic studies. In: *Protein Kinase C: Current Concepts and Future Perspectives,* edited by D. S. Lester and R. M. Epand, pp. 117–134. Ellis Horwood, New York.

O'Brian, C. A., and I. B. Weinstein. 1987. *In vitro* inhibition of rat brain protein kinase C by rhodamine 6G. Profound effects of the lipid cofactor on the inhibition of the enzyme. *Biochem. Pharmacol.* 36:1231–1235.

O'Brian, C. A., D. S. Lawrence, E. T. Kaiser, and I. B. Weinstein. 1984. Protein kinase C phosphorylates the synthetic peptide Arg-Arg-Lys-Ala-Ser-Gly-Pro-Pro-Val in the presence of phospholipid plus either Ca^{2+} or a phorbol ester tumor promoter. *Biochem. Biophys. Res. Commun.* 124:296–302.

O'Brian, C. A., R. M. Liskamp, D. H. Solomon, and I. B. Weinstein. 1985. Inhibition of protein kinase C by tamoxifen. *Cancer Res.* 45:2462–2465.

O'Brian, C. A., R. M. Liskamp, D. H. Solomon, and I. B. Weinstein. 1986. Triphenylethylenes: A new class of protein kinase C inhibitors. *J. Natl. Cancer Inst.* 76:1243–1246.

O'Brian, C. A., G. M. Housey, and I. B. Weinstein. 1987. Binding of protein kinase C to naphthalenesulfonamide- and phenothiazine-agarose columns: Evidence for direct interactions between protein kinase C and cationic amphiphilic inhibitors of the enzyme. *Biochem. Pharmacol.* 36:4179–4181.

O'Brian, C. A., N. E. Ward, and B. W. Anderson. 1988. Role of specific interactions between protein kinase C and triphenylethylenes in inhibition of the enzyme. *J. Natl. Cancer Inst.* 80:1628–1633.

O'Brian, C. A., N. E. Ward, R. M. Liskamp, D. B. de Bont, and J. H. van Boom. 1990a. N-myristyl-Lys-Arg-Thr-Leu-Arg: A novel protein kinase C inhibitor. *Biochem. Pharmacol.* 39:49–57.

O'Brian, C. A., C. G. Ioannides, N. E. Ward, and R. M. Liskamp. 1990b. Inhibition of protein kinase C and calmodulin by the geometric isomers *cis-* and *trans-*tamoxifen. *Biopolymers* 29:97–104.

O'Brian, C. A., N. E. Ward, R. M. Liskamp, D. B. de Bont, L. E. Earnest, J. H. van Boom, and D. Fan. 1991. A novel N-myristylated synthetic octapeptide inhibits protein kinase C activity and partially reverses murine fibrosarcoma cell resistance to Adriamycin. *Investig. New Drugs* 9:169–179.

O'Brian, C. A., N. E. Ward, K. Kundu, and D. Fan. 1993. Role of protein kinase C in multidrug resistance. In: *Cancer Treatment and Research,* edited by W. L. McGuire. Kluwer Acad. Publishers, New York, in press.

Ohta, H., T. Tanaka, and H. Hidaka. 1988. Putative binding site(s) of 1-(5-isoquinolinesulfonyl)-2-methylpiperazine (H-7) on protein kinase C. *Biochem. Pharmacol.* 37:2704–2706.

Oishi, K., R. L. Raynor, and J. F. Kuo. 1988a. Regulation of protein kinase C by lysophospholipids. Potential role in signal transduction. *J. Biol. Chem.* 263:6865–6871.

Oishi, K., B. Zheng, J. F. White, W. R. Vogler, and J. F. Kuo. 1988b. Inhibition of Na, K-ATPase and sodium pump by anticancer ether lipids and protein kinase C inhibitors ET-18-OCH$_3$ and BM 41.440. *Biochem. Biophys. Res. Commun.* 157:1000–1006.

Oishi, K., B. Zheng, and J. F. Kuo. 1990. Inhibition of Na, K-ATPase and sodium pump by protein kinase C regulators sphingosine, lysophosphatidylcholine, and oleic acid. *J. Biol. Chem.* 265:70–75.

Ono, Y., T. Fujii, K. Ogita, U. Kikkawa, K. Igarashi, and Y. Nishizuka. 1988. The structure, expression, and properties of additional members of the protein kinase C family. *J. Biol. Chem.* 263:6927–6932.

Rando, R. R., and Y. Kishi. 1992. The structural basis of protein kinase C activation by diacylglycerols and tumor promoters. In: *Protein Kinase C: Current Concepts and Future Perspectives,* edited by D. S. Lester and R. M. Epand, pp. 41–61. Ellis Horwood, New York.

Raynor, R. L., B. Zheng, and J. F. Kuo. 1991. Membrane interactions of amphiphilic polypeptides mastoparan, melittin, polymyxin B, and cardiotoxin. Differential inhibition of protein kinase C, Ca^{2+}/calmodulin-dependent protein kinase II and synaptosomal membrane Na,K-ATPase, and Na$^+$ pump and differentiation of HL60 cells. *J. Biol. Chem.* 266:2753–2758.

Raynor, R. L., Y.-S. Kim, B. Zheng, W. R. Vogler, and J. F. Kuo. 1992. Membrane interactions of mastoparan analogues related to their differential effects on protein kinase C, Na,K-ATPase and HL60 cells. *FEBS Lett.* 307:275–279.

Robinson, P. J. 1992. Potencies of protein kinase C inhibitors are dependent on the activators used to stimulate the enzyme. *Biochem. Pharmacol.* 44:1325–1334.

Rodriguez-Paris, J. M., M. Shoji, S. Yeola, D. Liotta, W. R. Vogler, and J. F. Kuo. 1989. Fluorimetric studies of protein kinase C interactions with phospholipids. *Biochem. Biophys. Res. Commun.* 159:495–500.

Rotenberg, S. A., S. Smiley, M. Ueffing, R. S. Krauss, L. B. Chen, and I. B. Weinstein. 1990. Inhibition of rodent protein kinase C by the anticarcinoma agent dequalinium. *Cancer Res.* 50:677–685.

Sako, T., A. I. Tauber, A. Y. Jeng, S. H. Yuspa, and P. M Blumberg. 1988. Contrasting actions of staurosporine, a protein kinase C inhibitor, on human neutrophils and primary mouse epidermal cells. *Cancer Res.* 48:4646–4650.

Senisterra, G. and R. M. Epand. 1992. Dual modulation of protein kinase C activity by sphingosine. *Biochem. Biophys. Res. Commun.* 187:635–640.

Shinomura, T., Y. Asaoka, M. Oka, K. Yoshida, and Y. Nishizxka. 1991. Synergistic action of diacylglycerol and unsaturated fatty acid for protein kinase C activation: Its possible implications. *Proc. Natl. Acad. Sci. U. S. A.* 88:5149–5153.

Shoji, M., R. L. Raynor, W. E. Berdel, W. R. Vogler, and J. F. Kuo. 1988. Effects of thioether phospholipid BM 41.440 on protein kinase C and phorbol ester-induced differentiation of human leukemia HL60 and KG-1 cells. *Cancer Res.* 48:6669–6673.

Shoji, M., R. L. Raynor, E. A. M. Fleer, H. Eibl, W. R. Vogler, and J. F. Kuo. 1991. Effects of hexadecylphosphocholine on protein kinase C and TPA-induced differentiation of HL60 cells. *Lipids* 26:145–149.

Soderling, T. R. 1990. Protein kinases. *J. Biol. Chem.* 265:1823–1826.

Strulovici, B., S. Daniel-Issakani, G. Baxter, J. Knopf, L. Sultzman, H. Cherwinski, J. Nestor, D. R. Webb, and J. Ranson. 1991. Distinct mechanisms of regulation of protein kinase C-ϵ by hormones and phorbol esters. *J. Biol. Chem.* 266:168–173.

Su, H.-D., G. J. Mazzei, W. R. Vogler, and J. F. Kuo. 1985. Effect of tamoxifen, a non-steroidal antiestrogen, on phospholipid/calcium-dependent protein kinase and phosphorylation of its endogenous substrate proteins from the rat brain and ovary. *Biochem. Pharmacol.* 34:3649–3653.

Su, H.-D., B. E. Kemp, R. S. Turner, and J. F. Kuo. 1986. Synthetic myelin basic protein peptide analogs are specific inhibitors of phospholipid/calcium-dependent protein kinase (protein kinase C). *Biochem. Biophys. Res. Commun.* 134:78–84.

Tamaoki, T., H. Nomoto, I. Takahashi, Y. Kato, M. Morimoto, and F. Tomita. 1986. Staurosporine, a potent inhibitor of phospholipid/Ca^{++} dependent protein kinase. *Biochem. Biophys. Res. Commun.* 135:397–402.

Tanaka, T., T. Ohmura, T. Yamakado, and H. Hidaka. 1982. Two types of calcium-dependent protein phosphorylations modulated by calmodulin antagonists. *Molec. Pharmacol* 22:408–412.

Tidwell, T., G. Guzman, and W. R. Vogler. 1981. The effect of alkyllysophospholipids on leukemic cell lines. I. Differential action on two human leukemic cell lines HL60 and K562. *Blood* 57:794–797.

Toullec, D., P. Pianetti, H. Coste, P. Bellevergue, T. Grand-Perret, M. Ajakane, V. Baudet, P. Boissin, E. Boursier, F. Loriolle, L. Duhamel, D. Charon, and J. Kirilovsky. 1991. The bisindolylmaleimide GF109203X is a potent and selective inhibitor of protein kinase C. *J. Biol. Chem.* 266:15771–15781.

Turner, R. S., B. E. Kemp, H.-D. Su, and J. F. Kuo. 1985. Substrate specificity of phospholipid/Ca^{2+}-dependent protein kinase as probed with synthetic peptide fragments of the bovine myelin basic protein. *J. Biol. Chem.* 260:11503–11507.

Ward, N. E. and C. A. O'Brian. 1992a. The intrinsic ATPase activity of protein kinase C is catalyzed at the active site of the enzyme. *Biochemistry* 31:5905–5911.

Ward, N. E. and C. A. O'Brian. 1992b. Kinetic analysis of protein kinase C inhibition by staurosporine: Evidence that inhibition entails inhibitor binding at a conserved region of the catalytic domain but not competition with substrates. *Molec. Pharmacol.* 41:387–392.

Ways, D. K., P. P. Cook, C. Webster, and P. J. Parker. 1992. Effect of phorbol esters on protein kinase C-ζ. *J. Biol. Chem.* 267:4799–4805.

Weinstein, I. B. 1990. The role of protein kinase C in growth control and the concept of carcinogenesis as a progressive disorder in signal transduction. In: *The Biology and Medicine of Signal Transduction,* edited by Y. Nishizuka et al., pp. 307–316. Raven Press, New York.

Whitehouse, S., J. R. Feramisco, J. E. Casnellie, E. G. Krebs, and D. A. Walsh. 1983. Studies on the kinetic mechanism of the catalytic subunit of the cAMP-dependent protein kinase. *J. Biol. Chem.* 258:3693–3701.

Wise, B. C. and J. F. Kuo. 1983. Modes of inhibition by acylcarnitines, adriamycin and trifluoperazine of cardiac phospholipid-sensitive calcium-dependent protein kinase. *Biochem. Pharmacol.* 32:1259–1265.

Zheng, B., K. Oishi, M. Shoji, H. Eibl, W. E. Berdel, J. Haidu, W. R. Vogler, and J. F. Kuo. 1990. Inhibition of protein kinase C, (sodium plus potassium)-activated adenosine triphosphatase, and sodium pump by synthetic phospholipid analogues. *Cancer Res.* 50:3025–3031.

5

Receptor-Regulated Phospholipases and Their Generation of Lipid Mediators, Which Activate Protein Kinase C

J. DAVID LAMBETH

PHOSPHOLIPASE C AND SIGNAL TRANSDUCTION

The Phosphoinositide Cycle, Diacylglycerol, and Protein Kinase C

The past decade has seen growing interest in the role of phospholipids in signal transduction, in particular with regard to the phosphoinositide cycle. The basis for the discovery of such a role came with the accidental observation (Hokin and Hokin, 1953) that acetylcholine stimulation of pancreatic slices resulted in a tenfold increase in the incorporation of $[^{32}P]$-P_i into phospholipids. The major glycerophospholipid class that was labeled was the phosphoinositides, which include phosphatidylinositol (PI), phosphatidylinositol 4-phosphate (PIP), and phosphatidylinositol 4,5-bisphosphate (PIP$_2$). By incorporating radiolabeled inositol into phospholipids during a prelabeling period, it was demonstrated that the regulated step was the hydrolysis of phosphoinositides, in particular PIP$_2$, and that the increased incorporation of $[32P]$-P_i into lipids was secondary to resynthesis of phosphoinositides from the breakdown products. The early history of this area is detailed in a 1980s review (Hokin, 1985).

By the mid-1970s, it was clear that in nearly all mammalian cell types a variety of receptors were coupled to phosphoinositide hydrolysis and turnover (Blank et al., 1991). As shown in Figure 5.1, the receptor-linked activation of phospholipase C (PLC) results in the hydrolysis of PIP$_2$, with the concomitant generation of

Supported by National Institutes of Health grants CA46508 and AI22809. I would like to thank David Perry, Eddie Bowman, and Isabel Lopez for their critical reading, discussion, and editorial comments on the manuscript.

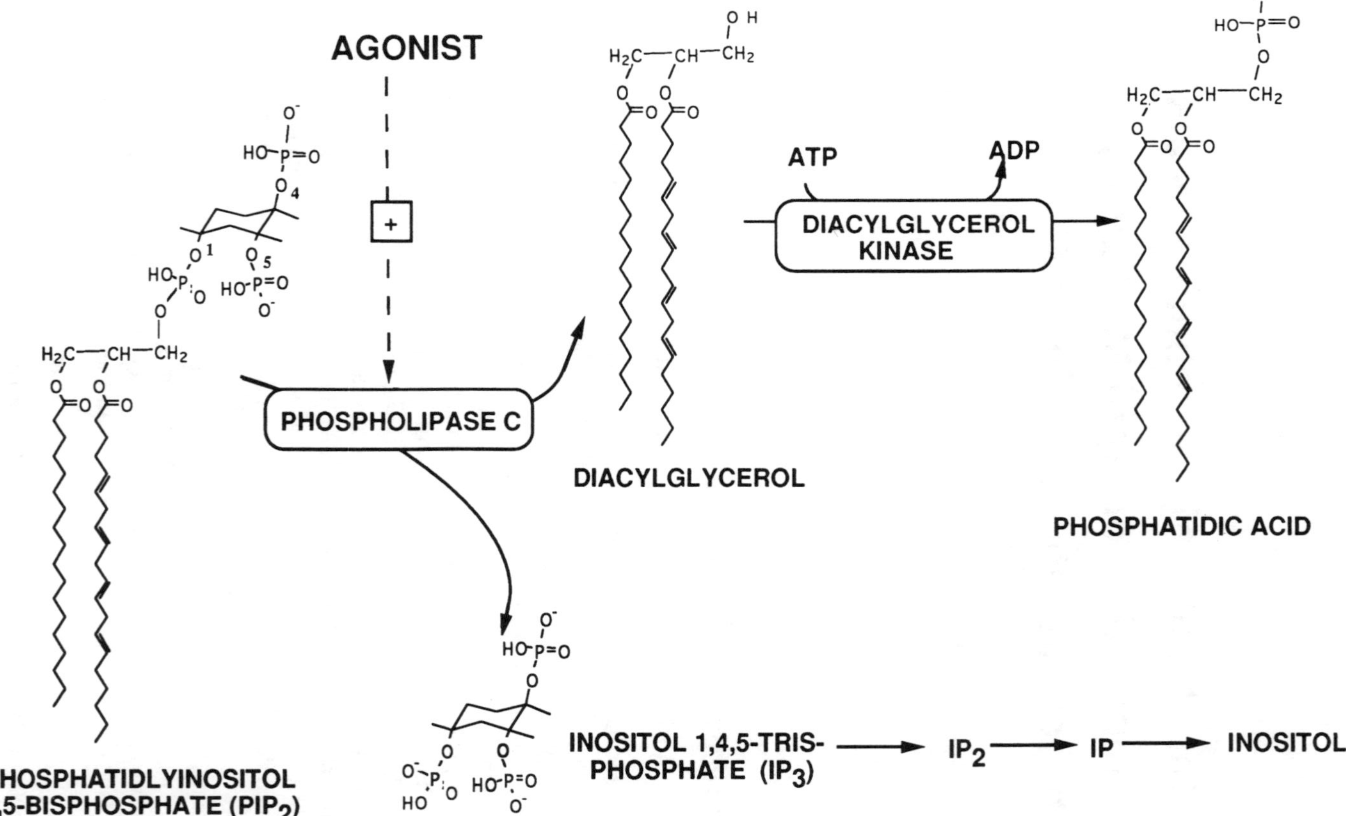

Figure 5.1   Phospholipase C hydrolysis of phosphatidylinositol 4,5-bisphosphate (PIP$_2$). Agonist activates phospholipase C, with consequent hydrolysis of PIP$_2$ to generate inositol 1,4,5-trisphosphate (IP$_3$) and diacylglycerol. Also shown is the subsequent metabolism of diacylglycerol to phosphatidic acid via the action of diacylglycerol kinase and the sequential metabolism of IP$_3$ to IP$_2$, IP, and inositol.

diacylglycerol and inositol 1,4,5-trisphosphate (IP_3). As summarized in recent reviews (Berridge, 1984; Nishizuka, 1984; Majerus et al., 1986), IP_3 acts on a "receptor" in the endoplasmic reticulum or related structures to open a calcium channel, releasing Ca^{2+} into the cytosol and transiently increases its concentration. Diacylglycerol functions in concert with calcium to activate protein kinase C (PKC), as detailed in other chapters. Thus, the hydrolysis of phosphoinositides can result in the activation of PKC as well as other Ca^{2+}-activated enzymes/processes.

Agonist-activated phosphoinositide hydrolysis occurs in a variety of tissues. This chapter will not chronicle the diverse agonists, tissues, and biological responses that have been associated with phosphoinositide hydrolysis, and the interested reader is referred to the above reviews. PIP_2 hydrolysis is activated in many cases to "serpentine" receptors, known to mediate their effects via heterotrimeric G proteins (e.g., receptors for α-adrenergic agonists, chemotactic peptides, angiotensin II, etc.) whereas in other cases hydrolysis is activated by growth factors (e.g., epidermal [EGF,] platelet-derived [PDGF] that function via tyrosine kinase-type receptors. The activation mechanisms are best understood by first considering the biochemistry of phosphoinositide-specific PLC (PI-PLC) isoenzymes.

The Biochemistry of PI-PLC: Discovery of Multiple Isoenzymes.

β, γ, and δ Isoenzymes

Much of what is currently known about the biochemistry of receptor-activated phospholipases C derives from pioneering studies by Rhee and colleagues (Ryu et al., 1987a,b) in which brain cytosol was chromatographically resolved into PLC isoforms. The original nomenclature designating forms I, II, and III has been replaced respectively by β, γ, and δ. The three forms were immunochemically distinct and differed in their apparent molecular weights by sodium dodecyl sulfate (SDS) gel electrophoresis (β = 150 kDa, γ = 145 kDa, and δ = 85 kDa), and in some of their kinetic and catalytic properties. All required calcium for activity toward PI and PIP_2, and at low calcium all preferred PIP_2 as substrate.

All three isoforms have been cloned and sequenced. The sequence of the γ form corresponds to a protein of molecular weight 148,431, and contains regions corresponding to homology regions SH2 and SH3 of the nonreceptor tyrosine kinase encoded by *src* (Stahl et al., 1988; Suh et al., 1988a). PLC isoforms β and δ were subsequently cloned (Katan et al., 1988; Suh et al., 1988b), and encode distinct proteins of molecular weights 138,255 and 85,840, respectively. The three isoforms showed an overall low degree of sequence homology. Nevertheless, significant homology was seen in two regions, one about 170 amino acids and the other about 260 residues, referred to as regions X and Y, respectively. The conserved regions were separated by nonconserved sequences that in the γ isoform contained the SH2 and SH3 homology domains. It has been proposed that the X and Y regions participate in the phosphoinositide binding and/or catalytic activity. Consistent with this suggestion, deletion of the X and Y domains revealed that they are essential for activity, whereas activity is retained in the absence of the SH2/SH3 domain.

Other Phospholipase C Isoforms

Other PLC isoforms have been isolated from either membranes or cytosol from various tissues. These include a 140-kDa form from sheep seminal vesicles (Hofman and Majerus, 1982), a 62-kDa form from guinea pig uterus cytosol (Bennett and Crook, 1987) and 140- and 154-kDa form from bovine brain membranes (Katan and Parker, 1987; Lee et al., 1987). All showed a substrate preference for phosphoinositides. An antibody to the 62-kDa uterus PLC was used to clone (Bennett et al., 1988) the "PLC α" gene from an HL-60 library. This form does not contain the X or Y homology regions, and it is uncertain whether the cloned gene actually encodes a PLC activity. A PI-specific 88-kDa PLC has also been isolated from bovine brain cytosol, but its relationship to the other brain enzymes is unclear.

In contrast to the phosphoinositide-specific enzymes described above, a neutral pH optimum PLC partially purified from canine myocardial cytosol showed a substrate preference for phosphatidylcholine (PC) and phosphatidylethanolamine (PE) (Wolf et al., 1985). In addition, the enzyme showed a preference for phospholipids containing a 1-O-alkenyl (plasmalogen) linkage rather than a 1-acyl linked fatty acid. Receptor-mediated hydrolysis of PC by PLC has been noted in some systems, for example MDCK cells (see Slivka et al., 1988), but the relevance of PC-specific PLC to signal transduction is less well understood than PI hydrolysis.

PLC β, γ, and δ Subfamilies

Additional members of the PLC superfamily have recently been identified, and are closely related to the originally described β, γ, and δ isoforms from brain (now designated phospholipases C-β1, -γ1, and -δ1). The new members were identified by low stringency hybridization to various cDNA libraries using DNA probes made to the X and Y conserved regions of the brain isoenzymes. Two new mammalian β-type sequences were obtained (Park et al., 1992a; Rhee and Choi, 1992), one from an HL-60 leukocyte library and one from a fibroblast library, and are designated PLC-β2 and PLC-β3, respectively. Additionally, two *Drosophila* genes, *norp*A (Bloomquist et al., 1988), which is defective in a visual mutant, and PLC-21 (Shortridge et al., 1991), which is encoded in the nervous system, are also members of the PLC-β subfamily. A second member of the PLC-γ family (PLC-γ2) was identified by hybridization methods from cDNA libraries from lymphocytes (Ohta et al., 1988), HL-60 cells (Kriz et al., 1990) and rat muscle (Emori et al., 1989). Two additional members of the PLC-δ subfamily have also been described. PLC-δ2 was purified from bovine brain (Meldrum et al., 1991), and a cDNA-encoding PLC-δ3 was identified from a fibroblast cDNA library (Rhee and Choi, 1992).

As described below, the β isoforms are activated by heterotrimeric G proteins, while the γ isoforms are stimulated by tyrosine phosphorylation. The δ isoforms are less well understood. It is possible that this subfamily is regulated in part by intracellular calcium, since these isoforms contain a fairly well-conserved EF-hand calcium-binding motif (Bairoch and Cox, 1990). The δ isoforms will not be considered further.

G Protein Activation of the PLC-β Subfamily

Indirect Evidence for G Protein Activation of PLC

In the mid-1980s several lines of evidence implicated G proteins in the regulation of PI-PLC. In a variety of systems including permeabilized cells and isolated membranes, the hydrolysis-resistant GTP analogue GTPγS and/or aluminum fluoride (both G protein activators) activated phosphoinositide hydrolysis and/or synergized with other activating factors. GTPγS activated phosphoinositide hydrolysis in permeabilized neutrophils (Cockcroft and Gomperts, 1985), liver plasma membranes (Uhing et al., 1985), cerebral cortical membranes (Claro et al., 1989), platelet membranes (Rock and Jackowski, 1987), and others. Aluminum fluoride had similar effects (e.g., see Blackmore and Exton, 1986; Deckmyn et al., 1986; Paris and Pouyssegun, 1987). In several systems including membranes from blowfly salivary glands (Litosch et al., 1985) and from neutrophil (Smith et al., 1986), agonist-activated phosphoinositide hydrolysis was reconstituted in the presence of GTP. In the latter system and in cytosolic fractions from platelet and brain (Deckmyn et al., 1986), the guanine nucleotide reduced the required calcium concentration by as much as 100-fold. In these systems, the G protein was thus proposed to activate by altering the sensitivity of PI-PLC toward calcium.

Another line of evidence implicating a heterotrimeric G protein was the finding that in some cell types, receptor-linked activation of phosphoinositide hydrolysis was blocked by pertussis toxin. Pertussis toxin catalyzes the transfer of the ADP-ribose moiety of NAD^+ to a specific residue on the G_i subfamily of G proteins, inhibiting their function. Toxin pretreatment of neutrophils (Verghese et al., 1985), HL-60 granulocytes (Kikuchi et al., 1986), mast cells (Nakamura and Ui, 1985), and some fibroblasts (Paris and Pouyssegur, 1986) resulted in loss of receptor-activated phosphoinositide hydrolysis, while other cell types (e.g., hepatocytes, astrocytoma cells, 3T3 cells) were not affected (Litosch and Fain, 1986). In pertussis toxin-inhibited HL-60 membranes, G_o and G_i (a mixture of isoforms isolated from brain) reconstituted f-Met-Leu-Phe receptor-regulated phosphoinositide hydrolysis (Kikuchi et al., 1986), but the low activity and the high activating concentrations of G proteins complicated the interpretation of these results.

Other approaches have demonstrated functional coupling and/or physical association between PLC-coupled receptors and heterotrimeric G proteins. In platelet membranes (Grandt et al., 1986; Houslay et al., 1986), agonists that activated PI hydrolysis also activated GTPase activity. In neutrophil membranes, f-Met-Leu-Phe stimulated the ADP-ribosylation of two 40 to 41-kDa proteins, identified as the α subunits of G_{i2} and G_{i3} (Gierschik and Jakobs, 1987; Gierschik et al., 1989). The f-Met-Leu-Phe receptor co-purified with a GTP-binding protein containing a distinct 40-kDa pertussis toxin substrate[1] (Polakis et al., 1988). Functional coupling has also been demonstrated with muscarinic receptor subtypes, based on the well-known observation that the affinity of G protein-linked receptors for their agonists is decreased when their associated G protein binds a GTP analogue. In this study (Bloomquist et al., 1988), GTPγS selectively decreased the binding of agonists for the muscarinic receptor subtypes that regulate PI hydrolysis.

Support for a role for G protein α subunits came initially from studies in which isolated $\beta\gamma$ subunits inhibited receptor- or G protein-activated PLC. According to this approach, the $\beta\gamma$ subunits are expected to combine with and inhibit the effector functions of free α subunits. *Xenopus* oocytes transfected with exogenous receptors activate phosphoinositide hydrolysis via both pertussis toxin-sensitive and -insensitive pathways, and the response can be assayed indirectly by monitoring a calcium-dependent Cl^- current (which is responsive to the IP_3-stimulated Ca^{2+} flux). Microinjection of G protein $\beta\gamma$ subunits from either brain or erythrocytes attenuated the agonist-stimulated response, and both pertussis toxin-sensitive and -insensitive pathways were affected, suggesting that both pathways are G protein regulated (Moriarty et al., 1988, 1989). In subsequent studies (Moriarty et al., 1990), microinjection of the α subunit of brain G_o resulted in a Cl current, but a variety of other G α subunits failed to produce the effect. Parallel inhibition of AIF_4^--stimulated PLC and adenylate cyclase activities by isolated $\beta\gamma$ subunits was also seen in turkey erythrocyte membranes (Boyer et al., 1989), implicating a G protein α subunit.

Activation of Phospholipase C-β1 by α Subunits of G_q/G_{11} Subfamily of G Proteins

A novel class of heterotrimeric G protein α subunits was recently identified (Strathmann et al., 1990), which currently contains four members, G_q, G_{11}, G_{14}, and G_{16}. $G\alpha_q$ and $G\alpha_{11}$ have an overall amino acid identity of 90 percent, being almost identical in their C-terminal 158 amino acid residues, a region expected to contain the effector enzyme binding region. They are widely distributed, with highest levels in lung and brain, and lack the predicted site for derivatization by pertussis toxin.

A combination of purification, reconstitution, and immunochemical approaches has shown that members of the G_q class of G protein α subunits activate PLC-β, in particular PLC-β1. A G protein α subunit preparation purified from liver membranes containing 42- and 43-kDa proteins activated (in the presence of aluminum fluoride) purified PLC-β1, but not PLC-γ1 or PLC-δ1 (Blank et al., 1991; Taylor et al., 1991). The activating G proteins were identified as member(s) of the $G\alpha_q$ class by immunochemical methods. Although activation required calcium, the G protein did not affect the affinity of the enzyme for this metal. In other studies (Pang and Sternweis, 1989; Smrcka et al., 1991), the $G\alpha_q$ subunit was isolated from solubilized beef brain membranes using a $\beta\gamma$ subunit affinity

---

[1]Nevertheless, it has not been possible to identify in this system a G protein α subunit that activates phospholipase C, a finding consistent with a novel G protein activation mechanism involving $\beta\gamma$ subunits, as described below.

column. The purified $G\alpha_q$ stimulated bovine brain PLC in the presence of aluminum fluoride but not GTPγS, consistent with the poor binding of this guanine nucleotide by the isolated $G\alpha_q$. Similarly, a PLC from turkey erythrocyte membranes was activated by a G protein preparation from the same source, which contained an α subunit that was immunochemically related to $G\alpha_{q/11}$ (Waldo et al., 1991). PLC-β2 was not stimulated by $G\alpha_q$ (Park et al., 1992a), indicating that G_q discriminates not only among the major PLC classes but also among the β subfamily members. The effector enzyme PLC-β1 stimulates the GTPase activity of isolated $G_{q/11}$ (Berstein et al., 1992a), providing direct evidence for the interaction between $G\alpha_{q/11}$ and its effector enzyme. The effector enzyme thus functions in a manner analogous to GTPase activating proteins (GAP), which stimulate the GTPase activity of *ras*-family small M_r G proteins, terminating their activated state.

Coupling of PI-PLC-activating receptors to $G\alpha_q$ and $G\alpha_{11}$ has been demonstrated using a combination of biochemical and immunochemical approaches. With use of antibodies to the C-terminal 12 amino acids common to $G\alpha_q$ and $G\alpha_{11}$ (part of the putative receptor interaction domain), agonist-activated phosphoinositide hydrolysis was blocked in membranes from 1321N1 astrocytoma cells, NG108–15 neuroblastoma × glioma cells, and rat liver (Gutowski et al., 1991). Similar studies (Shenker et al., 1991) documented the role of $G\alpha_{q/11}$ in thromboxane A_2-stimulated GTPase activity in platelet membranes. With use of a [^{32}P]-azidoanalido GTP photoaffinity analogue, $G\alpha_q$ and $G\alpha_{11}$ in liver were labeled in response to vasopressin (Gutowski et al., 1991). With use of all purified components, a complete receptor-activated, G_q-coupled phosphoinositide hydrolysis has been reconstituted (Berstein et al., 1992b); recombinant m1 muscarinic receptor was co-reconstituted into phospholipid vesicles with either hepatic $G_{q/11}$ or brain $\alpha_{q/11}$, together with PLC-β1. Agonist binding stimulated the rate of GTPγS binding to G_q up to 50-fold, and GTPγS binding paralleled phosphoinositide hydrolysis.

The specificity and coupling of the G_q class of G proteins both to their receptors and to PLC has been further clarified using recombinant DNA methodologies. Cos-7 cells were transfected with cDNA corresponding to the α subunits of G_q, G_{11}, G_{14}, and G_{16} (Lee et al., 1992). Membranes were then isolated, combined with purified PLC-β1 or PLC-β2, and GTPγS-stimulated phosphoinositide hydrolysis was monitored. All four α subunits stimulated PLC-β1, with α_q and α_{11} being the most efficient. $G\alpha_{16}$ provided the best activation of PLC-β2, with the other subunits showing little stimulation. Thus, members of the G_q family discriminate among PLC-β isoenzyme subforms. Transfection was also used to demonstrate specificity in the interaction of G_q α subunits with the three α1-adrenergic receptor subtypes (α1A, α1B, and α1C) (Wu et al., 1992a). $G\alpha_q$ and $G\alpha_{11}$ coupled all three receptor subtypes to PLC-β1, whereas $G\alpha_{14}$ and $G\alpha_{16}$ showed differences in specificity for the receptor subtypes.

Activation of Phospholipase C-β3 by G Protein βγ Subunits

Rather than serving merely as a docking site for α subunits, G protein βγ subunits now appear to have distinct regulatory roles. Receptor-stimulated GTP binding to α subunits causes dissociation of the α subunit from $\beta\gamma$ subunits, freeing them for their unique regulatory roles. Precedent for regulation by $\beta\gamma$ subunits has come from studies of the activation of isoforms of adenylate cyclase (Tang and Gilman, 1991) and of retinal PLA_2 (see later section in this chapter). Soluble fractions from HL-60 granulocytes contain a PI-PLC that is markedly stimulated by GTP analogues. G protein $\beta\gamma$ subunits obtained from transducin or from bovine brain markedly stimulated phosphoinositide hydrolysis, and this was inhibited by the transducin α subunit, presumably by combining with and inactivating the free $\beta\gamma$ subunits (Camps et al., 1992). A cytosolic PLC from liver was activated by purified G protein $\beta\gamma$ subunits that were shown by Western blots to be free of detectable α subunits (Blank et al., 1992). In a phospholipid vesicle-reconstituted system containing $G\alpha_{11}$ and a PLC from turkey erythrocytes, $\beta\gamma$ subunits had a biphasic effect on aluminum fluoride-activated phosphoinositide hydrolysis (Boyer et al., 1992). At low concentrations $\beta\gamma$ subunits inhibited, while at high concentrations they stimulated activity. Recently, marked activation of PLC-$\beta3$, and to a lesser extent -$\beta2$ by G protein $\beta\gamma$ subunits has been observed (Park et al., 1993). Thus, while coupling of G protein $\alpha_{q/11}$ subunits to PLC-$\beta1$ and -$\beta2$ accounts for pertussis toxin-*insensitive* activation, $\beta\gamma$ subunit regulation of phospholipase-$\beta3$ and perhaps -$\beta2$ can explain pertussis toxin-*sensitive* PI hydrolysis.

Tyrosine Kinases and Activation of Phospholipase C-γ

Receptor Tyrosine Kinases and Activation of Phospholipase C-γ

The receptors for EGF and PDGF, known protein tyrosine kinases, respond to receptor occupancy by activating phosphoinositide hydrolysis, whereas other receptor tyrosine kinases (e.g., the receptors for insulin, CSF-1, and fibroblast growth factor) fail to do so. Early studies, reviewed in Rhee and Choi (1992), showed that growth factor activation of PI hydrolysis differed in many respects from activation by G protein-coupled receptors, and was therefore unlikely to occur via the same sort of G protein mechanism. Mutant EGF or PDGF receptors that lacked tyrosine kinase activity were used to demonstrate that this activity was necessary to mediate PIP_2 hydrolysis (Escobedo et al., 1988; Moolenaar et al., 1988; Margolis et al., 1990).

Phosphorylation of PLC-γ by these receptor kinases occurs in response to receptor occupancy. Stimulation of A431 cells with EGF resulted in activation of phosphoinositide hydrolysis and increased retention of PLC activity on an immunoaffinity matrix comprising antiphosphotyrosine antibodies (Wahl et al., 1988). This suggested that PLC or a closely associated protein was tyrosine phosphorylated. In several cell types (Kumjian et al., 1989; Margolis et al., 1989; Meisenhelder et al., 1989; Wahl et al., 1989; Kim et al., 1991), EGF, PDGF, or nerve growth factor increased the phosphorylation of PLC-$\gamma1$ but not PLC-$\beta1$ or PLC-$\delta1$. Phosphorylation occurred on both serine and tyrosine residues, and was independent of receptor internalization. In PLC-$\gamma1$ obtained by immunoprecipitation, tyrosine phosphorylation was associated with altered kinetic parameters of the enzyme (Wahl et al., 1992). The K_m for PIP_2 decreased by 7.5-fold and changed from a sigmoidal to a hyperbolic saturation curve. Three major phosphorylation sites were seen with PDGF, and two of these were also seen with EGF (Meisenhelder et al., 1989); phosphorylation at these sites correlated with activation (Margolis et al., 1989). Two of the sites (tyrosines 771 and 1254) were identified in PLC-γ tryptic peptides (Wahl et al., 1990), and additional sites (472 and 783) were subsequently characterized (J. W. Kim et al., 1990). The first three sites are located in the SH2 and SH3 *src* homology regions of PLC-γ. To investigate the functional relevance of individual tyrosines, PLC-γ genes encoding phenylalanine rather than tyrosine at sites 771, 783, and 1254 were constructed and expressed in NIH 3T3 cells (H. K. Kim et al., 1991). The 783 mutation completely blocked PLC activation by PDGF, while that at 1254 was inhibitory. Thus, tyrosine phosphorylation is essential for intracellular activation of PLC-γ.

PLCs associate physically with EGF and PDGF receptors. PLC-γ1 co-immunoprecipitated with the EGF or PDGF receptors, using antibodies to either the receptor kinases or the phospholipase (Kumjian et al., 1989; Margolis et al., 1989; Meisenhelder et al., 1989; Wahl et al., 1989; Kitagawa et al., 1990). Tyrosine phosphorylation is important for physical association, since mutant PDGF and EGF receptors that lack tyrosine kinase activity also failed to co-immunoprecipitate (Margolis et al., 1990; Morrison et al., 1990). Reversal of the phosphorylated state of both proteins by phosphatases was associated with loss of physical association (Kumjian et al., 1991).

In addition to physical association with growth factor receptors, PLC-γ1 translocates from the cytosol to membrane fractions on EGF or PDGF stimulation, coincident with activation of PI hydrolysis (U. H. Kim et al., 1990; Todderud et al., 1990). The cytosolic PLC-γ1 contained more phosphotyrosine than the membrane-associated form. The translocated form may interact directly with growth factor receptors in an enzyme-substrate complex, or there may be additional membrane interaction sites such as cytoskeletal elements, as discussed below.

Nonreceptor Tyrosine Kinases and Activation of PLC-γ

Several cells of the immune system are regulated by cell surface receptors that do not themselves contain a kinase domain but that activate tyrosine phosphorylation and phosphoinositide hydrolysis. These receptors activate nonreceptor tyrosine kinases such as those related to the *src* family of enzymes. In several of these systems, PLC-γ1 becomes phosphorylated on tyrosine residues upon cell activation. B lymphocytes bind antigen through a membrane-bound IgM (mIgM). Ligation of mIgM with antigen activates a tyrosine kinase (perhaps the B-cell specific *blk* gene product, a member of the *src* family) and phosphoinositide hydrolysis. This is presumably mediated through PLC-γ1, which becomes rapidly phosphorylated on tyrosines (Carter et al., 1991). Tyrosine kinase inhibitors blocked both tyrosine phosphorylation of PLC-γ1 and phosphoinositide hydrolysis, implying a causal relationship between phosphorylation and lipase activation. Similarly, the T-cell receptor, CD3, is a multimeric complex that lacks tyrosine kinase activity, but may interact with and activate a nonreceptor tyrosine kinase such as *fyn* or *lck*, both members of the *src* family. Ligation of the T-cell receptor with antibodies resulted in activation of PI hydrolysis and tyrosine phosphorylation (June et al., 1990; Secrist et al., 1991; Weiss et al., 1991). PLC-γ1 was phosphorylated on the same tyrosine and serine residue as were seen in EGF- or PDGF-treated A431 cells. Thus, a nonreceptor tyrosine kinase is coupled to the T-cell CD3 complex, and acts on PLC-γ1. Phosphorylation of PLC-γ1 by nonreceptor tyrosine kinases has also been demonstrated with IgE receptors of rat basophilic leukemia cells (Park et al., 1991) and Fc receptors on a monocytic cell line (Liao et al., 1992).

Phospholipase C-γ and the Cytoskeleton

Enzymes of the phosphoinositide cycle including PLC, PI-3-kinase, and diacyl-glycerol kinase are associated with the cytoskeleton in platelets and in A431 cells, and enzyme association increased markedly with agonist stimulation (Grondin et al., 1991). It was proposed (Rhee and Choi, 1992) that tyrosine phosphorylation induces a conformational change that exposes cytoskeletal interaction domains such as the SH3 region on PLC-γ. Immunofluorescence techniques showed co-localization of actin and PLC-γ1 in rat embryo fibroblasts (McBride et al., 1991). The enzyme was evenly distributed along the length of actin microfilaments, and co-localized with PKC-α at focal contacts, where actin filaments are anchored to the membrane. PDGF caused tyrosine phosphorylation of both soluble and cyto-skeletal PLC-γ, and activation was associated with a detectable increase in cyto-skeleton-associated PLC-γ.

Specific interactions with cytoskeleton or cytoskeletal-associated proteins may have important functional consequences in modulating the activity of PLC-γ. Although EGF and PDGF stimulate phosphoinositide hydrolysis in responsive cells, reconstitution of the process with purified growth factor receptor plus PLC-γ has been complicated by the fact that the isolated PLC-γ is constitutively active, and phosphorylation on tyrosine residues fails to activate further. In a recent study (Goldschmidt-Clermont et al., 1991), the actin-binding protein profilin modulated PLC activity. Profilin binds to the substrate PIP_2 and inhibits its hydrolysis by nonphosphorylated PLC-γ1. Phosphorylation of the latter by the purified EGF receptor tyrosine kinase overcame the inhibition by profilin and resulted in effec-tive activation of PLC-γ1. Thus, the effect of phosphorylation may be to relieve inhibition by this or other proteins.

Signal Cross-Talk and Feedback Regulation of PLC.   There are numerous exam-ples of signaling cross-talk wherein a receptor or second-messenger system inhib-its the activation of phosphoinositide hydrolysis by an activating ligand, as reviewed by (Rhee and colleagues (1990). For example, thrombin-induced phos-phoinositide hydrolysis in platelets is inhibited by prostacyclin, a AMP-elevating agonist (Watson et al., 1984). In Swiss 3T3 cells, phorbol esters inhibited bom-besin-stimulated PI hydrolysis, and down-regulation of protein kinase C resulted in the loss of the ability to inhibit and produced elevated basal levels of inositol phosphates (Brown et al., 1987). Because phosphoinositide hydrolysis results in the activation of PKC, these studies provide evidence for a feedback inhibitory loop involving PKC. Thus, both signal cross-talk and feedback inhibition affect the activity of PLC.

The molecular mechanisms for down-regulation have been investigated. For PLC-β1, treatment of several cell types with phorbol ester stimulated the phosphorylation of serine residues on PLC-β1 (Ryu et al., 1990). *In vitro* phosphorylation of PLC-β1 by purified PKC resulted in stoichiometric incorporation of phosphate at serine 887, but was without any effect on phospholipase enzymatic activity. The authors proposed that rather than directly modulating enzymatic activity, phosphorylation inhibited the interaction with a G protein.

For PLC-γ1, treatment of cells with cAMP-elevating agents increased serine phosphorylation of PLC-γ1, but not the other isoforms (U.H. Kim et al., 1989; Olashaw et al., 1990). Purified cAMP-dependent protein kinase also selectively phosphorylated this isoenzyme. Nevertheless, phosphorylation failed to affect *in vitro* activity, and it was proposed that phosphorylation affected the interaction with other modulatory proteins. Consistent with this explanation, prior incubation of T cells with either PMA or forskoline (which elevates cAMP) inhibited both tyrosine phosphorylation of PLC-γ1 and phosphoinositide hydrolysis in response to stimulation of the T-cell receptor (Park et al., 1992b). Both treatments stimulated phosphorylation at serine 1248, the same site that becomes phosphorylated *in vitro* with use of purified protein kinase C or cAMP-dependent protein kinase. Activation of the T-cell receptor also caused phosphorylation at the same residue, implying that this phosphorylation is part of a normal feedback mechanism operating through PKC. The authors proposed that the serine phosphorylation decreases tyrosine phosphorylation either by inhibiting the functional interaction with the tyrosine kinase or by augmenting the interaction with a tyrosine phosphatase.

PHOSPHOLIPASE D, PHOSPHATIDIC ACID, AND SECONDARY GENERATION OF DIACYLGLYCEROL

Relevance of Phospholipase D Products to Activation of Protein Kinase C and Other Enzymatic Systems

Diacylglycerol

The action of PLD (e.g., on phosphatidylcholine) gives rise directly to phosphatidic acid (PA) plus the free headgroup (e.g., choline), as in Figure 5.2. The PA may have direct regulatory roles (see below), or can be metabolized to other bioactive molecules. The action of PA phosphohydrolase on PA generates a diradylglycerol, either diacylglycerol or in some cases 1-O-alkyl,2-acylglycerol, both of which may regulate PKC. In contrast to diacylglycerol, which activates, the 1-alkyl-linked diglyceride is variously reported as either an activator (Ford et al., 1989b) or an inhibitor (Daniel et al., 1988) of PKC. Whereas the onset of diacylglycerol generation from phosphatidylinositol-specific PLC is rapid and the effect is short-lived, diradylglycerol generation from PLD/PA phosphohydrolase frequently occurs more slowly and in a more sustained fashion (e.g., see Tyagi et al., 1989). In one study (P. Lin et al., 1992), agonist-activated diradylglycerol generation from PLD did not *initiate* PKC activation (perhaps because of a

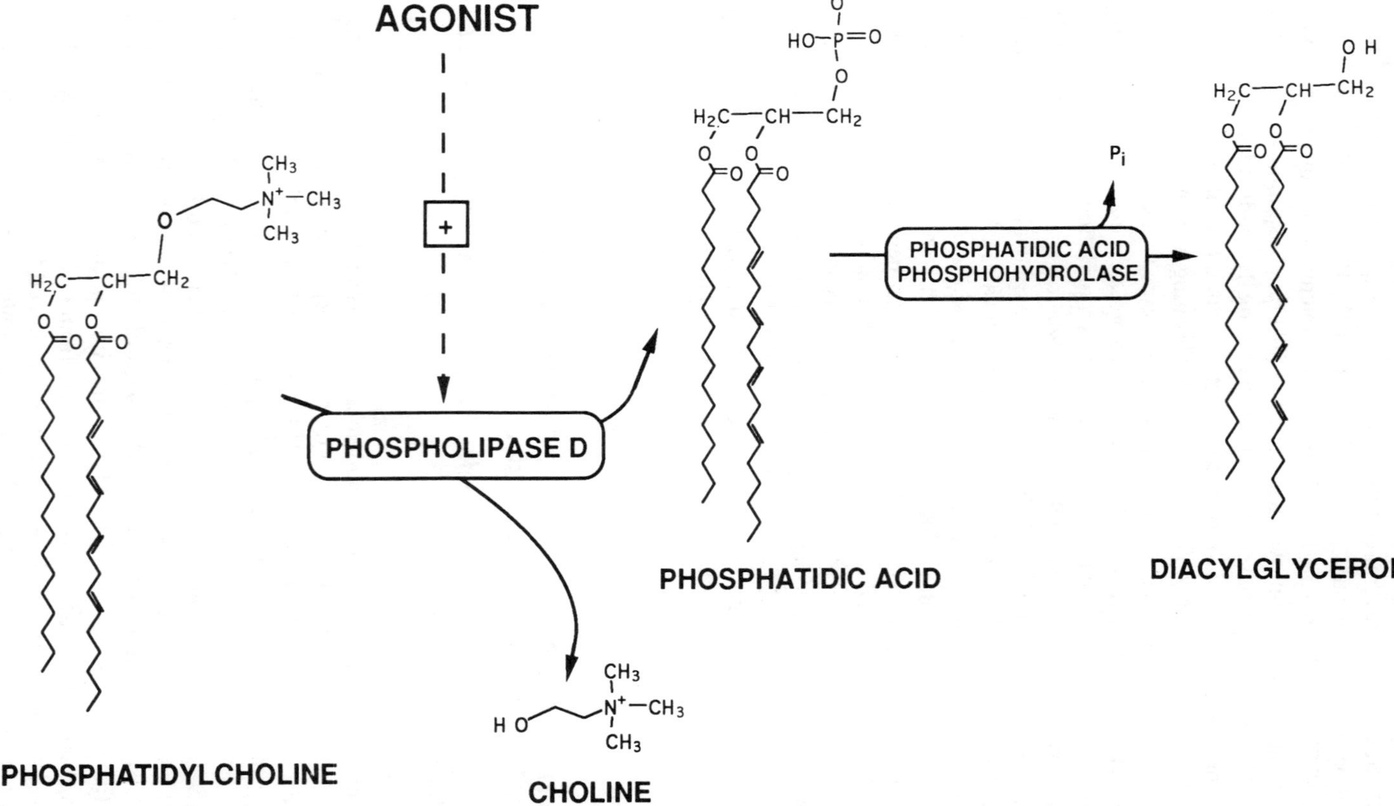

Figure 5.2 Phospholipase D hydrolysis of phosphatidylcholine. Agonist activates phospholipase D, with the hydrolysis of phosphatidylcholine to generate choline and phosphatidic acid. Also shown is the hydrolysis of phosphatidic acid to diacylglycerol via the action of phosphatidic acid phosphohydrolase.

requirement for calcium for translocation), but appeared to participate in its sustained activation.

Phosphatidic Acid and Lysophosphatidic Acid

PA is a direct product of the action of PLD (Figure 5.2) and can be further metabolized by a PA-specific PLA_2 (Billah et al., 1981), to give lysophosphatidic acid (LPA) and a free fatty acid. PA and LPA have been implicated as direct regulators of a number of enzymes. In one study, PA had an inhibitory effect on PKC selectively in the presence of diacylglycerol (Epand and Stafford, 1990). PA also activates several other important regulatory enzymes including PI 4-kinase (Moritz et al., 1992), PI-PLC (Jackowski and Rock, 1989), adenylate cyclase (Dunlop and Larkins, 1989), the neutrophil NADPH-oxidase (Peveri and Curnutte, 1990; Agwu et al., 1991b), and a novel protein kinase (Bocckino et al., 1991).

PA and/or LPA added to cells or generated in the outer leaflet of the plasma membrane by added PLD has diverse cellular effects, including stimulation of insulin release (Metz and Dunlop, 1990), activation of superoxide generation in neutrophils (Ohtsuka et al., 1989), and activation of mitogenesis (Moolenaar et al., 1986; van Corven et al., 1989; Moolenaar, 1991; Kondo et al., 1992a). Exogenous PA increases calcium permeability in a number of cells (Putney et al., 1980; Salmon and Honeyman, 1980; Ohsako and Deguchi, 1981; Altin and Bygrave, 1987), and has a calcium ionophoretic effect in model membranes (Serhan et al., 1982; Tyson et al., 1992). Effects could also be related to LPA binding to a putative receptor, recently labeled using a photoreactive LPA analog (van der Bend et al., 1992).

Alternatively, it may be that PA or LPA in the outer leaflet is further metabolized to exert its biological effects. With use of a fluorescent PA, movement to internal membranes required hydrolysis of the phosphate ester (Pagano and Longmuir, 1985). Recently we found that a nonhydrolyzable phosphonate analogue of PA failed to activate the respiratory burst in the neutrophil, and demonstrated that the action of a novel *ecto*-PA phosphohydrolase on the outer leaflet of the plasma membrane was necessary for the flip-flop and subsequent intracellular function of the diradylglycerol moiety (D. Perry and D. Lambeth, unpublished). Exogenous PA was in effect functioning as a "timed-release diacylglycerol," and its actions appeared to be PKC mediated, on the basis of inhibitor studies.

Importance of the Phospholipase D Pathway in Cellular Signaling

Studies Correlating Phospholipase D Activation and Cellular Responses

In contrast to situations in which PA is added or generated exogenously, activation of PLD by receptor-coupled mechanisms presumably results in PA on the inner leaflet of the plasma membrane where it may exert regulatory effects without further metabolism. Several studies have therefore aimed to correlate PA with various cellular responses. In neutrophils primed with tumor necrosis factor-α, superoxide generation correlated with PA but not with diacylglycerol levels (Bauldry et al., 1991). In another study using cytochalasin-primed cells (Perry et al., 1992), activity correlated with diacylglycerol but not with PA. Thus, PA effects in these cells may be complex, and could depend on the experimental or physiological setting.

Regarding mitogenic effects, receptor-linked activation of PLD often correlates with a mitogenic response, but it is difficult to differentiate whether other systems are also activated. In recent studies, sphingosine and, to a greater extent sphingosine-1-phosphate, increased cell proliferation and DNA synthesis (Zhang et al., 1991), and these factors activated PLD and PA generation (Desai et al., 1992). Nevertheless, in a study in which cells were transfected to express either muscarinic receptors or EGF receptor, there was no correlation between PLD activation and the mitogenic response (McKenzie et al., 1992). The authors concluded that activation of this enzyme was neither necessary nor sufficient for cell growth and DNA synthesis. Thus, it remains unclear whether PA has direct regulatory roles in these systems, and additional approaches are needed.

Pharmacological Evaluation of the Role of Receptor-Activated Phospholipase D

Two important pharmacological approaches have been developed to investigate both the occurrence and the biological role of the PLD/PA phosphohydrolase pathway. In one, PLD-mediated generation of PA (and subsequent metabolites) is inhibited with use of primary alcohols such as ethanol or butanol. The basis of this inhibition is the transphosphatidylation reaction (Yang et al., 1967) in which the PA moiety of the phospholipid is transferred to the primary alcohol rather than to water, yielding the corresponding phosphatidylalcohol (see Figure 5.3). This reaction provides a unique marker of PLD activity, and the phosphatidylalcohol is relatively inert metabolically, in contrast to PA, which is readily degraded. At 1.5-percent ethanol, inhibition of neutrophil PA generation is essentially complete, corresponding to the increase in phosphatidylethanol. With this approach, parallel inhibition of PA generation and f-Met-Leu-Phe-activated superoxide generation in neutrophils (Bosner et al., 1989) and of IgE-activated exocytosis in mast cells (Gruchalla et al., 1990; P. Lin et al., 1991) was seen. Nevertheless, this approach fails to distinguish between direct regulatory effects of PA and effects due to its metabolism to diradylglycerol.

In another approach, inhibition of PA phosphohydrolase permits accumulation of PA, but prevents its further metabolism to diradylglycerol. Propranolol is frequently used for this purpose (Pappu and Hauser, 1983). In addition, sphingosine inhibits this enzyme (Mullmann et al., 1991), and we have recently characterized a series of PA phosphohydrolase inhibitors that vary widely in their potencies (Perry et al., 1992). In the unprimed neutrophil (which normally has a very low PLD activity), inhibition of PA phosphohydrolase resulted in an enhanced f-Met-Leu-Phe-activated respiratory burst (Bellavite et al., 1988), consistent with an activating or priming role of PA. Using a series of inhibitors, we showed that diradylglycerol derived from the PLD/PA phosphohydrolase pathway is critical for both f-Met-Leu-Phe and phorbol ester activation of the respiratory burst in cytochalasin-primed neutrophils (Perry et al., 1992). Propranolol also inhibits IgE-induced exocytosis in mast cells (P. Lin et al., 1991), supporting a functional

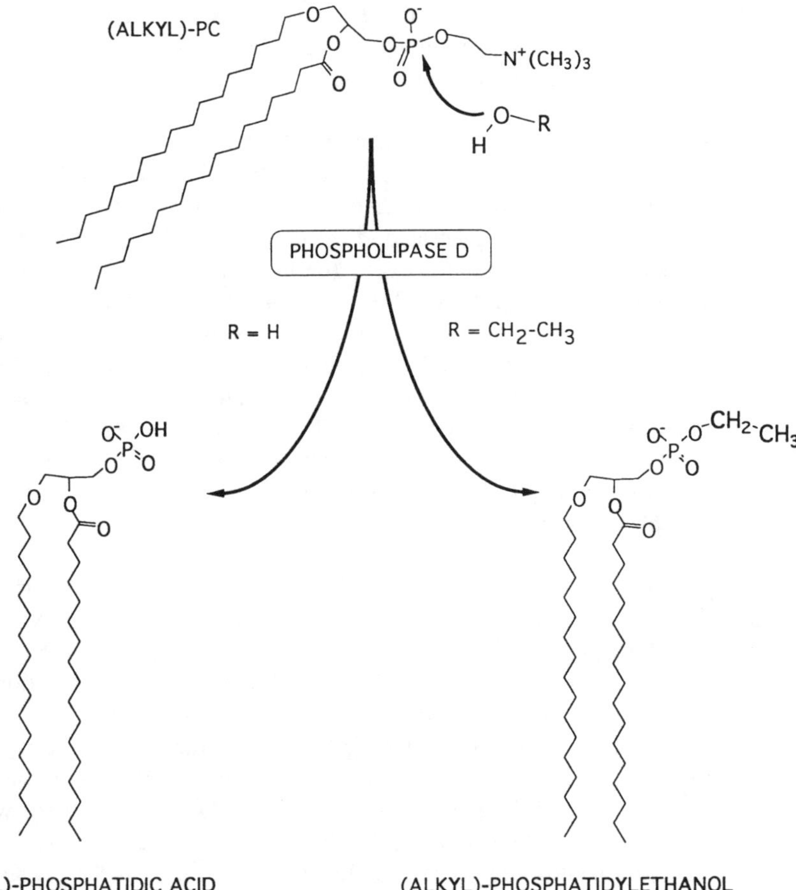

Figure 5.3   Phospholipase D catalysis of phosphatidylcholine hydrolysis versus trans-phosphatidylation. See text for details.

role for diradylglycerol derived from the PLD pathway. Thus, both correlational and pharmacological approaches have indicated that the receptor-activated PLD pathway is important in cellular regulation, and both PA and diacylglycerol have been implicated.

The Discovery of Receptor-Activated Phospholipase D

Historical Perspective

Phospholipase D was discovered in plants almost 50 years ago (Hanahan and Chaikoff, 1947). The early history of this enzyme, comprising mostly studies in plants and bacteria, is reviewed in Heller (1978) and will not be considered here. The focus of the signal transduction field on PI hydrolysis during the 1970s and 1980s provided a paradigm for phospholipid-derived second messengers. Nevertheless, although several mammalian PLD activities had been described in the 1970s and early 1980s, the enzyme was not generally appreciated to be relevant to mammalian signal transduction until the mid-1980s.

Agonist-Activated Phosphatidylcholine Hydrolysis

Several early studies led to the conclusion that in a variety of cells, PI hydrolysis by PLC could not account for the quantity, kinetics, or molecular composition of the diradylglycerol that was generated in response to signals, and it became necessary to postulate an additional pathway (either PLC or PLD). Early studies in agonist-activated mouse fibroblasts (Mufson et al., 1981) had shown choline and phosphocholine release in the absence of inositol phosphate generation. In liver cells activated by vasopressin and other agonists, two peaks of diradylglycerol generation were seen, and fatty acid analysis revealed that the diglyceride was derived in part from sources other than PI (Bocckino et al., 1985). Phorbol esters and growth factors were subsequently shown to activate diacylglycerol and PA generation corresponding to PC breakdown (Besterman et al., 1986), as was also seen in liver plasma membranes (Irving and Exton, 1987) and MDCK cells (Daniel et al., 1986).

The quantitative importance of alternative (non-PI hydrolysis) pathways has been demonstrated by quantifying the molecular mass of diradylglycerols. In neutrophils, PMA priming (which enhances the subsequent activation in response to receptor-linked agonists such as f-Met-Leu-Phe) caused an inhibition of PI hydrolysis, but markedly enhanced diradylglycerol mass (Tyagi et al., 1988). In some cells, the chemical linkage at the 1 position of diradylglycerol can serve as a marker of the parent phospholipid. In neutrophil, nearly 50 percent of the PC is 1-O-alkyl linked and 70 percent of the PE is 1-O-alkenyl linked, whereas phosphatidylinositol contains only acyl linkages (Mueller et al., 1982; Tencé et al., 1985). When neutrophils were stimulated with f-Met-Leu-Phe (Tyagi et al., 1989), PMA (Rider et al., 1988), or fluoride (Olson et al., 1990), nearly half of the diradylglycerol was found to be 1-O-alkyl linked, implying that PC hydrolysis was quantitatively the major source of diradylglycerol in these cells. No 1-alkenyl linkages were seen, suggesting that PE hydrolysis did not occur to a significant degree. Molecular species analysis has also identified PC as a quantitatively important source of diradylglycerol in several cell types, including platelet (Takamura et al., 1987), hepatocyte (Augert et al., 1989), and mast cells (Kennerly, 1990). In this methodology, agonist-generated diacylglycerol is usually characterized as to its fatty acid/alcohol composition, and this is compared with that in the major cellular phospholipids. The molecular composition thus serves as a ''fingerprint'' linking the diradylglycerol to its parent phospholipid.

The quantitative importance of PC hydrolysis by PLD in diacylglycerol generation has been further evaluated using pharmacological approaches. Studies in which PA phosphohydrolase was inhibited with propranolol in neutrophils indicated that the PLD pathway was the major (greater than 90%) source of diradylglycerol (Billah et al., 1989a). In astrocytoma cells (Martinson et al., 1989), PMA-activated diacylglycerol generation occurred almost exclusively from PC, while carbachol activation resulted in about 20 percent generation from PI, with the remainder from PC. In platelets activated with thrombin, 13 percent of the PA was generated from PLD, while the remainder was from a PC-specific PLC (Huang et al., 1991). In a comparative study using three different cell types (R. Huang et al., 1990), diradylglycerol generation occurred through both PLC and PLD pathways, but the quantity depended on the cell type. In MDCK cells the predominant pathway was PLD, whereas in arterial smooth muscle cells the major pathway was PLC. In pulmonary arterial endothelial cells, both pathways participated. Thus, the quantitative significance of PC hydrolysis and the participation of PLD in diradylglycerol generation depends on the cell type and the agonist used.

Discovery of Receptor-Coupled PLD

Several early studies pointed to the possibility of receptor-activated PLD. In acetylcholine-stimulated pancreas, release of inositol but not inositol phosphate was seen (Hokin-Neaverson et al., 1975). In platelets (Lapetina and Cuatrecasas, 1979) and neutrophils (Cockcroft, 1984), agonist-induced PA production had been observed, and in the latter system the PA was shown to contain a significantly lower [^{32}P]-specific activity than cellular [^{32}P]-ATP in [^{32}P]-P$_i$ prelabeled cells, consistent with direct generation of PA via PLD (Figure 5.2) rather than indirect generation via the sequential action of PLC and DG kinase (Figure 5.1). Despite these prophetic studies, it was not until a study published several years later (Bocckino et al., 1987) that a mammalian receptor-activated PLD became generally accepted. In this study, calcium-linked hormones stimulated PA generation in isolated rat hepatocytes without affecting DG levels. Plasma membranes isolated from these cells generated PA in response to GTPγS, without incorporating radioactivity from [γ-^{32}P]-ATP, and the fatty acid composition of the product suggested an origin from PC. Proof for agonist-activated PLD activity in an intact cell came with studies in neutrophil using double-labeled PC as a substrate (Billah et al., 1989a). 1-O-alkyl,2-lysophosphatidylcholine (lysoPAF) was [^{32}P] labeled in the phosphate and [^{3}H] labeled in the alkyl chain. During a preincubation, lysoPAF is taken up by the cell and acylated, thus creating a double-labeled PC pool. PA that originates directly from a PLD pathway (Figure 5.2) will contain the same isotopic ratio as the parent PC, whereas that derived from PLC followed by the action of DG kinase (utilizing unlabeled ATP in the cells, see Figure 5.1) will have a low ratio of [^{32}P] to [^{3}H]. The PA formed in response to f-Met-Leu-Phe was found to have the same isotopic ratio as the parent phospholipid, thus proving the direct pathway.

Since 1987 there has been a veritable explosion of literature documenting the occurrence of receptor-coupled PLD in a variety of cell types. Some of these studies are summarized in Table 5.1. As is shown, PLD activity has been demonstrated in a large number of cell types, ranging from neural tissue to blood cells to fibroblasts. Activation occurs in response to a large number of receptor-coupled agonists, including various hormones, inflammatory mediators, and neurotransmitters.

Substrate Specificity of Receptor-activated Phospholipase(s) D

Most evidence indicates that PLD is specific for PC. Evidence comes from choline headgroup release, from observation of a predominance of 1-O-alkyl-linked dir-

Table 5.1   Tissue Distribution and Activator Specificity of Agonist-Activated Phospholipase D

Tissue/cell	Activator	Reference
Hepatocytes	Vasopressin, angiotensin II, EGF, ATP, ADP, A23187	Bocckino et al., 1987
Neutrophil/HL-60	fMLP	Billah et al., 1989a
	Complement C5a	Mullmann et al., 1990b
	ATP (P_2-purinergic agonists)	Xie and Dubyak, 1991
	Fluoride	Olson et al., 1990
	PMA	Mullmann et al., 1990a; Chabot et al., 1992
	A23187, Diacylglycerol[a]	Billah et al., 1989b
	Phosphatidic acid[a]	D. Perry, unpublished
Mast cells/lines	IgE[a]	Gruchalla et al., 1990; P. Lin et al., 1992
	PMA[b]	Yamada et al., 1991
U937 monocytes	PAF, PMA	Balsinde and Mollinedo, 1991
Eosinophils	C5a, A23187, PMA	Minnicozzi et al., 1990
Erythroleukemia cells	Thrombin, PAF, A23187, PMA	Halenda and Rehm, 1990
	PGE	Wu et al., 1992b
Platelet	Thrombin	Rubin, 1988
Lymphocytes	PMA	Cao et al., 1990
Natural killer lymphocytes	Anti Kp43 Ab	Balboa et al., 1992
HeLa cells	Ionomycin	Plein et al., 1992
	EGF[c]	Kaszkin et al., 1992
Endothelial cells	P_{2Y} purinergics	Martin and Michaelis,
	PMA[a]	Martin et al., 1990
A431 cells	EGF[c]	Kaszkin et al., 1992
NIH 3T3 cells	Bradykinin	Fu et al., 1992
Brain		
Canine	Cholinergic agonists	Qian and Drewes, 1989
Rat	α1-adrenergic,[b] PMA[a]	Llahi and Fain, 1992
Hippocampal cells	Glutamate receptor agonists	Boss and Conn, 1992
Sciatic nerve	PMA[a]	Eggen and Eichberg, 1992
PC12 cells	Bradykinin,[b] PMA[a]	Horwitz, 1991; Horwitz and Ricanatik, 1992
	ATP, UTP (purinergic agonists)	Murrin and Boarder, 1992

Tissue/cell	Activator	Reference
NG108-15 (neuroblastoma × glioma hybrid	PMA,[a,d] diC8[a,d] Sphingosine	Liscovitch, 1989 Lavie and Liscovitch, 1990
Neuroblastoma (LA-N-2) cells	Carbachol(muscarin),[b] PMA[a]	Sandmann and Wurtman, 1991
1321N1 astrocytoma	Carbacho, PMA	Martinson et al., 1989
Pituitary gonadotrope (αT3-1) cells	GnRH	Netiv et al., 1991
Vascular smooth muscle	Angiotensin II PMA, PDGF	Kondo et al., 1992b Konishi et al., 1991
Rat embryo fibroblasts	Vasopressin, PMA, diC8	Cabot et al., 1988; Huang and Cabot, 1992
Swiss 3T3 fibroblasts	Bombesin, vasopressin, PGF2α, PMA PDGF	Cook and Wakelam, Plevin et al., 1991
Fibroblasts (human foreskin	Bradykinin,[a] PMA[a]	van Blitterswijk et al., 1991
Corneal epithelium	Ionomycin, PMA	Akhtar and Choi, 1992
BAC1.2F5 macrophages	ATP	El-Moatassim and Dubyak, 1992
Adrenal glomerulosa cells	Angiotensin II	Bollag et al., 1990
Ovarian granulosa cells	GnRH,[b] PMA[a]	Liscovitch and Amsterdam, 1989
Leydig cells	PMA,[a,d] vasopressin[d]	Vinggaard and Hansen, 1991
Kidney cells	PMA	C. Huang and Cabot, 1990
Sperm (sea urchin)	Fucose sulfate glycoconjugate (egg)	Domino et al., 1989

[a]Inhibited by staurosporine, H-7, or other protein kincase C inhibitors.
[b]Not inhibited by protein kinase C inhibitors.
[c]Not inhibited by chronic phorbol ester down-regulation of protein kinase C.
[d]Inhibited by chronic phorbol ester down-regulation of protein kinase C.

adylglycerol or PA in some inflammatory cells, and from molecular species analysis of DG and PA. For example, in PC12 cells stimulated with PMA or bradykinin (Holbrook et al., 1992), the molecular species of the phosphatidylethanol derived from transphosphatidylation was identical to that of the cellular PC, but differed from that of other phospholipids. Using an endogenous labeling method to tag cellular phospholipids (Mohn et al., 1992), rat brain PLD had an absolute specificity for PC.

Despite this evidence, at least in some cases mammalian PLD hydrolyzes other phospholipids, although the quantitative significance is not yet clear. Early studies in pancreas (Hokin-Neaverson et al., 1975) suggested PI hydrolysis by PLD, the basis of on release of free inositol. Subsequently, a soluble, neutral-pH optimum PLD activity was described in neutrophil (Balsinde et al., 1989) that showed a preference for PI over PC. In MDCK cells, PMA elicited PC hydrolysis, while bradykinin stimulated PI hydrolysis, both accompanied by ethanol-dependent transphosphatidylation (C. Huang et al., 1992). The activities were partially characterized, and two PLD activities were seen. A membrane-bound, PC-selective enzyme was seen that was calcium independent. A cytosolic form was PI selective and required calcium. Evidence has also accrued for a PLD acting on PE. A membrane-associated, PE-specific PLD activity was described in rat heart (Schmid et al., 1983). In ethanolamine-prelabeled cells, agonist-activated ethanolamine release was seen (Kiss and Anderson 1989, 1990). Thus, while most of the evidence points to PC as the major substrate for PLD in many cell types, there also may be isoforms that act on other phospholipids and whose quantitative importance may be cell type and agonist specific.

The Mechanism of Activation of Receptor-Coupled Phospholipase D

Overview of Activation Mechanisms

The activation mechanisms for PLD are probably complex, involving several different and perhaps interacting signaling systems. For example, in neutrophil several agonists (receptor-coupled, phorbol ester, A23187) were used in various combinations with inhibitors (staurosporine, wortmannin) to distinguish at least three separable activation pathways or mechanisms (Reinhold et al., 1990). Currently there is evidence for the participation of G proteins, calcium fluxes, and phosphorylation by protein kinase C and/or tyrosine kinases. Also at issue is whether PLD is activated independently of or secondary to PLC. The activation mechanisms have not yet been described at a molecular level, and the proof for these mechanisms must await experiments utilizing defined components. The following sections survey available information for putative activation pathways, but it should be kept in mind that some or all of these mechanisms may be linked sequentially or synergistically in an as-yet unknown manner.

G Protein Activation of Phospholipase D

Exton and colleagues focused attention on G protein regulation of PLD by demonstrating GTPγS-stimulated PA generation in isolated liver plasma membranes (Bocckino et al., 1987), and a similar link to G proteins has subsequently been found in many systems. In intact neutrophils, PLD is activated by fluoride (Olson et al., 1990; English et al., 1991) and receptor activation is blocked by pertussis toxin (Kanaho et al., 1991). In permeabilized NG108-15 cells, GTPγS activated PLD (Liscovitch and Eli, 1991). In synaptosomes from canine brain, GTP analogues, fluoride, and cholera toxin stimulated PLD, and a low concentration of GTP synergized with cholinergic agonists, supporting receptor-PLD coupling via a G protein (Qian and Drewes, 1989). In neutrophils (Kanaho et al., 1992b), high concentrations of staurosporine activated PLD but not PLC in a pertussis toxin-sensitive manner (the effect was unrelated to inhibition of PKC), and also activated the GTPase activity of several purified pertussis toxin-sensitive G proteins (G_{i1}, G_{i2}, and G_o). Transfection of known G protein-coupled receptors into embryonic kidney cells (Sandmann et al., 1991) or Rat-1 fibroblasts (MacNulty et al., 1992) rendered PLD sensitive to the appropriate agonists, and in the latter system, transfection was coupled to agonist activation of a membrane GTPase that was inhibited in parallel with PLD by pertussis toxin.

We have investigated activation of PLD by GTP and its hydrolysis-resistant analogues in a cell-free system consisting of isolated plasma membrane plus cytosol (Olson et al., 1991). Activation required a low concentration of calcium and occurred independently of ATP, indicating that the GTP effect did not require the participation of a protein kinase (I. Lopez and D. Lambeth, unpublished). By preincubating cell fractions with guanine nucleotide analogues and then reisolating fractions away from the free nucleotides, the G protein has been localized to the plasma membrane (E. Bowman and D. Lambeth, unpublished data). The type (heterotrimeric versus small) and identity of the G protein remain to be established.

Protein Kinase C Activation of Phospholipase D

As shown in Table 5.1 (Billah and Anthes, 1990), activation of PLD by phorbol esters is a nearly universal response in a variety of cell types. Nevertheless, as indicated in the table, results using protein kinase C inhibitors in intact cells are variable and may be tissue, agonist, and/or inhibitor specific. In particular, although PMA activation of PLD is usually blocked by PKC inhibitors or PKC down-regulation, inhibition of receptor-linked activation is less consistently seen.

In some cells, receptor activation of PLD seems to be secondary to activation of PLD/PKC. In NG108-15 cells (Liscovitch, 1989), Leydig cells (Vinggaard and Hansen, 1991), and fibroblasts (van Blitterswijk et al., 1991) inhibitors of PKC blocked activation by PMA and in the latter case by a receptor-coupled agonist. In Leydig cells (Vinggaard and Hansen, 1991), mast cells (P. Lin and Gilfillan, 1992), human fibroblasts (van Blitterswijk et al., 1991), Swiss 3T3 fibroblasts (Cook and Wakelam, 1991), and endothelial cells (Martin et al., 1989), down-regulation of PKC resulted in a loss of activation of PLD by both PMA and several receptor-binding agonists, suggesting that both PMA- and receptor-linked activation was mediated by PKC. In another study in which cells were transfected with either the EGF receptor or a muscarinic receptor subtype (McKenzie et al., 1992), the ability of a given agonist to stimulate PI-PLC correlated with its ability to activate PLD. In addition, in PKC-down-regulated cells, receptor-binding agonists activated PLC but not PLD. Overexpression of PKC-B1 in rat fibroblasts caused enhanced PLD activity and DG generation in response to phorbol esters, with no enhanced PI turnover (Pai et al., 1991). Thus, these studies support a model in which PLD is activated via the prior activation of PI-PLC and PKC.

In contrast, other studies have given rise to a more complicated activation picture involving both PKC-dependent and -independent mechanisms. In some cases, agonists activate PLD without detectable PI hydrolysis, calcium mobilization, or PKC activation. This result was reported with interleukin-1-stimulated T lymphocytes (Rosoff et al., 1988), interleukin-3-stimulated mast cells (Duronio et al., 1989), EGF-stimulated IIC9 fibroblasts (Wright et al., 1988), and P_{2z} receptor agonist-stimulated macrophages (El-Moatassim and Dubyak, 1992). In PC-12 cells (Horwitz and Ricanatik, 1992), bradykinin and PMA both activated PLD, but only the PMA activation was blocked by staurosporine, suggesting separate activation pathways. In granulocytes (Billah et al., 1989b), lymphocytes (Cao et al., 1990), and a mast cell line (Yamada et al., 1991), PKC inhibitors blocked PMA-dependent phosphorylation, but failed or only partially inhibited PLD. The authors concluded that activation occurred by both PKC-dependent and -independent mechanisms.

A role for PKC has been further investigated in cell-free PLD systems. In an unprecedented result, PMA-activated PLD in fibroblast plasma membranes required added PKC, but was *independent* of ATP or phosphorylation (Conricode et al., 1992). While such a result may seem heretical, it could explain the lack of inhibition of PMA-activated PLD in some systems by PKC inhibitors such as H-7 and staurosporine, which compete for ATP binding to the kinase. Nevertheless, in the PMA-activated cell-free system from neutrophil, depletion of cytosolic ATP by dialysis or size exclusion chromatography permits the demonstration of an absolute requirement for ATP (Olson et al., 1991) and (I. Lopez and D. Lambeth, unpublished studies). Thus, at least in granulocytes, PKC plays a conventional role.

In addition to having a direct effect on PLD, PKC may synergize with G protein-mediated activation mechanisms. In permeabilized granulocytes, phorbol esters synergized with GTPγS and calcium to activate PLD (Geny and Cockcroft, 1992). In platelets, (van der Meulen and Haslam, 1990), pretreatment of the cells with PMA markedly increased the ability of GTPγS to stimulate PLD, and activation was further enhanced by 1 μM calcium. Thus, these studies suggest an interplay among PKC, a G protein, and calcium to achieve full activation.

Calcium and Phospholipase D Activation

As summarized in Table 5.1, many of the stimuli that activate PLD are either Ca^{2+} ionophores or Ca^{2+} mobilizing agents. This has suggested a role for Ca^{2+} in PLD activation. PLD activation by receptor-mediated stimuli and Ca^{2+} ionophores (but not by phorbol esters) is almost always partially or completely dependent on both extracellular and intracellular Ca^{2+} (Cockcroft, 1984; Augert et al., 1989; Billah et al., 1989b; Kessels et al., 1991; Kanaho et al., 1992a; P. Lin et al., 1992; Wu et al., 1992b). In contrast (*vide infra*), PI hydrolysis does not require extracellular Ca^{2+}. In electropermeabilized neutrophils (Kessels et al., 1991), f-Met-Leu-Phe failed to activate at 100 nM (a resting Ca^{2+} level), but activated at 500 nM (corresponding to levels achieved after agonist stimulation). In PGE-stimulated erythroleukemia cells (Wu et al., 1992b), PLD activity correlated with Ca^{2+} fluxes. In this study, an inhibitor of PLC blocked both the Ca^{2+} flux and PLD activation, and the authors proposed that the Ca^{2+} flux mediated PLD activation.

Although Ca^{2+} seems necessary for optimal activation, it remains unclear whether it activates directly or permits some other factor to act. Consistent with the latter interpretation, in the neutrophil cell-free PLD system Ca^{2+} was required for activation by either GTPγS or by protein kinase C (Olson et al., 1991). Ca^{2+} alone, however, failed to activate even at nonphysiologically high concentrations.

Thus, in neutrophil Ca^{2+} appears to act permissively rather than as an activator. Indeed, in some systems calcium greater than 1 μM is inhibitory (Liscovitch and Eli, 1991).

Tyrosine Kinases and Phospholipase D Activation

Several growth factors that act via protein tyrosine kinase receptors also activate PLD (Table 5.1). These include PDGF (Plevin et al., 1991) and EGF (Wright et al., 1988). In addition, transformation of cells with *v-src,* which encodes a soluble tyrosine kinase, resulted in increased PLD activity with a resulting increase in DG (Wright et al., 1988). EGF activation of PLD occurred independently of PI hydrolysis (Wright et al., 1988) and was unaffected by down-regulation of PKC (Kaszkin et al., 1992). In neutrophils (Uings et al., 1992), tyrosine kinase inhibitors blocked chemoattractant receptor but not PMA activation of PLD, and an inhibitor of phosphotyrosine phosphatase enhanced both tyrosine phosphorylation and PLD activity. These data suggest a role for a tyrosine phosphorylation in PLD activation by some receptors.

Sphingolipids and Activation of Phospholipase D

Signal transduction roles for sphingolipids and their metabolites represents an emerging area (Hannun and Bell, 1989). Several studies have noted increased PLD activity on addition of sphingosine to cells (Kiss and Anderson, 1990; Desai et al., 1992). Sphingosine-1-phosphate was subsequently shown to activate more rapidly and in lower doses (Desai et al., 1992), and it was suggested that earlier effects reported for sphingosine may have required its metabolism to sphingosine phosphate.

Cyclic AMP, Signal Cross-talk, and Regulation of Phospholipase D

In neutrophils, treatments that elevated cyclic AMP blocked activation of PLD by receptor-linked agonists such as f-Met-Leu-Phe, but failed to inhibit PMA activation (Agwu et al., 1991a; Tyagi et al., 1991), suggesting signal ''cross-talk'' between cAMP and PLD activation pathways. In contrast, in endothelial cells, elevation of cAMP resulted in increased PLD activity in response to thrombin. Thus, modulation of signaling pathways by cAMP differs depending on the cell type.

The Molecular Nature of Receptor-Activated Phospholipase D

Are There Isoenzymic Forms of PLD?

To date, no eukaryotic intracellular PLD has been purified or its cDNA cloned. Sufficient data have accumulated, described in part above, to suggest strongly that by analogy with PI-PLC, PLD will constitute a family of isoenzymic forms with distinct activation mechanisms. Several activities have been partially characterized (e.g., pH optima, size by gel exclusion chromatography), and in some cases partially purified. Some of these represent constitutive or detergent-activated activities that are not known to participate in signal transduction, while more recent evidence has partially characterized forms that are likely to participate in signal transduction.

Constitutive or Detergent-activated PLD Activities

Among the first PLD activities reported in mammals were acid-pH optima enzymes: a membrane-associated brain enzyme (Saito and Kanfer, 1975), and a soluble activity from eosinophil cytosol (Kater et al., 1976). The latter had an apparent M_r of 60,000, whereas the former was 200,000 (Taki and Kanfer, 1979). The brain form absolutely required detergent, and a similar activity was subsequently reported in lung (Chalifour and Kanfer, 1980). The brain enzyme may be the same as a neutral-pH optimum synaptosomal enzyme described recently (Chalifa et al., 1990), since the buffer used in the earlier Kanfer studies inhibited at neutral pH. The synaptosomal activity required the amphiphile oleate, and was strongly stimulated by 1 mM Mg^{2+} or 0.25 mM Ca^{2+}. A membrane-associated form was also described from endothelial cells (Martin, 1988). This neutral-pH optimum enzyme required Triton X-100 and was selective for PC, with no activity toward PI and weak activity toward sphingomyelin and PE. In contrast, a Triton X-100-activated, neutral-pH optimum enzyme from rat heart showed greatest specificity for PE and was Ca^{2+} and Mg^{2+} independent (Schmid et al., 1983).

In addition to the eosinophil enzyme, other soluble PLD activities have been described. A cytosolic, neutral-pH optimum form was from neutrophil and monocyte (Balsinde et al., 1989); it was calcium dependent and selective for PI. The form was stimulated by deoxycholate but not Triton X-100, and eluted on gel filtration columns as a 400,000-kDa peak. A soluble, acidic-pH optimum PLD was recently characterized from lung cytosol (P. Wang et al., 1991), with similar activities seen in brain, spleen, heart, kidney, and thymus. This form was constitutively active, but increased its activity up to 20-fold on fractionation by anion exchange chromatography, apparently due to the removal of an inhibitor. The enzyme showed specificity for both PC and PE, but was less active toward PI. Two interconvertible activity peaks of apparent molecular weights 30,000 and 80,000 were seen on gel exclusion chromatography, suggesting dimerization. Calcium was not necessary, but stimulated activity 1.5- to 2-fold at high concentrations, and the enzyme was inhibited by detergents. Thus, a variety of PLD activities with different molecular properties exist in mammalian systems, but their relevance to signal transduction is unclear.

Regulated PLD Activities

Studies in isolated liver plasma membranes (Bocckino et al., 1987) first defined a GTPγS-activatable PLD with selectivity for PC, but its molecular and other catalytic properties are unknown. Another GTPγS-dependent activity was recently described in plasma membranes from Swiss 3T3 cells (Kiss and Anderson, 1990). The enzyme was capable of hydrolyzing PE.

When we attempted to investigate a similar activity in human neutrophils, we found that although GTPγS-stimulated activity was readily observed in neutrophil homogenates, it was lost when isolated plasma membranes were used. Activity was restored when plasma membranes were recombined with cytosol (Anthes et al., 1991; Olson et al., 1991), and thermal and proteolytic inactivation demonstrated activating protein factors in both fractions (Olson et al., 1991). Activation was also achieved using phorbol esters in the presence of ATP and activity was inhibited by staurosporine (Olson et al., 1991; I. Lopez, unpublished data). Isolated brain PKC failed to replace cytosol in the assay. As described above, we have found that the GTP-dependent factor is associated with the membrane. The cytosolic protein, which we believe to be the PLD-catalytic moiety, has an apparent molecular weight of 50,000 by gel exclusion chromatography (E. Bowman, unpublished data). Despite the partial characterization of this activity, the enzyme has resisted extensive attempts at purification by at least two laboratories, and its molecular nature remains undefined. Its activation by both guanine nucleotides and PKC is consistent with a variety of *in vivo* studies, described above, that suggested multiple activation pathways.

PHOSPHOLIPASE A2, CIS-UNSATURATED FATTY ACIDS, AND LYSOPHOSPHOLIPIDS

Sources of *cis*-Unsaturated Fatty Acids and Lysophospholipids

Phospholipase A_2 (PLA_2) catalyzes the hydrolysis of a fatty acid from the *sn*-2 position of phospholipids, releasing the fatty acid plus the lysophospholipid. Typically the *sn*-2 position of glycerophospholipids contains an unsaturated or polyunsaturated fatty acid such as arachidonate. For example, in neutrophils arachidonate is distributed among PE (60%), PC (18%), and PI (18%). Released arachidonate may participate directly as a mediator, or may be metabolized via cyclooxygenase and lipoxygenase pathways to form other biologically active lipids such as the prostaglandins and leukotrienes.

cis-Unsaturated fatty acid generation may also occur via other pathways. For example, diglyceride lipase acts on diacylglycerol to release its fatty acids (Tao et al., 1989; Balsinde et al., 1991). In polymorphonuclear leukocytes, the diacylglycerol kinase inhibitor R59022 increased agonist-stimulated diacylglycerol levels and potentiated arachidonate release, consistent with the participation of PLC and diglyceride lipase in fatty acid release. In addition, a PLA_2 activity that is specific for PA has been described in platelets (Billah et al., 1981). Thus, release of arachidonate in some cases could require the participation of two or more phospholipases (e.g., PLD followed by a phosphatidate-specific PLA_2). Other potential pathways are considered in a recent review (Dennis, 1987). The current discussion will focus on PLA_2.

Relevance of Fatty Acids and Lysophospholipids to Protein Kinase C Activation

Fatty acids are implicated in the activation of PKC. In a detergent extract from human neutrophils, unsaturated fatty acids stimulated PKC by increasing its sensitivity to calcium and phosphatidylserine (McPhail et al., 1984). In an *in vitro* system, *cis*-unsaturated fatty acids potentiated diacylglycerol- and phorbol ester-stimulated PKC activity, permitting activation at physiological levels of calcium (Shinomura et al., 1991). Types I, II, and III protein kinase C all responded to some extent to synergistic activation by *cis* fatty acids plus diacylglycerol, but type III was the most sensitive (Chen and Murakami, 1992). In platelets, *cis*-unsaturated fatty acids acted synergistically with diacylglycerol or phorbol ester to enhance PKC-dependent phosphorylation of a 47-kDa protein (Yoshida et al., 1992). The effect could be attributed in part to increased calcium sensitivity of PKC. Fatty acids may also activate PKC in different cellular compartments than does diacylglycerol. In a recent study, *cis*-unsaturated fatty acids activated soluble PKC, whereas diacylglycerol activated membrane-associated PKC (Khan et al., 1992).

With regard to lysophospholipids, Kuo and co-workers (Oishi et al., 1988) observed that lysophosphatidylcholine (LPC) exerts biphasic effects on PKC, either stimulatory or inhibitory depending on the calcium concentration. Subsequent studies (Marquardt and Walker, 1991) showed that LPC added to mouse mast cells induced translocation of PKC to the membranes, and synergized with an antigen to stimulate secretion. These authors attributed the effect to an elevation of intracellular calcium. In T lymphocytes, however, LPC potentiated cell activation by diacylglycerol (or a tumor promoter) in the presence of a calcium ionophore, indicating that its effects cannot be attributed solely to an elevation in cellular calcium.

Receptor-Coupled Phospholipase A$_2$

Phospholipases A$_2$ include secreted or extracellular forms such as the pancreatic and snake venom enzymes (Dennis, 1987), reported ectoenzyme forms in the outer leaflet of the plasma membrane (Edgar et al., 1980; Kennedy and Becker, 1987), and intracellular forms. The intracellular forms are the best candidates for physiologically regulated generation of lipid-derived mediators, although in certain pathological conditions, the secreted and/or ecto forms may be important. The intracellular forms differ from the secreted forms in several respects. Extracellular/secreted forms are stabilized by cysteinyl disulfide linkages and are therefore sensitive to dithiothreitol. In addition, the extracellular forms show an absolute dependence on calcium for catalytic activity, whereas the intracellular forms, while utilizing calcium for membrane interactions, are not absolutely dependent on calcium (*vide infra*). A great deal of information is available about the secreted forms, as reviewed in Dennis (1987). This chapter will be limited to consideration of intracellular forms.

Evidence in many cell types demonstrates hormonal or other agonist activation of arachidonate release and/or generation of prostaglandins and leukotrienes. A partial list includes chemoattractants in neutrophils and HL-60 cells (Bokoch and Gilman, 1984; Nakashima et al., 1988; Cockcroft, 1991; Nielson et al., 1991); α-adrenergic agonists in FRTL-5 thyroid cells (Burch et al., 1986), MDCK cells (Parker et al., 1987; Slivka and Insel, 1987), and the pineal (Ho and Klein, 1987); light in rod outer segments (Jelsema, 1987; Jelsema and Axelrod, 1987); H-1 histamine in platelets (Nakashima et al., 1987a,b); bradykinin in Swiss 3T3 fibroblasts (Burch and Axelrod, 1987; Gil et al., 1991); arginine vasopressin, epidermal growth factor, and lipopolysaccharide in rat mesangial cells (Gronich et al., 1988; Wang et al., 1988; Goldberg et al., 1990); angiotensin II in rabbit aorta smooth muscle (Ford and Gross, 1989); and erythropoietin in erythroid cells (Mason-Garcia et al., 1992). Thus, direct or indirect agonist activation through receptors of PLA_2 is a widespread phenomenon and is likely to be of general regulatory importance.

On a Role for G Proteins in Activation of Phospholipase A_2

Evidence from a variety of systems implicates G proteins in activation of PLA_2. However, a critical issue has been whether the activation occurs directly via a G protein coupled to PLA_2 or is secondary to other processes that are activated via the G protein. In particular, calcium and protein phosphorylation via PKC may play a role, as detailed below. Thus, in some cases it has been proposed that PLA_2 is activated secondary to activation PI-PLC. In addition, the lipocortin/annexin family of proteins inhibit PLA_2, and are themselves substrates for phosphorylation by tyrosine and serine/threonine kinases (Russo-Marie, 1991). Thus, this protein may participate in PLA_2 regulation, although this has been an area of controversy. Multiple isoenzymic forms of intracellular PLA_2 with distinct activation mechanisms may also exist.

G protein participation in PLA_2 activation has been inferred from studies involving G protein activators and inhibitors. In permeabilized cells or isolated membrane fractions, GTPγS and other hydrolysis-resistant GTP analogues and in some cases fluoride activate or synergize with the normal agonist in stimulating arachidonate release. Systems that have been characterized in this manner include platelets (Nakashima et al., 1987a, b; Kajiyama et al., 1989; Silk et al., 1989; Murayama et al., 1990), neutrophils and HL-60 cells (Bokoch and Gilman, 1984; Nakashima et al., 1988; Cockcroft, 1991; Nielson et al., 1991; Cockcroft, 1991), FRTL-5 thyroid cells (Burch et al., 1986), Swiss 3T3 fibroblasts (Burch and Axelrod, 1987; Gil et al., 1991), mast cells (Okano et al., 1987; Churcher et al., 1990), rod outer segments (Jelsema, 1987; Jelsema and Axelrod, 1987), and *Aplysia* sensory neurons (Volterra and Siegelbaum, 1988). In all cases, activation by the normal agonist is blocked by pretreating the cells with pertussis toxin. Thus, a heterotrimeric G protein of the G_i/G_o class has been implicated. In some cells (e.g., neutrophil and HL-60 cells), pertussis toxin also blocks activation of PLC, so that a direct versus an indirect pathway cannot be readily distinguished. However, in several systems, pertussis toxin inhibits PLA_2 but not PLC (e.g., see Burch et al., 1986; Nakashima et al., 1987a; Axelrod et al., 1988), demonstrating independent G protein-associated pathways.

Other pharmacological and kinetic evidence has also led to a clear distinction between PLA_2 and PI-PLC/PKC. In mastoparan-stimulated Swiss 3T3 fibroblasts, arachidonic acid release was stimulated without release of inositol phosphates (Gil et al., 1991). In PMA-pretreated fibroblasts, bradykinin-stimulated PI hydrolysis was inhibited but PLA_2 activation remained intact (Burch and Axelrod, 1987). In ATP-depleted permeabilized neutrophils, fMLP plus GTPγS activated PLA_2, suggesting that PKC or another protein kinase was not involved (Cockcroft, 1992). In MDCK cells (Slivka and Insel, 1987, 1988), arachidonate release occurred more rapidly than PI hydrolysis. The former required extracellular calcium, whereas the latter did not. Neomycin, an inhibitor of PLC, blocked inositol phosphate but not arachidonate release. Similar studies have been carried out with FRTL5 thyroid cells wherein neomycin had no effect on arachidonate release but blocked IP_3 generation (Burch et al., 1986; Axelrod et al., 1988). These studies support a model wherein in some cases PLA_2 activation occurs independently of PLC/PKC.

Other studies have directly implicated a G protein in PLA_2 activation. In a novel approach (Gupta et al., 1990), chimeric DNAs encoding portions of the α subunits of the heterotrimeric G proteins G_a and G_i were constructed and trans-fected into CHO cells. One of these, which encoded the last 36 residues from α_i with the remainder from α_s, proved to be functional in activating adenylate cyclase but inhibited the activation of PLA_2 by thrombin and P_2 purinergic recep-tor agonists. The mutant G protein may bind to the effector enzyme in a nonpro-ductive manner and block its interaction with the normal $G_i\alpha$. In dark-adapted retinal rod outer segments (Jelsema, 1987; Jelsema and Axelrod, 1987; Axelrod et al., 1988), GTPγS mimicked the activation by light of PLA_2. Pertussis and cholera toxins (both of which catalyze ADP ribosylation of the α subunit of the retinal G protein transducin) blocked light activation both of cGMP phosphodi-esterase and of arachidonate release. GTPγS- and light-stimulated activation of PLA_2 was lost on removal of transducin by hypotonic washes, and phospholipase activity was partially restored on addition of exogenous holo-transducin. Unex-pectedly, when experiments were carried out using transducin separated into the α and $\beta\gamma$ subunits, it was found that the α subunit activated cGMP phosphodi-esterase but not PLA_2, while the converse was seen with $\beta\gamma$ subunits. The $\beta\gamma$ subunits may bind to and stimulate PLA_2 directly, or might activate by combining with an inhibitor protein. Thus, there is considerable evidence for the participation of G proteins in the activation of PLA_2, but direct evidence for G protein-effector enzyme interaction is currently lacking.

Activation of Phospholipase A_2 by Tyrosine Kinases

Both EGF and PDGF activate arachidonate release in some systems (Hasegawa-Saskai, 1985; Bonventre et al., 1990; Goldberg et al., 1990). In mesangial cells, EGF activates PLA_2 independently of PLC, since EGF-activated inositol phos-phate release is not seen (Bonventre et al., 1990). Growth factor-stimulated arach-idonate release differs from other activation mechanisms, since it is insensitive to pertussis toxin and to down-regulation of PKC. In cells transfected with either a normal or a mutant form of the EGF receptor lacking tyrosine kinase activity, EGF stimulated arachidonate release only when the normal receptor was present. Following EGF or PDGF, stable activation of cytosolic PLA_2 occurs, suggesting a posttranslational modification such as phosphorylation. While tyrosine phos-phorylation of PLA_2 is an exciting possibility, to date this has not been observed. Rather, in Rat-2 cells stimulated with PDGF, an immunoprecipitated cytosolic PLA_2 is phosphorylated on serine but not on tyrosine residues (L. Lin et al., 1992). Thus, growth factor receptor activation of PLA_2 may be mediated indirectly by a serine kinase(s).

Activation of Phospholipase A$_2$ by Protein Kinase C

PKC has been proposed to either directly activate or act synergistically with other factors to activate PLA$_2$ in a variety of cells. PKC activators (phorbol myristate acetate, mezerein, and diacylglycerols), inhibitors (staurosporine, H-7, and sphingosine), and in some cases PKC down-regulation, have implicated this enzyme in PLA$_2$ activation in MDCK cells (Parker et al., 1987; Godson et al., 1990), rat aorta (Slivka and Insel, 1987), human neutrophils (McIntyre et al., 1987), rat thymic epithelium (Liu et al., 1992), and platelets (Lapetina and Crouch, 1989). Following PMA treatment of MDCK cells, small increases in cytosolic PLA$_2$ activity were seen, perhaps due to phosphorylation of the enzyme (Clark et al., 1991), and transfection approaches showed that the PKC-α activates (directly or indirectly) PLA$_2$ in this cell (Godson et al., 1990). As discussed below in another context, there is now direct evidence for phosphorylation of PLA$_2$ by PKC. However, a recent study (see below) has demonstrated that PKC-dependent phosphorylation of cytosolic PLA2 fails to affect its activity.

Activation of Phospholipase A$_2$ Mitogen-Activated Protein Kinase

The sequence of a recently described 110-kDa cytosolic PLA$_2$ (c-PLA$_2$), which has been implicated as a receptor-regulated enzyme (see below), was recently identified as containing a consensus sequence for MAP (mitogen-activated protein) kinase. A purified form of the latter was shown by Davis, Knopf, and colleagues (L. Lin et al., 1993) to phosphorylate PLA$_2$. Phosphorylation was accompanied by a shift in migration to a larger M_r on Western blots, which was also seen with physiologically relevant receptor-linked activation of the enzyme. In the presence of calcium, phosphorylation resulted in activation of the enzyme. Because MAP kinase lies downstream of both protein kinase C and tyrosine kinases, the participation of MAP kinase appears to provide a unifying mechanism for the activation of PLA$_2$ by phorbol esters and by growth factors linked to tyrosine kinase receptors.

Activation of Phospholipase A$_2$ by Calcium

In some cases, PKC activators failed to activate or only weakly activated arachidonate release by themselves, by synergized with agents or hormones that elevate intracellular calcium. In neutrophils (McIntyre et al., 1987), platelets (Banga et al., 1991), and pulmonary endothelium (Chakraborti et al., 1992), PLA$_2$ was stimulated by calcium ionophores, and this was augmented by PKC activation, whereas activation of PKC alone failed to activate. In rat aorta (Jeremy and Dandona, 1987) and pineal (Ho and Klein, 1987), phorbol ester- and other agonist-stimulated arachidonate release was blocked by calcium channel blockers and/or EGTA, indicating a calcium requirement.

Although calcium is necessary, it is unlikely that this agent can act alone as a PLA$_2$ activator under normal cellular conditions. Even with use of a variety of methods and cell types, calcium levels fail to correlate with activation. In neutrophils, activation by a series of purinergic agonists caused similar elevations in calcium levels but resulted in different levels of arachidonate release (Cockcroft and Stutchfield, 1989; Tao et al., 1989). IP$_3$ added to permeabilized platelets induced thromboxane A$_2$ synthesis, implying calcium was sufficient to activate PLA$_2$ (Lapetina and Crouch, 1989). However, with other agonists, a correlation with calcium levels failed to hold up. A possible explanation, as discussed below, is that calcium may regulate the association of PLA$_2$ with the membrane, and that once there other processes (e.g., phosphorylation, G proteins) activate the enzyme. Calcium-dependent membrane association of PLA$_2$ has been observed in both cellular systems (e.g., see Channon and Leslie, 1990) and *in vitro* systems (see below).

Purification, Characterization, and Cloning of Isozymes of Phospholipase A$_2$

A High Molecular Weight Hormonally Regulated Form

A species with an apparent size of 110 kDa by gel exclusion chromatography was isolated from rat kidney, and similar forms were seen in monocytes and brain (Gronich et al., 1990). The kidney enzyme hydrolyzed arachidonate from PC and PE, but not from PI, and was stimulated by physiological concentrations of calcium, with an EC$_{50}$ of 500 nM. A similar or identical form was isolated from U937 cells and its cDNA was cloned and sequenced (Clark et al., 1991). Interestingly, reduced stringency hybridization failed to detect related members of a PLA$_2$ gene family. The enzyme expressed from the cDNA selectively cleaved arachidonate from phospholipid vesicles, and translocated to artificial phospholipid membranes in response to physiologically relevant changes in calcium concentration. An amino terminal 140 residue region also translocated to membranes in response to calcium. This region contained a 45 amino acid sequence that was homologous to the C2 region of PKC, p65 (a protein that binds acidic phospholipids in a calcium-dependent manner and may function in calcium-induced fusion of synaptic vesicles with the plasma membrane), GAP (which also binds acidic phospholipids such as PIP$_2$), and PLC-γ1. This region, proposed to be the calcium-binding, phospholipid-interaction motif, contained a conserved threonine that the authors speculated may serve as a regulatory phosphorylation site. As noted above, the sequence also contains a MAP kinase consensus site.

Recent evidence demonstrates definitively that this form of PLA$_2$ is indeed regulated by receptors (L. Lin et al., 1992). In transfected CHO cells overexpressing this PLA$_2$, thrombin and ATP (a purinergic agonist) increased arachidonic acid release compared with that seen in the parental cells or cells expressing a secreted form of PLA$_2$. Lysates from ATP-treated cells contained increased PLA$_2$ activity, which was sensitive to phosphatase treatment. Immunoprecipitated PLA$_2$ showed that activation was associated with phosphorylation on serine residues, and pretreatment with the protein kinase inhibitor staurosporine inhibited both phosphorylation and activation. Thus, the response involves not only calcium-dependent translocation, but also an activating phosphorylation.

A Low Molecular Weight Ischemia-activated Phospholipase A$_2$

Myocardial ischemia results in the accumulation of lysophospholipids, due to activation of a PLA$_2$ that is selective for arachidonoyl-containing plasmenyle-thanolamine. The latter predominates in the sarcolemmal membrane (Hazen et al., 1990). A 40-kDa enzyme has been purified to homogeneity on the basis of its ability to bind specifically and reversibly to adenine nucleotide triphosphate agarose affinity matrices (Hazen and Gross, 1991). It differs from secreted and other intracellular forms of PLA$_2$ because of its calcium independence and its reactivity with a halenol lactone mechanism-based inhibitor (Hazen et al., 1991). The enzyme was selective for *sn*-2 arachidonoyl phospholipids and showed the following rank order of substrate preference: plasmenylcholine > 1-O-alkyl PC > PC. The mechanism by which ischemia activates this calcium-independent enzyme remains obscure, and it is not clear whether this enzyme should be considered an agonist-activated PLA$_2$. It should be noted, however, that ATP and its hydrolysis-resistant analogues augment the rate of phospholipid hydrolysis by this enzyme (Hazen and Gross, 1991).

Platelet Forms of Phospholipase A$_2$

A large molecular weight form of PLA$_2$ was purified from rabbit platelets (Burstein et al., 1992). This form was similar in size, substrate preference, and calcium stimulation to the large molecular weight forms seen in U937 monoblast and in kidney cells, described above, and may represent the same isoenzyme. A small molecular weight PLA$_2$ representing the major activity in sheep platelets has also been isolated, characterized, and cloned (Zupan et al., 1991; 1992). The purified enzyme was activated by calcium, but a high concentration of NaCl also supported activity. These results underscore the fundamental difference between this intracellular PLA$_2$ and secreted forms, since calcium is obligatory in the latter where it is utilized in the catalytic mechanism to polarize the *sn*-2 carbonyl during hydrolysis. The cDNA for the platelet enzyme predicted a 30-kDa polypeptide that showed 73 percent homology to a family of proteins originally isolated from brain and known collectively as the 14-3-3 proteins. Hybridization studies revealed the presence of at least three related genes, suggesting a gene family. Unlike secreted forms of PLA$_2$ wherein the polarized carbonyl carbon is attacked by water to release arachidonic acid, this enzyme catalyzed arachidonate release by forming an enzyme thioester-arachidonoyl intermediate.

Summary and Perspectives of Phospholipase A$_2$

The foregoing discussion establishes a number of signal transduction paradigms by which agonists activate one or more PLA$_2$s. These include activation by G protein-coupled agonists, tyrosine kinase-related growth factor receptors, PKC, and calcium. Since at least some of these pathways do not seem to operate through common mechanisms, it seems reasonable to suggest that the different signal transduction mechanisms may utilize different isoenzymic forms of PLA$_2$. Consistent with this hypothesis, several isoenzymic forms of this enzyme have been isolated and some of these have been cloned. Although the relationship of some of these isoforms to signal transduction pathways remains to be established, activation of one of these—the U937 enzyme—has clearly been shown to be receptor mediated, with a mechanism that involves calcium-dependent translocation and activation by serine phosphorylation. Further studies on the relationship of signal transduction pathways to phospholipases C, D, and A_2 should prove fruitful areas of research.

REFERENCES

Agwu, D. E., C. E. McCall, and L. C. McPhail. 1991a. Regulation of phospholipase D-induced hydrolysis of choline-containing phosphoglycerides by cyclic AMP in human neutrophils. *J. Immunol.* 146:3895–3903.

Agwu, D. E., L. C. McPhail, S. Sozzani, D. A. Bass, and C. E. McCall. 1991b. Phosphatidic acid as a second messenger in human polymorphonuclear leukocytes: Effects on activation of NADPH oxidase. *J. Clin. Invest.* 88:531–539.

Akhtar, R. A., and M. W. Choi. 1992. Stimulation of phospholipase D by phorbol esters and ionomycin in bovine corneal epithelial cells. *Curr. Eye Res.* 11:553–564.

Altin, J. G., and F. L. Bygrave. 1987. Phosphatidic acid and arachidonic acid each interact synergistically with glucagon to stimulate Ca^{++} influx in the perfused rat liver. *Biochem. J.* 247:613–619.

Anthes, J. C., P. Wang, M. I. Siegel, R. W. Egan, and M. M. Billah. 1991. Granulocyte phospholipase D is activated by a guanine nucleotide dependent protein factor. *Biochem. Biophys. Res. Commun.* 175:236–243.

Augert, G., S. B. Bocckino, P. F. Blackmore, and J. H. Exton. 1989. Hormonal stimulation of diacylglycerol formation in hepatocytes: Evidence for phosphatidylcholine breakdown. *J. Biol. Chem.* 264:21689–21698.

Axelrod, J., R. M. Burch, and C. L. Jelsema. 1988. Receptor-mediated activation of phospholipase A_2 via GTP-binding proteins: Arachidonic acid and its metabolites as second messengers. *TINS* 11:113–123.

Bairoch, A., and J. A. Cox. 1990. EF-hand motifs in inositol phospholipid-specific phospholipase C. *FEBS Lett.* 269:454–456.

Balboa, M. A., J. Balsinde, J. Aramburu, F. Mollinedo, and M. Lopez-Botet. 1992. Phospholipase D activation in human natural killer cells through the Kp43 and CD16 surface antigens takes place by different mechanisms. Involvement of the phospholipase D pathway in tumor necrosis factor alpha synthesis. *J. Exp. Med.* 176:9–17.

Balsinde, J., and F. Mollinedo. 1991. Platelet-activating factor synergizes with phorbol myristate acetate in activating phospholipase D in the human promonocytic cell line U937: Evidence for different mechanisms of activation. *J. Biol. Chem.* 266:18726–18730.

Balsinde, J., E. Diez, B. Fernandez, and F. Mollinedo. 1989. Biochemical characterization of phospholipase D activity from human neutrophils. *Eur. J. Biochem.* 186:717–724.

Balsinde, J., E. Diez, and F. Mollinedo. 1991. Arachidonic acid release from diacylglycerol in human neutrophils: Translocation of diacylglycerol-deacylation enzyme activities from an intracellular pool to plasma membrane upon cell activation. *J. Biol. Chem.* 266:15638–15643.

Banga, H. S., S. P. Halenda, and M. B. Feinstein. 1991. Potentiation of arachidonic acid release by phorbol myristate acetate in platelets is not due to inhibition of arachidonic acid uptake or incorporation into phospholipids. *Biochim. Biophys. Acta* 1091:115–119.

Bauldry, S. A., D. A. Bass, S. L. Cousart, and C. E. McCall. 1991. Tumor necrosis factor α priming of phospholipase D in human neutrophils: Correlation between phosphatidic acid production and superoxide generation. *J. Biol. Chem.* 266:4173–4179.

Bellavite, P., F. Corso, S. Dusi, M. Grzeskowiak, V. Della-Bianca, and F. Rossi. 1988. Activation of NADPH-dependent superoxide production in plasma membrane extracts of pig neutrophils by phosphatidic acid. *J. Biol. Chem.* 263:8210–8214.

Bennett, C. F., and S. T. Crook. 1987. Purification and characterization of phosphoinositide-specific phospholipase C from guinea pig uterus: Phosphorylation by protein kinase C *in vivo*. *J. Biol. Chem.* 262:13789–13797.

Bennett, C. F., J. M. Balcarek, A. Varrichio, and S. T. Crooke. 1988. Molecular cloning and complete amino acid sequence of form-I phosphoinositide-specific phospholipase. C. *Nature* 344:268–270.

Berridge, M. J. 1984. Inositol trisphosphate and diacylglycerol as second messengers. *Biochem. J.* 220:345–360.

Berstein, G., J. L. Blank, D. Jhon, J. H. Exton, S. G. Rhee, and E. M. Ross. 1992a. Phospholipase C-β1 is a GTPase-activating protein for $G_{q/11}$, its physiologic regulator. *Cell* 70:411–418.

Berstein, G., J. L. Blank, A. V. Smrcka, T. Higashijima, P. C. Sternweis, J. H. Exton, and E. M. Ross. 1992b. Reconstitution of agonist-stimulated phosphatidylinositol 4,5-bisphosphate hydrolysis using purified m1 muscarinic receptor, $G_{q/11}$, and phospholipase C-β1. *J. Biol. Chem.* 267:8081–8088.

Besterman, J. M., V. Duronio, and P. Cuatrecasas. 1986. Rapid formation of diacylglycerol from phosphatidylcholine: A pathway for generation of a second messenger. *Proc. Nat. Acad. Sci. U.S.A.* 83:6785–6789.

Billah, M., and J. C. Anthes. 1990. The regulation and cellular functions of phosphatidylcholine hydrolysis. *Biochem. J.* 269:281–291.

Billah, M. M., E. G. Lapetina, and P. Cuatrecasas. P. 1981. Phospholipase A_2 activity specific for phosphatidic acid: A possible mechanism for the production of arachidonic acid in platelets. *J. Biol. Chem.* 256:5399–5403.

Billah, M. M., S. Eckel, T. J. Mullmann, R. W. Egan, and M. I. Siegel. 1989a. Phosphatidylcholine hydrolysis by phospholipase D determines phosphatidate and diglyceride levels in chemotactic peptide-stimulated human neutrophils: Involvement of phosphatidate phosphohydrolase in signal transduction. *J. Biol. Chem.* 264:17069–17077.

Billah, M. M., J. Pai, T. J. Mullmann, R. W. Egan, and M. I. Siegel. 1989b. Regulation of phospholipase D in HL-60 granulocytes. *J. Biol. Chem.* 264:9069–9076.

Blackmore, P. F., and J. H. Exton. 1986. Studies on the hepatic calcium-mobilizing activity of aluminum fluoride and glucagon: Modulation by cAMP and phorbol myristate acetate. *J. Biol. Chem.* 261:11056–11063.

Blank, J. L., A. H. Ross, and J. H. Exton. 1991. Purification and characterization of two G-proteins that activate the β1 isozyme of phosphoinositide-specific phospholipase C: Identification as members of the G_q class. *J. Biol. Chem.* 266:18206–18216.

Blank, J. L., K. A. Brattain, and J. H. Exton. 1992. Activation of cytosolic phosphoinositide phospholipase C by G-protein βγ subunits. *J. Biol. Chem.* 267:23069–23075.

Bloomquist, B. T., R. D. Shortridge, S. Schneuwly, M. Perdew, C. Montell, H. Steller, G. Rubin, and W. L. Pak. 1988. Isolation of putative phospholipase C gene of *Drosophila*, norpA, and its role in phototransduction. *Cell* 54:723–733.

Bocckino, S. B., P. F. Blackmore, and J. H. Exton. 1985. Stimulation of 1,2-diacylglycerol accumulation in hepatocytes by vasopressin, epinepherine, and angiotensin II. *J. Biol. Chem.* 260:14201–14207.

Bocckino, S. B., P. F. Blackmore, P. B. Wilson, and J. H. Exton. 1987. Phosphatidate accumulation in hormone-treated hepatocytes via a phospholipase D mechanism. *J. Biol. Chem.* 262:15309–15315.

Bocckino, S. B., P. B. Wilson, and J. H. Exton. 1991. Phosphatidate-dependent protein phosphorylation. *Proc. Natl. Acad. Sci. U.S.A.* 88:6210–6213.

Bokoch, G. M., and A. G. Gilman. 1984. Inhibition of receptor-mediated release of arachidonic acid by pertussis toxin. *Cell* 39:301–308.

Bollag, W. B., P. Q. Barrett, C. M. Isales, M. Liscovitch, and H. Rasmussin. 1990. A potential role for phospholipase D in the angiotensin II-induced stimulation of aldosterone secretion from bovine adrenal glomerulosa cells. *Endocrinology* 127:1436–1443.

Bonventre, J. V., J. H. Gronich, and R. A. Nemenoff. 1990. Epidermal growth factor enhances glomerular mesangial cell soluble phospholipase A_2 activity. *J. Biol. Chem.* 265:4934–4938.

Bosner, R. W., N. T. Thompson, R. W. Randall, and L. G. Garland. 1989. Phospholipase D activation is functionally linked to superoxide generation in the human neutrophil. *Biochem. J.* 264:617–620.

Boss, V., and P. J. Conn. 1992. Metabotropic excitatory amino acid receptor activation stimulates phospholipase D in hippocampal slices. *J. Neurochem.* 59:2340–2343.

Boyer, J. L., G. L. Waldo, T. Evans, J. K. Northrup, C. P. Downes, and T. K. Harden. 1989. Modification of AIF_4^--and receptor-stimulated phospholipase C activity by G-protein beta gamma subunits. *J. Biol. Chem.* 264:13917–13922.

Boyer, J. L., G. L. Waldo, and T. K. Harden. 1992. βγ-subunit activation of G-protein-regulated phospholipase C. *J. Biol. Chem.* 267:25451–25456.

Brown, K. D., D. M. Blakeley, M. H. Hamon, M. S. Laurie, and A. N. Corps. 1987. Protein kinase C-mediated negative feedback inhibition of unstimulated and bombesin-stimulated polyphosphoinositide hydrolysis in Swiss-mouse 3T3 cells. *Biochem. J.* 245:631–639.

Burch, R. M., and J. Axelrod. 1987. Dissociation of bradykinin-induced prostaglandin formation from phosphatidylinositol turnover in Swiss 3T3 fibroblasts: Evidence for G protein regulation of phospholipase A_2. *Proc. Natl. Acad. Sci. U.S.A.* 84:6374–6378.

Burch, R. M., A. Luini, and J. Axelrod. 1986. Phospholipase A_2 and phospholipase C are activated by distinct GTP binding proteins in response to α1-adrenergic stimulation in FRTL5 thyroid cells. *Proc. Natl. Acad. Sci. U.S.A.* 83:7201–7205.

Burstein, E. S., W. H. Brondyk, and I. G. Macara. 1992. Amino acid residues in the Ras-like GTPase Rab3A that specify sensitivity to factors that regulate the GTP/GDP cycling of Rab3A. *J. Biol. Chem.* 267:22715–22718.

Cabot, M. C., C. J. Welsh, H. Cao, and H. Chabbott. 1988. The phosphatidylcholine pathway of diacylglycerol formation stimulated by phorbol diesters occurs via phospholipase D activation. *FEBS Lett.* 233:153–157.

Camps, M., C. Hou, D. Sidiropoulos, J. B. Stock, K. H. Jakobs, and P. Gierschik. 1992. Stimulation of phospholipase C by guanine-nucleotide-binding protein βγ subunits. *Eur. J. Biochem.* 206:821–831.

Cao, Y. Z., C. C. Reddy, and A. M. Mastro. 1990. Evidence for protein kinase C independent activation of phospholipase D by phorbol esters in lymphocytes. *Biochem. Biophys. Res. Commun.* 171:955–962.

Carter, R. H., D. J. Park, S. G. Rhee, and D. T. Fearson. 1991. Tyrosine phosphorylation of phospholipase C induced by membrane immunoglobulin in B lymphocytes. *Proc. Natl. Acad. Sci. U.S.A.* 88:2745–2749.

Chabot, M. C., L. C. McPhail, R. L. Wykle, D. A. Kennerly, and C. E. McCall. 1992. Comparison of diglyceride production from choline-containing phosphoglycerides in human neutrophils stimulated with *N*-formylmethionyl-leucylphenylalanine, ionophore A23187 or phorbol 12-myristate 13-acetate. *Biochem. J.* 286:693–699.

Chakraborti, S., J. R. Michael, and T. Sanyal. 1992. Defining the role of protein kinase C in calcium-ionophore-(A23187)-mediated activation of phospholipase A_2 in pulmonary endothelium. *Eur. J. Biochem.* 206:965–972.

Chalifa, V., H. Mohn, and M. Liscovitch. 1990. A neutral phospholipase D activity from rat brain synaptic plasma membranes: Identification and partial characterization. *J. Biol. Chem.* 265:17512–17519.

Chalifour, R. J., and J. N. Kanfer. 1980. Microsomal phospholipase D of rat brain and lung tissues. *Biochem. Biophys. Res. Commun.* 96:742–747.

Channon, J. Y., and C. C. Leslie. 1990. A calcium-dependent mechanism for associating a soluble arachidonoyl-hydrolyzing phospholipase A_2 with membrane in the macrophage cell line RAW264.7. *J. Biol. Chem.* 265:5409–5413.

Chen, S. G., and K. Murakami. 1992. Synergistic activation of type III protein kinase C by *cis*-fatty acid and diacylglycerol. *Biochem. J.* 282:33–39.

Churcher, Y., D. Allan, and B. D. Gomperts. 1990. Relationship between arachidonate generation and exocytosis in permeabilized mast cells. *Biochem. J.* 266:157–163.

Clark, J. D., L. Lin, R. W. Kriz, C. S. Ramesha, L. A. Sultzman, A. Y. Lin, N. Milona, and J. L. Knopf. 1991. A novel arachidonic acid-selective cytosolic PLA_2 contains a Ca^{2+}-dependent translocation domain with homology to PKC and GAP. *Cell* 65:1043–1051.

Claro, E., M. A. Wallace, H. M. Lee, and J. N. Fain. 1989. Carbachol in the presence of guanosine 5′-O-(3-thiotriphosphate) stimulates the breakdown of exogenous phosphatidylinositol 4,5-biphosphate, phosphatidylinositol 4-phosphate, and phosphatidlyinositol by rat brain membranes. *J. Biol. Chem.* 264:18288–18295.

Cockcroft, S. 1984. Ca^{2+}-dependent conversion of phosphatidylethanolamine to phosphatidate in neutrophils stimulated with fMet-Leu-Phe or ionophore A23187. *Biochim. Biophys. Acta* 795:37–46.

Cockcroft, S. 1991. Relationship between arachidonate release and exocytosis in permeabilized human neutrophils stimulated with formylmethionyl-leucyl-phenylalanine (fMetLeuPhe), guanosine 5′-[γ-thio]triphosphate (GTP[S]) and Ca^{2+}. *Biochem. J.* 275:127–131.

Cockcroft, S. 1992. G-protein-regulated phospholipases C, D and A_2-mediated signalling in neutrophils. *Biochim. Biophys. Acta* 1113:135–160.

Cockcroft, S., and B. D. Gomperts. 1985. Role of guanine nucleotide binding protein in the activation of polyphosphoinositide phosphodiesterase. *Nature* 314:534–536.

Cockcroft, S., and J. Stutchfield. 1989. The receptors for ATP and fMet-Leu-Phe are independently coupled to phospholipases C and A_2 via G protein(s): Relationship between phospholipase C and A_2 activation and exocytosis in HL60 cells and human neutrophils. *Biochem. J.* 263:715–723.

Conricode, K. M., K. A. Brewer, and J. H. Exton. 1992. Activation of phospholipase D by protein kinase C: Evidence for a phosphorylation-independent mechanism. *J. Biol. Chem.* 267:7199–7202.

Cook, S. J., and M. J. O. Wakelam. 1991. Hydrolysis of phosphatidylcholine by phospholipase D is a common response to mitogens which stimulate inositol lipid hydrolysis in Swiss 3T3 fibroblasts. *Biochim. Biophys. Acta* 1092:265–272.

Daniel, L. W., M. Waite, and R. L. Wykle. 1986. A novel mechanism of diglyceride formation: 12-O-tetradecanoylphorbol-13-acetate stimulates the cyclic breakdown and resynthesis of phosphatidylcholine. *J. Biol. Chem.* 261:9128–9132.

Daniel, L. W., G. W. Small, and J. D. Schmitt. 1988. Alkyl-linked diglycerides inhibit protein kinase C activation by diacylglycerols. *Biochem. Biophys. Res. Commun.* 151:291–297.

Deckmyn, H., S.-M. Tu, and P. W. Majerus. 1986. Guanine nucleotides stimulate soluble phosphoinositide-specific phospholipase C in the absence of membranes. *J. Biol. Chem.* 261:16553–16558.

Dennis, E. A. 1987. Regulation of eicosanoid production: Role of phospholipases and inhibitors. *Bio/Technology* 5:1294–1300.

Desai, N. N., H. Zhang, A. Olivera, M. E. Mattie, and S. Spiegel. 1992. Sphingosine-1-phosphate, a metabolite of sphingosine, increases phosphatidic acid levels by phospholipase D activation. *J. Biol. Chem.* 267:23122–23128.

Domino, S. E., S. B. Bocckino, and D. L. Garbers. 1989. Activation of phospholipase D by the fucose-sulfate glycoconjugate that induces an acrosome reaction in spermatozoa. *J. Biol. Chem.* 264:9412–9419.

Dunlop, M. E., and R. G. Larkins. 1989. Effects of phosphatidic acid on islet cell phosphoinositide hydrolysis, Ca^{++}, and adenylate cyclase. *Diabetes* 38:1187–1192.

Duronio, V., L. Nip, and S. L. Pelech. 1989. Interleukin 3 stimulates phosphatidylcholine turnover in a mast/megakaryocyte cell line. *Biochem. Biophys. Res. Commun.* 164: 804–808.

Edgar, A. D., L. Freysz, P. Mandel, and L. A. Horrocks. 1980. Phospholipid metabolism in low and high density C6 cells. *Trans. Am. Soc. Neurochem.* 11:100.

Eggen, B. J. L., and J. Eichberg. 1992. Phorbol ester-mediated stimulation of phospholipase D activity in sciatic nerve from normal and diabetic rats. *J. Neurochem.* 59: 1467–1473.

El-Moatassim, C., and G. R. Dubyak. 1992. A novel pathway for the activation of phospholipase D by P_{2z} purinergic receptors in BAC1.2F5 macrophages. *J. Biol. Chem.* 267:23664–23673.

Emori, Y., Y. Homma, H. Sorimachi, H. Kawasaki, O. Naikanishi, K. Suzuki, and T. Takenawa. 1989. A second type of rat phosphoinositide-specific phospholipase C containing a *src*-related sequence not essential for phosphoinositide-hydrolyzing activity. *J. Biol. Chem.* 264:21885–21890.

English, D., G. Taylor, and J. G. N. Garcia. 1991. Diacylglycerol generation in fluoride-treated neutrophils: Involvement of phospholipase D. *Blood* 77:2746–2756.

Epand, R. M., and A. R. Stafford. 1990. Counter-regulatory effects of phosphatidic acid on protein kinase C activity in the presence of calcium and diolein. *Biochem. Biophys. Res. Comm.* 171:487–490.

Escobedo, J. A., P. J. Barr, and L. T. Williams. 1988. Role of tyrosine kinase and membrane-spanning domains in signal transduction by the platelet-derived growth factor receptor. *Molec. Cell. Biol.* 8:5126–5131.

Ford, D. A., and R. W. Gross. 1989. Plasmenylethanolamine is the major storage depot for arachidonic acid in rabbit vascular smooth muscle and is rapidly hydrolyzed after angiotensin II stimulation. *Proc. Natl. Acad. Sci. U.S.A.* 86:3479–3483.

Ford, D. A., R. Miyake, P. E. Glaser, and R. W. Gross. 1989. Activation of protein kinase C by naturally occurring ether-linked diglycerides. *J. Biol. Chem.* 264:13818–13824.

Fu, T., Y. Okano, and Y. Nozawa. 1992. Differential pathways (phospholipase C and phospholipase D) of bradykinin-induced biphasic 1,2-diacylglycerol formation in non-transformed and K-*ras*-transformed NIH-3T3 fibroblasts: Involvement of intracellular Ca^{2+} oscillations in phosphatidylcholine breakdown. *Biochem. J.* 283: 347–354.

Geny, B., and S. Cockcroft. 1992. Synergistic activation of phospholipase D by protein kinase C- and G-protein-mediated pathways in streptolysin O-permeabilized HL60 cells. *Biochem. J.* 284:531–538.

Gierschik, P., and K. H. Jakobs. 1987. Receptor-mediated ADP-ribosylation of a phospholipase C-stimulating G protein. *FEBS Lett.* 224:219–223.

Gierschik, P., D. Sidiropoulos, and K. H. Jakobs. 1989. Two distinct G_i-proteins mediate formyl peptide receptor signal transduction in human leukemia (HL-60) cells. *J. Biol. Chem.* 264:21470–21473.

Gil, J., T. Higgins, and E. Rozengurt. 1991. Mastoparan, a novel mitogen for Swiss 3T3 cells, stimulates pertussis toxin-sensitive arachidonic acid release without inositol phosphate accumulation. *J. Cell. Biol.* 113:943–950.

Godson, C., B. A. Weiss, and P. A. Insel. 1990. Differential activation of protein kinase C-α is associated with arachidonate release in Madin-Darby canine kidney cells. *J. Biol. Chem.* 265:8369–8372.

Goldberg, H. J., M. M. Viegas, B. L. Margolis, J. Schlessinger, and C. L. Skorecki. 1990. The tyrosine kinase activity of the epidermal-growth-factor receptor is necessary for phospholipase A_2 activation. *Biochem. J.* 267:461–465.

Goldschmidt-Clermont, P. J., J. W. Kim, L. M. Machesky, S. G. Rhee, and T. D. Pollard. 1991. Regulation of phospholipase C-γ1 by profilin and tyrosine phosphorylation. *Science* 251:1231–1233.

Grandt, R., K. Aktories, and K. H. Jacobs. 1986. Evidence for two GTPases activated by thrombin in membranes of human platelets. *FEBS Lett.* 196:279–283.

Grondin, P., M. Plantavid, C. Sultan, M. Breton, G. Mauco, and H. Chap. 1991. Interaction of pp60^{c-src}, phospholipase C, inositol-lipid, and diacylglycerol kinases with the cytoskeletons of thrombin-stimulated platelets. *J. Biol. Chem.* 266:15705–15709.

Gronich, J. H., J. V. Bonventre, and R. A. Nemenoff. 1988. Identification and characterization of a hormonally regulated form of phospholipase A_2 in rat renal mesangial cells. *J. Biol. Chem.* 263:16645–16651.

Gronich, J. H., J. V. Bonventre, and R. A. Nemenoff. 1990. Purification of a high-molecular-mass form of phospholipase A_2 from rat kidney activated at physiological calcium concentrations. *Biochem. J.* 271:37–43.

Gruchalla, R. S., T. T. Dinh, and D. A. Kennerly. 1990. An indirect pathway of receptor-mediated 1,2-diacylglycerol formation in mast cells: IgE receptor-mediated activation of phospholipase D. *J. Immunol.* 144:2334–2342.

Gupta, S. K., E. Diez, L. E. Heasley, S. Osawa, and G. L. Johnson. 1990. A G protein mutant that inhibits thrombin and purinergic receptor activation of phospholipase A_2. *Science* 249:662–666.

Gutowski, S., S. Smrcka, L. Nowak, D. Wu, M. Simon, and P. Sternweis. 1991. Antibodies to the αq subfamily of G protein α subunits attenuate activation of phosphatidylinositol 4,5-bisphosphate hydrolysis by hormones. *J. Biol. Chem.* 266:20519–20524.

Halenda, S. P., and A. G. Rehm. 1990. Evidence for the calcium-dependent activation of phospholipase D in thrombin-stimulated human erythroleukemia cells. *Biochem. J.* 267:479–483.

Hanahan, D. J., and I. L. Chaikoff. 1947. Phosphorus-containing lipids of carrot. *J. Biol. Chem.* 168:233–240.

Hannun, Y. A., and R. M. Bell. 1989. Functions of sphingolipids and sphingolipid breakdown products in cellular regulation. *Science* 243:500–508.

Hasegawa-Saskai, H. 1985. Early changes in inositol lipids and their metabolites induced by platelet-derived growth factor. *Biochem. J.* 232:99–109.

Hazen, S. L., and R. W. Gross. 1991. ATP-dependent regulation of rabbit myocardial cytosolic calcium-independent phospholipase A_2. *J. Biol. Chem.* 266:14526–14534.

Hazen, S. L., R. J. Stuppy, and R. W. Gross. 1990. Purification and characterization of canine myocardial cytosolic phospholipase A_2: A calcium-independent phospholipase with absolute *sn*-2 regiospecificity for diradyl glycerophospholipids. *J. Biol. Chem.* 265:10622–10630.

Hazen, S. L., L. A. Zupan, R. H. Weiss, D. P. Getman, and R. W. Gross. 1991. Suicide inhibition of canine myocardial cytosolic calcium-independent phospholipase A_2: Mechanism-based discrimination between calcium-dependent and -independent phospholipases A_2. *J. Biol. Chem.* 266:7227–7232.

Heller, M. 1978. Phospholipase D. *Adv. Lip. Res.* 16:267–326.

Ho, A. K., and D. C. Klein. 1987. Activation of α1-adrenoceptors, protein kinase C, or treatment with intracellular free Ca^{++} elevating agents increases pineal phospholipase A_2 activity: Evidence that protein kinase C may participate in Ca^{++}-dependent α1-adrenergic stimulation of pineal phospholipase A_2 activity. *J. Biol. Chem.* 262:11764–11770.

Hofman, S. L., and P. W. Majerus. 1982. Identification and properties of two distinct phosphatidylinositol-specific phospholipase C enzymes from sheep seminal vesicular glands. *J. Biol. Chem.* 257:6461–6489.

Hokin, L. E. 1985. Receptors and phosphoinositide-generated second messengers. *Annu. Rev. Biochem.* 54:205–235.

Hokin, M. R., and L. E. Hokin, 1953. Enzyme secretion and the incorporation of ^{32}P into phospholipids of pancreatic slices. *J. Biol. Chem.* 203:967–977.

Hokin-Neaverson, M., K. Sadeeghian, A. L. Majumder, and Jr. F. Eisenberg 1975. Inositol is the water-soluble product of acetylcholine stimulated breakdown of phosphatidylinositol in mouse pancrease. *Biochem. Biophys. Res. Commun.* 67:1537–1544.

Holbrook, P. G., L. K. Pannell, Y. Murata, and J. W. Daly. 1992. Molecular species analysis of a product of phospholipase D activation: Phosphatidylethanol is formed from phosphatidylcholine in phorbol ester- and bradykinin-stimulated PC12 cells. *J. Biol. Chem.* 267:16834–16840.

Horwitz, J. 1991. Bradykinin activates a phospholipase D that hydrolyzes phosphatidylcholine in PC12 cells. *J. Neurochem.* 56:509–517.

Horwitz, J., and S. Ricanatik. 1992. Bradykinin and phorbol dibutyrate activate phospholipase D in PC12 cells by different mechanisms. *J. Neurochem.* 59:1474–1480.

Houslay, M. D., D. Bojanic, and A. Wilson. 1986. Platelet activating factor and U44069 stimulate a GTPase activity in human platelets which is distinct from the guanine nucleotide regulatory proteins, Ns and Ni. *Biochem. J.* 234:737–740.

Huang, C., and M. C. Cabot. 1990. Phorbol diesters stimulate the accumulation of phosphatidate, phosphatidylethanol, and diacylglycerol in three cell types: Evidence for the indirect formation of phosphatidylcholine-derived diacylglycerol by a phospholipase D pathway and direct formation of diacylglycerol by a phospholipase C pathway. *J. Biol. Chem.* 265:14858–14863.

Huang, C., and M. C. Cabot. 1992. Phospholipase D activity in nontransformed and transformed fibroblasts. *Biochim. Biophys. Acta* 1127:242–248.

Huang, C., R. L. Wykle, L. W. Daniel, and M. C. Cabot. 1992. Identification of phosphatidylcholine-selective and phosphatidylinositol-selective phospholipases D in Madin-Darby canine kidney cells. *J. Biol. Chem.* 267:16859–16865.

Huang, R., G. L. Kucera, and S. E. Rittenhouse. 1991. Elevated cytosolic Ca^{2+} activates phospholipase D in human platelets. *J. Biol. Chem.* 266:1652–1655.

Irving, H. R., and J. H. Exton. 1987. Phosphatidylcholine breakdown in rat liver plasma membrane. *J. Biol. Chem.* 262:3440–3443.

Jackowski, S., and C. O. Rock. 1989. Stimulation of phosphatidylinositol 4,5-bisphosphate phospholipase C activity by phosphatidic acid. *Arch. Biochem. Biophys.* 268:516–524.

Jelsema, C. L. 1987. Light activation of phospholipase A_2 in rod outer segments of bovine retina and its modulation by GTP-binding proteins. *J. Biol. Chem.* 262:163–168.

Jelsema, C. L., and J. Axelrod. 1987. Stimulation of phospholipase A_2 activity in bovine rod outer segments by the βγ subunits of transducin and its inhibition by the α subunit. *Proc. Nat. Acad. Sci. U.S.A.* 84:3623–3627.

Jeremy, J. Y., and P. Dandona. 1987. The role of the diacylglycerol-protein kinase C system in mediating adrenoreceptor-prostacyclin synthesis coupling in the rat aorta. *Eur. J. Pharm.* 136:311–316.

June, C. H., M. C. Fletcher, J. A. Ledbetter, and L. E. Samelson. 1990. Increases in tyrosine phosphorylation are detectable before phospholipase C activation after T cell receptor stimulation. *J. Immunol.* 144:1591–1599.

Kajiyama, Y., T. Murayama, and Y. Nomura. 1989. Pertussis toxin-sensitive GTP-binding proteins may regulate phospholipase A_2 in response to thrombin in rabbit platelets. *Arch. Biochem. Biophys.* 274:200–208.

Kanaho, Y., H. Kanoh, and Y. Nozawa. 1991. Activation of phospholipase D in rabbit neutrophils by fMet-Leu-Phe is mediated by a pertussis toxin-sensitive GTP-binding protein that may be distinct from a phospholipase C-regulating protein. *FEBS Lett.* 279:249–252.

Kanaho, Y., A. Nishida, and Y. Nozawa. 1992a. Calcium rather than protein kinase C is the major factor to activate phospholipase D in FMLP-stimulated rabbit peritoneal neutrophils: Possible involvement of calmodulin/myosin L chain kinase pathway. *J. Immunol.* 149:622–628.

Kanaho, Y., K. Takahashi, U. Tomita, T. Iiri, T. Katada, M. Ui, and Y. Nozawa. 1992b. A protein kinase C inhibitor, staurosporine, activates phospholipase D via a pertussis toxin-sensitive GTP-binding protein in rabbit peritoneal neutrophils. *J. Biol. Chem.* 267:23554–23559.

Kaszkin, M., L. Seidler, R. Kast, and V. Kinzel. 1992. Epidermal-growth-factor-induced production of phosphatidylalcohols by HeLa cells and A431 cells through activation of phospholipase D. *Biochem. J.* 287:51–57.

Katan, M., and P. J. Parker. 1987. Purification of phosphoinositide-specific phospholipase C from a particulate fraction of bovine brain. *Eur. J. Biochem.* 168:413–418.

Katan, M., R. W. Kriz, N. Totty, R. Philp, E. Meldrum, R. A. Aldape, J. L. Knopf, and P. J. Parker. 1988. Determination of the primary structure of PLC-154 demonstrates diversity of phosphoinositide-specific phospholipase C activities. *Cell* 54:171–177.

Kater, L. A., E. J. Goetzl, and K. F. Austen. 1976. Isolation of human eosinophil phospholipase D. *J. Clin. Invest.* 57:1173–1180.

Kennedy, S. P., and E. L. Becker. 1987. Ectophospholipase A_2 activity of the rabbit peritoneal neutrophil. *Int. Arch. Aller. Appl. Immunol.* 83:238–246.

Kennerly, D. A. 1990. Phosphatidylcholine is a quantitatively more important source of increased 1,2-diacylglycerol than is phosphatidylinositol in mast cells. *J. Immunol.* 144:3912–3919.

Kessels, G. C. R., D. Roos, and A. J. Verhoeven. 1991. fMet-Leu-Phe-induced activation of phospholipase D in human neutrophils: Dependence on changes in cytosolic free Ca^{2+} concentration and relation with respiratory burst activation. *J. Biol. Chem.* 266:23152–23156.

Khan, W. A., G. C. Blobe, and Y. A. Hannun. 1992. Activation of protein kinase C by oleic acid: Determination and analysis of inhibition by detergent micelles and physiologic membranes: Requirement for free oleate. *J. Biol. Chem.* 267:3605–3612.

Kikuchi, A., O. Kozawa, K. Kaibuchi, T. Katada, M. Ui, and Y. Takai. 1986. Direct evidence for involvement of a guanine nucleotide-binding protein in chemotactic peptide-stimulated formation of inositol bisphosphate and trisphosphate in differentiated human leukemic (HL-60) cells: Reconstitution with G_i or G_o of the plasma membranes of ADP-ribosylated by pertussis toxin. *J. Biol. Chem.* 261:11558–11562.

Kim, H. K., J. W. Kim, A. Zilberstein, B. Margolis, J. G. Kim, J. Schlessinger, and S. G. Rhee. 1991. PDGF stimulation of inositol phospholipid hydrolysis requires PLC-gamma 1 phosphorylation on tyrosine residues 783 and 1254. *Cell* 65:435–441.

Kim, J. W., S. S. Sim, U. H. Kim, S. Nishibe, M. I. Wahl, G. Carpenter, and S. G. Rhee. 1990. Tyrosine residues in bovine phospholipase C-(gamma) phosphorylated by the epidermal growth factor receptor *in vitro. J. Biol. Chem.* 265:3940–3943.

Kim, U. H., J. W. Kim, and S. G. Rhee. 1989. Phosphorylation of phospholipase C-γ by cAMP-dependent protein kinase. *J. Biol. Chem.* 264:20167–20170.

Kim, U. H., H. S. Kim, and S. G. Rhee. 1990. Epidermal growth factor and platelet-derived growth factor promote translocation of phospholipase C-gamma from cytosol to membrane. *FEBS Lett.* 270:33–36.

Kim, U. H., D. Fink Jr., H. S. Kim, D. J. Park, M. L. Contreras, G. Guroff, and S. G. Rhee. 1991. Nerve growth factor stimulates phosphorylation of phospholipase C gamma in PC12 cells. *J. Biol. Chem.* 266:1359–1362.

Kiss, Z., and W. B. Anderson. 1989. Phorbol ester stimulates the hydrolysis of phosphatidylethanolamine in leukemic HL-60, NIH 3T3, and baby hamster kidney cells. *J. Biol. Chem.* 264:1483–1487.

Kiss, Z., and W. B. Anderson. 1990. ATP stimulates the hydrolysis of phosphotidylethanolamine in NIH 3T3 cells: Potentiating effects of guanosine triphosphates and sphingosine. *J. Biol. Chem.* 265:7345–7350.

Kitagawa, M., J. A. Williams, and R. C. De Lisle. 1990. Amylase release from streptolysin O-permeabilized pancreatic acini. *Am. J. Physiol.* 259:G157–G164.

Kondo, T., H. Inui, F. Konishi, and T. Inagami. 1992a. Phospholipase D mimics platelet-derived growth factor as a competence factor in vascular smooth muscle cells. *J. Biol. Chem.* 267:23609–23616.

Kondo, T., F. Konishi, H. Inui, and T. Inagami. 1992b. Diacylglycerol formation from phosphatidylcholine in angiotensin II-stimulated vascular smooth muscle cells. *Biochem. Biophys. Res. Commun.* 187:1460–1465.

Konishi, F., T. Kondo, and T. Inagami. 1991. Phospholipase D in cultured rat vascular smooth muscle cells and its activation by phorbol ester. *Biochem. Biophys. Res. Commun.* 179:1070–1076.

Kriz, R., L. L. Lin, L. Sultzman, C. Ellis, C. H. Heldin, T. Pawson, and J. Knopf. 1990. Phospholipase C isozymes: Structural and functional similarities. *Ciba Found. Symp.* 150:112–123.

Kumjian, D. A., M. I. Wahl, S. G. Rhee, and T. O. Daniel. 1989. Platelet-derived growth factor (PDGF) binding promotes physical association of PDGF receptors with phospholipase C. *Proc. Natl. Acad. Sci. U.S.A.* 86:8232–8236.

Kumjian, D. A., A. Barnstein, S. G. Rhee, and T. O. Daniel. 1991. Phospholipase C gamma complexes with ligand-activated platelet-derived growth factor receptors. An intermediate implicated in phospholipase activation. *J. Biol. Chem.* 266:3973–3980.

Lapetina, E. G., and M. F. Crouch. 1989. The relationship between phospholipases A_2 and C in signal transduction. *Ann. N. Y. Acad. Sci.* 3:155–171.

Lapetina, E. G., and P. Cuatrecasas. 1979. Stimulation of phosphatidic acid production in platelets precedes the formation of arachidonate and parallels the release of serotonin. *Biochim. Biophys. Acta* 573:394–402.

Lavie, Y., and M. Liscovitch. 1990. Activation of phospholipase D by sphingoid bases in NG108-15 neural-derived cells. *J. Biol. Chem.* 265:3868–3872.

Lee, C. H., D. Park, D. Wu, S. G. Rhee, and M. I. Simon. 1992. Members of the G_q α subunit gene family activate phospholipase C B isozymes. *J. Biol. Chem.* 267:16044–16047.

Lee, K. Y., S. H. Ryu, P. G. Suh, W. C. Choi, and S. G. Rhee. 1987. Phospholipase C associated with particulate fractions of bovine brain. *Proc. Natl. Acad. Sci. U.S.A.* 84:5540–5554.

Liao, F., H. S. Shin, and S. G. Rhee. 1992. Tyrosine phosphorylation of phospholipase C-gamma 1 induced by cross-linking of the high-affinity or low-affinity receptor for IgG in U937 cells. *Proc. Natl. Acad. Sci. U.S.A.* 89:3659–3663.

Lin, L., A. Y. Lin, and J. L. Knopf. 1992. Cytosolic phospholipase A_2 is coupled to hormonally regulated release of arachidonic acid. *Proc. Natl. Acad. Sci. U.S.A.* 89:6147–6151.

Lin, L., M. Wartmann, A. Y. Lin, J. L. Knopf, A. Seth, and R. J. Davis. 1993. CPLA2 Is phosphorylated by MAP kinase. *Cell* 72:268–278.

Lin, P., and A. M. Gilfillan. 1992. The role of calcium and protein kinase C in the IgE-dependent activation of phosphatidylcholine-specific phospholipase D. *Eur. J. Biochem.* 207:163–168.

Lin, P., G. A. Wiggan, and A. M. Gilfillan. 1991. Activation of phospholipase D in a rat mast (RBL2H3) cell line: A possible unifying mechanism for IgE-dependent degranulation and arachidonic acid metabolite release. *J. Immunol.* 146:1609–1616.

Lin, P., W. C. Fung, and A. M. Gilfillan. 1992. Phosphatidylcholine-specific phospholipase D-derived 1,2-diacylglycerol does not initiate protein kinase C activation in the RBL 2H3 mast-cell line. *Biochem. J.* 287:325–331.

Liscovitch, M. 1989. Phosphatidylethanol biosynthesis in ethanol-exposed NG108-15 neuroblastoma x glioma hybrid cells: Evidence for activation of a phospholipase D phosphatidyl transferase activity by protein kinase C. *J. Biol. Chem.* 264:1450–1456.

Liscovitch, M., and A. Amsterdam. 1989. Gonadotropin-releasing hormone activates phospholipase D in ovarian granulosa cells: Possible role in signal transduction. *J. Biol. Chem.* 265:11762–11767.

Liscovitch M., and Y. Eli. 1991. Ca^{2+} inhibits guanine nucleotide-activated phospholipase D in neural-derived NG108-15 cells. *Cell Reg.* 2:1011–1019.

Litosch, I., and J. N. Fain. 1986. Regulation of phosphoinositide breakdown by guanine nucleotides. *Life Sci.* 39:187–194.

Litosch, I., C. Wallis, and J. N. Fain. 1985. 5-Hydroxytryptamine stimulates inositol phosphate production in a cell-free system from blowfly salivary glands. Evidence for a role of GTP in coupling receptor activation to phosphoinositide breakdown. *J. Biol. Chem.* 260:5464–5471.

Liu, P., L. Sun, M. Wen, and J. Hayashi. 1992. Stimulation of prostaglandin production in rat thymic epithlial cells by protein kinase C mediated activation of phospholipase A₂. *Biochem. Int.* 27:931–938.

Llahi, S., and J. N. Fain. 1992. α₁-Adrenergic receptor-mediated activation of phospholipase D in rat cerebral cortex. *J. Biol. Chem.* 267:3679–3685.

MacNulty, E. E., S. J. McClue, I. C. Carr, T. Jess, M. J. O. Wakelam, and G. Milligan. 1992. α₂-C10 adrenergic receptors expressed in rat 1 fibroblasts can regulate both adenylylcyclase and phospholipase D-mediated hydrolysis of phosphatidylcholine by interacting with pertussis toxin-sensitive guanine nucleotide-binding proteins. *J. Biol. Chem.* 267:2149–2156.

Majerus, P. W., T. M. Connolly, H. Deckmyn, T. S. Ross, T. E. Bross, H. Ishii, V. S. Bansal, and D. B. Wilson. 1986. The metabolism of phosphoinositide-derived messenger molecules. *Science* 234:1519–1526.

Margolis, B., S. G. Rhee, S. Felder, M. Mervic, R. Lyall, A. Levitzki, A. Ullrich, A. Zilberstein, and J. Schlessinger. 1989. EGF induces tyrosine phosphorylation of phospholipase C-II: A potential mechanism for EGF receptor signaling. *Cell* 57: 1101–1107.

Margolis, B., F. Bellot, A. M. Honegger, A. Ullrich, J. Schlessinger, and A. Zilberstein. 1990. Tyrosine kinase activity is essential for the association of phospholipase C-gamma with the epidermal growth factor receptor. *Molec. Cell. Biol.* 10:435–441.

Marquardt, D. L., and L. L. Walker. 1991. Lysophosphatidylcholine induces mast cell secretion and protein kinase C acivation. *J. Aller. Clin. Immunol.* 5:721–730.

Martin, T. W. 1988. Formation of diacylglycerol by a phospholipase D-phosphatidate phosphatase pathway specific for phosphatidylcholine in endothelial cells. *Biochim. Biophys. Acta* 962:282–296.

Martin, T. W., and K. Michaelis. 1989. P₂-purinergic agonists stimulate phosphodiesteratic cleavage of phosphatidylcholine in endothelial cells: Evidence for activation of phospholipase D. *J. Biol. Chem.* 264:8847–8856.

Martin, T. W., D. R. Feldman, K. E. Goldstein, and J. R. Wagner. 1989. Long-term phorbol ester treatment dissociates phospholipase D activation from phosphoinositide hydrolysis and prostacyclin synthesis in endothelial cells stimulated with bradykinin. *Biochem. Biophys. Res. Commun.* 165:319–326.

Martin, T. W., D. R. Feldman, and K. C. Michaelis. 1990. Phosphatidylcholine hydrolysis stimulated by phorbol myristate acetate is mediated principally by phospholipase D in endothelial cells. *Biochim. Biophys. Acta* 1053:162–172.

Martinson, E. A., D. Goldstein, and J. H. Brown. 1989. Muscarinic receptor activation of phosphatidylcholine hydrolysis. Relationship to phosphoinositide hydrolysis and diacylglycerol metabolism. *J. Biol. Chem.* 264:14748–14754.

Mason-Garcia, M., S. Clejan, J. Tou, and B. S. Beckman. 1992. Signal transduction by the erythropoietin receptor: Evidence for the activation of phospholipases A₂ and C. *Am. J. Physiol. Cell Physiol.* 262:C1197–C1203.

McBride, K., S. G. Rhee, and S. Jaken. 1991. Immunocytochemical localization of phospholipase C-gamma in rat embryo fibroblasts. *Proc. Natl. Acad. Sci. U.S.A.* 88: 7111–7115.

McIntyre, T. M., S. L. Reinhold, S. M. Prescott, and G. A. Zimmerman. 1987. Protein kinase C activity appears to be required for the synthesis of platelet-activating factor and leukotriene B4 by human neutrophils. *J. Biol. Chem.* 262:15370–15376.

McKenzie, F. R., K. Seuwen, and J. Pouyssegur. 1992. Stimulation of phosphatidylcholine breakdown by thrombin and carbachol but not by tyrosine kinase receptor ligands in cells transfected with M1 muscarinic receptors: Rapid desensitization of phosphocholine-specific (PC) phospholipase D but sustained activity of PC-phospholipase C. *J. Biol. Chem.* 267:22759–22769.

McPhail, L. C., C. C. Clayton, and R. Snyderman. 1984. A potential second messenger role for unsaturated fatty acids: Activation of a Ca^{++}-dependent protein kinase. *Science* 224:622–625.

Meisenhelder, J., P. G. Suh, S. G. Rhee, and T. Hunter. 1989. Phospholipase C-γ is a substrate for the PDGF and EGF receptor protein-tyrosine kinases in vivo and in vitro. *Cell* 57:1109–1122.

Meldrum, E., R. W. Kriz, N. Totty, and P. J. Parker. 1991. A second gene product of the inositol-phospholipid-specific phospholipase C delta subclass. *Eur. J. Biochem.* 196:159–165.

Metz, S. A., and M. Dunlop. 1990. Stimulation of insulin release by phospholipase D. *Biochem. J.* 270:427–435.

Minnicozzi, M., J. C. Anthes, M. I. Siegel, M. M. Billah, and R. W. Egan. 1990. Activation of phospholipase D in normodense human eosinophils. *Biochem. Biophys. Res. Commun.* 170:540–547.

Mohn, H., V. Chalifa, and M. Liscovitch. 1992. Substrate specificity of neutral phospholipase D from rat brain studied by selective labeling of endogenous synaptic membrane phospholipids *in vitro. J. Biol. Chem.* 267:11131–11136.

Moolenaar, W. H. 1991. Mitogenic action of lysophosphatidic acid. *Adv. Can. Res.* 57: 87–102.

Moolenaar, W. H., W. Kruijer, B. C. Tilly, I. Verlaan, A. J. Bierman, and S. W. de Latt. 1986. Growth factor-like action of phosphatidic acid. *Nature* 323:171–173.

Moolenaar, W. H., A. J. Bierman, B. C. Tilly, I. Verlaan, L. H. K. Defize, A. M. Honegger, A. Ullrich, and J. Schlessinger. 1988. A point mutation in the ATP binding site of the EGF receptor abolishes signal transduction. *EMBO J.* 7:707–710.

Moriarty, T. M., B. Gillo, D. J. Carty, R. T. Premont, E. M. Landau, and R. Lyengar. 1988. Beta gamma subunits of GTP-binding proteins inhibit muscarinic receptor stimulation of phospholipase C. *Proc. Natl. Acad. Sci. U.S.A.* 85:8865–8869.

Moriarty, T. M., S. C. Sealfon, D. J. Carty, J. L. Roberts, R. Lyengar, and E. M. Landau. 1989. Coupling of exogenous receptors to phospholipase C in *Xenopus* oocytes through pertussis toxin-sensitive and -insensitive pathways. Cross-talk through heterotrimeric G-proteins. *J. Biol. Chem.* 264:13524–13530.

Moriarty, T. M., E. Padrell, D. J. Carty, G. Omri, E. M. Landau, and R. Lyengar. 1990. G_o protein as signal transducer in the pertussis toxin-sensitive phosphatidylinositol pathway. *Nature* 343:79–82.

Moritz, A., P. N. E., De Graan, W. H., Gispen, and K. W. A. Wirtz. 1992. Phosphatidic acid is a specific activator of phosphatidylinositol-4-phosphate kinase. *J. Biol. Chem.* 267:7207–7210.

Morrison, D. K., D. R. Kaplan, S. G. Rhee, and L. T. Williams. 1990. Platelet-derived growth factor (PDGF)-dependent association of phospholipase C-gamma with the PDGF receptor signaling complex. *Mole. Cell. Biol.* 10:2359–2366.

Mueller, H. W., J. T. O'Flaherty, and R. L. Wykle. 1982. Ether lipid content and fatty acid distribution in rabbit polymorphonuclear neutrophil phospholipids. *Lipids* 17: 72–77.

Mufson, R. A., E. Okin, and I. B. Weinstein. 1981. Phorbol esters stimulate the rapid release of choline from prelabelled cells. *Carcinogenesis* 2:1095–1102.

Mullmann, T. J., M. I. Siegel, R. W. Egan, and M. M. Billah. 1990a. Phorbol-12-myristate-

13-acetate activation of phospholipase D in human neutrophils leads to the production of phosphatides and diglycerides. *Biochem. Biophys. Res. Commun.* 170: 1197–1202.

Mullmann, T. J., M. I. Siegel, R. W. Egan, and M. M. Billah. 1990b. Complement C5a activation of phospholipase D in human neutrophils: A major route to the production of phosphatidates and diglycerides. *J. Immunol.* 144:1901–1908.

Mullmann, T. J., M. I. Siegel, R. W. Egan, and M. M. Billah. 1991. Sphingosine inhibits phosphatidate phosphohydrolase in human neutrophils by a protein kinase C-independent mechanism. *J. Biol. Chem.* 266:2013–2016.

Murayama, T., Y. Kajiyama, and Y. Nomura. 1990. Histamine-stimulated and GTP-binding protein-mediated phospholipase A_2 activation in rabbit platelets. *J. Biol. Chem.* 265:4290–4295.

Murrin, R. J., and M. R. Boarder. 1992. Neuronal ''nucleotide'' receptor linked to phospholipase C and phospholipase D? Stimulation of PC12 cells by ATP analogues and UTP. *Molec. Pharmacol.* 41:561–568.

Nakamura, T., and M. Ui. 1985. Simultaneous inhibitions of inositol phospholipid breakdown, arachidonic acid release, and histamine secretion in mast cells by islet activating protein, pertussis toxin. *J. Biol. Chem.* 260:3584–3593.

Nakashima, S., H. Hattori, L. Shirato, A. Takenaka, and Y. Nozawa. 1987a. Differential sensitivity of arachidonic acid release and 1,2-diacylglycerol formation to pertussis toxin, GDPβS and NaF in saponin-permeabilized human platelets: Possible evidence for distinct GTP-binding proteins involving phospholipase C and A_2 activation. *Biochem. Biophys. Res. Commun.* 148:971–978.

Nakashima, S., T. Tohmatsu, H. Hattori, A. Suganuma, and Y. Nozawa. 1987b. Guanine nucleotides stimulate arachidonic acid release by phospholipase A_2 in saponin-permeabilized human platelets. *J. Biochem.* 101:1055–1058.

Nakashima, S., K.-I. Nagata, K. Ueeda, and Y. Nozawa. 1988. Stimulation of arachidonic acid release by guanine nucleotide in saponin-permeabilized neutrophils: Evidence for involvement of GTP-binding protein in phospholipase A_2 activation. *Arch. Biochem. Biophys.* 261:375–382.

Netiv, E., M. Liscovitch, and Z. Naor. 1991. Delayed activation of phospholipase D by gonadotropin-releasing hormone in a clonal pituitary gonadotrope cell line (alpha T3-1). *FEBS Lett.* 295:107–109.

Nielson, C. P., J. Stutchfield, and S. Cockcroft. 1991. Chemotactic peptide stimulation of arachidonic acid release in HL60 cells, an interaction between G protein and phospholipase C mediated signal transduction. *Biochim. Biophys. Acta* 1095:83–89.

Nishizuka, Y. 1984. Turnover of inositol phospholipids and signal transduction. *Science* 225:1365–1370.

Ohsako, S., and T. Deguchi. 1981. Stimulation by phosphatidic acid of calcium influx and cyclic GMP synthesis in neuroblastoma cells. *J. Biol. Chem.* 256:10945–10948.

Ohta, S., A. Matsui, Y. Nazawa, and Y. Kagawa. 1988. Complete cDNA encoding a putative phospholipase C from transformed human lymphocytes. *FEBS Lett.* 242: 31–35.

Ohtsuka, T., M. Ozawa, N. Okamura, and S. Ishibashi. 1989. Stimulatory effects of a short chain phosphatidate on superoxide anion production in guinea pig polymorphonuclear leukocytes. *J. Biochem.* 106:259–263.

Oishi, K., R. L. Raynor, P. A. Charp, and J. F. Kuo. 1988. Regulation of protein kinase C by lysophospholipids: Potential role in signal transduction. *J. Biol. Chem.* 263: 6865–6871.

Okano, Y., K. Yamada, K. Yano, and Y. Nozawa. 1987. Guanosine 5′-(γ-thio)triphosphate stimulates arachidonic acid liberation in permeabilized rat peritoneal mast cells. *Biochem. Biophys. Res. Commun.* 145:1267–1275.

Olashaw, N. E., S. G. Rhee, and W. J. Pledger. 1990. Cyclic AMP agonists induce the phosphorylation of phospholipase C-γ and of a 76 kDa protein co-precipitated by anti-(phospholipase C-γ) monoclonal antibodies in BALB/c-3T3 cells. *Biochem. J.* 272:297–303.

Olson, S. C., S. R. Tyagi, and J. D. Lambeth. 1990. Fluoride activates diradylglycerol and superoxide generation in human neutrophils via PLD/PA phosphohydrolase-dependent and -independent pathways. *FEBS Lett.* 272:19–24.

Olson, S. C., E. P. Bowman, and J. D. Lambeth. 1991. Phospholipase D activation in a cell-free system from human neutrophils by phorbol 12-myristate 13-acetate and guanosine 5′-*O*-(3-thiotriphosphate): Activation is calcium dependent and requires protein factors in both the plasma membrane and cytosol. *J. Biol. Chem.* 266: 17236–17242.

Pagano, R. E., and K. J. Longmuir. 1985. Phosphorylation, transbilayer movement, and facilitated intracellular transport of diacylglycerol are involved in the uptake of a fluorescent analog of phosphatidic acid by cultured fibroblasts. *J. Biol. Chem.* 260: 1909–1916.

Pai, J.-K., J. A. Pachter, I. B. Weinstein, and W. R. Bishop. 1991. Overexpression of protein kinase C β1 enhances phospholipase D activity and diacylglycerol formation in phorbol ester-stimulated rat fibroblasts. *Proc. Natl. Acad. Sci. U.S.A.* 88: 598–602.

Pang, I. H., and P. C. Sternweis. 1989. Purification of unique α-subunits of GTP-binding regulatory proteins by affinity chromatography with immobilized βγ-subunits. *J. Biol. Chem.* 265:18707–18712.

Pappu, A. S., and G. Hauser. 1983. Propranolol-induced inhibition of rat brain cytoplasmic phosphatidate phosphohydrolase. *Neurochem. Res.* 8:1565–1575.

Paris, S., and J. Pouyssegur. 1986. Pertussis toxin inhibits thrombin-induced activation of phosphoinositide hydrolysis and Na^+/H^+ exchange in hamster fibroblasts. *EMBO J.* 5:55–60.

Paris, S., and J. Pouyssegur. 1987. Further evidence for a phospholipase C-coupled G protein in hamster fibroblasts: Induction of inositol phosphate formation by fluoroaluminate and vanadate and inhibition by pertussis toxin. *J. Biol. Chem.* 262: 1970–1976.

Park, D., D. Y. Jhon, C. W. Lee, K. H. Lee, and S. G. Rhee. 1993. Activation of phospholipase C isozymes by G protein βγ subunits. *J. Biol. Chem.* 268:4573–4576.

Park, D. J., H. K. Min, and S. G. Rhee. 1991. IgE-induced tyrosine phosphorylation of phospholipase C-gamma 1 in rat basophilic leukemia cells. *J. Biol. Chem.* 266: 24237–24240.

Park, D. J., D. Jhon, R. Kriz, J. Knopf, and S. G. Rhee. 1992a. Cloning, sequencing, expression, and G_q-independent activation of phospholipase C-β2. *J. Biol. Chem.* 267:16048–16055.

Park, D. J., H. K. Min, and S. G. Rhee. 1992b. Inhibition of CD3-linked phospholipase C by phorbol ester and by cAMP is associated with decreased phosphotyrosine and increased phosphoserine contents of PLC-γ1. *J. Biol. Chem.* 267:1496–1501.

Parker, J., L. W. Daniel, and M. Waite. 1987. Evidence of protein kinase C involvement in phorbol diester-stimulated arachidonic acid release and prostaglandin synthesis. *J. Biol. Chem.* 262:5385–5393.

Perry, D. K., W. L. Hand, D. E. Edmondson, and J. D. Lambeth. 1992. Role of phospholipase D-derived diradylglycerol in the activation of the human neutrophil respiratory burst oxidase: Inhibition by phosphatidic acid phosphohydrolase inhibitors. *J. Immunol.* 149:2749–2758.

Peveri, P., and J. T. Curnutte. 1990. Phosphatidic acid may be a physiologic activator of human neutrophil NADPH oxidase. *Blood* 76:751 (Abs.).

Plein, P., M. Kaszkin, and V. Kinzel. 1992. Accumulation of ester- and ether-linked phosphatidates by HeLa cells in response to ionophore A23187 through activation of phospholipase D. *Biol. Chem. Hoppe Selyer* 373:151–157.

Plevin, R., S. J. Cook, S. Palmer, and M. J. Wakelam. 1991. Multiple sources of *sn*-1,2-diacylglycerol in platelet-derived-growth-factor-stimulated Swiss 3T3 fibroblasts: Evidence for activation of phosphoinositidase C and phosphatidylcholine-specific phospholipase D. *Biochem. J.* 279:559–565.

Polakis, P. G., R. J. Uhing, and R. Snyderman. 1988. The formylpeptide chemoattractant receptor copurifies with a GTP-binding protein containing a distinct 40-kDa pertussis toxin substrate. *J. Biol. Chem.* 263:4969–4976.

Putney, J. W., S. J. Weiss, C. M. Van de Walle, and R. A. Haddas. 1980. Is phosphatidic acid a calcium ionophore under neutrohumoral control? *Nature* 284:345–347.

Qian, Z., and L. R. Drewes. 1989. Muscarinic acetylcholine receptor regulated phosphatidylcholine phospholipase D in canine brain. *J. Biol. Chem.* 264:21720–21724.

Reinhold, S. L., S. M. Prescott, G. A. Zimmerman, and T. M. McIntyre. 1990. Activation of human neutrophil phospholipase D by three separable mechanisms. *FASEB J.* 4:208–214.

Rhee, S. G., and K. D. Choi. 1992. Multiple forms of phospholipase C isozymes and their activation mechanisms. *Adv. Sec. Mess. Phosphoprot. Res.* 26:35–61.

Rhee, S. G., U. H. Kim, and J. W. Kim. 1990. Cross-talk between cellular signalling cascade pathways suggested by cAMP-induced phosphorylation of phospholipase C-γ. In: *The Biology and Medicine of Signal Transduction,* edited by Y. Nishizuka, pp. 164–169. Raven Press, New York.

Rider, L. G., R. W. Dougherty, and J. E. Niedel. 1988. Phorbol diesters and dioctanoylglycerol stimulate accumulation of both diacylglycerols and alkylacylglycerols in human neutrophils. *J. Immunol.* 140:200–207.

Rock, C. O., and S. Jackowski. 1987. Thrombin- and nucleotide-activated phosphatidylinositol 4,5-bisphosphate phospholipase C in human platelet membranes. *J. Biol. Chem.* 262:5492–5498.

Rosoff, P. M., N. Savage, and C. A. Dinarello. 1988. Interleukin-1 stimulates diacylglycerol production in T lymphocytes by a novel mechanism. *Cell* 54:73–81.

Rubin, R. 1988. Phosphatidylethanol formation in human platelets: Evidence for thrombin-induced activation of phospholipase D. *Biochem. Biophys. Res. Commun.* 156:1090–1096.

Russo-Marie, F. 1991. Lipocortins: An update. *Prost. Leuk. Ess. Fatty Acids* 42:83–89.

Ryu, S. H., K. S. Cho, K.-Y. Lee, P.-G. Suh, and S. G. Rhee. 1987a. Purification and characterization of two immunologically distinct phosphoinositide-specific phospholipase C from bovine brain. *J. Biol. Chem.* 262:12511–12518.

Ryu, S. H., P.-G. Suh, K. S. Cho, K.-Y. Lee, and S. G. Rhee. 1987b. Bovine brain cytosol contains three immunologically distinct forms of inositol phospholipid-specific phospholipase C. *Proc. Natl. Acad. Sci. U.S.A.* 84:6649–6653.

Ryu, S. H., U. H. Kim, M. I. Wahl, A. B. Brown, G. Carpenter, K.-P. Huang, and S. G. Rhee. 1990. Feedback regulation of phospholipase C-β by protein kinase C. *J. Biol. Chem.* 265:17941–17945.

Saito, M., and J. Kanfer. 1975. Phosphatidohydrolase activity in a solubilized preparation from rat brain particulate fractions. *Arch. Biochem. Biophys.* 169:318–323.

Salmon, D. M., and T. W. Honeyman. 1980. Proposed mechanism of cholinergic action in smooth muscle. *Nature* 284:344–345.

Sandmann, J., and R. J. Wurtman. 1991. Stimulation of phospholipase D activity in human neuroblastoma (LA-N-2) cells by activation of muscarinic acetylcholine receptors or by phorbol esters: Relationship to phosphoinositide turnover. *J. Neurochem.* 56:1312–1319.

Sandmann, J., E. G. Peralta, and R. J. Wurtman. 1991. Coupling of transfected muscarinic acetylcholine receptor subtypes to phospholipase D. *J. Biol. Chem.* 266:6031–6034.

Schmid, P. C., P. V. Reddy, V. Natarajan, and H. H. O. Schmid. 1983. Metabolism of N-acylethanolamine phospholipids by a mammalian phosphodiesterase of the phospholipase D type. *J. Biol. Chem.* 258:9302–9306.

Secrist, J. P., L. Karnitz, and R. T. Abraham. 1991. T-cell antigen receptor ligation induces tyrosine phosphorylation of phospholipase C-gamma 1. *J. Biol. Chem.* 266:12135–12139.

Serhan, C. N., J. Fridovich, E. J. Goetzl, P. B. Dunham, and G. Weissmann. 1982. Leukotriene B4 and phosphatidic acid are calcium ionophores: Studies using arsenazo III in liposomes. *J. Biol. Chem.* 257:4746–4752.

Shenker, A., P. Goldsmith, C. G. Unson, and A. M. Spiegel. 1991. The G-protein coupled to the thromboxane A_2 receptor in human platelets is a member of the novel G_q family. *J. Biol. Chem.* 266:9309–9313.

Shinomura, T., Y. Asaoka, M. Oka, K. Yoshida, and Y. Nishizuka. 1991. Synergistic action of diacylglycerol and unsaturated fatty acid for protein kinase C activation: Its possible implications. *Proc. Natl. Acad. Sci. U.S.A.* 88:5149–5153.

Shortridge, R. D., J. Yoon, C. R. Lending, B. T. Bloomquist, M. Perdew, and W. L. Pak. 1991. A *Drosophila* phospholipase C gene that is expressed in the central nervous system. *J. Biol. Chem.* 266:12474–12480.

Silk, S. T., S. Clejan, and K. Witkom. 1989. Evidence of GTP-binding protein regulation of phospholipase A_2 activity in isolated human platelet membranes. *J. Biol. Chem.* 264:21466–21469.

Slivka, S. R., and P. A. Insel. 1987. α1-Adrenergic receptor-mediated phosphoinositide hydrolysis and prostaglandin E_2 formation in Manin-Darby canine kidney cells: Possible parallel activation of phospholipase C and phospholipase A_2. *J. Biol. Chem.* 262:4200–4207.

Slivka, S. R., and P. A. Insel. 1988. Phorbol ester and neomycin dissociate bradykinin receptor-mediated arachidonic acid release and polyphosphoinositide hydrolysis in Madin-Darby canine kidney cells. Evidence that bradykinin mediates noninterdependent activation of phospholipases A_2 and C. *J. Biol. Chem.* 263:14640–14647.

Slivka, S. R., K. E. Meier, and P. A. Insel. 1988. α1-Adrenergic receptors promote phosphatidylcholine hydrolysis in MDCK-D1 cells: A mechanism for rapid activation of protein kinase C. *J. Biol. Chem.* 263:12242–12246.

Smith, C. D., C. C. Cox, and R. Snyderman. 1986. Receptor-coupled activation of phosphoinositide-specific phospholipase C by an N protein. *Science* 232:97–100.

Smrcka, A. V., J. R. Helper, K. O. Brown, and P. C. Sternweis. 1991. Regulation of polyphosphoinositide-specific phospholipase C by purified G_q. *Science* 251:804–807.

Stahl, M. L., C. R. Ferenz, K. L. Kelleher, R. W. Kriz, and J. L. Knopf. 1988. Sequence similarity of phospholipase C with the non-catalytic region of src. *Nature* 332:269–272.

Strathmann, M., and M. Simon. 1990. G protein diversity: A distinct class of α subunits is present in vertebrates and invertebrates. *Proc. Natl. Acad. Sci. U.S.A.* 87:9113–9117.

Suh, P. G., S. H. Ryu, K. H. Moon, H. W. Suh, and S. G. Rhee 1988a. Inositol phospholipid-specific phospholipase C: Complete cDNA and protein sequences and sequence homology to tyrosine kinase-related oncogene products. *Proc. Natl. Acad. Sci. U.S.A.* 85:5419–5423.

Suh, P. G., S. H. Ryu, K. H. Moon, H. W. Suh, and S. G. Rhee. 1988b. Cloning and sequence of multiple forms of phopholipase C. *Cell* 54:161–169.

Takamura, H., H. Narita, H. J. Park, K. Tanaka, T. Matsuura, and M. Kito. 1987. Differential hydrolysis of phospholipid molecular species during activation of human platelets with thrombin and collagen. *J. Biol. Chem.* 262:2262–2269.

Taki, T., and J. N. Kanfer. 1979. Partial purification and properties of a rat brain phospholipase D. *J. Biol. Chem.* 254:9761–9765.

Tang, W., and A. G. Gilman. 1991. Type-specific regulation of adenylyl cyclase by G protein βγ subunits. *Science* 254:1500–1503.

Tao, W., T. F. P. Molski, and R. I. Sha'afi. 1989. Arachidonic acid release in rabbit neutrophils. *Biochem. J.* 257:633–637.

Taylor, S. J., H. Z. Chae, S. G. Rhee, and J. H. Exton. 1991. Activation of the β1 isozyme of phospholipase C by α subunits of the G_q class of G proteins. *Nature* 350:516–518.

Tence, M., E. Jouvin-Marche, G. Bessou, M. Record, and J. Benveniste. 1985. Etherphospholipid composition in neutrophils and platelets. *Thromb. Res.* 38:207–214.

Todderud, G., M. I. Wahl, S. G. Rhee, and G. Carpenter. 1990. Stimulation of phospholipase C-gamma 1 membrane association by epidermal growth factor. *Science* 249:296–298.

Tyagi, S. R., M. Tamura, D. N. Burnham, and J. D. Lambeth. 1988. Phorbol myristate acetate augments chemoattractant-induced diglyceride generation in human neutrophils, but inhibits phosphoinositide hydrolysis: Implications for the mechanism of PMA priming of the respiratory burst. *J. Biol. Chem.* 263:13191–13198.

Tyagi, S. R., D. N. Burnham, and J. D. Lambeth. 1989. On the biological occurrence and regulation of 1-acyl and 1-O-alkyl-diglycerides in human neutrophils: Selective destruction of 1-acyl species using rhizopus lipase. *J. Biol. Chem.* 264:12977–12982.

Tyagi, S. R., S. C. Olson, D. N. Burnham, and J. D. Lambeth. 1991. Cyclic AMP-elevating agents block chemoattractant activation of diradylglycerol generation by inhibiting phospholipase D activation. *J. Biol. Chem.* 266:3498–3504.

Tyson, C. A., H. Vande Zande, and D. E. Green. 1992. Phospholipids as ionophores. *J. Biol. Chem.* 1976:1326–1332.

Uhing, R. J., H. Jiang, V. Prpic, and J. H. Exton. 1985. Regulation of a liver plasma membrane phosphoinositide phosphodiesterase by guanine nucleotides and calcium. *FEBS. Lett.* 188:317–320.

Uings, I. J., N. T. Thompson, R. W. Randall, G. D. Spacey, R. W. Bonser, A. T. Hudson, and L. G. Garland. 1992. Tyrosine phosphorylation is involved in receptor coupling to phospholipase D but not phospholipase C in the human neutrophil. *Biochem. J.* 281:597–600.

van Blitterswijk, W., H. Hilkmann, J. de Widt, and R. van der Bend. 1991. Phosopholipid metabolism in bradykinin-stimulated human fibroblasts. II. Phosphatidylcholine breakdown by phospholipases C and D; involvement of protein kinase C. *J. Biol. Chem.* 266:10344–10350.

van Corven, E. J., A. Groenink, K. Jalink, T. Eichholtz, and W. H. Moolenaar. 1989.

Lysophosphatidate-induced cell proliferation: Identification and dissection of signaling pathways mediated by G proteins. *Cell* 59:45–54.

van der Bend, R. L., J. Brunner, K. Jalink, E. J. van Corven, W. H. Moolenaar, and W. J. van Blitterswijk. 1992. Identification of a putative membrane receptor for the bioactive phospholipid, lysophosphatidic acid. *EMBO J.* 11:2495–2501.

van der Meulen, J., and R. J. Haslam. 1990. Phorbol ester treatment of intact rabbit platelets greatly enhances both the basal and guanosine 5'-[γ-thio]triphosphate-stimulated phospholipase D activities of isolated platelet membranes: Physiological activation of phospholipase D may be secondary to activation of phospholipase C. *Biochem. J.* 271:693–700.

Verghese, M. W., C. D. Smith, and R. Snyderman. 1985. Potential role for a guanine nucleotide regulatory protein in chemoattractant receptor mediated polyphosphoinositide metabolism, Ca^{++} mobilization and cellular responses by leukocytes. *Biochem. Biophys. Res. Comm.* 127:450–457.

Vinggaard, A. M., and H. S. Hansen. 1991. Phorbol ester and vasopressin activate phospholipase D in Leydig cells. *Molec. Cell. Endocrinol.* 79:157–165.

Volterra, A., and S. A. Siegelbaum. 1988. Role of two different guanine nucleotide-binding proteins in the antagonistic modulation of the S-type K^+ channel by cAMP and arachidonic acid metabolites in *Aplysia* sensory neurons. *Proc. Natl. Acad. Sci. U.S.A.* 85:7810–7814.

Wahl, M. I., T. O. Daniel, and G. Carpenter. 1988. Antiphosphotyrosine recovery of phospholipase C activity after EGF treatment of A-431 cells. *Science* 241:968–970.

Wahl, M. I., S. Nishibe, P. G. Suh, S. G. Rhee, and G. Carpenter. 1989. Epidermal growth factor stimulates tyrosine phosphorylation of phospholipase C-II independently of receptor internalization and extracellular calcium. *Proc. Natl. Acad. Sci. U.S.A.* 86: 1568–1572.

Wahl, M. I., S. Nishibe, J. W. Kim, H. Kim, S. G. Rhee, and G. Carpenter 1990. Identification of two epidermal growth factor-sensitive tyrosine phosphorylation sites of phospholipase C-(gamma) in intact HSC-1 cells. *J. Biol. Chem.* 265:3944–3948.

Wahl, M. I., G. A. Jones, S. Nishibe, S. G. Rhee, and G. Carpenter. 1992. Growth factor stimulation of phospholipase C-gamma 1 activity. Comparative properties of control and activated enzymes. *J. Biol. Chem.* 267:10447–10456.

Waldo, G. L., J. L. Boyer, A. J. Morris, and T. K. Harden. 1991. Purification of an AlF_{4-} and G-protein βγ-subunit-regulated phospholipase C-activating protein. *J. Biol. Chem.* 266:14217–14225.

Wang, J., M. Kester, and M. J. Dunn. 1988. Involvement of a pertussis toxin-sensitive G-protein-coupled phospholipase A_2 in lipopolysaccharide-stimulated prostaglandin E_2 synthesis in cultured rat mesangial cells. *Biochim. Biophys. Acta* 963:429–435.

Wang, P., J. C. Anthes, M. I. Siegel, R. W. Egan, and M. M. Billah. 1991. Existence of cytosolic phospholipase D: Identification and comparison with membrane-bound enzyme. *J. Biol. Chem.* 266:14877–14880.

Watson, S. P., R. T. McConnell, and E. G. Lapetina. 1984. The rapid formation of inositol phosphates in human platelets by thrombin is inhibited by prostacyclin. *J. Biol. Chem.* 259:13199–13203.

Weiss, A., G. Koretzky, R. C. Schatzman, and T. Kadlecek. 1991. Functional activation of the T-cell antigen receptor induces tyrosine phosphorylation of phospholipase C-gamma 1. *Proc. Natl. Acad. Sci. U.S.A.* 88:5484–5488.

Wolf, R. A., and R. W. Gross 1985. Identification of neutral active phospholipase C which hydrolyzes choline glycerophospholipids and plasmalogen selective phospholipase A_2 in canine myocardium. *J. Biol. Chem.* 260:7295–7303.

Wright, T. M., L. A. Rangan, H. S. Shin, and D. M. Raben. 1988. Kinetic analysis of 1,2-diacylglycerol mass levels in cultured fibroblasts: Comparison of stimulation by α-thrombin and epidermal growth factor. *J. Biol. Chem.* 263:9374–9380.

Wu, D., A. Katz, C. H. Lee, and M. I. Simon. 1992a. Activation of phospholipase C by α_1-adrenergic receptors is mediated by the α subunits of G_q family. *J. Biol. Chem.* 267:25798–25802.

Wu, H., M. R. James-Kracke, and S. P. Halenda. 1992b. Direct relationship between intracellular calcium mobilization and phospholipase D activation in prostaglandin E-stimulated human erythroleukemia cells. *Biochemistry* 31:3370–3377.

Xie, M., and G. R. Dubyak. 1991. Guanine-nucleotide- and adenine-nucleotide-dependent regulation of phospholipase D in electropermeabilized HL-60 granulocytes. *Biochem. J.* 278:81–89.

Yamada, K., Y. Kanaho, K. Miura, and Y. Nozawa. 1991. Antigen-induced phospholipase D activation in rat mast cells is independent of protein kinase C. *Biochem. Biophys. Res. Commun.* 175:159–164.

Yang, S. F., S. Freer, and A. A. Benson. 1967. Transphosphatidylation by phospholipase D. *J. Biol. Chem.* 242:477–484.

Yoshida, K., Y. Asaoka, and Y. Nishizuka. 1992. Platelet activation by simultaneous actions of diacylglycerol and unsaturated fatty acids. *Proc. Natl. Acad. Sci. U.S.A.* 89:6443–6446.

Zhang, H., N. N. Desai, A. Olivera, T. Seki, G. Brooker, and S. Spiegel. 1991. Sphingosine-1-phosphate, a novel lipid, involved in cellular proliferation. *J. Cell Biol.* 114:155–167.

Zupan, L. A., K. K. Kruszka, and R. W. Gross. 1991. Calcium is sufficient but not necessary for activation of sheep platelet cytosolic phospholipase A_2. *FEBS Lett.* 284:27–30.

Zupan, L. A., D. L. Steffens, C. A. Berry, M. Landt, and R. W. Gross. 1992. Cloning and expression of a human 14-3-3 protein mediating phospholipolysis: Identification of an arachidonoyl-enzyme intermediate during catalysis. *J. Biol. Chem.* 267:8707–8710.

6

Protein Kinase C in Multidrug Resistance, Neoplastic Transformation, and Differentiation

ROBERT I. GLAZER

Since the discovery of Ca^{2+}- and phospholipid-dependent protein kinase (protein kinase C, PKC) more than a decade ago by Nishizuka and co-workers (Kikkawa et al., 1982) and its identification as a major receptor for phorbol esters (Ashendel et al., 1983; Kikkawa et al., 1983), a multitude of physiological signaling mechanisms have been ascribed to PKC. The intense interest in PKC stems from its unique ability to be activated *in vitro* by diacylglycerol (and its phorbol ester mimetics), an effector whose formation is coupled to phospholipid turnover induced by the action of growth and differentiation factors. PKC comprises eight isoforms termed α, β_1, β_2, γ, δ, ϵ, ζ, and η (Figure 6.1), and thus many of its functions are likely to be both isoform and tissue specific (Kikkawa et al., 1989; Parker et al., 1989). The α, β, and γ isoforms are Ca^{2+}, phospholipid, and diacylglycerol dependent, while the more recently discovered δ, ϵ, ζ, and η isoforms all lack the Ca^{2+}-binding C2 domain and are activated by phospholipid and diacylglycerol. Because of these features as well as its diversity, PKC has been proposed to serve a highly specialized signal transduction function in response to physiological stimuli (Nishizuka, 1989). From a pharmacological standpoint, PKC has also served as a focal point for the design of drugs with possible therapeutic efficacy in the treatment of cancer (Gescher and Dale, 1989), although at the present time it is not clear what influence PKC has on the proliferation of malignant cells, as discussed elsewhere in this chapter.

This review will describe some of the newer developments in cancer biology

The work described was supported, in part, by grants from the Bristol-Myers Squibb Company and the National Institutes of Health.

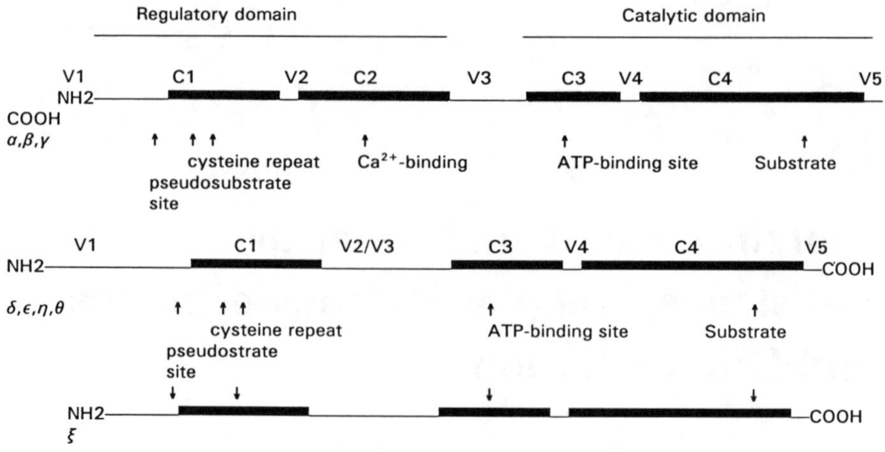

Figure 6.1 Structural organization of the genes encoding the isoforms of protein kinase C. The protein kinase C gene is divided into a regulatory and a catalytic domain encompassing five variable (V1–V5) and four constant (C1–C4) regions. The Ca^{2+}- and phospholipid-dependent isoforms (α, βI, βII, and γ) differ from the Ca^{2+}-independent and phospholipid-dependent forms (δ, ϵ, ξ and η) by the presence of a Ca^{2+}-binding (C2) domain.

that relate directly to PKC, particularly in the fields of multi-drug resistance, transformation, and differentiation. Because of the voluminous literature on PKC, this overview will necessarily be restricted to the more recent developments in these fields. Evidence will be presented that directly implicates PKC as a modulator of drug resistance through a complex pattern of regulation that involves the drug efflux pump, P-glycoprotein (PGP), as a target substrate as well as proteins associated with the proliferative process such as topoisomerases and transcription factors. The regulation of these phosphoproteins by PKC may provide both short- and long-term regulation of drug resistance, proliferation, and differentiation through transcriptional and posttranslational mechanisms.

PROTEIN KINASE C AND MULTIDRUG RESISTANCE

The precise relationship of PKC activity to the multidrug-resistant (MDR) phenotype may be related to differences in the expression (activity and isoform) and regulation (intracellular distribution and activation) of PKC in various MDR cell lines. Cells selected for multidrug resistance generally contain increased levels of PKC (Aquino et al., 1988, 1990a; Fine et al., 1988; Lee et al., 1992; Melloni et al., 1989; O'Brian et al., 1989; Palayoor et al., 1987; Posada et al., 1989a, Pb), and in several instances the elevated expression of PKC has been attributed to a specific isoform such as PKC α (Blobe et al., 1992; Lee et al., 1992; O'Brian et al., 1991; Posada et al., 1989a; Yu et al., 1991), PKC β (Fan et al., 1992; Melloni et al., 1989) and PKC γ (Aquino et al., 1990a). The appearance of PKC γ in

Adriamycin (ADR)-resistant HL-60 cells (Bhalla et al., 1985) that do not express PGP, an isoform normally absent in wild-type cells (Aquino et al., 1990a), suggests that PKC may also have a specific function in multidrug resistance independent of PGP. This was demonstrated for PKC βI in PGP-negative rat fibroblasts, where overexpression of this isoform led to a low level of resistance to ADR, *Vinca* alkaloids, and actinomycin D as well as reduced drug accumulation (Fan et al., 1992). An isolate of Friend erythroleukemia cells lacking PGP contained elevated PKC ε activity (originally misidentified as PKC β; see Powell et al., 1992) and exhibited a low level of resistance to vincristine and reduced drug accumulation that was antagonized by verapamil (Richon et al., 1991). The elevation in PKC ε activity in vincristine-resistant Friend cells also correlated with their enhanced responsiveness to the differentiating agent, hexamethylene bisacetamide (Michaeli et al., 1990). In contrast, no changes in drug sensitivity were noted for MCF-7 breast carcinoma cells stably transfected with PKC α (Yu et al., 1991) or for BC-19 cells (MCF-7 cells transfected with the *MDR*1 gene; Fairchild et al., 1990) transfected with PKC γ (Ahmad et al., 1992). Thus, there is some evidence that the βI, and ε isoforms but not the α and γ isoforms of PKC may confer low levels of multidrug resistance independently of the expression of PGP. Alternatively, PKC βI or ε may activate a resistance gene or gene product other than the *MDR*1 gene, such as the recently discovered *MRP* transporter gene that confers multidrug resistance in non-PGP expressing cells (Cole et al., 1992).

Since the majority of the isoforms within the PKC family lack the C2 calcium-binding domain, it was of interest to determine if they were differentially expressed in wild-type and MDR MCF-7 cells (Ahmad et al., 1992; Ahmad and Glazer, 1993). The only non-C2 isoform expressed to any significant degree in BC-19 cells was PKCε, and its level was reduced in cells transfected with PKC γ (Ahmad et al., 1992). Conversely, PKC ε levels were increased in ADR-resistant MCF-7 cells (MCF-7/ADR) after transfection with the antisense PKC α cDNA (Ahmad and Glazer, 1993). MCF-7/ADR cells were also found to contain reduced PKC ε and elevated PKC α in comparison with the parental cell line (Blobe et al., 1992). Thus there is some evidence that the α and ε isoforms of PKC exist in a closely regulated steady state where up-or down-regulation of one isoform results in a compensatory change in the other isoform. These observations are also consistent with studies in rat fibroblasts transfected with the c-H-*ras* oncogene where transformed isolates that expressed higher levels of PKC α showed a concomitant reduction in PKC ε (Borner et al., 1990). Whether these changes result from an autoregulatory mechanism in their respective promoter regions through phorbol ester or PKC-dependent *trans*-activation is not known.

Protein Kinase C and P-Glycoprotein

Carlsen and colleagues (1977) were the first to note that PGP existed as a phosphoprotein in colchicine-resistant, but not in revertant, cells. This finding has been corroborated in a variety of MDR cell lines (Center, 1983, 1985; Hamada et al., 1987; Mellado and Horwitz, 1987; Myers et al., 1989; Posada et al., 1989a; Schurr et al., 1989). A clue to the origin of the protein kinase(s) responsible for this

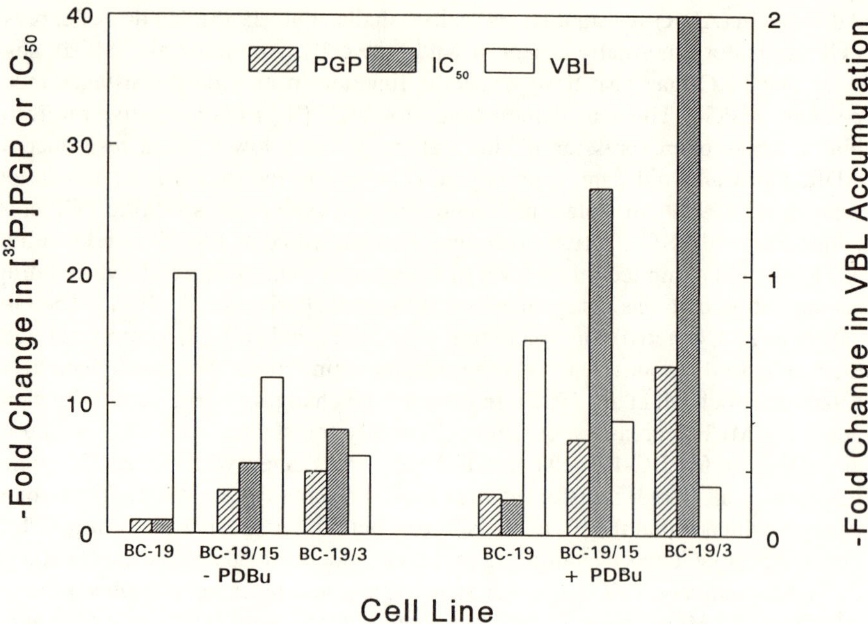

Figure 6.2 Changes in [^{32}P]P-glycoprotein, Adriamycin cytotoxicity, and vinblastine accumulation in the absence and presence of phorbol dibutyrate (PDBu). The -fold changes in BC-19/15 and BC-19/3 cells in the absence or presence of PDBu are relative to untreated BC-19 cells = 1. [^{32}P]P-glycoprotein (*pgp*) represents the radioactivity in two phorbol ester-sensitive tryptic phosphopeptides of P-glycoprotein. Adriamycin toxicity is based on its IC$_{50}$, and vinblastine (*VBL*) accumulation reflects drug retention over a 2-hour period. (From Yu et al., 1991.)

effect was provided by studies showing that phorbol ester activators of PKC could stimulate the phosphorylation of PGP *in vivo* (Hamada et al., 1987; Chambers et al., 1990b; Yu et al., 1991; Bates et al., 1992; Chambers et al., 1992). The involvement of PKC in this process was further supported by studies demonstrating that PGP is a substrate for PKC *in vitro* (Chambers et al., 1990a) and that stimulation of PGP phosphorylation by phorbol esters *in vivo* leads to reduced drug accumulation in vincristine-resistant KB cells (Chambers et al., 1990b, 1992).

To directly address the question of whether PKC could modulate multidrug resistance, BC-19 cells were stably transfected with the cDNA-encoding rabbit brain PKC α (Ohno et al., 1987). BC-19 cells contained the same level of PGP as MCF-7/ADR cells (Fairchild et al., 1990), but were 30-fold less resistant to ADR and contained 30-fold less PKC activity (Yu et al., 1991). When two stable drug-resistant isolates of BC-19 cells were examined, it was found that the isolate with the highest PKC activity was an order of magnitude more resistant than BC-19 cells (Yu et al., 1991). The isolates of BC-19 cells transfected with PKC α expressed 20- to 30-fold greater PKC activity than wild-type MCF-7 or parental BC-19 cells and possessed greater endogenous phosphorylation of PGP than BC-19 cells, which was stimulated further by phorbol dibutyrate (PDBu) (Yu et al.,

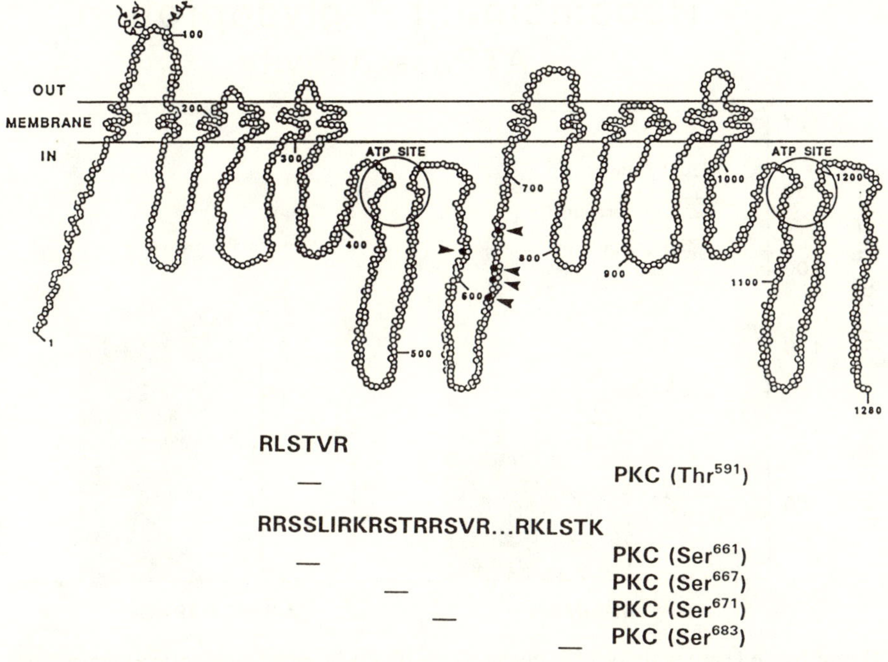

RLSTVR

—   PKC (Thr^{591})

RRSSLIRKRSTRRSVR...RKLSTK

—   PKC (Ser^{661})
—   PKC (Ser^{667})
—   PKC (Ser^{671})
—   PKC (Ser^{683})

Figure 6.3 Putative phosphorylation sites for protein kinase C in P-glycoprotein. *Arrowheads* indicate phosphorylation consensus sequences for PKC in the intervening linker region between the two ATP-binding domains. A recent study has confirmed Ser-661 and Ser-671 as sites of phosphorylation by PKC *in vitro* (Chambers et al., 1993).

1991). Thus, overexpression of PKC α produced a positive regulatory effect on resistance characterized by increased phorbol ester-dependent resistance, increased PGP phosphorylation, and reduced drug accumulation (Figure 6.2). In contrast to these results, transfection of BC-19 cells with the γ isoform of PKC that is not expressed in sensitive or resistant MCF-7 cells did not lead to increased multidrug resistance (Ahmad et al., 1992). Conversely, transfection of MCF-7/ADR cells with the antisense cDNA to PKC α led to reduced PGP phosphorylation and increased sensitivity to ADR in proportion to the reduction in PKC α (Ahmad and Glazer, 1993).

On the basis of the phosphorylation consensus sequence for PKC, there are several potential phosphorylation sites in the intervening region between the two ATP-binding sites in PGP (Figure 6.3). A recent study using membrane vesicles and purified mixed isoforms of PKC *in vitro* has identified Ser-661 and Ser-671 and one or more serines at positions 667, 675, and 683 as PKC phosphorylation sites in the linker region between the ATP-binding domains (Chambers et al., 1993). Therefore, it can be envisioned that PKC isoform-specific phosphorylation of PGP leads to conformational changes in the cytoplasmic domain with accompanying changes in ATP and drug - binding. Although there is precedent for this model within the large ATPase-binding cassette (ABC) protein family, for example in the regulation of the activity of the cystic fibrosis transmembrane regulator

Recombinant P-glycoprotein
ATPase Activity

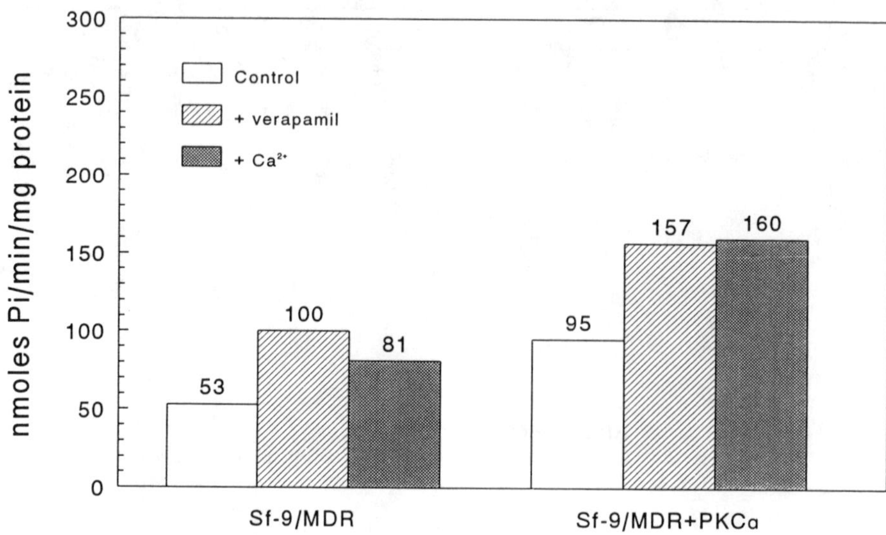

Figure 6.4 ATPase activity of P-glycoprotein (PGP) in membrane vesicle preparations from Sf-9 insect cells expressing either PGP or PGP and PKC α. Membrane vesicles were prepared by nitrogen cavitation and assayed for vanadate-sensitive ATPase activity. Assays were carried out under linear kinetics for 20 minutes at 37°C with 10 μg of membrane vesicle protein. In some assays, activity was measured in the presence of either 10 μM verapamil or 0.5 mM calcium.

(Ames and Lecar, 1992), phosphorylation has yet to be proven essential for the drug efflux and/or drug-binding activities of PGP. Recent studies with baculo-virus-expressed PGP have shown that its ATPase activity is increased about two-fold when co-expressed with PKC α in insect cells in comparison with PGP alone (Figure 6.4). With the elucidation of the phosphorylation sequences in PGP phos-phorylated by PKC and the use of site-directed mutagenesis, the validity or irrel-evance of this model should be forthcoming.

Protein Kinase C Inhibitors and Activators and Multidrug Resistance

The majority of studies assessing the role of PKC in multidrug resistance or other processes have utilized pharmacological agents that are known *in vitro* to either stimulate or inhibit PKC. It is assumed in all of these studies that the inhibition or activation of PKC is specific, and therefore that the effects observed are directly related to PKC, an assumption that is not necessarily true.

Protein Kinase Inhibitors

The inhibitors staurosporine and H-7, now widely regarded as nonspecific protein kinase inhibitors, decreased resistance to ADR in S180 cells (Posada et al., 1989a) and fibrosarcoma cells (O'Brian et al., 1989), respectively. H-7 reversed chemoresistance in colon carcinoma cell line KM12L4a and antagonized the reduction in drug accumulation induced by phorbol esters and diacylglycerol (Dong et al., 1991). In wild-type or MCF-7/ADR cells, staurosporine had no effect on ADR toxicity (Schwartz et al., 1991). A possible reason for the latter effect in some MDR cell lines is that there may be a high endogenous rate of proteolysis of PKC to its catalytic fragment, M-kinase, which is insensitive to staurosporine and other inhibitors of PKC (Sauvage et al., 1991). An alternative explanation is that staurosporine-sensitive kinases are not involved in multi-drug resistance in this cell line.

One major obstacle to the use of inhibitors such as staurosporine and H-7, apart from their lack of specificity for PKC (Meyer et al., 1989), is that they produce cytotoxicity that may be unrelated to inhibition of protein kinase activity (Smith et al., 1988). A second drawback to their use in MDR cells is that staurosporine and its analogues (Sato et al., 1990; Miyamoto et al., 1992), as well as H-7 and other isoquinoline sulfonamide analogs (Wakusawa et al., 1992), compete for drug-binding sites on PGP *in vitro*. Thus, the effects of inhibitors with regard to multidrug resistance and PGP phosphorylation may present a misleading and confusing picture of the specific role of PKC in this process. The ability of the serine-protein phosphatase inhibitor, okadaic acid, to increase PGP phosphorylation without affecting drug accumulation or drug sensitivity in MDR KB cells (Chambers et al., 1992), further attests to the complexity of PGP phosphorylation in multidrug resistance, and indicates that conclusions drawn from these types of studies should be interpreted with caution.

Protein Kinase Activators

Other studies have utilized phorbol esters as activators of PKC (Castagna et al., 1982) to elucidate the role of this enzyme in resistance. These studies can be generally categorized into those that assume that short-term treatment with phorbol esters will activate PKC, and those that assume that long-term treatment will down-regulate PKC.

Short-Term Phorbol Ester Treatment. Exposure of wild-type or MDR B-cell leukemia cell line MOLT-3 for 90 minutes to 12-O-tetradecanoylphorbol-13-acetate (TPA) increased membrane-associated PKC activity and increased trimetrexate (a lipophilic antifolate) resistance (Schwartz et al., 1991). TPA mediated a rapid increase in resistance to colchicine in human B-cell leukemia cells (O'Connor, 1985), and similar results were obtained after wild-type and MCF-7/ADR cells (Fine et al., 1988), BC-19 cells (Ahmad et al., 1992), or BC-19 cells transfected with PKC α (Yu et al., 1991) were treated for 2 hours with PDBu. Treatment of ADR-resistant colon carcinoma cell line SW620 with TPA resulted in increased PGP phosphorylation that counteracted the decrease in PGP phosphorylation and increase in drug accumulation mediated by sodium butyrate (Bates et al., 1992). Except for one instance in which treatment of wild-type S180 cells for 30 minutes with TPA resulted in reduced resistance to ADR coincident with enhanced PKC activity (Posada et al., 1989b), short-term exposure to TPA in most instances stimulated the phosphorylation of PGP and reduced drug accumulation in MDR cells.

Long-Term Phorbol Ester Treatment.   The rationale behind long-term exposure to phorbol esters is that the sustained translocation of PKC to the plasma membrane will result in its proteolysis and inactivation and hence the term "down-regulation." It would be naive to assume that this is always the case, particularly since the various isoforms of PKC respond differently to phorbol esters. In addition, phorbol esters produce a variety of transcriptional effects that may subsequently lead to a physiological response quite unrelated to the initial effects on PKC (Karin, 1992).

Long-term exposure to TPA increased ADR resistance in wild-type and ADR-resistant S-180 cells (Posada et al., 1989b); however, a similar regime of TPA sensitized HeLa and A253 carcinoma cells to cisplatin (Basu et al., 1990). Since cisplatin is not a substrate for PGP, and a prolonged incubation with TPA is required to elicit this effect, sensitization to cisplatin may reflect transcriptional activation of the target gene involved in the cytotoxicity produced by this drug or in the down-regulation of a gene that attenuates its activity.

The majority of these studies suggest that PKC is a positive regulator of multidrug resistance. However, if one accepts the current "activation" paradigm for PKC (Nishizuka, 1989), it is paradoxical that short-term exposure to phorbol esters produces the same effect as prolonged exposure unless the species of PKC affected under the latter conditions is not highly susceptible to down-regulation. There is some evidence to substantiate the latter premise as shown for PKC α in BC-19 cells (Ahmad et al., 1992) and for PKC ξ which is refractory to both translocation and down-regulation following prolonged TPA treatment (Kiley et al., 1990; Liyanage et al., 1992). Alternatively, phorbol esters may not affect the same target after short- and long-term exposure, and thereby produce their effects under these diverse regimens by entirely different mechanisms. It is possible that transcriptional activation by prolonged phorbol ester treatment results in the synthesis of a new set of PKC substrates that may or may not require Ca^{2+} or diacylglycerol for phosphorylation. Alternatively, the commonly accepted model for PKC activation (Nishizuka, 1989) may not, in fact, pertain to either the prolonged or short-term effects of phorbol esters. It has been proposed, on the basis of experimental evidence in the literature, that the "activation" of PKC by phorbol esters may actually be a reflection of its endogenous substrate, lipid environment, and diacylglycerol-mediated changes in membrane fluidity, rather than a direct activation by Ca^{2+}, diacylglycerol, and phospholipid (Zidovetski and Lester, 1992).

PROTEIN KINASE C AND ATYPICAL DRUG RESISTANCE

Some forms of multidrug resistance are unrelated to expression of PGP and have been termed atypical multidrug resistance (Beck, 1987). This type of resistance generally refers to alterations in the activity or structure of topoisomerases (Beck, 1987, 1989). These nuclear enzymes are required for breaking and rejoining single (topoisomerase I) or double-stranded (topoisomerase II) DNA to facilitate transcription and DNA replication. Topoisomerase I is a specific target for the anticancer drug camptothecin, whereas topoisomerase II is believed to be a major target for ADR and etoposide (Ross, 1986). In camptothecin-resistant tumor cells, the level as well as the sensitivity of topoisomerase I to camptothecin is reduced (Sugimoto et al., 1990). While the mechanism for this effect is unknown, two studies have suggested that this enzyme is regulated through phosphorylation. Durban and colleagues (1983) were the first to demonstrate that topoisomerase I is activated by phosphorylation *in vitro*. A recent study by Pommier and co-workers (1990) has shown that both the 100-kDa and 68-kDa forms of topoisomerase I from 9-hydroxyellipticine-resistant Chinese hamster cells are totally inactivated on dephosphorylation but are fully reactivated by PKC. The reactivated forms of topoisomerase I behaved like the native enzyme with respect to the processivity of DNA relaxation and sequence selectivity of DNA cleavage in the presence of camptothecin. That PKC may indeed be a regulator of topoisomerase I activity was also suggested by the stimulation in its activity in HL-60 cells treated with TPA (Gorsky et al., 1989) as well as in 3T3 cells where TPA increased the phosphorylation of topoisomerase I fourfold within 2 hours of treatment (Samuels and Shimizu, 1992). It remains to be determined whether reduced PKC-mediated phosphorylation or down-regulation of PKC is responsible for the decreased activity of topoisomerase I in camptothecin-resistant cells.

Topoisomerase II is also a substrate for PKC *in vitro* (Sahyoun et al., 1986; De Vore et al., 1992) and is phosphorylated *in vivo* (Heck et al., 1989; Kroll and Rowe, 1991). However, treatment of HL-60 cells with TPA rapidly inhibits rather than stimulates etoposide-mediated DNA cleavage (Zwelling et al., 1990). Etoposide resistance is also associated with reduced PKC activity in P388 cells (Ido et al., 1987) but no connection between PKC and topoisomerase activity was made. Mutant forms of topoisomerase II from etoposide-resistant KB cells that have higher activities also contained levels of serine phosphorylation in intact cells that were 14- to 18-fold higher than for the wild-type enzyme (Takano et al., 1991). The kinase responsible for the latter effect was not identified, although it was stated that phorbol esters did not increase phosphorylation. It is still premature to conclude that topoisomerase II is regulated directly by PKC.

PROTEIN KINASE C IN TRANSFORMATION

It is becoming increasingly evident that PKC may play a role in cell cycle progression in an isoform- and phenotype-specific manner. Rat fibroblasts overexpressing PKC β_1 exhibited increased growth in soft agar and anchorage-independent growth (Housey et al., 1988), became more susceptible to transformation by H-*ras* (Hsiao et al., 1989), and responded to TPA with increased diacylglycerol formation (Pai et al., 1991). This effect was similar to the increased growth and tumorigenicity observed for NIH 3T3 mouse fibroblasts transfected with PKC γ (Persons et al., 1988). The role of PKC α in transformation has yielded mixed results. No transformation (growth in soft agar) of 3T3 fibroblasts was observed after overexpression of PKC α (Eldar et al., 1990) or PKC α mutated in the regulatory domain (Borner et al., 1991). Other studies utilizing 3T3 cells transfected with both wild-type (Finkenzeller et al., 1992) and mutated PKC α (Megidish and Mazurek, 1989) have demonstrated increased growth in soft agar as well as solid tumor formation in nude mice (Megidish and Mazurek, 1989), and factor-independent growth and disappearance of epidermal growth factor receptor number (Eldar et al., 1990). Some of these inconsistencies may be a result of the degree of expression of PKC α or possibly changes in other endogenous isoforms of PKC. CHO cells overexpressing PKC δ did not exhibit changes in growth but were arrested in G2/M phase after treatment with TPA (Watanabe et al., 1992). TPA-arrested cells were predominantly dikaryons indicating that cytokinesis rather than mitosis was inhibited by phorbol ester treatment. The positive proliferative effect of PKC α and δ in fibroblasts is contrary to findings in regenerating liver where there is complete down-regulation of nuclear PKC α and a 50-percent elevation in nuclear PKC δ 4 hours after partial hepatectomy (Alessenko et al., 1992); however, changes in PKC isoforms during later periods of regeneration during the peak of cell division were not measured.

Similar studies with tumor cells have yielded quite different results. No differences were observed in the growth of breast carcinoma cells transfected with either PKC γ (Ahmad et al., 1992) or PKC α (Yu et al., 1991), but retroviral infection of human colon carcinoma cell line HT-29 with PKC β_1 resulted in reduced growth and tumorigenicity (Choi et al., (1990). Many of the effects of overexpression of various isoforms of PKC may be related to the level of expression of the enzyme as well as the particular phenotypic characteristics of the cell line studied. For example, expression of the antisense cDNA of PKC α in MCF-7/ADR cells did not alter its growth (Ahmad and Glazer, 1993), but similar levels of expression of PKC α in glioblastoma cells resulted in an increased doubling time and reduced tumorigenicity in nude mice (Figure 6.5). The latter effects were related to interference with growth factor-dependent growth and the induction of a more differentiated phenotype. Factors such as the degree of differentiation of the tumor, its ability to undergo further differentiation, and its growth factor dependence may account for the differences between fibroblasts and transformed cells.

PROTEIN KINASE C AND DIFFERENTIATION

The role of PKC in differentiation has been addressed predominantly in the human promyelocytic leukemia cell line HL-60 and therefore has centered on hematopoietic cell maturation, with few studies addressing other differentiated cell types. PKC activity is elevated in HL-60 cells undergoing differentiation in response to the granulocytic differentiating agents retinoic acid and dimethylsulfoxide or the monocytic differentiating agents 1,25-dihydroxyvitamin D_3 and phorbol ester (Zylber-Katz & Glazer, 1985; Zylber-Katz et al., 1986). Dimethylsulfoxide and retinoic acid increased the levels of PKC α and β, as well as γ (Makowske et al., 1988). Increased PKC β activity was also noted after HL-60 cells were treated with retinoic acid (Hashimoto et al., 1990). PKC γ, an isoform that is not expressed in HL-60 cells, was abundant in an ADR-resistant variant of HL-60

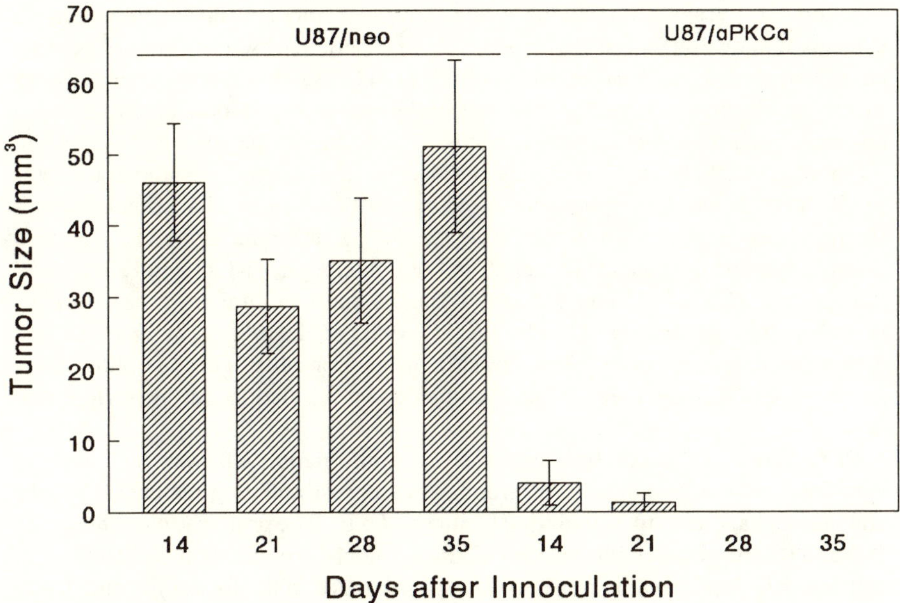

Figure 6.5 The effect of antisense PKC α on the tumorigenicity of glioblastoma cells. Human glioblastoma cell line U87 was transfected with the antisense cDNA to PKC α and selected for *neo* resistance with G418. Cells were inoculated subcutaneously into nude mice and tumor size was measured periodically.

cells that is also resistant to retinoic acid, DMSO, and 1,25-dihydroxyvitamin D_3, but not to TPA (Aquino et al., 1990a). Treatment of HL-60 cells with 1,25-dihydroxyvitamin D_3 increased PKC α and β mRNA levels (Obeid et al., 1990; Solomon et al., 1991), although there is disagreement about whether this is caused by a transcriptional (Obeid et al., 1990) or posttranslational mechanism (Solomon et al., 1991). In monocytic cell line U937, TPA treatment resulted in translocation of PKC α to the particulate fraction and induction of PKC γ (Strulovici et al., 1989) and induced elevations in PKC α and β at 0.1 to 1 nM TPA and down-regulation of PKC α, β, and ε at higher concentrations (Ways et al., 1992b). PKC α was more resistant to down-regulation by TPA than PKC β and ε, a result similar to findings in HL-60 cells (Aihara et al., 1991). Interestingly, complete down-regulation of PKC α was not required for the differentiation of HL-60 cells to macrophage-like cells (Aihara et al., 1991), a result consistent with earlier studies by Zylber-Katz and colleagues (1986). A recent study has confirmed this finding, and in fact found that induction of the macrophage phenotype in HL-60 cells with TPA increased the level of PKC α and reduced the level of PKC β (Edashige et al., 1992).

Although numerous studies have described phorbol ester-dependent and Ca^{2+}- and phospholipid-dependent phosphorylation in myeloid cells, few have described an association between a specific endogenous PKC substrate and its function in differentiation. Studies with neutrophils from normal and hypertensive subjects demonstrated that the superoxide burst elicited by chemotactic peptide correlated directly to the level of protein kinase C (Salamino et al., 1991). Other investigations have described PKC-dependent phosphorylation of two surface antigens in hematopoietic cells that may give a clue to the role of PKC during differentiation. May and colleagues (1984) reported that the transferrin receptor in HL-60 cells was hyperphosphorylated by PDBu, an effect that coincided with the rapid down-regulation of this receptor during differentiation. TPA induced phosphorylation of the CD34 receptor in both myelocytic (KG-1 and KG-1a) and lymphocytic (MOLT-13 and RPMI 8402) cells that was similar to the phospho-peptide pattern observed *in vitro* by PKC (Fackler et al., 1980). The latter antigen is expressed only on early hematopoietic progenitor cells, and thus phosphory-lation of CD34 may precede its eventual down-regulation during differentia-tion.

PKC may also be involved in the signal transduction pathway mediated by cytokines such as granulocyte/macrophage colony-stimulating factor (GM-CSF) and macrophage colony-stimulating factor (M-CSF). The protein kinase inhibitors H-7 and staurosporine inhibited GM-CSF-dependent growth of myeloid leukemia cell line KG-1 (Katayama et al., 1992) and the PKC activators bryostatin 1 and TPA synergized with GM-CSF in promoting GM colony formation from normal myeloid progenitor cells (Grant et al., 1991; McCrady et al., 1991). M-CSF stim-ulated PKC activity coincident with diacylglycerol formation in normal mono-cytes (Imamura et al., 1990) and an antisense oligodeoxynucleotide against PKC ϵ, but not against PKC α, blocked the GM-CSF-dependent glycolytic burst eryth-roleukemia cell line TF-1, which is dependent on GM-CSF for proliferation (Bax-ter et al., 1992a). Therefore, cytokine-mediated differentiation of myeloid pro-genitor cells may involve a common pathway downstream from multiple CSF receptors that comprises, in part, one or more isoforms of PKC.

These data are also consistent with the differentiation pattern in other cell types. Melanoma cells transfected with PKC α exhibited diminished anchorage-depen-dent growth, less tumorigenicity, and a more differentiated phenotype (Gruber et al., 1992). Embryonal carcinoma cells treated with retinoic acid exhibited increased PKC activity that correlated with differentiation to the parietal endo-derm phenotype (Kraft and Anderson, 1983). Thus, the consensus of studies examining the expression of PKC during myeloid differentiation is that elevation of PKC α is required for both granulocytic and monocytic differentiation. It is still not clear what role PKC α or other PKC isoforms play in this process at either a molecular or functional level.

Resistance to Differentiation by Phorbol Esters.

Down-regulation of PKC in 3T3 cells after prolonged phorbol ester treatment resulted in resistance to the proliferative effects of PDBu that could be remedied by microinjection of these cells with PKC (Pasti et al., 1986). Although this study provided compelling evidence that PKC was the primary target of long-term exposure to phorbol esters, subsequent investigations of phorbol ester-resistant cells have not produced a consensus as to the underlying cause and role of PKC in this phenomenon. Phorbol ester-resistant variants of HL-60 cells do not undergo differentiation to a macrophage-like phenotype on exposure to phorbol esters, although they remain responsive to other monocytic differentiating agents such as 1,25-dihydroxyvitamin D_3 (Homma et al., 1986; Nishikawa et al., 1990). TPA-resistant isolates of HL-60 cells generally contain a reduction in cytosolic PKC activity with little or no change in membrane-associated activity (Perella et al., 1986; Nishikawa et al., 1990; Tonetti et al., 1992). The major isoform affected was PKC β (Nishikawa et al., 1990; Tonetti et al., 1992), and in one instance this was attributed to reduced PKC β mRNA (Tonetti et al., 1992). While other investigators did not find differences in the activity of PKC α or β in TPA-resistant HL-60 cells, they did find that the up-regulation of PKC β mRNA as well as its transcription in wild-type cells in response to TPA was completely ablated in resistant cells (McSwine-Kennick et al., 1991). These studies suggest that attenuation of downstream processes that regulate the transcriptional activation of PKC β (McSwine-Kennick et al., 1991), and perhaps other genes as well, for example the early response genes such as c-*jun* and c-*fos* (Tonetti et al., 1992), are important for conferring resistance. Since 1,25-dihydroxyvitamin D_3-induced monocytic differentiation of HL-60 cells results in induction of both PKC α and β (Obeid et al., 1990; Solomon et al., 1991) and phorbol ester-resistant cells remain sensitive to differentiation by this agent, PKC α would be implicated as playing a major role in monocytic differentiation, whereas reduction in PKC β expression appears to be the predominant underlying change in phorbol ester-resistant HL-60 cells. These results are also consistent with the greater resistance of PKC α to phorbol ester-induced down-regulation (Aihara et al., 1991).

Phorbol ester resistance has also been investigated in nonmyeloid cells. In TPA-resistant EL4 thymoma cells, there was 50 percent less PKC α mRNA, the major isoform of PKC in this cell line, 85 percent less PKC ε mRNA, and no discernible PKC ε protein in comparison with sensitive cells (Homan et al., 1991). In TPA-resistant T-cell leukemia Jurkat cells, PKC α was the predominant isoform that was reduced with little or no change in PKC β and γ (Isakov et al., 1990; Tchou-Wong and Weinstein, 1992). While no changes in PKC α and β mRNAs or growth rate was noted in one study (Tchou-Wong and Weinstein, 1992), other investigators found that PKC α and β mRNA levels were lower in TPA-resistant Jurkat cells (Isakov et al., 1990). In the latter study, there was no up-regulation of PKC α and β mRNA levels after TPA treatment of resistant cells, and one of two TPA-resistant clones that failed to show an increase in PKC α mRNA in response to TPAalso exhibited a reduced growth and dependence on TPA for proliferation (Isakov et al., 1990). The latter findings are also consistent with the observations that PKC α may play a role in cell cycle progression. Therefore, resistance to phorbol esters may involve desensitization in phorbol ester responsiveness at the transcriptional level, which may involve inactivation of transcription factors (through dephosphorylation/phosphorylation?) directly involved in the up-regulation of PKC genes. Similar studies of PKC isoform expression in cells resistant to differentiating agents other than phorbol esters appear to be lacking.

PROTEOLYTIC ACTIVATION OF PROTEIN KINASE C BY PHORBOL ESTERS

The effects mediated by phorbol esters may involve not only direct activation of PKC, but also activation of PKC by proteolysis to a Ca^{2+}- and phospholipid-independent form, termed M-kinase (Inoue et al., 1977; Takai et al., 1977; Kishimoto et al., 1983). In some cell lines there is evidence that the actual effector of phorbol ester action may be M-kinase per se (Melloni et al., 1986; Murray et al., 1987). HL-60/ADR cells contain high constitutive levels of M-kinase (Aquino et al., 1988), as well as Ca^{2+}-dependent protease activity (Aquino et al., 1990b). Vincristine-resistant Friend erythroleukemia cells contain high levels of PKC ϵ compared with parental cells and exhibit higher and more sustained levels of M-kinase than do wild-type cells after short-term exposure to TPA (Melloni et al., 1989). Small-cell lung carcinoma cell line NCI N417 contains high levels of PKC ϵ and expresses constitutively the active catalytic fragment of PKC ϵ whose levels are up-regulated by TPA or gastrin-releasing peptide (Baxter et al., 1992b). Proteolytic activation of PKC ϵ in this cell line is phenotype specific, since treatment of $CD4^+/CD8^+$ mouse thymocytes with TPA did not induce a similar down-regulation of PKC ϵ (Strulovici et al., 1991).

The differential sensitivity of various isoforms of PKC to Ca^{2+}-dependent neutral protease (calpain) also suggests that regulation of PKC via proteolysis may be physiologically relevant (Huang et al., 1989; Kishimoto et al., 1989). The phorbol ester-responsiveness of sensitive and resistant cells may be a function of the predominant isoform of PKC that is expressed in the cell and its sensitivity to proteolysis. PKC β and γ were 6- and 50-fold more sensitive, respectively, than PKC α to proteolysis by calpain I *in vitro* (Kishimoto et al., 1989). Other studies have found that calpain produced similar rates of proteolysis of PKC α and β to a catalytically stable form of M-kinase, whereas neutral serine protease destroyed PKC activity without generating M-kinase (Pontremoli et al., 1990a). Treatment of human neutrophils with TPA or chemotactic peptide maximally generated M-kinase within 10 minutes (Pontremoli et al., 1990b), a process that was abolished by the calpain inhibitor leupeptin (Melloni et al., 1986). In several cell lines treated with phorbol esters, PKC α was less susceptible to degradation than PKC β (Ase et al., 1988; Huang et al., 1989; Pontremoli et al., 1990b). Endogenous Ca^{2+}- and phospholipid-dependent PKC activity (mostly PKC α) was virtually unchanged after PDBu treatment of BC-19 cells that overexpressed PKC γ, whereas PKC γ was rapidly degraded within 10 minutes (Ahmad et al., 1992). Positive effectors of PKC activity such as diacylglycerol and phospholipid increased proteolysis of PKC by trypsin (Huang et al., 1989; Newton and Koshland, 1989) and calpain (Imajoh et al., 1986; Kishimoto et al., 1989), and increased the susceptibility of PKC α to trypsin (Huang et al., 1989). These results explain why membrane-associated PKC α is more sensitive to trypsin digestion than the soluble form of the enzyme (Newton and Koshland, 1989) and provide a rationale for the posttranslational effects of phorbol esters, that is, that rapid translocation of PKC to the plasma membrane results in subsequent proteolysis to a catalytically active fragment on short-term exposure to phorbol esters and proteolytic inactivation on long-term treatment. Interestingly, autophosphorylation of PKC α is a prerequisite for proteolytic cleavage (Ohno et al., 1990), and therefore its activation by membrane association (Zidovetski and Lester, 1992) alone might generate M-kinase. Therefore it is possible that the numerous effects attributed to phorbol esters may actually be mediated by M-kinase arising from one or more isoforms of PKC. However, one should also note that not all isoforms of PKC are down-regulated by phorbol esters. Two notable examples are PKC η, which is specifically localized in the nucleus of lung and epithelial cells and is unaffected by TPA treatment (Greif et al., 1992), and PKC ξ, which is found in many hematopoietic cell lines and is neither translocated to the plasma membrane nor down-regulated by chronic exposure to TPA (Ways et al., 1992a).

PROTEIN KINASE C AND TRANSCRIPTIONAL ACTIVATION

The regulation of eukaryotic transcription occurs through specific *cis*-acting regulatory elements termed promoters or enhancers, which are composed of distinctive DNA sequence motifs (Johnson and McKnight, 1989; Mitchell and Tjian, 1989). Regulatory proteins interact with these elements in a highly specific and sometimes cooperative manner and determine which genes will be transcriptionally active. Many of the regulatory proteins are inducible factors (Maniatis et al., 1987) that respond to such diverse factors as heavy metal ions (Karin et al., 1984; Searle et al., 1985), heat shock (Pelham, 1983), glucocorticoids (Chandler et al., 1983), phorbol esters (Angel et al., 1987), and anticancer drugs (Kohno et al., 1989; Chin et al., 1990).

In many instances, transcription factor activity is modulated by phosphorylation/dephosphorylation that may involve PKC (Hunter and Karin, 1992; Karin, 1992). A prime example of this is the activation of NFκB by the cytosolic inhibitor IκB (Bauerle and Baltimore, 1988; Lenardo and Baltimore, 1989) where phosphorylation of IκB by PKC leads to dissociation of the complex and subsequent activation and translocation of NFκB to the nucleus (Shirakawa and Mizel, 1989). (Figure 6.6) *Trans*-activation by PKC γ is also involved in the tumorigenicity of fibroblasts overexpressing PKC γ (Persons et al., 1988). These cells exhibit enhanced *trans*-activation of a retroviral enhancer element in a manner similar to that caused by TPA (Persons et al., 1991).

Nuclear Localization of Protein Kinase C

Regulation of transcription factor activity by phosphorylation would be expected to be facilitated by the presence of the kinase near or within the nucleus. Translocation of PKC or its catalytic fragment to the nucleus would represent a unique

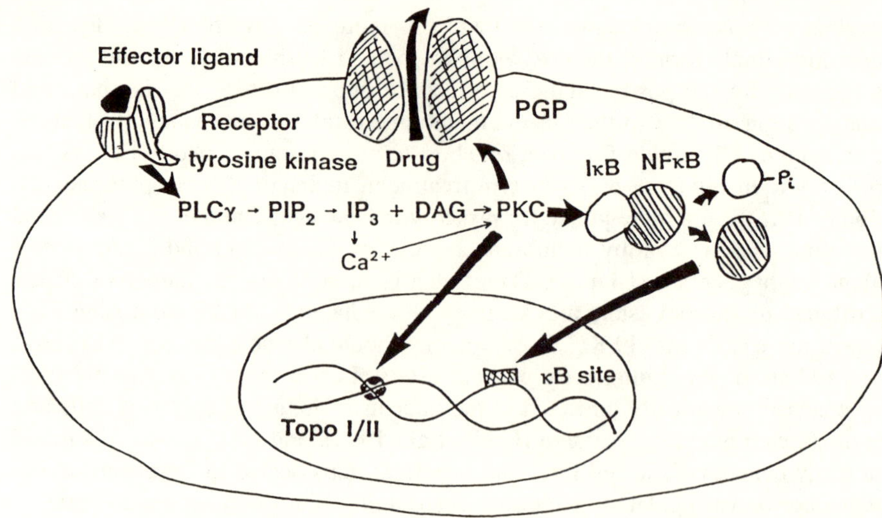

Figure 6.6 Summary of the pleiotropic cellular effects of protein kinase C. PKC acts as a central transducer of the activity of growth factors through their receptor tyrosine kinase-dependent activation of phospholipase Cγ that in turn yields increased levels of PKC cofactors such as diacylglycerol (DAG) and Ca^{2+} through hydrolysis of phosphatidylinositol 4,5-bisphosphate (PIP_2). Activated PKC can directly activate P-glycoprotein (PGP) through phosphorylation of the inter-ATP binding domain. Activation of PKC may also result in activation of transcription through direct phosphorylation of topoisomerase I (Topo I) and/or topoisomerase II (Topo II) or by activation of transcription factor NFκB by phosphorylation and inactivation of its inhibitor IκB.

isoform-specific means of regulating transcription by phorbol esters. In this regard, PKC α is rapidly and specifically localized in the nuclear envelope of 3T3 cells after TPA treatment (Thomas et al., 1988; Leach et al., 1989), and transfection of cos cells with PKC α M-kinase resulted in its exclusive localization in the nucleus (Eldar et al., 1992; James and Olson, 1992), an effect also seen after transfection with the full length PKC α cDNA following treatment with TPA (Eldar et al., 1992) or PDBu (James and Olson, 1992). Similar results were seen when human neutrophils were treated with TPA, where a selective increase in nuclear PKC βII but not PKC α activity occurred. In contrast, treatment of HL-60 cells with phorbol esters resulted in plasma membrane but not nuclear translocation of PKC α and βII (Thomas et al., 1988; Hocevar and Fields, 1991), although treatment with bryostatin caused selective association of PKC βII with the nuclear membrane fraction (Hocevar and Fields, 1991). Some isoforms such as PKC η that are not down-regulated by phorbol esters do not require phorbol ester activation for nuclear localization, as shown in keratinocyte and squamous cell carcinoma cells (Greif et al., 1992). Therefore, there is substantial evidence that some PKC isoforms can translocate to the nucleus either as the holoenzyme or as the catalytic fragment. Interestingly, it is M-kinase and not the PKC holoenzyme that can activate TPA-responsive elements in the collagenase (Hata et al., 1989) and c-*fos* (Muramatsu et al., 1989) promoters. These data suggest that translocation of some species of M-kinase to the nucleus may selectively activate nuclear factors that differentially regulate transcription in resistant cells.

The Promoter Region of Protein Kinase C

The immediate 5′-flanking region to exon 1 in the genomic sequences of PKC β and γ contain consensus-binding elements for transcription factors AP-1, AP-2, and Sp1 (Chen et al., 1990; Niino et al., 1992; Obeid et al., 1992). Both AP-1 and AP-2 activity are stimulated by phorbol esters (Imagawa et al., 1987; Lee et al., 1987). Transcription factor Sp1 also exists as a phosphoprotein, but it is unclear what protein kinase is involved or how phosphorylation regulates its activity (Jackson et al., 1990). Therefore, part of the basal promoter activity for at least two PKC genes may be autoregulatory. Some evidence for this proposal has been obtained in yeast, where cells resistant to staurosporine exhibited increased activity of an AP-1-like transcription factor concomitant with a tenfold increase in protein kinase activity (Toda et al., 1991). In HL-60/ADR cells that contain elevated levels of PKC γ (Aquino et al., 1990a), the level and DNA-binding activity of transcription factors Sp1 (Borellini et al., 1990), CREB, and AP-1 (Rohlff et al., 1993) are greatly enhanced in comparison with sensitive cells. The human *MDR*1 gene also contains a major downstream promoter region that contains binding elements for Sp1 and AP-1 (Ueda et al., 1987).

Therefore, it is conceivable that changes in PKC activity could affect the activity of nuclear transcription factors that are responsible for the initiation of the resistant phenotype as well as preneoplastic hyperplasia. Various subpopulations of cells expressing higher transcription factor activity would in turn have a selective advantage during induction of drug resistance or transformation and would account for the heterogeneity in response of cell populations with emerging resistance or neoplasia.

CONCLUSION

There is mounting evidence that specific isoforms of PKC are involved in the regulation of neoplastic growth, differentiation, and drug resistance. Drug resistance may arise from the activation of transcription factors by phosphorylation via PKC or M-kinase to induce the expression of MDR-specific genes such as *MDR*1 and/or a direct stimulatory effect on the activity of the *MDR*1 gene product, PGP (Figure 6.6). In a similar context, the onset of neoplasia may involve a cell-cycle-specific role of PKC in the transcriptional regulation of proliferation-associated genes. Understanding the multifactorial role of PKC in cellular transformation and drug resistance at both the transcriptional and posttranslational levels should help define their etiology, as well as lead to a possible molecular target that can be exploited therapeutically.

REFERENCES

Ahmad, S., and R. I. Glazer. 1993. Expression of the antisense cDNA to protein kinase Cα attenuates resistance in doxorubicin-resistant MCF-7 breast carcinoma cells. *Molec. Pharmacol.* 43:858–862.

Ahmad, S., J. B. Trepel, S. Ohno, K. Suzuki, T. Tsuruo, and R. I. Glazer. 1992. The role of protein kinase C in the modulation of multidrug resistance: Expression of the atypical gamma isoform of protein kinase C does not confer increased resistance to doxorubicin. *Molec. Pharmacol.* 42:1004–1009.

Aihara, H., Y. Asaoka, K. Yoshida, and Y. Nishizuka. 1991. Sustained activation of protein kinase C is essential to HL-60 cells differentiation to macrophage. *Proc. Natl. Acad. Sci. U.S.A.* 88:11062–11066.

Allesenko, A., W. A. Khan, W. C. Wetsel, and Y. A. Hannun. 1992. Selective changes in protein kinase C isoenzymes in rat liver nuclei during liver regeneration. *Biochem. Biophys. Res. Commun.* 182:1333–1339.

Ames, G. F.-L., and H. Lecar. 1992. ATP-dependent bacterial transporters and cystic fibrosis: Analogy between channels and transporters. *FASEB J.* 6:2660–2666.

Angel, P. M. Imagawa, R. Chiu, B. Stein, R. J. Imbra, H. J. Rahmsdorf, C. Jonat, P. Herrlich, and M. Karin. 1987. Phorbol ester-inducible genes contain a common *cis*-element recognized by a TPA-modulated trans-acting factor. *Cell* 49:729–739.

Aquino, A., K. D. Hartman, M. C. Knode, K.-P. Huang, C.-H. Niu, and R. I. Glazer. 1988. Role of protein kinase C in phosphorylation of vinculin in Adriamycin-resistant HL-60 leukemia cells. *Cancer Res.* 48:3324–3329.

Aquino, A., B. Warren, J. Omichinski, K. D. Hartman, and R. I. Glazer. 1990a. Protein kinase C-γ is present in Adriamycin-resistant HL-60 leukemia cells. *Biochem. Biophys. Res. Commun.* 166:723–728.

Aquino, A., M. Johnson-Thompson, and R. I. Glazer. 1990b. Enhanced Ca^{2+}-dependent proteolysis associated with Adriamycin-resistant HL-60 cells. *Cancer Commun.* 2:243–247.

Ase, K., N. Berry, U. Kikkawa, A. Kishimoto, and Y. Nishizuka. 1988. Differential down-regulation of protein kinase C subspecies in KM3 cells. *Fed. Eur. Biochem. Soc. Lett.* 236:396–400.

Ashendel, C. L., J. M. Staller, and R. K. Boutwell. 1983. Protein kinase activity associated with a phorbol ester receptor purified from mouse brain. *Cancer Res.* 43:4333–4337.

Basu, A., B. A. Teicher, and J. S. Lazo. 1990. Involvement of protein kinase C in phorbol ester-induced sensitization of HeLa cells to *cis*-diamminechloroplatinum (II). *J. Biol. Chem.* 265:8451–8457.

Bates, S. E., S. J. Currier, M. Alvarez, and A. T. Fojo. 1992. Modulation of P-glycoprotein phosphorylation and drug transport by sodium butyrate. *Biochemistry* 31:6366–6372.

Baeuerle, P. A., and D. Baltimore. (1988). IηB: A specific inhibitor of the NFηB transcription factor. *Science* 242:540–546.

Baxter, G., D. L. Miller, R. C. Kuo, H. G. Wada, and J. C. Owicki. 1992a. PKCε is involved in granulocyte-macrophage colony-stimulating factor signal transduction: Evidence from microphysiometry and antisense oligonucleotide experiments. *Biochemistry* 31:10950–10954.

Baxter, G., E. Oto, S. Daniel-Issakani, and B. Strulovici. 1992b. Constitutive expression of a catalytic fragment of protein kinase Cε in a small cell lung carcinoma cell line. *J. Biol. Chem.* 267:1910–1917.

Beck, W. T. 1987. The cell biology of multiple drug resistance. *Biochem. Pharmacol.* 36: 2879–2887.

Beck, W. T. 1989. Unknotting the complexities of multidrug resistance: The involvement of DNA topoisomerases in drug action and resistance. *J. Natl. Cancer Inst.* 81: 1683–1685.

Bhalla, K., A. Hindenburg, R. N. Traub, and S. Grant. 1985. Isolation and characterization of anthracycline-resistant human leukemic cell line. *Cancer Res.* 45:3657–3662.

Blobe, G. C., C. W. Sachs, W. A. Khan, D. Fabbro, S. Stabel, W. C. Wetsel, L. M. Obeid, R. L. Fine, and Y. A. Hannun. 1993. Selective regulation of expression of protein kinase C (PKC) isoenzymes in multidrug-resistant MCF-7 cells. *J. Biol. Chem.* 268:658–664.

Borellini, F., A. Aquino, S. F. Josephs, and R. I. Glazer. 1990. Increased expression and DNA-binding activity of transcription factor Sp1 in doxorubicin-resistant HL-60 leukemia cells. *Molec. Cell. Biol.* 10:5541–5547.

Borner, C., S. N. Guadagno, L.-L. Hsieh, W.-L. W. Hsiao, and I. B. Weinstein. 1990. Transformation by a *ras* oncogene causes increased expression of protein kinase C-α and decreased expression of protein kinase C-ε. *Cell Growth Differen.* 1:653–660.

Borner, C., I. Filipuzzi, I. B. Weinstein, and R. Imber. 1991. Failure of wild type of a mutant form of protein kinase C-α to transform fibroblasts. *Nature* 353:78–80.

Carlsen, S. A. J. E. Till, and V. Ling. 1977. Modulation of drug permeability in Chinese hamster ovary cells. Possible role for phosphorylation of surface glycoproteins. *Biochem. Biophys. Acta* 467:238–250.

Castagna, M, Y. Takai, K. Kaibuchi, K. Sano, U. Kikkawa, and Y. Nishizuka Y. 1982. Direct activation of calcium-activated, phospholipid-dependent protein kinase by tumor-promoting phorbol esters. *J. Biol. Chem.* 257:7847–7851.

Center, M. S. 1983. Evidence that Adriamycin resistance in Chinese hamster lung cells is regulated by phosphorylation of a plasma membrane glycoprotein. *Biochem. Biophys. Res. Commun.* 115:159–166.

Center, M. S. 1985. Mechanisms regulating cell resistance to Adriamycin. Evidence that drug accumulation in resistant cells is modulated by phosphorylation of a plasma membrane glycoprotein. *Biochem. Pharmacol.* 34:1471–1476.

Chambers, T. C., E. M. McAvoy, J. W. Jacobs, and G. Eilon. 1990a. Protein kinase C phosphorylates P-glycoprotein in multidrug resistant human KB carcinoma cells. *J. Biol. Chem.* 265:7679–7686.

Chambers, T., I. Chalikonda, and G. Eilon. 1990b. Correlation of protein kinase C translocation, P-glycoprotein phosphorylation and reduced drug accumulation in multidrug resistant human KB cells. *Biochem. Biophys. Res. Commun.* 169:253–259.

Chambers, T., B. Zheng, and J. F. Kuo. 1992. Regulation by phorbol ester and protein kinase C inhibitors, and by a protein phosphatase inhibitor (okadaic acid), of P-glycoprotein phosphorylation and relationship to drug accumulation in multidrug-resistant human KB cells. *Molec. Pharmacol.* 41:1008–1015.

Chambers, T. C., J. Pohl, R. L. Raynor, and J. F. Kuo. 1993. Identification of specific sites in human P-glycoprotein phosphorylated by protein kinase C. *J. Biol. Chem.* 268: 4592–4595.

Chandler V. L. B. A. Maler, and K. R. Yamamoto. 1983. DNA sequences bound specifically by glucocorticoid receptor in vitro render a heterologous promoter responsive in vivo. *Cell* 33:489–499.

Chen, K.-H., S. G. Widen, S. H. Wilson, and K.-P. Huang. 1990. Characterization of the

5'-flanking region of the rat protein kinase Cγ gene. *J. Biol. Chem.* 265:19961–19965.

Chin, K.-V., S. S. Chauhan, I. Pastan, and M. M. Gottesman. 1990. Regulation of *mdr* RNA levels in response to cytotoxic drugs in rodent cells. *Cell Growth Differen.* 1:361–365.

Choi, P. M., K.-M. Wong, and I. B. Weinstein. 1990. Overexpression of protein kinase C in HT29 colon cancer cells causes growth inhibition and tumor suppression. *Molec. Cell. Biol.* 10:4650–4657.

Cole, S. P. C., G. Bhardwaj, J. H. Gerlach, J. E. Mackie, C. E. Grant, K. C. Almquist, A. J. Stewart, E. U. Kurz, A. M. V. Duncan, and R. G. Deeley. 1992. Overexpression of a transporter gene in a multidrug-resistant human lung cancer cell line. *Science* 258:1650–1654.

Croop, J. M., P. Gros, and D. E. Housman. 1988. Genetics of multidrug resistance. *J. Clin. Invest.* 81:1303–1309.

DeVore, R. F., A. H. Corbett, and N. Osheoff. 1992. Phosphorylation of topoisomerase II by casein kinase II and protein kinase C: Effects on enzyme-mediated DNA cleavage-religation and sensitivity to the antineoplastic drugs etoposide and 4'-(9-acridinylamino)methane-sulfon-*m*-anisidide. *Cancer Res.* 52:2156–2161.

Dickson, R. B., and M. M. Gottesman. 1990. Understanding of the molecular basis of drug resistance in cancer reveals new targets for chemotherapy. *Trends Pharmacol. Sci.* 11:305–307.

Dong, Z., N. E. Ward, D. Fan, K. P. Gupta, and C. A. O'Brian. 1991. *In vitro* model for intrinsic drug resistance: Effects of protein kinase C activators on the chemosensitivity of cultured human colon cancer cells. *Molec. Pharmacol.* 39:563–569.

Durban, E., J. S. Mills, D. Roll, and H. Busch. 1983. Phosphorylation of purified Novikoff hepatoma topoisomerase I. *Biochem. Biophys. Res. Commun.* 111:897–905.

Edashige, K., E. F. Sato, K. Akimaru, M. Kasai, and K. Utsumi. 1992. Differentiation of HL-60 cells by phorbol ester is correlated with up-regulation of protein kinase C-α. *Arch. Biochem. Biophys.* 299:205–209.

Eldar, H., Y. Zisman, A. Ullrich, and E. Livneh. 1990. Overexpression of protein kinase C α-subtype in Swiss/3T3 fibroblasts causes loss of both high and low affinity receptor numbers for epidermal growth factor. *J. Biol. Chem.* 265:13290–13296.

Eldar, H., J. Ben-Chaim, and E. Livneh. 1992. Deletions in the regulatory or kinase domains of protein kinase C-α cause association with the cell nucleus. *Exp. Cell Res.* 202:259–266.

Fackler, M. J., C. I. Civin, D. R. Sutherland, M. A. Baker, and W. S. May. 1990. Activated protein kinase C directly phosphorylates the CD34 antigen on hematopoietic cells. *J. Biol. Chem.* 265:11056–11061.

Fairchild, C. R., J. S. Moscow, E. E. O'Brien, and K. H. Cowan. 1990. Multidrug resistance in cells transfected with human genes encoding a variant P-glycoprotein and glutathione-*S*-transferase-π. *Molec. Pharmacol.* 37:801–809.

Fan, D., I. J. Fidler, N. E. Ward, C. Seid, L. E. Earnest, G. M. Housey, and C. A. O'Brian. 1992. Stable expression of a cDNA encoding rat brain protein kinase C-βI confers a multidrug-resistant phenotype on rat fibroblasts. *Anticancer Res.* 12:661–668.

Fine, R. L., J. Patel, and B. A. Chabner. 1988. Phorbol esters induce multidrug resistance in human breast cancer cells. *Proc. Natl. Acad. Sci. U.S.A.* 85:582–586.

Finkenzeller, G., D. Marme, and H. Hug. 1992. Inducible over expression of human protein kinase Cα in NIH 3T3 fibroblasts results in growth abnormalities. *Cell. Signal.* 4:163–177.

Gescher, A., and L. L. Dale. 1989. Protein kinase C—a novel target for rational anticancer drug design? *Anticancer Drug Design* 4:93–105.

Gorsky, L. D., S. M. Cross, and M. J. Morin. 1989. Rapid increase in the activity of DNA topoisomerase I, but not topoisomerase II, in HL-60 promyelocytic leukemia cells treated with a phorbol diester. *Cancer Commun.* 1:83–92.

Grant, S., G. R. Pettit, C. Howe, and C. McCrady. 1991. Effect of protein kinase C activating agent bryostatin 1 on the clonogenic response of leukemic blast progenitors to recombinant granulocyte-macrophage colony-stimulating factor. *Leukemia* 5: 393–398.

Greif, H., J. Ben-Chaim, T. Shimon, E. Bechor, H. Eldar, and E. Livneh. The protein kinase C-related PKC-L(η) gene product is localized in the cell nucleus. 1992. *Molec. Cell. Biol.* 12:1304–1311.

Gruber, J. R., S. Ohno, and R. M. Niles. 1992. Increased expression of protein kinase Cα plays a key role in retinoic acid-induced melanoma differentiation. *J. Biol. Chem.* 267:13356–13360.

Hamada, H., K.-I. Hagiwara, I. Nakajima, and T. Tsuruo. 1987. Phosphorylation of the M_r 170,000 to 180,000 glycoprotein specific to multidrug resistant cells: Effects of verapamil, trifluoperazine, and phorbol esters. *Cancer Res.* 47:2860–2865.

Hashimoto, K., A. Kishimoto, H. Aihara, I. Yasuda, K. Mikawa, and Y. Nishizuka. 1990. Protein kinase C during differentiation of human promyelocytic leukemia cell line, HL-60. *Fed. Eur. Biol. Soc. Lett.* 263:31–34.

Hata, A., Y. Akita, Y. Konno, K. Suzuki, and S. Ohno. 1989. Direct evidence that the kinase activity of protein kinase C is involved in transcriptional activation through a TPA-responsive element. *Fed. Eur. Biochem. Soc. Lett.* 252:144–146.

Heck, M. M. S., W. N. Hittelman, and W. C. Earnshaw. 1989. *In vivo* phosphorylation of the 170-kDa form of eukaryotic DNA topoisomerase II. *J. Biol. Chem.* 264:15161–15164.

Hocevar, B. A., and A. P. Fields. 1991. Selective translocation of β_{II}-protein kinase C to the nucleus of human promyelocytic (HL60) leukemia cells. *J. Biol. Chem.* 266: 28–33.

Homan, E. C., D. E. Jensen, and J. J. Sando. 1991. Protein kinase C isozyme expression in phorbol ester-sensitive and -resistant EL4 thymoma cells. *J. Biol. Chem.* 266: 5676–5681.

Homma, Y., C. B. Henning-Chubb, and E. Huberman. 1986. Translocation of protein kinase C in human leukemia cells susceptible or resistant to differentiation induced by phorbol 12-myristate 13- acetate. *Proc. Natl. Acad. Sci. U.S.A.* 83:7316–7319.

Housey, G. M., M. D. Johnson, W. L. W. Hsiao, C. A. O'Brian, J. P. Murphy, P. Kirschmeier, and I. B. Weinstein. 1988. Overproduction of protein kinase C causes disordered growth control in rat fibroblasts. *Cell* 52:343–354.

Hsiao, W.-L. W., G. M. Housey, M. D. Johnson, and I. B. Weinstein. 1989. Cells that overproduce protein kinase C are more susceptible to transformation by an activated H-*ras* oncogene. *Molec. Cell. Biol.* 9:2641–2647.

Huang, F. L., Y. Yoshida, J. R. Cunha-Melo, M. A. Beaven, and K.-P. Huang. 1989. Differential down-regulation of protein kinase C isozymes. *J. Biol. Chem.* 264: 4238–4243.

Hunter, T., and M. Karin. 1992. The regulation of transcription by phosphorylation. *Cell* 70:375–387.

Ido, M., K. Sato, M. Sakurai, M. Inagaki, M. Saitoh, M. Watanabe, and H. Hidaka. 1987. Decreased phorbol ester receptor and protein kinase C in P388 murine leukemic cells resistant to etoposide. *Cancer Res.* 47:3460–3463.

Imagawa, M., R. Chiu, and M. Karin. 1987. Transcription factor AP-2 mediates induction by two different signal-transduction pathways: protein kinase C and cAMP. *Cell* 51:251–260.

Imajoh, S., H. Kawasaki, and K. Suzuki. 1986. The amino-terminal hydrophobic region of the small subunit of calcium-activated neutral protease (CANP) is essential for its activation by phosphatidylinositol. *J. Biochem.* (Tokyo) 99:1281–1284.

Imamura, K., A. Dianoux, T. Nakamura, and D. Kufe. 1990. Colony-stimulating factor 1 activates protein kinase C in human monocytes. *EMBO J.* 9:2423–2429.

Inoue, M., A. Kishimoto, Y. Takai, and Y. Nishizuka. 1977. Studies on a cyclic nucleotide-independent protein kinase and its proenzyme in mammalian tissues. II. Proenzyme and its activation by calcium-dependent protease from rat brain. *J. Biol. Chem.* 252:7610–7616.

Isakov, N., P. McMahon, and A. Altman. 1990. Selective post-transcriptional down-regulation of protein kinase C isoenzymes in leukemic T cells chronically treated with phorbol ester. *J. Biol. Chem.* 265:2091–2097.

Jackson, S. P., J. J. MacDonald, S. Lees-Miller, and R. Tjian. 1990. GC box binding induces phosphorylation of Sp1 by a DNA-dependent protein kinase. *Cell* 63:155–165.

James, G., and E. Olson. 1992. Deletion of the regulatory domain of protein kinase Cα exposes regions in the hinge and catalytic domains that mediate nuclear targeting. *J. Cell Biol.* 116:863–874.

Johnson, P. F., and S. L. McKnight. 1989. Eukaryotic transcriptional regulatory proteins. *Annu. Rev. Biochem.* 58:799–839.

Karin, M. 1992. Signal transduction from cell surface to nucleus in development and disease. *FASEB J.* 6:2581–2590.

Karin, M., A. Haslinger, H. Holtgreve, R. I. Richards, P. Krauter, H. M. Westphal, and M. Beato. 1984. Characterization of DNA sequences through which cadmium and glucocorticoid hormones induce human metallothionein-II$_A$ gene. *Nature* 308:513–519.

Katayama, N., M. Nishizuka, N. Shimizu, F. Komada, T. Sekine, N. Minami, and S. Shirakawa. 1992. A role for protein kinase C in the growth of human erythroid progenitor cells. *Leukemia Res.* 16:145–151.

Kikkawa, U., Y. Takai, R. Minakuchi, S. Inohara, and Y. Nishizuka. 1982. Calcium-activated, phospholipid-dependent protein kinase from rat brain. Subcellular distribution, purification, and properties. *J. Biol. Chem.* 257:13341–13348.

Kikkawa, U., Y. Takai, Y. Tanaka, R. Miyake, R and Y. Nishizuka. 1983. Protein kinase C as a possible receptor protein of tumor-promoting phorbol esters. *J. Biol. Chem.* 258:11442–11445.

Kikkawa, U., A. Kishimoto, and Y. Nishizuka. 1989. The protein kinase C family: Heterogeneity and its implications. *Annu. Rev. Biochem.* 58:31–44.

Kiley, S., D. Schaap, P. Parker, L.-L. Hsieh, and S. Jaken. 1990. Protein kinase C heterogeneity in GL$_4$C$_1$ rat pituitary cells. Characterization of a Ca^{2+}-independent phorbol ester receptor. *J. Biol. Chem.* 265:15704–15712.

Kishimoto, A., N. Kajikawa, M. Shiota, and Y. Nishizuka. 1983. Proteolytic activation of calcium-activated, phospholipid-dependent protein kinase by calcium-dependent neutral protease. *J. Biol. Chem.* 258:1156–1164.

Kishimoto, A., K. Mikawa, K. Hashimoto, I. Yasuda, S.-I. Tanaka, M. Tominaga, T. Kuroda, and Y. Nishizuka. 1989. Limited proteolysis of protein kinase C subspecies by calcium-dependent neutral protease (calpain). *J. Biol. Chem.* 264:4088–4092.

Kohno, K., S.-I. Sato, H. Takano, K.-I. Matsuo, and M. Kuwano. 1989. The direct activation of human multidrug resistance gene (MDR1) by anticancer agents. *Biochem. Biophys. Res. Commun.* 165:1415–1421.

Kraft, A. S., and W. B. Anderson. 1983. Characterization of cytosolic calcium-activated phospholipid-dependent protein kinase activity in embryonal carcinoma cells. *J. Biol. Chem.* 258:9178–9183.

Kroll, D. J., and T. C. Rowe. 1991. Phosphorylation of DNA topoisomerase II in a human tumor cell line. *J. Biol. Chem.* 266:7957–7961.

Leach K. L., E. A. Powers, V. A. Ruff, S. Jaken, and S. Kaufmann. 1989. Type 3 protein kinase C localization to the nuclear envelope of phorbol ester-treated NIH 3T3 cells. *J. Cell Biol.* 109:686–695.

Lee, A. A., J. W. Karaszkiewicz, and W. B. Anderson. 1992. Elevated level of protein kinase C in multidrug-resistant MCF-7 human breast carcinoma cells. *Cancer Res.* 52:3750–3759.

Lee, W., P. Mitchell, and R. Tjian. 1987. Purified transcription factor AP-1 interacts with TPA-inducible enhancer elements. *Cell* 49:741–752.

Lenardo, M. J., and D. Baltimore. 1989. NFκB: A pleiotropic mediator of inducible and tissue-specific gene control. *Cell* 58:227–229.

Liyanage, D., D. Firth, E. Livneh, and S. Stabel. 1992. Protein kinase C group B members PKC-δ, -ε, -ξ and PKC-L (η). Comparison of properties of recombinant proteins *in vitro* and *in vivo. Biochem. J.* 283:781–787.

Makowske, M., R. Ballester, Y. Cayre, and O. M. Rosen. 1988. Immunochemical evidence that three protein kinase C isozymes increase in abundance during HL-60 differentiation induced by dimethyl sulfoxide and retinoic acid. *J. Biol. Chem.* 263:3402–3410.

Maniatis, T., S. Goodbourn, and J. A. Fischer. 1987. Regulation of inducible and tissue-specific gene expression. *Science* 236:1237–1244.

May, W. S., S. Jacobs, and P. Cuatrecasas. 1984. Association of phorbol ester-induced hyperphosphorylation and reversible regulation of transferrin membrane receptors in HL60 cells. *Proc. Natl. Acad. Sci. U.S.A.* 81:2016–2020.

McCrady, C. W., J. Staniswalis, G. R. Pettit, C. Howe, and S. Grant. 1991. Effect of pharmacologic manipulation of protein kinase C by phorbol dibutyrate and bryostatin 1 on the clonogenic response of human granulocyte-macrophage progenitors to recombinant GM-CSF. *Brit. J. Haematol.* 77:5–15.

McSwine-Kennick, R. L., E. M. McKeegan, M. D. Johnson, and M. J. and Morin. 1991. Phorbol ester-induced alterations in the expression of protein kinase C isozymes and their mRNAs. Analysis in wild-type and phorbol diester-resistant HL-60 cell clones. *J. Biol. Chem.* 266:15135–15143.

Megidish, T., and N. Mazurek. 1989. A mutant protein kinase C that can transform fibroblasts. *Nature* 342:807–811.

Mellado, W., and S. B. Horwitz. 1987. Phosphorylation of the multidrug resistance associated glycoprotein. *Biochemistry* 26:6900–6904.

Melloni, E., S. Pontremoli, M. Michetti, O. Sacco, B. Sparatore, and B. L. Horecker. 1986. The involvement of calpain in the activation of protein kinase C in neutrophils stimulated by phorbol myristic acid. *J. Biol. Chem.* 261:4101–4105.

Melloni, E., S. Pontremoli, P. L. Viotti, M. Patrone, P. A. Marks, and R. A. Rifkind. 1989. Differential expression of protein kinase C isozymes and erythroleukemia cell differentiation. *J. Biol. Chem.* 264:18414–18418.

Meyer, T., U. Regenass, D. Fabbro, E. Alteri, J. Rosel, M. Muller, G. Caravatti, and A. Matter. 1989. A derivative of staurosporine (CGP 41 251) shows selectivity for protein kinase C inhibition and *in vitro* anti-proliferative as well as *in vivo* anti-tumor activity. *Intl. J. Cancer* 43:851–856.

Michaeli, J., Y. B. Lebedev, V. M. Richon, Z.-X. Chen, P. A. Marks, and R. A. Rifkind. 1990. Conversion of differentiation inducer resistance to differentiation inducer sensitivity in erythroleukemia cells. *Molec. Cell. Biol.* 10:3535–3540.

Mitchell, P. J., and R. Tjian. 1989. Transcriptional regulation in mammalian cells by sequence-specific DNA binding proteins. *Science* 245:371–378.

Miyamoto, K. I., S. Wakusawa, K. Inoko, K. Takagi, and M. Koyama. 1992. Reversal of vinblastine resistance by a new staurosporine derivative, NA-382, in P388/ADR cells. *Cancer Lett.* 177–183.

Muramatsu, M.-A., K. Kaibuchi, and K.-I. Arai. 1989. A protein kinase C cDNA without the regulatory domain is active after transfection *in vivo* in the absence of phorbol ester. *Molec. Cell. Biol.* 9:831–836.

Murray, A. W., A. Fournier, and S. J. Hardy. 1987. Proteolytic activation of protein kinase C: A physiological reaction? *Trends Biol. Sci.* 12:53–54.

Myers, M. B., L. Rittmann-Grauer, J. P. O'Brien, and A. R. Safa. 1989. Characterization of monoclonal antibodies recognizing a M_r 180,000 P-glycoprotein: Differential expression of the M_r 180,000 and M_r 170,000 P-glycoproteins in multidrug-resistant human tumor cells. *Cancer Res.* 49:3209–3214.

Newton, A. C., and D. E. Koshland Jr. 1989. High cooperativity, specificity, and multiplicity in the protein kinase C-lipid interaction. *J. Biol. Chem.* 264:14909–14915.

Niino, Y. S., S. Ohno, and K. Suzuki. 1992. Positive and negative regulation of the transcription of the human protein kinase Cβ gene. *J. Biol. Chem.* 267:6158–6163.

Nishikawa, M., F. Komada, Y. Uemura, H. Hidaka, and S. Shirakawa. 1990. Decreased expression of type II protein kinase C in HL-60 variant cells resistant to induction of cell differentiation by phorbol diester. *Cancer Res.* 50:621–626.

Nishizuka, Y. 1989. Studies and prospectives of the protein kinase C family for cellular recognition. *Cancer* 10:1892–1903.

Obeid, L. M., T. Okazaki, L. A. Karolak, and Y. A. Hannun. 1990. Transcriptional regulation of protein kinase C by 1,25 dihydroxyvitamin D_3 in HL-60 cells. *J. Biol. Chem.* 265:2370–2374.

Obeid, L. M., G. C. Blobe, L. A. Karolak, and Y. A. Hannun. 1992. Cloning and characterization of the major promoter of the human protein kinase Cβ gene. *J. Biol. Chem.* 267:20804–20810.

O'Brian, C. A., D. Fan, N. E. Ward, C. Seid, and I. J. Fidler. 1989. Level of protein kinase C activity correlates directly with resistance to Adriamycin in murine fibrosarcoma cells. *Fed. Eur. Biochem. Soc. Lett.* 246:78–82.

O'Brian, C. A., D. Fan, N. E. Ward, Z. Dong, L. Iwamoto, K. P. Gupta, L. E. Earnest, and I. J. Fidler. 1991. Transient enhancement of multidrug resistance by the bile acid deoxycholate in murine fibrosarcoma cells *in vitro*. *Biochem. Pharmacol.* 41:797–806.

O'Connor, T. W. E. 1985. Phorbol ester-induced loss of colchicine ultrasensitivity in chronic lymphocytic leukaemia lymphocytes. *Leukemia Res.* 9:885–895.

Ohno, S., H. Kawasaki, S. Imajoh, and K. Suzuki, K. 1987. Tissue-specific expression of three distinct types of rabbit protein kinase C. *Nature* 325:161–166.

Ohno, S., Y. Konno, Y. Akita, A. Yano, and K. Suzuki. 1990. A point mutation at the putative ATP-binding site of protein kinase Cα abolishes the kinase activity and renders it down-regulation insensitive. *J. Biol. Chem.* 265:6296–6300.

Pai, J.-K., J. A. Pachter, I. B. Weinstein, and W. R. Bishop. 1991. Overexpression of protein kinase C β1 enhances phospholipase D activity and diacylglycerol formation in phorbol ester-stimulated rat fibroblasts. *Proc. Natl. Acad. Sci. U.S.A.* 88: 598–602.

Palayoor, S. T., J. M. Stein, and W. N. Hait. 1987. Inhibition of protein kinase C by antineoplastic agents: Implications for drug resistance. *Biochem. Biophys. Res. Commun.* 148:718–725.

Parker, P. J., G. Kour, R. M. Marais, F. Mitchell, C. Pears, D. Schaap, S. Stabel, and C. Webster. 1989. Protein kinase C—a family affair. *Molec. Cell. Endrocrinol.* 65:1–11.

Pasti, G., J.-C. Lacal, B. S. Warren, S. A. Aaronson, and P. M. Blumberg. 1986. Loss of mouse fibroblast cell response to phorbol esters restored by microinjected protein kinase *C. Nature* 324:375–377.

Pelham, H. R. B. 1982. Regulatory upstream promoter element in the Drosophila Hsp 70 heat shock gene. *Cell* 30:517–528.

Perletti, G., A. Ghessi, E. Raffaldoni, and F. Piccinini. 1991. The activity of a beta subtype of protein kinase C purified from nuclei of human neutrophils is enhanced by treatment with phorbol 12-myristate 13-acetate. *Biochem. Biophys. Res. Commun.* 181:348–352.

Perrella, F. W., B. D. Hellmig, and L. Diamond. 1986. Regulation of the phorbol ester receptor-protein kinase C in HL-60 variant cells. *Cancer Res.* 46:567–572.

Persons, D. A., W. O. Wilkison, R. M. Bell, and O. J. Finn. 1988. Altered growth regulation and enhanced tumorigenicity of NIH 3T3 fibroblasts transfected with protein kinase C-I cDNA. *Cell* 52:447–458.

Persons, D. A., R. D. Downs, M. C. Ostrowski, and O. J. Finn. 1991. Protein kinase Cγ expression mimics phorbol ester-induced transcriptional activation of a murine VL30 enhancer element. *Cell Growth Differen.* 2:7–14.

Pommier, Y., D. Kerrigan, K. D. Hartman, and R. I. Glazer. 1990. Phosphorylation of mammalian DNA topoisomerase I and activation by protein kinase C. *J. Biol. Chem.* 265:9418–9422.

Pontremoli, S., E. Melloni, B. Sparatore, M. Michetti, F. Salamino, and B.L. Horecker. 1990a. Isozymes of protein kinase C in human neutrophils and their modification by two endogenous proteinases. *J. Biol. Chem.* 265:706–712.

Pontremoli, S., M. Michetti, E. Melloni, B. Sparatore, F. Salamino, and B. L. Horecker. 1990b. Identification of the proteolytically activated form of protein kinase C in stimulated human neutrophils. *Proc. Natl. Acad. Sci. U.S.A.* 87:3705–3707.

Posada, J. A., E. M. McKeegan, K. F. Worthington, M. J. Morin, S. Jaken, and T. R. Tritton. 1989a. Human multidrug resistant KB cells overexpress protein kinase C: Involvement in drug resistance. *Cancer Commun.* 1:285–292.

Posada, J., P. Vichi, and T. R. Tritton. 1989b. Protein kinase C in Adriamycin action and resistance in mouse sarcoma 180 cells. *Cancer Res.* 49:6634–6639.

Powell, C. T., L. Leng, L. Dong, H. Kiyokawa, X. Busquets, K. O'Driscoll, P. A. Marks, and R. A. Rifkind. 1992. Protein kinase C isozymes ε and α in murine erythroleukemia cells. *Proc. Natl. Acad. Sci. U.S.A.* 89:147–151.

Richon, V. M., N. Weich, L. Leng, H. Kiyokawa, L. Ngo, R. A. Rifkind, and P. A. Marks. 1991. Characteristics of erythroleukemia cells selected for vincristine resistance that have accelerated inducer-mediated differentiation. *Proc. Natl. Acad. Sci. U.S.A.* 88:1666–1670.

Rohlff, C., B. Safa, A. Rahman, Y. S. Cho-Chung, R. W. Klecker, and R. I. Glazer. 1993. Reversal of resistance of Adriamycin by 8-C1-cAMP in Adriamycin-resistant HL-60 leukemia cells is associated with reduction of type I cyclic AMP-dependent protein kinase and CREB and Sp1 DNA-binding activities. *Molec. Pharmacol.* 43, in press.

Ross, W. E. 1986. DNA topoisomerases as targets for cancer therapy. *Biochem. Pharmacol.* 34:4191–4195.

Safa, A. R., R. K. Stern, K. Choi, M. Agresti, T. Tamai, N. D. Mehta, and I. B. Roninson. 1990. Molecular basis of preferential resistance to colchicine in multidrug-resistant human cells conferred by Gly-185→Val-185 substitution in P-glycoprotein. *Proc. Natl. Acad. Sci. U.S.A.* 87:7225–7229.

Sahyoun, N., M. Wolf, J. Besterman, T.-S. Hsieh, M. Sander, H. LeVine III, K.-J. Chang, and P. Cuatrecasas. 1986. Protein kinase C phosphorylates topoisomerase II: Topoisomerase activation and its possible role in phorbol ester-induced differentiation of HL-60 cells. *Proc. Natl. Acad. Sci. U.S.A.* 83:1603–1607.

Salamino, F., B. Spartore, R. DeTullio, P. Mengotti, E. Melloni, and S. Pontremoli. 1991. Respiratory burst in activated neutrophils is directly correlated to the intracellular level of protein kinase C. *Eur. J. Biochem.* 200:573–577.

Samuels, D. S., and N. Shimizu. 1992. DNA topoisomerase I phosphorylation in murine fibroblasts treated with 12-*O*-tetradecanoylphorbol-13-acetate and *in vitro* by protein kinase C. *J. Biol. Chem.* 267:11156–11162.

Sato, W., K. Yusa, M. Naito, and T. Tsuruo. 1990. Staurosporine, a potent inhibitor of C-kinase, enhances drug accumulation in multidrug-resistant cells. *Biochem. Biophys. Res. Commun.* 173:1252–1257.

Sauvage, C., C. B. Cash, and M. Maitre. 1991. Isolation of human brain protein kinase C: Evidence for kinase C catalytic fragment modulates G protein-GTPase activity. *Biochem. Biophys. Res. Commun.* 174:593–599.

Schurr, E., M. Raymond, J. C. Bell, and P. Gros. 1989. Characterization of the multidrug resistance protein expressed in cell clones stably transfected with the mouse *mdr1* cDNA. *Cancer Res.* 49:2729–2734.

Schwartz, G. K., H. Arkin, J. F. Holland, and T. Ohnuma. 1991. Protein kinase C activity and multidrug resistance in MOLT-3 human lymphoblastic leukemia cells resistant to trimetrexate. *Cancer Res.* 51:55–61.

Searle, P. F., G. W. Stuart, and R. D. Palmiter. 1985. Building a metal-responsive promoter with synthetic regulatory elements. *Molec. Cell. Biol.* 5:1480–1489.

Shirakawa, F., and S. B. Mizel. 1989. In vitro activation and nuclear translocation of NFκB catalyzed by cyclic AMP-dependent protein kinase and protein kinase C. *Molec. Cell. Biol.* 9:2424–2430.

Smith, C. D., J. F. Glickman, and K.-J. Chang. 1988. The antiproliferative effects of staurosporine are not exclusively mediated by inhibition of protein kinase C. *Biochem. Biophys. Res. Commun.* 156:1250–1256.

Solomon, D. H., K. O'Driscoll, G. Sosne, I. B. Weinstein, and Y. E. Cayre. 1991. 1α,25-Dihydroxyvitamin D_3-induced regulation of protein kinase C gene expression during HL-60 cell differentiation. *Cell Growth Differen.* 2:187–194.

Staats, J., D. Marquardt, and M. Center. 1990. Characterization of a membrane-associated protein kinase of multidrug-resistant HL60 cells which phosphorylates P-glycoprotein. *J. Biol. Chem.* 265:4084–4090.

Strulovici, B., S. Daniel-Issakami, E. Oto, J. Nester Jr., H. Chan, and A.-P. Tsou. 1989. Activation of distinct protein kinase C isozymes by phorbol esters: Correlation with induction of interleukin 1β gene expression. *Biochemistry* 28:3569–3576.

Strulovici, B., S. Daniel-Issakami, G. Baxter, J. Knopf, L. Sultzman, H. Cherwinski, J. Nester Jr., D. R. Webb, and J. Ransom. 1991. Distinct mechanisms of regulation of protein kinase Cε by hormones and phorbol esters. *J. Biol. Chem.* 266:168–173.

Sugimoto, Y., I. B. Roninson, and T. Tsuruo, T. 1987. Decreased expression of the amplified *mdr*1 gene in revertants of multidrug-resistant human myelogenous leukemia K562 occurs without loss of amplified DNA. *Molec. Cell. Biol.* 7:4549–4552.

Sugimoto, Y., S. Tsukahara, T. Oh-hara, T. Isoe, and T. Tsuruo. 1990. Decreased expression of topoisomerase I in camptothecin-resistant tumor cell lines as determined by a monoclonal antibody. *Cancer Res.* 50:6925–6930.

Takai, Y., A. Kishimoto, M. Inoue, and Y. Nishizuka. 1977. Studies on a cyclic nucleotide-independent protein kinase and its proenzyme in mammalian tissues. I. Purification and characterization of an active enzyme from bovine cerebellum. *J. Biol. Chem.* 252:7603–7609.

Takano, H., K. Kohno, M. Ono, Y. Uchida, and M. Kuwano. 1991. Increased phosphorylation of DNA topoisomerase II in etoposide-resistant mutants of human cancer KB cells. *Cancer Res.* 51:3951–3957.

Tchou-Wong, K.-M., and I. B. Weinstein. 1992. Altered expression of protein kinase C, lck, and CD45 in a 12-*O*-tetradecanoylphorbol-13-acetate-dependent leukemic T-cell variant that expresses a high level of interleukin-2 receptor. *Molec. Cell. Biol.* 12:394–401.

Thomas, T. P., H. S. Talwar, and W. B. Anderson. 1988. Phorbol ester-mediated association of protein kinase C to the nuclear fraction in NIH 3T3 cells. *Cancer Res.* 48: 1910–1919.

Toda, T., M. Shimanuki, and M. Yanagida. 1991. Fission yeast genes that confer resistance to staurosporine encode an AP-1-like transcription factor and a protein kinase related to the mammalian ERK1/MAP2 and budding yeast *FUS3* and *KSS1* kinases. *Genes Devel.* 5:60–73.

Tonetti, D. A., M. Horio, F. R. Collart, and E. Huberman. 1992. Protein kinase Cβ gene expression is associated with susceptibility of human promyelocytic leukemia cells to phorbol ester-induced differentiation. *Cell Growth Differen.* 3:739–745.

Ueda, K., I. Pastan, and M. M. Gottesman. 1987. Isolation and sequence of the promoter region of the human multidrug-resistance (P-glycoprotein) gene. *J. Biol. Chem.* 262:17432–17436.

Wakusawa, S., S. Nakamura, K. Tajima, K. I. Miyamoto, M. Hagiwara, and H. Hidaka. 1992. Overcoming of vinblastine resistance by isoquinolinesulfonamide compounds in Adriamycin-resistant leukemia cells. *Molec. Pharmacol.* 41:1034–1038.

Watanabe, T., Y. Ono, Y. Taniyama, K. Hazama, K. Igarashi, K. Ogita, U. Kikkawa, and Y. Nishizuka. 1992. Cell division arrest induced by phorbol ester in CHO cells overexpressing protein kinase C-δ subspecies. *Proc. Natl. Acad. Sci. U.S.A.* 89: 10159–10163.

Ways, D. K., P. P. Cook, C. Webster, and P. J. Parker. 1992a. Effect of phorbol esters on protein kinase C-ξ. *J. Biol. Chem.* 267:4799–4805.

Ways, D. K., B. R. Messer, T. O. Garris, W. Qin, P. P. Cook, and P. J. Parker. 1992b. Modulation of protein kinase C-ϵ by phorbol esters in the monoblastoid U937 cell. *Cancer Res.* 52:5604–5609.

Yu, G., S. Ahmad, A. Aquino, C. R. Fairchild, J. B. Trepel, S. Ohno, K. Suzuki, T. Tsuruo, K. H. Cowan, and R. I. Glazer. 1991. Transfection with protein kinase Cα confers increased multidrug resistance to MCF-7 cells expressing P-glycoprotein. *Cancer Commun.* 3:181–189.

Zidovetski, R., and D. S. Lester. 1992. The mechanism of activation of protein kinase C: A biophysical perspective. *Biochim. Biophys. Acta* 1134:261–272.

Zwelling, L. A., M. Hinds, D. Chan, E. Altschuler, J. Mayes, and T. F. Zipf. 1990. Phorbol ester effects on topoisomerase II activity and gene expression in HL-60 human leukemia cells with different proclivities toward monocytoid differentiation. *Cancer Res.* 50:7116–7122.

Zylber-Katz, E., and R. I. Glazer. 1985. Phospholipid- and Ca^{2+}-dependent protein kinase activity and protein phosphorylation patterns in the differentiation of human promyelocytic leukemia cell line HL-60. *Cancer Res.* 45:5159–5164.

Zylber-Katz, E., M. C. Knode, and R. I. Glazer. 1986. Retinoic acid promotes phorbol ester-initiated macrophage differentiation in HL-60 leukemia cells without disappearance of protein kinase C. *Leukemia Res.* 10:1425–1432.

7

Protein Kinase C in the Nervous System

P. JEFFREY CONN
J. DAVID SWEATT

It has long been established that activation of neurotransmitter receptors can induce rapid transient changes in the membrane potential of effector neurons that last on the order of milliseconds. Such short-lived changes, known as postsynaptic potentials, can be inhibitory (IPSP) or excitatory (EPSP) and result from changes in the conductances of ligand-gated ion channels such as the nicotinic acetylcholine receptor (AChR) or $GABA_A$ receptor chloride channel. In recent years it has become increasingly clear that brief activation of neurotransmitter receptors can also induce changes in neuronal function that last minutes, hours, or even days or months. Such long-lasting changes in neuronal function play a critical role in regulating animal behavior, and can include changes in a cell's electrical properties or in important cellular processes such as neurotransmitter synthesis, neurotransmitter release, or gene expression. Generally, changes in neuronal function that persist after the neurotransmitter has been inactivated are brought about by second-messenger-mediated activation of specific protein kinases. The mammalian central nervous system (CNS) contains a wide variety of different families of protein kinases. Of these, members of the protein kinase C (PKC) family have been some of the best characterized in terms of distribution in the CNS and specific roles in regulating neuronal function.

PKC is highly enriched in the nervous systems of many different animal species (Kuo et al., 1980; Girard et al., 1985) and the brain is the most abundant source of PKC in terms of both quantity of enzyme and number of PKC subspecies expressed (Shinomura et al., 1992). Indeed, two PKC subspecies (ϵ and γ) may be expressed exclusively in the CNS and have not been detected in other tissues (Shinomura et al., 1992). Increasing evidence suggests that PKC plays a key role in regulating a variety of aspects of CNS function, including neuronal excitability, neurotransmitter release, and synaptic transmission.

REGULATION OF PKC ACTIVITY IN THE CNS

For the most part, the mechanisms involved in regulation of PKC activity in the CNS are the same as in other tissues (see Chapter 6). Receptors coupled to virtually all of the transduction systems involved in activating PKC have been identified in the CNS. These include receptors coupled to phosphoinositide hydrolysis (Fisher and Agranoff, 1987), phospholipase D activation (Boss and Conn, 1992; Llahi and Fain, 1992), and arachidonic acid release (Axelrod et al., 1988). However, there are also other mechanisms for activation of PKC that may be unique to the CNS and excitable cells in the periphery. For instance, activation of ligand-gated cation channels, such as the glutamate receptors, can result in activation of PKC that is secondary to calcium influx. This can be due to activation of calcium-permeable ligand-gated channels (such as the N-methyl-D-aspartate subtype of glutamate receptor)(Etoh et al., 1991a,b; Vaccarino et al., 1991) or activation of ligand-gated channels that are impermeable to calcium but that cause cell depolarization and subsequent activation of voltage-dependent calcium channels (Vaccarino et al., 1991; Wakade et al., 1991).

One of the most interesting aspects of PKC with regard to its role in regulating neuronal function is that it has a number of properties that make it a prime candidate for being an enzyme involved in synaptic plasticity. Neurobiologists have long been interested in the mechanisms by which a transient stimulus can induce a change in synaptic function that persists long after the initiating stimulus has been removed. Such lasting changes in neuronal function are likely to play a critical role in a number of normal physiological processes, such as learning and memory, as well as various pathological disorders. One potential mechanism by which such lasting changes could occur is that transient activation of a neurotransmitter receptor could result in persistent activation of a protein kinase. As with other second messengers, second messengers that activate PKC generally are rapidly metabolized and remain elevated only a few seconds after receptor activation. However, because of the unique properties of PKC, the effects of transient formation of these second messengers may persist long after their levels have returned to baseline. For example, PKC was first described as protein kinase from rat brain that could be activated by specific proteolysis by a calcium-activated protease. Such proteolysis produces a catalytic fragment of PKC, termed protein kinase M (PKM)(see Murray et al., 1987, for review), that is constitutively active in the absence of calcium and phospholipid. Such a reaction occurring *in vivo* would result in a long-lasting increase in PKC activity that would diminish with a time course related to the biological half-life of PKM. Recently, a second mechanism for production of a constitutively active form of PKC was proposed. Activation of PKC is normally associated with translocation of the enzyme from a cytosolic to a membrane fraction of the cell. Using phospholipid vesicles and phospholipid monolayers, Bazzi and Nelsestuen (1988a,b,c) found that calcium can induce translocation of the enzyme such that it remains bound to the membrane in an irreversible manner; that is, it appears that the enzyme has been inserted into the membrane. In its membrane-inserted form, PKC is tonically active and cannot be further activated by addition of calcium or phorbol esters.

As discussed in detail below, formation of constitutively active forms of PKC may play an important role in long-term potentiation (LTP) of synaptic efficacy and other forms of neuronal plasticity.

REGIONAL DISTRIBUTION AND DEVELOPMENTAL REGULATION OF PKC ISOFORMS

The members of the PKC family of isozymes are widely distributed in mammalian tissues. Of the eight well-characterized isoforms, all but the η subtype are represented in the CNS, and overall the brain is highly enriched in PKC (see Nishizuka, 1989; Stabel and Parker, 1991, for reviews). In fact, the γ and ε subtypes are selectively expressed in the CNS. The regional distribution of the α, β, and γ isoforms in the CNS has been well documented.

Distribution of PKC α

Localization of PKC α was determined using an antiserum directed against a unique carboxy-terminal sequence in the enzyme (Ito et al., 1990). PKC α appeared to be largely localized to neurons, and was present throughout the cell, except immunostaining was lacking from the nucleus. Glia contained very little immunoreactivity. Notable areas of high PKC α immunoreactivity include the olfactory bulb, lateral septal nucleus, substantia nigra pars compacta, inferior olive, and hippocampus/dentate gyrus. More modest staining was observed in layers II–III of the cerebral cortex, and very little immunostaining was observed in the cerebellar cortex. In the same study the mRNA expression for PKC α was found to correlate well with the immunoreactivity.

Distribution of PKC βI

Localization of PKC βI has been determined using an approach similar to that described above for PKC α (Hosada et al., 1989). Similar to the PKC α immunoreactivity, PKC βI immunoreactivity was localized to neurons but not glia, and the subcellular distribution appeared to be throughout the cell with the exception of the nucleus. Notable areas of high PKC βI immunoreactivity were the septal nucleus and pontine nuclei. Interestingly, there was very little overlap in the distribution of reactivity for PKC γ versus PKC βI (see below).

Distribution of PKC βII

The distribution of PKC βII is distinct from that of PKC βI and PKC γ (Saito et al., 1989). It was found to be highly represented in olfactory nucleus, olfactory tubercle, amygdala, caudate nucleus/putamen, nucleus accumbens, lateral septal nucleus, the CA1 region of the hippocampus, the cerebral cortex, and various spinal and brainstem nuclei. Like PKC α and PKC βI, PKC βII was largely localized to neurons, and was present throughout the cell. Interestingly, PKC βII was found to be uniquely associated with the Golgi complex.

Distribution of PKC γ

PKC γ is located exclusively in the brain and spinal cord (Saito et al., 1988). In the brain it is found in high levels in the hippocampus, cerebral cortex, amygdala, and cerebellum. The localization within the cerebellum is confined largely to the cell bodies, dendrites, and axons of Purkinje cells. In the hippocampus, this isoform of PKC appears to be localized to postsynaptic elements of pyramidal neurons.

Functional Consequences of Selective Distribution of PKC Isoforms

Linden and colleagues (1992) have described an interesting functional correlate of the selective cellular distribution of PKC γ versus PKC β. In the cerebellum, Purkinje cells express PKC γ, whereas granule cells express predominantly PKC β. PKC γ can be activated by *cis-* unsaturated fatty acids such as oleic acid, whereas PKC β is largely insensitive to this class of compounds. Both Purkinje cells and granule cells in the cerebellum possess large voltage-gated potassium currents of the delayed rectifier type, which can be attenuated by PKC activation. Linden and co-workers (1992) showed that oleic acid could selectively decrease delayed rectifier potassium currents in Purkinje cells and not in granule cells, providing direct physiological evidence for a functional consequence of the discrete distribution of PKC isoforms in the cerebellum.

Developmental Regulation of PKC Expression

Sposi (1989) and Hashimoto (1988) and their colleagues have investigated the ontogeny of the expression of the genes for α, β, and γ PKC in rat and mouse brain. PKC α and β genes were expressed in early stages of embryonic development, showing a progressive increase in expression into adult stages. In contrast, PKC γ was not expressed in embryos; expression began in the perinatal period, and substantial expression occurred only in the adult. In general these results are consistent with a role for PKC α and PKC β in early brain development.

PKC-INDUCED MODULATION OF CALCIUM CHANNELS

Studies of the modulation of calcium channels by PKC and other second-messenger/protein kinase systems is of particular interest to many investigators because of the wide variety of functions that these channels have in regulating cellular function (see Scott et al., 1991; Tsien et al., 1991, for reviews). Like other ion channels, calcium channels can be involved in generating electrical signals. In many invertebrate neurons, such as the *Aplysia* bag cell neurons discussed below, calcium is the primary carrier of inward current during an action potential. Likewise, calcium is a primary charge carrier during action potentials in many excitable cells in vertebrates, including cardiac and smooth muscle cells. In most vertebrate neurons, sodium is the primary carrier of inward current during action potentials, but calcium currents play an important role in generating pacemaker depolarizations and paroxysmal depolarizing shifts that occur during seizure activity. In addition to giving rise to electrical signals, calcium serves as an intracellular second messenger. Increases in intracellular calcium concentrations are important for a number of cellular processes including neurotransmitter release, regulation of potassium channels and other ion channels, and activation of specific enzymes, including proteases, phosphatases, and protein kinases.

Modulation of Calcium Channels in Bag Cell Neurons of *Aplysia*

The nervous system of marine mollusk *Aplysia* provides an excellent model system for studying regulation of neuronal excitability by second messengers and protein kinases. One particularly useful population of neurons in *Aplysia* is a group of cells called the bag cell neurons, which control egg-laying behavior in this animal (see Conn and Kaczmarek, 1989, for review). Normally the bag cell neurons are electrically silent. However, brief stimulation of the pleuroabdominal nerve (a major afferent input to these cells) or application of certain neuroactive peptides elicits a long-lasting afterdischarge of spontaneously firing calcium-dependent action potentials. The afterdischarge lasts approximately 30 minutes and results in release of several neuroactive peptides that act on various target cells to initiate egg-laying behavior. Intracellular recordings from individual bag cell neurons reveal that about 1 minute into an afterdischarge the bag cell action potentials become significantly enhanced in both height and width. It is likely that this enhancement of bag cell action potentials results in greater calcium influx during each action potential and thereby maximizes the efficiency of neurotransmitter release during the afterdischarge.

A great deal of effort has been focused on determining the cellular mechanisms involved in generation of the afterdischarge and the enhancement of bag cell action potentials. The onset of the afterdischarge is accompanied by an increase in phosphoinositide hydrolysis (Fink et al., 1988). Furthermore, the bag cell neurons are a rich source of PKC (DeReimer et al., 1985a). Thus, PKC is a prime candidate for a protein kinase involved in bringing about the excitability changes that are characteristic of the afterdischarge. The first evidence that this is the case was supplied by DeReimer and colleagues (1985b), who found that exposure of cultured bag cell neurons to a cell-permeable DAG or the PKC-activating phorbol ester TPA causes a marked potentiation of the height of evoked action potentials, similar to that occurring during an afterdischarge in the intact abdominal ganglion. This effect was mimicked by intracellular injection of purified PKC or by addition of other phorbol esters that activate PKC but not by phorbol esters that do not activate PKC (DeReimer et al., 1985b). In addition, the effect of TPA on action potentials is inhibited if the protein kinase inhibitors, sphinganine (erythro dihydro-sphingosine) or H-7 are added 15 minutes prior to addition of TPA (Conn et al., 1989a,b). This suggests that these actions of TPA are mediated by activation of PKC.

Whole-cell patch clamp studies suggest that the enhancement of action potentials is mediated by an increase in calcium current in these cells. Thus, PKC activators have no effect on the major potassium currents in bag cell neurons but induce an approximate doubling of voltage dependent calcium current (DeReimer et al., 1985b). As with the effect of PKC activators on action potentials, this effect is not mimicked by phorbol esters that do not activate PKC (DeReimer et al., 1985b) and is blocked by prior incubation with the PKC inhibitors spinganine or H-7 (Conn et al., 1989a,b).

Cell-attached patch clamp techniques were used to determine the mechanism by which PKC activators enhance calcium current in bag cell neurons at the single channel level (Strong et al., 1987). In control bag cell neurons, a single species of calcium channel is present. This channel has a relatively small unitary conductance ($\approx$12 pS) (picosiemens) and is distributed in clusters on the cell membrane. After treatment of bag cell neurons with PKC activators, the 12-pS channel is present in the same density as in control cells, and the properties of this channel (e.g., mean open time, frequency of openings.) remain unchanged. However, in TPA-treated cells a new larger conductance ($\approx$24 pS) channel is present that is never seen in control cells. In contrast to the small-conductance channel, the 24-pS channel is distributed evenly on the cell surface and tends to open in bursts of high-frequency openings. Since PKC activation does not alter the density of the 12-pS channel, it is likely that the large-conductance calcium channel is a novel species of channel that is actively recruited to the cell membrane on activation of PKC. This hypothesis is supported by the distinct spatial distributions, distinct unitary conductances, and distinct bursting kinetics of the two channels. Thus, although the possibility that the large-conductance channel results from conversion of the small-conductance channel cannot be completely ruled out, evidence suggests that activation of PKC brings about the acute recruitment of a previously covert species of calcium channel.

Although it remains to be determined whether the large-conductance calcium channel is expressed during the afterdischarge, the increase in phosphoinositide hydrolysis during the afterdischarge is likely to result in activation of PKC and lead to expression of the covert calcium channel. However, studies in cultured cells give no indication of the role, if any, that the covert calcium channel plays in bringing about the excitability changes that occur during the afterdischarge. One way to address this issue would be to use selective inhibitors of the 24-pS calcium channel to see what effect blockade of this current would have on expression of the afterdischarge. Unfortunately, the pharmacology of the 12-pS and the 24-pS channels are very similar and there are no known selective blockers of the 24-pS channel. However, consistent with their effect on whole-cell calcium current, sphinganine and H-7 prevent expression of the covert calcium channel when added 15 minutes before addition of TPA. Thus, the availability of these cell-permeable PKC inhibitors allows blockade of expression of the covert calcium channel and thereby allows direct assessment of the contribution of PKC and the covert calcium channel to the afterdischarge. When the afterdischarge is monitored with extracellular recordings, there are no obvious differences between afterdischarges in control ganglia and ganglia treated with PKC inhibitors (Conn et al., 1989b). This suggests that PKC and the covert calcium channel are not obligatory for initiation and maintenance of a bag cell afterdischarge. However, intracellular recordings from individual neurons within the cluster reveal that the normal enhancement of action potential height and width does not occur during the afterdischarge if ganglia are treated with PKC inhibitors (Conn et al., 1989b). Furthermore, PKC inhibitors reduce the release of neuropeptides during the afterdischarge (Loechner et al., 1992). Thus, it is likely that PKC and the covert calcium channel are not required for initiation or maintenance of the afterdischarge but may contribute to spike enhancement during the afterdischarge and thereby enhance neuropeptide release.

One particularly interesting aspect of this mechanism of enhancing calcium current in bag cell neurons is that expression of a new species of calcium channels could provide a mechanism for induction of a rapid change in neuronal excitability that greatly outlasts the initiating stimulus. As discussed above, the stimulus to the pleuroabdominal nerve that initiates an afterdischarge is brief, lasting only seconds. However, the spike enhancement that is brought about by expression of the covert calcium channel persists for the duration of the afterdischarge. Interestingly, although addition of sphinganine or H-7 15 minutes before addition of TPA results in complete inhibition of TPA-induced spike enhancement, these compounds do not reverse the effect of TPA on action potential height if they are added 20 minutes after addition of TPA (after action potentials have become fully enhanced) (Conn et al., 1989b). This suggests that transient activation of PKC can act as a ''trigger'' for inducing expression of the covert calcium channel and that, once expressed, the channel activity is autonomous and does not depend on PKC activity for maintenance. The mechanism of the persistence of PKC-activated calcium current is not known at present. One possible mechanism is that transient activation of the enzyme results in formation of one of the constitutively active forms of PKC (see above). However, this mechanism is unlikely since such an enzyme would be expected to be sensitive to PKC inhibitor H-7. H-7 inhibits PKC by competing with ATP for the catalytic site and inhibits PKM by acting at the same site (Hidaka et al., 1984). Thus, H-7 would be expected to reverse any effects of TPA that are mediated by formation of this constitutively active form of PKC. Another possible mechanism is that PKC induces insertion of the covert channel into the cell membrane. If this is the case, once the covert channel is expressed it may require normal membrane recycling for inactivation.

PKC activators have also been shown to increase calcium current in neurons from *Helix aspersa* (Hill-Venning and Cottrell, 1992) and *H. pomatia* (Kostyuk et al., 1992), suggesting that enhancement of calcium current may be a common effect of PKC in molluscan neurons. However, the exact mechanisms by which PKC enhances calcium current in these cells has not been determined.

Modulation of Voltage-sensitive Calcium Currents in Vertebrate Cells

Classification of Vertebrate Calcium Channels
There is a great deal of diversity among voltage-dependent calcium channels in the vertebrate CNS. Thus, any simple taxonomy for classifying voltage-dependent

calcium channels will be an oversimplification. However, if rigid adherence to such a classification is avoided, such systems for classifying calcium channels can provide a useful framework for studying modulation of voltage-dependent calcium channels by PKC and other second-messenger/protein kinase systems. One of the most commonly accepted nomenclatures for discussing voltage-dependent calcium channels in vertebrate neurons was proposed by Nowycky and colleagues (1985) on the basis of studies originally done with chick dorsal root ganglion (DRG) neurons. These investigators divided calcium channels in DRG neurons into T, N, and L subtypes. It is now clear that these three channel subtypes are present on a variety of neuronal and nonneuronal cells, often coexisting on a single cell. L-, N-, and T-type channels can be distinguished on the basis of many factors, including sensitivity to specific channel blockers and activators and differences in conductance and gating properties. In addition, these channels have distinct distributions and each is likely to play a unique role in regulating cell function (see Tsien et al., 1988; for review).

The L-type channel has the largest unitary conductance of the three channel subtypes ($\approx$25 pS unitary conductance with 110 mM barium as the charge carrier) and requires relatively strong depolarizations for activation. The activation range for L-type channels in DRG neurons is positive to -10 mV, but this range can vary considerably in different cell types. The L-type channels are very slowly inactivating over a period of several hundred milliseconds of strong depolarization. Because they require strong depolarization for activation and do not completely inactivate at relatively depolarized potentials, L-currents can be evoked from holding potentials as high as -40 to -10 mV. One of the distinguishing characteristics of the L-type channels is their sensitivity to a group of compounds called dihydropyridines (DHPs). The DHP calcium channel agonist, Bay K 8644, greatly enhances L-type calcium currents but has no significant effect on N- or T-type currents (Tsien et al., 1988). In addition, DHP calcium channel antagonists selectively block L-type calcium channels relative to N- and T-type channels. This high sensitivity of L-type channels to DHPs has provided an excellent tool for probing the function of this channel subtype and for purifying and cloning the DHP-binding subunit of the channel (Tanabe et al., 1987). Evidence suggests that the L-type channel is a substrate for PKC *in vitro* since PKC phosphorylates DHP receptors from rabbit skeletal muscle (Nastainczyk et al., 1987; O'Callahan et al., 1988).

The characteristics of the T-type calcium channel are in many ways opposite to those of the L-type channel. T-type channels have the smallest unitary conductance of the three channel subtypes ($\approx$8 pS, 110 mM barium conductance), are activated at relatively negative potentials (positive to -0 mV), and can only be measured from hyperpolarized holding potentials ($\approx$-100 mV). Also, unlike L-type channels, T-type channels are rapidly inactivating and are resistant to DHPs.

N-type channels have characteristics in common with both T- and L-type channels. Like the T-type channel, they are insensitive to DHPs and they inactivate at positive potentials. However, the kinetics of inactivation of N-type channels are considerably slower than those of T-type channels. In common with L-type channels, N-type channels require fairly strong depolarizations for activation (positive to -20 mV). Also, the pharmacology of N-type channels with respect to inorganic calcium channel blockers is more similar to that of L-type channels than to that of T-type channels. One of the most distinguishing characteristics of the N-type calcium channels is that they are selectively blocked by ω-conotoxin. Interestingly, the N-type channels have only been identified in neurons, suggesting that they may serve a neuron-specific function. Also, N-type channels from rat brain have been shown to serve as substrates from PKC (Ahlijanian et al., 1991), suggesting that this enzyme may play a role in regulating N-type channel function.

It is now clear that there is diversity within each of these three classes of calcium channels and that many calcium channels do not fit comfortably into any of these subclasses. One of the best examples of this is a voltage-dependent calcium channel originally described in cerebellar Purkinje cells that is referred to as the P-type channel (P for Purkinje) (Tsien et al., 1991; Llinas et al., 1992). This channel has a high threshold, shows little or no inactivation, and is insensitive to dihydropyridines or ω-conotoxin. P-type channels are selectively blocked by a polyamine referred to as FTX (funnel web spider toxin) and a peptide from the venom of *A. aperta* (Ω-AGA-IVA). P-type calcium channels have now been shown be present in a number of vertebrate and invertebrate neurons. Since the P-type calcium currents have only recently been characterized, most of the studies in which the effects of PKC activators on calcium currents were investigated did not consider the possibility that this current was present in the cells being examined. It is possible that some previously measured high threshold calcium currents, thought to be composed of L-type and/or N-type currents, also included a significant P-type current component. This should be kept in mind in interpreting earlier studies of PKC-induced modulation of these currents.

Inhibition of Calcium Current in Dorsal Root Ganglion Neurons

The first report that activators of PKC modulate voltage-dependent calcium channels was from experiments performed with chick DRG neurons, the same cell type used for the experiments in which calcium channels were originally classified into N, L, and T subtypes. The effects of the cell-permeable DAG 1,2-oleoylacetylglycerol (OAG) and the phorbol ester 12-deoxyphorbol 13-isobutyrate (DPB) on whole-cell calcium currents were measured in DRG neurons using the whole-cell patch clamp technique (Holz et al., 1986; Rane and Dunlap, 1986; Rane et al., 1989). Cells were dialyzed with cesium and TEA to block outward potassium currents. Puffer application of either OAG or DPB, through a blunt microelectrode placed in close proximity to the cell, caused a rapid decrease in the peak amplitude of calcium current (mean decrease = 55%). The effect was seen within 10 seconds of application and was rapidly reversible on cessation of the puffer application. A phorbol ester that does not activate PKC did not alter calcium current in any of six cells tested. Norepinephrine (NE) causes a similar reduction in calcium current in these cells and it was found that NE has no further effect when added in the presence of a maximally effective concentration of OAG. Furthermore, dialysis of the cells with specific peptide inhibitors of PKC inhibits

NE- and OAG-induced suppression of calcium currents (Rane et al., 1989). The effects of PKC activators on calcium currents in chick DRG neurons are at least partially mediated by inhibition of the T- and L-type currents (Marchetti and Brown, 1988). The effect of PKC activators on N-type calcium current in these cells has not been investigated.

These data suggest that NE inhibits calcium currents in chick DRG neurons by activating PKC. Recently, Hockberger and colleagues (1989) reported data similar to those discussed above. However, these investigators did additional experiments that indicate that under their experimental conditions PKC activators may have effects on calcium currents that are not mediated by PKC. For instance, they found that, in addition to PKC-activating phorbol esters, an inactive phorbol ester inhibits calcium currents in DRG neurons. They then incubated the cells with PKC inhibitors before application of phorbol esters and DAGs and found that concentrations of inhibitors many times higher than the respective inhibition constants (K_i values) had no effect on DAG or phorbol ester-induced inhibition of calcium currents. In addition, they used a rapid delivery system to show that inhibition of calcium currents occurs within 50 milliseconds of application of OAG or TPA to the outside of the cell. This is too rapid to be consistent with diffusion of these compounds across the cell membrane and suggests that the action may be at an extracellular site. Consistent with this, application of phorbol esters or OAG to the inside of the cell had no effect on calcium current. These data suggest that the effect of DAGs and phorbol esters on calcium current in DRG neurons is not mediated by activation of PKC but may be mediated by action on the outside of the DRG neurons.

The effects reported by Rane and co-workers were distinct from those reported by Hockberger and colleagues (1989) in many respects and it is unlikely that they are mediated by this non-PKC-mediated mechanism. This has been discussed in detail by Rane and co-workers (1989). Most important, the effects reported by Rane were blocked by selective PKC inhibitors. Furthermore the kinetics of onset of the effects of PKC activators and recovery following washout reported by Rane and Dunlap (1986) are consistent with an intracellular site of action. It is important to note, however, that injection of peptide inhibitors of PKC completely inhibits the effect of NE on calcium currents but inhibits only 78 percent of the effect of OAG (Rane et al., 1989). Although there are other potential explanations of these data, it is possible that the residual inhibition of calcium current by OAG in the presence of maximal concentrations of PKC inhibitors is mediated by the non-specific effects of these compounds reported by Hockberger and colleagues (1989).

Evidence suggests that PKC also reduces voltage-dependent calcium currents in mammalian DRG neurons. Rat DRG neurons respond to application of the neuroactive peptides, neuropeptide Y (NPY) and bradykinin (BK), with a marked decrease in both sustained and transient components of whole-cell calcium current (measured with whole-cell patch clamp) (Ewald et al., 1988; Walker et al., 1988). This is accompanied by a marked decrease in potassium-induced release of substance P, another neuroactive peptide. Indirect evidence suggesting that the effect of NPY may be mediated by activation of PKC came with the finding that down-regulation of PKC in DRG neurons diminished the effect of NPY on the sustained (but not transient) portion of the calcium current (Ewald et al., 1988). More direct evidence for PKC-induced modulation of calcium currents in mammalian DRG neurons was reported by Boland and co-workers (1991), who found that BK stimulates DAG formation and reduces calcium currents in a DRG cell line (F11-B9 cells). The effect of BK on calcium currents was mimicked by exogenously applied DAGs that activate PKC but not by inactive analogs. Furthermore, the response to both BK and DAG was partially blocked by a selective peptide inhibitor of PKC and the nonselective protein kinase inhibitor staurosporine. Taken together, these studies suggest that PKC reduces calcium currents in F11-B9 cells. Under the conditions used in this study, the current measured was likely composed of both N- and L-type components, and studies were not performed to determine if either of these components is reduced selectively. In mouse neurons, evidence suggests that PKC activators (OAG and phorbol dibutyrate [PDBu]) reversibly inhibit N-type calcium currents with little or no effect on T- or L-type currents (Werz and Macdonald, 1987a,b; Gross and Macdonald, 1989). However, it is important to note that Schroeder and colleagues (1990) reported that the effects of PKC-activating phobol esters on calcium currents in primary cultures of rat DRG neurons are temperature dependent and that at temperatures above 29°C phorbol esters selectively reduced T-type calcium currents without effects on high threshold calcium currents. Thus, although all studies to date support a role for PKC in reducing voltage-dependent calcium currents in mammalian DRG neurons, it is not entirely clear which of the major types of calcium currents are reduced under physiological conditions.

Modulation of Calcium Current in Other Vertebrate Cells

Inhibitory effects of PKC activators, similar to those described in DRG neurons, have also been observed when examining calcium currents in a variety of other vertebrate neurons and other cell types. These include mouse cerebral hemisphere neurons (Werz and Macdonald, 1987a,b), rat sympathetic neurons (Plummer et al., 1991; Abrahams and Schofield, 1992), mouse neuroblastoma cells (Linden and Routtenberg, 1989), GH$_3$ cells (Marchetti and Brown, 1988), the pituitary cell line AtT-20 (Lewis and Weight, 1988), and a variety of other cell types. In cultured and acutely dissociated hippocampal neurons, PKC activators inhibit N-type and, to a lesser extent, L-type calcium currents when measurements are made using whole-cell patch clamp techniques. This effect can be seen with a variety of PKC activators, including PKC-activating phorbol esters (Doerner et al., 1988, 1990; Doerner and Alger, 1992), DAGs (Doerner and Alger, 1992), and arachidonic acid (Keyser and Alger, 1990). The response to each of these PKC activators is blocked by selective and nonselective PKC inhibitors. High concentrations of an inactive phorbol ester also reduce calcium currents in these cells, but the response to these compounds requires 100-fold higher concentrations of phorbol esters and is not blocked by protein kinase inhibitors (Doerner et al., 1990).

In contrast to the effects of PKC activation on whole-cell calcium currents in hippocampal neurons, cell-attached patch clamp recordings reveal that phorbol esters increase rather than decrease the unitary channel activity through N- (Madison, 1989) and L-type channels (Madison, 1989; Doerner and Alger, 1992) in cultured hippocampal neurons. This discrepancy between data from whole-cell and single-channel recordings might suggest that intracellular dialysis during whole-cell patch clamp recordings may remove some intracellular factor that is important for the PKC-induced enhancement of N- and L-type channel activity. However, perforated patch clamp recordings prevent extensive intracellular dialysis, yet PKC activation also reduces whole-cell calcium currents in hippocampal neurons when this technique is used (Doerner and Alger, 1992). At present the reason for the seemingly opposite effects of PKC activation on calcium currents in hippocampal neurons when using whole-cell versus cell-attached patch clamp techniques is not known.

Evidence from other cells also suggests that PKC may increase calcium current in some vertebrate cell types. For instance, injection of *Xenopus* oocytes with messenger RNA from chick forebrain results in expression of a slowly inactivating voltage-dependent calcium current. The peak amplitude of this calcium current is increased by incubation with PKC-activating phorbol esters and OAG but not by inactive phorbol esters (Leonard et al., 1987; Sigel and Baur, 1988). The effect of PKC activators on this calcium current is blocked by incubation with a PKC inhibitor. These data suggest that PKC may enhance calcium current in some cells in vertebrate brain. Such an effect has not yet been observed in voltage-clamped neurons. However, indirect evidence suggests that activation of PKC enhances L-type calcium current in cortical (Adamson et al., 1989) or spinal cord (Gandhi and Jones, 1992) synaptosomes. Furthermore, evidence suggests that PKC activators may have a similar effect on L-type calcium channels in vertebrate smooth (Forder et al., 1985; Gleason and Flaim, 1986; Vivaudou et al., 1988), skeletal (Navarro, 1987), and cardiac (Lacerda et al., 1988) muscle cells. Interestingly, activation of PKC increases the density of L-type calcium channels in cultured skeletal muscle cells (Navarro, 1987) and may have this effect on PC12 cells (Messing et al., 1990). Thus, it is possible that PKC increases calcium current in vertebrate cells by increasing the number of active channels on the cell membrane. This mechanism would be similar to that described above for enhancement of calcium current in *Aplysia* neurons.

In summary, as would be expected, the effect of activation of PKC on calcium current is cell-type specific, both in terms of the subtype of calcium channel modulated and in the direction of the effect (increase or decrease).

MODULATION OF POTASSIUM CONDUCTANCES BY PKC

The most diverse group of voltage-sensitive ion channels, in terms of both properties and function, are the potassium channels. There are at least seven, and probably more, major classes of voltage-dependent potassium channels in neurons, and each of these serves a unique role in regulating cell excitability (Levitan and Kaczmarek, 1987). For instance, some potassium conductances are important for repolarization of individual sodium or calcium action potentials. The classic example of this type of current is the delayed rectifier potassium current ($I_{K(V)}$), which is responsible for repolarization of action potentials in squid giant axons. A primary potassium current that contributes to spike repolarization in some vertebrate neurons is a calcium- and voltage-dependent potassium current termed I_C. Other potassium currents, because of their slow activation kinetics, may not be significantly activated during a single action potential. However, if such currents are noninactivating, cumulative activation can occur during a series of repetitive spikes so that these currents can serve to limit repetitive firing. Still other potassium currents show cumulative inactivation after a series of spikes and thereby lead to enhanced action potential duration during repetitive firing. This would be expected to increase calcium influx and cause a greater amount of neurotransmitter release during the latter action potentials in a train. Other currents, such as the A current (I_A), are important in determining the rate at which a cell can fire, and potassium currents that are significantly active at negative potentials may be important for determining the cell's resting potential. Modulation of the latter type of potassium conductance by neurotransmitters and second-messenger/protein kinase systems may be involved in generation of slow postsynaptic potentials.

To a large extent, the unique electrical properties of a given neuron are defined by the relative abundances of various species of potassium channels. For instance, the resting potential, the ability or inability to fire repetitively, the pattern of firing, and the shape of action potentials in a given cell are largely determined by what potassium channels the cell expresses and in what proportions. By modulating specific potassium currents that serve distinct functions, neurotransmitters and second messengers can fine tune any number of the basic electrical properties of the cell. Thus it is not surprising that potassium conductances come under a great deal of modulatory control by PKC and other protein kinases.

Inhibition of Potassium Currents that Dampen Repetitive Firing

A common response of many vertebrate and invertebrate neurons to prolonged (500–800 msec) injection of depolarizing current is a brief burst of action potentials that slow and stop during the depolarizing pulse. This process of limiting repetitive firing in response to an excitatory stimulus has been termed spike frequency adaptation or spike accommodation. A number of studies have suggested that a calcium-dependent potassium conductance termed I_{AHP} may play a key role in mediating spike accommodation in a variety of invertebrate (Ottoson and Swerup, 1981; Lewis and Wilson, 1982) and vertebrate (Grafe et al., 1980; Buchert-Rau and Sonnhof, 1982; Madison and Nicoll, 1984) neurons. Baraban and colleagues (1985) and Malenka and co-workers (1986a) independently found that PKC-activating phorbol esters dramatically reduce spike accommodation in hippocampal neurons by reducing the amplitude of I_{AHP}. This effect of phorbol esters is blocked by PKC inhibitors (Doerner et al., 1988; Gerber et al., 1992; Sim et al., 1992) and mimicked by cell-permeable DAGs and agonists of phosphoinositide hydrolysis-linked receptors (Cole and Nicoll, 1983; Charpak et al., 1990; Desai and Conn, 1991; Desai et al., 1992). However, although agonists of muscarinic acetylcholine receptors (Cole and Nicoll, 1983) and metabotropic glutamate receptors (mGluRs)(Charpak et al., 1990; Desai and Conn, 1991; Desai et al., 1992) activate phosphoinositide hydrolysis and inhibit I_{AHP} in hippocampal slices, evidence suggests that the responses to these agonists is not mediated by PKC activation. Thus, nonspecific protein kinase inhibitors do not inhibit the effect of muscarinic or mGluR receptor agonists on I_{AHP} at concentrations that block phorbol ester-induced reduction in this current (Gerber et al., 1992; Sim et al., 1992). However, it is interesting to note that the ionotropic glutamate receptor agonist NMDA also reduces I_{AHP} in hippocampal pyramidal cells, and this response is inhibited by the protein kinase inhibitor H-7 (Baskys et al., 1990). As discussed above, activation of NMDA receptors activates PKC in neuronal preparations in a manner that is dependent on calcium influx through the NMDA receptor cation channel.

Another potassium current that may participate in spike accommodation in vertebrate neurons is a noninactivating time- and voltage-dependent potassium current termed the M current (I_M)(Adams et al., 1982). In addition to being involved in accommodation, I_M is significantly active at resting membrane potentials of some cells and contributes to the cell's resting potential (Jones and Adams, 1987). PKC activators (but not inactive phorbol esters) partially suppress I_M in NG108-15 cells (Higashida and Brown, 1986; Brown and Higashida, 1988), rat cervical ganglion neurons (Brown et al., 1989), and frog sympathetic ganglion neurons (Brown and Adams, 1987; Tsuji et al., 1987; Pfaffinger et al., 1988; Bosma and Hille, 1989), suggesting that PKC may be involved in regulating I_M in these cell types. However, studies by Bosma and Hille (1989) and Pfaffinger and colleagues (1988) suggest that, although luteinizing hormone-releasing hormone (LHRH) and substance P stimulate phosphoinositide hydrolysis and inhibit I_M in frog sympathetic ganglion cells, the effect of these peptides on I_M is not mediated by activation of PKC. For instance, the effect of PKC activators in these cells is only partial relative to that of LHRH, and the effect of LHRH is not occluded when it is added in the presence of a maximally active concentration of PDBu. More important, PKC inhibitors block the effect of PKC activators on I_M in these cells, but have no effect on I_M modulation by LHRH or substance P. Thus, although activation of PKC can lead to inhibition of I_M in frog sympathetic neurons, this enzyme is apparently not involved in modulation of this current by agonists that stimulate phosphoinositide hydrolysis. This is reminiscent of the studies related to regulation of I_{AHP} by agonists of mGluR and muscarinic acetylcholine receptors described above.

Modulation of Other Potassium Currents by PKC

In addition to modulating I_M and I_{AHP}, PKC has been suggested to play a role in modulation of a number of other potassium channels. In most neurons studied, PKC has an inhibitory effect on voltage-dependent potassium currents. PKC inhibits I_A (Farley and Auerbach, 1986; Alkon et al., 1986; Colby and Blaustein, 1988; Alkon et al., 1986; Grega et al., 1987; Werz and MacDonald, 1987a), $I_{K(V)}$ (Grega et al., 1987; Werz and Macdonald, 1987a; Doerner et al., 1988), and calcium-dependent potassium currents (Freedman and Aghajanian, 1987; Sawada et al., 1989a,b; Critz and Byrne, 1992) in a number of neurons. Interestingly, PKC inhibits some species of calcium-activated potassium currents that are activated in response to Ins[1,4,5]P$_3$-induced calcium mobilization. Thus, two second messengers formed in response to activation of the same transduction system can have opposing effects on a single ion channel. For the most part the exact physiological or behavioral relevance of PKC-induced inhibition of potassium currents is unknown. However, a noteworthy exception to this comes from studies in the marine mollusc *Hermissenda crassicornis,* where PKC-induced inhibition of a potassium conductance is likely to play an important role in classical conditioning (Etcheberrigaray et al., 1992).

The exact subtypes of PKC involved in inhibition of each of the potassium currents in various cell types have not yet been determined. However, evidence to date suggests that modulation of $I_{K(v)}$ and I_A may be a general effect of a variety of PKC subtypes. For instance, intracellular injection of PKC subtype I (γ), subtype II (βI + βII), or subtype III (α) reduces A-type potassium channels expressed in *Xenopus* oocytes (Lotan et al., 1990). Likewise, studies of modulation of $I_{K(v)}$ with subtype-selective PKC activators in cerebellar neurons suggest that both PKC subtype I and subtype II reduce this current (Linden et al., 1992). Effects of other PKC subtypes on I_A and $I_{K(v)}$ are not known.

Although the effect of PKC on potassium conductances is inhibitory in the vast majority of cases, there are some notable examples of increases in potassium conductances induced by PKC. For instance, activators of PKC enhance $I_{K(v)}$ in isolated ventricular cells (Walsh and Kass, 1988), and potentiate calcium-dependent potassium currents induced by Ins[1,4,5]P$_3$ in NG108-15 cells (Brown and Higashida, 1988).

MODULATION OF VOLTAGE-DEPENDENT SODIUM AND CHLORIDE CURRENTS

Voltage-dependent sodium channels are the primary channels responsible for the inward current involved in initiation and propagation of action potentials in most vertebrate neurons. Thus, modulation of these channels could have a dramatic impact on neuronal excitability and action potential propagation. Until recently there was little or no direct evidence for second-messenger-induced regulation of voltage-dependent sodium channels. However, in recent years progress on determining the effects of PKC on voltage-dependent sodium channels from mammalian brain has been extremely rapid, and these are now perhaps the most well characterized channels in terms of their modulation by PKC. The mammalian brain voltage-dependent sodium channel is a heterotrimeric protein that consists of a large α subunit and two smaller β subunits (βI and β_{II}). Four distinct subtypes of α subunits exist, which have been termed types I, II, IIA, and III. Expression of the α subunit alone in *Xenopus* oocytes or mammalian cells is sufficient for formation of functional voltage-dependent sodium channels. The α subunit of the voltage-dependent sodium channel is an excellent substrate for PKC (Costa and

Caterall, 1984), and recent mapping of tryptic digests suggest that the α subunit is phosphorylated on three distinct serine residues (Murphy and Caterall, 1992).

The first direct evidence for functional modulation of sodium channels by PKC came when Sigel and Baur (1988) found that activators of PKC markedly reduce the peak amplitude of sodium currents in *Xenopus* oocytes that have been injected with mRNA from rat brain. Similar effects can be seen in *Xenopus* oocytes injected with chick brain mRNA (Lotan et al., 1990). Interestingly, this effect is mimicked by intracellular injection of PKC subtype II or subtype III but not subtype I (Lotan et al., 1990). Thus, sodium channels may provide the first clear example of ion channels that are selectively modulated by only certain subtypes of PKC.

More recently, Numann and colleagues (1991) reported that PKC reduces peak sodium currents and slows sodium current inactivation in rat brain neurons and Chinese hamster ovary cells transfected with type IIA sodium channel subunits. Pharmacological studies indicate that this effect is mediated by PKC, since it can be elicited with either DAG or phorbol ester activators of PKC or by direct application of PKC to the cytoplasmic surface of inside-out membrane patches. Furthermore, this response was blocked by specific peptide inhibitors of PKC. Single-channel studies suggest that the PKC-induced reduction of sodium channel inactivation is mediated by an increase in the mean open time of sodium channels and an increase in the probability of reopening of sodium channels during prolonged depolarizations. The mechanism by which PKC reduces the peak sodium current is less clear. However, PKC does not decrease the single-channel conductance. Thus, it is likely that the decrease in the peak sodium current is due to a decrease in the number of active channels or a decrease in the probability of channel opening during the peak current (Numann et al., 1991).

The α subunit of voltage-dependent sodium channels has four major membrane-spanning regions, designated by the Roman numerals I–IV. The intracellular loop between domains III and IV plays an important role in inactivation of the sodium channel and has a consensus sequence for phosphorylation by PKC at Ser[1506]. This prompted West and colleagues (1991) to test the hypothesis that phosphorylation of this serine residue is involved in PKC-induced modulation of sodium channel function. They performed a series of studies in which they found that the functional properties of mutant sodium channels lacking this site are virtually identical to those of wild-type sodium channels in unstimulated cells but are not sensitive to modulation by PKC. These studies suggest that phosphorylation of Ser[1506] is required for modulation of brain sodium channels by PKC. However, they do not rule out an additional role for other PKC phosphorylation sites in modulation of these channels.

PKC activators have also been shown to turn off a voltage-dependent chloride conductance that is normally active at rest (Madison et al., 1986). The exact physiological consequences of PKC inhibition of this current are not yet fully understood. However, indirect evidence suggests that this conductance is localized on dendritic membranes. Thus, blockade of the channel by PKC could increase dendritic membrane resistance and thereby increase the sensitivity of the dendrite to excitatory synaptic input.

PKC-INDUCED MODULATION OF LIGAND-REGULATED ION CHANNELS

Examples of modulation of ligand-regulated ion channels by PKC or other protein kinases are not as numerous as are examples of modulation of voltage-sensitive ion channels. However, in recent years there has been a dramatic increase in our understanding of the structure and function of ligand-gated ion channels and regulation of these channels by protein kinases. Studies of protein kinase-mediated modulation of ligand-gated ion channels has recently been reviewed in detail (Swope et al., 1992).

Modulation of the Nicotinic Acetylcholine Receptor

The nicotinic acetylcholine receptor (AChR) is a ligand-gated cation channel found at the neuromuscular junction and at some synapses in the CNS. The AChR in *Torpedo* electroplaque has served as a model for studying the AChR and consists of four types of subunits in a stoichiometry of $\alpha_2\beta\gamma\delta$. The δ subunit of AChR is an excellent substrate for PKC (Huganir, 1987) and PKC activators reduce the sensitivity of cultured chick myotubes to ACh (Eusebi et al., 1985). This is apparently mediated by a number of effects on AChR cation channels (Eusebi et al., 1987). These include a decrease in the single-channel conductance (from about 54 pS to about 38 pS), an increase in decay time constants, and an increase in mean channel-closed time. The latter finding suggests that PKC activators either decrease the number of channels capable of opening or lengthen the time that individual channels stay closed. These effects are apparently mediated by PKC, since they are mimicked by various PKC activators (but not by inactive phorbol esters) and inhibited by PKC inhibitors. PKC may also decrease the sensitivity of the AChR in sympathetic ganglion neurons. In these cells PKC activators (but not inactive phorbol esters) enhance the rate of desensitization of the AChR during exposure to ACh (Downing and Role, 1987).

Modulation of GABA$_A$ Receptors

GABA, the major inhibitory neurotransmitter in mammalian brain, exerts many of its effects through activation of ligand-gated chloride channels referred to as the GABA$_A$ receptors. Like the AChR, GABA$_A$ receptors consist of multiple subunits that can be divided into four classes (α, β, γ, and δ). Multiple subtypes of α, β, and γ subunits have been identified and there may be alternative splice variants of the different subtypes (Swope et al., 1992). Protein kinase C phosphorylates both α and β subunits of GABA$_A$ receptors purified from bovine heart (Browning et al., 1990). Interestingly, this preparation consists of two separate subtypes of the β subunit, and only one of these serves as a substrate for PKC. In addition, Whiting and colleagues (1990) and Kofuji and co-workers (1991) independently reported that two alternative splice variants of the γ_2 subunit exist in mammalian brain. One of these forms contains an eight amino acid insert that includes a consensus phosphorylation site for PKC. Differential expression of different β subunits or differentially spliced forms of the γ_2 subunit may provide a mechanism for expression of distinct $GABA_A$ receptor subtypes that have differential sensitivity to modulation by PKC.

Electrophysiological experiments are beginning to give insight into the physiological consequences of phosphorylation of $GABA_A$ receptors by PKC. PKC activators reduce the peak amplitude of current through $GABA_A$ receptors expressed from rat brain mRNA injected into *Xenopus* oocytes (Sigel and Baur, 1988; Sigel et al., 1991, Leidenheimer et al., 1992). However, to date there is no direct evidence that PKC reduces currents through $GABA_A$ receptors in neurons. Indeed, Ticku and Mehta (1990) found that PKC activators have no effect on GABA-induced $^{36}Cl^-$ influx into cultured mammalian spinal cord neurons. These data may suggest that $GABA_A$ receptors in these cells do not contain the β or γ_2 subunits that are phosphorylated by PKC. However, future studies will be needed to determine the subunit composition of $GABA_A$ receptors in these cells and to determine the effects of PKC on $GABA_A$ receptor function in neurons expressing $GABA_A$ receptors with subunit compositions predicted to be sensitive to modulation by PKC.

PKC-induced Enhancement of NMDA Receptor Currents

The primary excitatory neurotransmitter in the CNS is glutamate. Glutamate activates two major families of ligand-gated cation channels. These have been classified and named on the basis of selective agonists at each site and are referred to as the N-methyl-D-aspartate (NMDA) receptors and the kainate/AMPA receptors. As with the AChR and $GABA_A$ receptors, there is a great deal of heterogeneity of subunits that comprise these receptor families. Although little is known about phosphorylation of specific sites on these channels by PKC, analysis of amino acid sequences of these receptor subunits reveals a number of potential phosphorylation sites. Furthermore, activation of PKC (or intracellular application of PKC) induces a marked increase in the peak amplitude of NMDA receptor currents in spinal cord neurons (Chen and Huang, 1991), hippocampal neurons (Aniksztejn et al., 1992; but also see Markram and Segal, 1992), and in *Xenopus* oocytes injected with rat brain mRNA (Urushihara et al., 1992) or RNA encoding specific NMDA receptor subunits (Durand et al., 1992). This effect can be elicited by agonists of μ-opioid receptors in spinal cord neurons (Chen and Huang, 1991) and metabotropic glutamate receptors in hippocampal neurons (Aniksztejn et al., 1992). The response to either PKC activators or receptor agonists can be blocked by a selective peptide inhibitor of PKC. Interestingly, expression of splice variants of NR1 (one of the cloned NMDA receptor subunits) in *Xenopus* oocytes suggests that the NR1b form of this receptor is much more sensitive to modulation by PKC than is NR1a (Durand et al., 1992). This is similar to the differential sensitivity of alternatively spliced forms of the γ subunit of $GABA_A$ receptors to phosphorylation by PKC (see above).

Single-channel and whole-cell patch clamp recordings suggest that PKC potentiates the response to NMDA receptor activation by increasing the probability of channel openings and by reducing a voltage-dependent block of this channel by magnesium (Chen and Huang, 1992). As discussed below, PKC-induced enhancement of NMDA receptor currents may play an important role in induction of long-term potentiation of synaptic efficacy in the hippocampus (Ben-Ari et al., 1992).

A MAJOR BRAIN-SPECIFIC PKC SUBSTRATE MAY BE INVOLVED IN NEURONAL GROWTH AND DIFFERENTIATION

PKC is likely to be involved in the control of process outgrowth in developing neurons. This inference is based largely on the observation that a major substrate for PKC is the protein neuromodulin, which is enriched in neuronal growth cones (reviewed in Skene, 1989). Neuromodulin is also known as GAP-43, which refers to the fact that it is a growth-associated protein of 43,000 apparent molecular weight on polyacrylamide gels. Neuromodulin is also known as B-50 and F1. The redundant nomenclature is an effect of the protein having been independently discovered in several contexts, it having been characterized as a calmodulin-binding protein (Andreason et al., 1983), a growth-associated protein (Skene and Willard, 1981), and a PKC substrate related to synaptic plasticity (Nelson and Routtenberg, 1985; Dekker et al., 1989). We have chosen for the purposes of this review to refer to the protein as neuromodulin, because this moniker was used in the first description of the protein's sequence and because the majority of the work describing the protein as a PKC substrate uses this nomenclature. However, the term GAP-43 is generally used when referring to the protein in a neurodevelopmental context, and the terms F1 and B-50 are generally used in the context of synaptic plasticity.

Neuromodulin is a substrate for PKC in vitro, and is phosphorylated by PKC at serine 41 in its sequence (Apel et al., 1990; Nielander et al., 1990). Neuromodulin is phosphorylated by PKC with a K_m of around 1 μM, and is not phosphorylated by the cAMP-dependent protein kinase or calcium/calmodulin-dependent protein kinases (Apel et al., 1990). Neuromodulin can be phosphorylated by casein kinase II in vitro, but this site does not appear to be phosphorylated in intact cells or rat brain (Spencer et al., 1992).

Although the precise function of neuromodulin in neurons is unknown, neuromodulin binds calmodulin in the absence of calcium (Andreason et al., 1983), and calcium elevation or phosphorylation of the protein by PKC decreases its affinity for calmodulin (Apel et al., 1990). Neuromodulin has also been implicated in regulating polyphosphoinositide metabolism (Van Hooff et al., 1988), and can induce process outgrowth when expressed even in nonneuronal cells (Zuber et al., 1989). Neuromodulin can also regulate the function of the GTP-binding protein G_o (Strittmatter et al., 1990).

A role for neuromodulin in the control of neuronal growth processes is strongly suggested by its selective localization in developing axonal growth cones (Goslin et al., 1988), and its increased expression in regenerating axons (Skene and Willard, 1981; Meiri et al., 1988; Doster et al., 1991). However, at this point insufficient information is available for formulating a detailed model of the role of the meuromodulin/PKC system in neuronal development.

INVOLVEMENT OF PKC IN REGULATION OF NEUROTRANSMITTER RELEASE

Any of the PKC-induced changes in ion channel function described above could ultimately influence neurotransmitter release. For instance, since exocytotic release of neurotransmitters is normally a calcium-dependent process (see Zucker, 1987, for review), increases or decreases in calcium conductances could increase or decrease depolarization-induced release respectively. Furthermore, inhibition of a potassium conductance could enhance cell firing and thereby increase calcium influx and augment neurotransmitter release. In addition, studies in a number of neuronal and nonneuronal cell types indicate that activators of PKC may directly influence exocytotic release of neurotransmitters by a mechanism that is independent of modulation of ion channels (Knight, 1987; Kikkawa et al., 1988). For instance, in various types of electropermeabilized cells, PKC can act synergistically with calcium to induce neurotransmitter release and thereby greatly reduce the calcium requirement for the process. In addition, in many cell types PKC can induce basal release of neurotransmitters under conditions in which resting levels of intracellular calcium would not be expected to rise or even in cells where PKC activators decrease calcium influx (Harris et al., 1986; Knight, 1987). In some instances, down-regulation of PKC (Matthies et al., 1987) or incubation with PKC inhibitors (Allgaier and Hertting, 1986) markedly inhibits depolarization-induced neurotransmitter release, suggesting that PKC is normally involved in release that occurs in response to a depolarizing stimulus. There are now a great number of examples of PKC-induced increase in both basal and evoked neurotransmitter release, and these studies suggest that the effects of PKC can involve any number of neurotransmitters and secretory products from various neuronal and nonneuronal cells. In vertebrate neurons alone, PKC has been suggested to be involved in modulating release of norepinephrine (Allgaier and Hertting, 1986; Matthies et al., 1987; Nichols et al., 1987), LHRH (Negro-Vilar et al., 1986), serotonin (Feurstein et al., 1987; Nichols et al., 1987; Wang and Friedman, 1987); dopamine (Nichols et al., 1987), GABA (Nichols et al., 1987; Bartmann et al., 1989; Weiss et al., 1989), acetylcholine (Nichols et al., 1987; Allgaier et al., 1988), aspartate (Nichols et al., 1987; Dekker et al., 1989), and glutamate (Malenka et al., 1987).

The mechanism of PKC action on neurotransmitter release is unknown. However, immunocytochemical studies indicate that PKC is present in high density in presynaptic terminals of various neuronal types in rat brain (Wood et al., 1986) and in rat Purkinje cells, the γ species of PKC is highly enriched in axonal synaptic vesicles. Thus, it is likely that PKC phosphorylates synaptic proteins that are important in regulating exocytosis. The phosphorylation of the synaptic protein neuromodulin is closely associated with transmitter release from hippocampal slices (Dekker et al., 1989) or rat brain synaptosomes (Dekker et al., 1990) and anti-neuromodulin antibodies inhibit calcium-dependent neuromodulin phosphorylation and neurotransmitter release (Dekker et al., 1990). However, the exact role of neuromodulin phosphorylation by PKC in regulation of neurotransmitter release is not yet known. Studies related to the role of PKC in regulating transmitter release have recently been reviewed in detail (De Graan et al., 1991; Dekker et al., 1991).

PKC PLAYS AN IMPORTANT ROLE IN VARIOUS FORMS OF SYNAPTIC PLASTICITY

The activity of PKC has been implicated as playing a role in a wide variety of forms of synaptic plasticity, including long-term depression of synaptic transmission in the cerebellum (Linden and Connor, 1991), facilitation of neurotransmitter release in *Aplysia* sensory neurons (Kandel and Schwartz, 1982; Ghiradi et al., 1992; Sossin and Schwartz, 1992), kindling-induced epileptogenesis (Chen et al., 1992), and altered synaptic coupling in *Hermissenda* (Etcheberrigaray et al., 1992). However, the system in which PKC has received the most attention is long-term potentiation of synaptic efficacy in the hippocampus.

LTP is manifest as an increase in evoked EPSPs in an individual neuron and occurs after brief repetitive stimulation of the afferent input of any one of a number of hippocampal synaptic connections (Bliss and Gardner-Medwin, 1973; Bliss and Lomo, 1973; Anderson et al., 1977; Harris and Cotman, 1986). Because it occurs in the hippocampus, which is know to be involved in learning and memory, and is a robust synaptic change, the phenomenon has immediate appeal as a potential cellular mechanism for memory. LTP also has several additional properties that make it an appealing candidate as a memory mechanism. LTP exhibits a stimulus threshold (McNaughton et al., 1978; Lee, 1983), and is usually homosynaptic; that is, the stimulated input is selectively potentiated (Anderson et al., 1977; McNaughton et al., 1978). LTP can be produced in an associative fashion. Pairing a weak (subthreshold) stimulus with a strong stimulus can produce LTP in the weak input (Levy and Steward, 1979; Barrinuevo and Brown, 1983). The postsynaptic neuron is the probable site for the association, as depolarization of the postsynaptic neuron can substitute for the strong synaptic input (Kelso and Brown, 1984; Kelso et al., 1986).

The capacity to study LTP in individual neurons using intracellular recording has allowed the definition of certain requisites for its induction. Depolarization of the postsynaptic cell is necessary (Kelso et al., 1986; Malinow and Miller, 1988), which demonstrates a role for the postsynaptic cell in the induction process. Additional work has suggested that postsynaptic calcium influx is a necessary step in LTP induction, and injection of a Ca^{2+} chelator into the postsynaptic cell blocked induction of LTP (Lynch et al., 1983). In addition, it is known that APV, an antagonist of the NMDA subtype of glutamate receptor, can block induction of LTP in the CA1 hippocampal subfield without altering normal synaptic transmission (Collingridge et al., 1983). How does LTP-inducing repetitive stimulation lead to the activation of NMDA receptors, and is this linked to the Ca^{2+} influx that appears necessary for the induction of LTP? Available evidence suggests that repetitive stimulation causes release of quantities of glutamate sufficient to lead to a high degree of depolarization of the postsynaptic cell. A high level of depolarization is sufficient to cause a loss of the voltage-dependent Mg^{2+} blockade of the NMDA receptor ion channel (MacDermott et al., 1986; Nowak et al., 1984; Jahr and Stevens, 1987; Mayer and Westbrook, 1987). The channel, once opened, allows the influx of Ca^{2+} (and other ions) and the induction of LTP. It should be noted, however, that while NMDA receptor-mediated Ca^{2+} influx may be necessary for the induction of LTP, a number of studies have shown that activation of this receptor is not sufficient for LTP induction (Kauer et al., 1988a; Malenka et al., 1988; Thibault et al., 1989).

Cellular Localization

How does the long-term increase in synaptic strength occur? Two possible sites of potentiation are readily apparent. LTP could be due to increased neurotransmitter (glutamate at these synapses) release presynaptically or increased responsiveness of glutamate receptors on the postsynaptic neuron. Though it is known that events occurring in the postsynaptic cell, specifically NMDA receptor activation and Ca^{2+} influx, are required for LTP induction, an interesting question remains concerning the possible involvement of the presynaptic cell as a locus for the maintenance of LTP. To date this issue remains controversial. Using a push-pull cannula technique, workers from Bliss's laboratory have observed increases in glutamate release in the maintenance phase of LTP (Bliss et al., 1986), suggesting involvement of a presynaptic mechanism in LTP. Nicoll and co-workers have presented evidence suggesting an exclusively postsynaptic locus for LTP maintenance, based on studies of postsynaptic ionic currents (Kauer et al., 1988a). Collingridge and co-workers have presented data that suggest that LTP is maintained by both presynaptic and postsynaptic mechanisms, each with different time courses. They demonstrated that sensitivity of neurons in hippocampal slices to ionophoretically applied quisqualate receptor ligands increases slowly after induction of LTP. This provided evidence for a functional postsynaptic change and also implied that presynaptic mechanisms also contribute to the early maintenance of LTP (Davies et al., 1989). More recent data have come out in strong support of a presynaptic locus for the maintenance of LTP. Statistical analyses of neurotransmission before and after LTP have shown a presynaptic change in neurotransmitter release to be associated with LTP (Bekkers and Stevens, 1990; Kullmann and Nicoll, 1992; Malgaroli and Tsien, 1992). Obviously, resolution of the relative contributions of presynaptic versus postsynaptic mechanisms for LTP awaits further experimentation. However at this point it seems clear that there is at a minimum some contribution of presynaptic changes to LTP maintenance.

PKC Activity Is Required for LTP Induction

Interestingly, certain kinase inhibitors can block the induction of LTP (Malinow et al., 1988, 1989). Sphingosine, an inhibitor acting at the DAG-regulator domain of PKC, can block induction but not maintenance of LTP. H7, an inhibitor of

PKC and other kinases that acts at the catalytic domain of the enzymes, can block both induction and maintenance of LTP. These data strongly implicate a role for PKC in LTP induction. However, a role for PKC in the induction of LTP is not definitively established, as inhibitors of C kinase are notoriously nonspecific agents and a viable alternative interpretation is that H7 is blocking the activity of cAMP-PK or CaM kinase. In addition, two groups have reported that H7 nonspecifically interferes with synaptic transmission. Also, although activators of PKC can cause facilitation of synaptic transmission in hippocampus, at least two groups have now shown that application of these agents does not in fact produce persistent potentiation. (Malenka et al., 1986b; Muller et al., 1988). Therefore, while PKC activity may be necessary, it is certainly not sufficient for the induction of LTP.

More recent studies using more selective peptide inhibitors have implicated a role for postsynaptic PKC activity and calmodulin activation in the initiation of LTP in the CA1 region of hippocampus. Work by Malinow and colleagues (1989) have shown that injection of peptide inhibitors selective for calmodulin-regulated enzymes of Ca^{2+}/calmodulin-dependent protein kinase into the postsynaptic cell can block induction of LTP. In addition, Malinow and colleagues (1989) showed that injection of selective peptide inhibitors of PKC could block induction of LTP. These studies complement and extend previous studies using less selective kinase inhibitors such as H7 and sphingosine. However, as Soderling has emphasized (Smith et al., 1990), in experiments in which peptides are injected into cells one does not know the final concentration of inhibitor achieved inside the cell. Given the fact that even selective peptide inhibitors cause some degree of nonspecific kinase inhibition, these experiments must be interpreted with caution.

PKC in the Maintenance of LTP

LTP has generally been divided into induction (early) and maintenance (late, i.e. approximately 45 minutes after tetanus) phases, on the basis of the time course of the effect. Less is known concerning the molecular basis for the maintenance of LTP than is known about its induction. From the viewpoint of understanding mechanisms for information storage in neurons, understanding the mechanisms for maintenance of LTP is a critical issue.

A variety of evidence suggests a role for persistent kinase activation in the maintenance of LTP. Application of H7, a nonspecific protein kinase inhibitor, can reversibly block the maintenance of LTP (Malinow et al., 1988). Other, more selective, PKC inhibitors can also reverse the maintenance of LTP (Colley et al., 1990). Given that activators of PKC cause potentiation of synaptic transmission in the hippocampus (Malenka et al., 1986b; Muller et al., 1988), these data suggest the model that persistent activation of PKC (or other kinases) could underlie the maintenance phase of LTP.

The idea that persistent PKC activation could underlie LTP is also appealing in that biochemical mechanisms exist whereby PKC can be persistently activated. PKC was initially characterized as a kinase activatable through proteolytic cleavage. Subsequent work has shown that PKC can also be persistently activated through autophosphorylation or translocation to membranes. Thus it is appealing to think that these mechanisms for establishing long-lasting biochemical changes might underlie long-lasting physiological changes such as occur in LTP.

Various studies have been aimed at more directly addressing the role of PKC in LTP maintenance. LTP is associated with long-term changes in the subcellular distribution of PKC (Akers et al., 1986), these data support the model that PKC is persistently activated in LTP and suggest that membrane translocation could underlie the effect. Interpretation of these studies is complicated due to the fact that the PKC could be translocated in an inactive form. As mentioned above, Routtenberg and co-workers and Gispen and co-workers have also reported that LTP is associated with an increased phosphorylation of the PKC substrate, neuromodulin (GAP-43/B-50/F1) (Lovinger et al., 1986; Linden et al., 1988; Dekker et al., 1989). These data must be interpreted with caution because the hippocampal homogenate was fractionated before being assayed, and phosphorylation of an endogenous substrate was measured. Therefore the effect could be due to translocation of either the substrate or the kinase, or perhaps due to a change in the phosphorylation state of the substrate that occurred before the assay was conducted. In addition, it is important to consider in experiments of this type that the altered phosphorylation could be due to a change in phosphatase activity. However, in an interesting series of experiments Gianotti and colleagues (1992) demonstrated that the phosphorylation of neuromodulin was increased *in situ* in association with LTP. Finally, Wang and Feng (1992) have shown that postsynaptic injection of PKC inhibitor peptides can reverse expression of LTP in the maintenance phase.

Despite the limitations to the interpretation of these various studies, PKC is overall a very appealing candidate for a kinase whose activity is modulated in a long-term fashion during LTP. The known properties of C kinase allow the formulation of a hypothesis to explain its possible persistent activation. Nelsestuen's group has described a mechanism whereby PKC can be inserted into the lipid bilayer in a persistently activated form (Bazzi and Nelsestuen, 1988a,b,c). Their data indicate that Ca^{2+} can cause translocation of PKC to the membrane in a partially active form. Over time and in the presence of Ca^{2+} the kinase is inserted into the membrane, where it is constitutively active, even in the absence of Ca^{2+} or other second messengers. The membrane-inserted form is tightly bound to the bilayer, as evidenced by the fact that the kinase activity fractionates with the membranes on centrifugation. While the PKC activator TPA can increase the efficacy of insertion into the membrane, it has no effect on the activity of the kinase once the kinase is inserted into the membrane.

Consideration of this evidence suggests a working model for how tetanic stimulation might elicit persistent activation of PKC. Tetanic stimulation is known to cause NMDA receptor activation and Ca^{2+} influx into the postsynaptic cell. Tetanic stimulation also presumably gives rise to large increases in Ca^{2+} influx in the presynaptic neuron. This Ca^{2+} influx could lead to both acute activation of PKC and generation of the membrane-inserted form of PKC. The membrane-inserted form, which is independent of Ca^{2+} and DAG, would then serve to maintain an increased phosphorylation of substrate proteins at these synapses. Of course other mechanisms for PKC activation such as proteolysis or autophosphorylation could also contribute to the persistent activation of PKC in LTP.

A recent series of experiments by Klann and co-workers (1991) have been performed to directly test the hypothesis that there is a long-lasting elevation in PKC activity associated with the maintenance phase of LTP, that is 45 minutes to 1 hour into the maintenance phase of LTP. These studies indicate that there is an increase in PKC activity associated with the maintenance phase of LTP in the CA1 region of rat hippocampus. Klann and colleagues (1991) observed an LTP-associated elevation of protein kinase activity in homogenates prepared from the CA1 region of a single hippocampal slice, using myosin light chain (MLC) or synthetic S6 peptide as kinase substrates *in vitro*. These substrates can be phosphorylated by a variety of protein kinases, including PKC. The increase in substrate phosphorylation is clearly a consequence of increased kinase activity, as opposed to decreased phosphatase activity, as the effect is not attenuated by addition of the general phosphatase inhibitor sodium pyrophosphate.

The increase in kinase activity is seen in homogenates prepared in buffer containing 1 mM EGTA, indicating that the elevated kinase activity is independent of altered Ca^{2+} levels. In addition, no LTP-associated change is observed for maximal Ca^{2+}-stimulated kinase activity, or in the maximal stimulation of kinase activity by phosphatidylserine plus diacylglycerol plus Ca^{2+}.

The increased kinase activity is selectively induced by LTP-inducing tetanic stimulation. Low-frequency stimulation of the same inputs, which does not cause LTP, elicits no alteration in kinase activity. This indicates that the elevated kinase activity is not a nonspecific effect of synaptic activity. In addition, application of 50 μM APV, which blocks induction of LTP, also blocks the ability of tetanic stimulation to induce the elevated kinase activity. These findings suggest that the elevation of kinase activity is selectively associated with stimuli that induce LTP. The finding that AVP blocks induction of the elevated kinase activity has an additional important implication. This finding suggests that activation of NMDA receptors is necessary for the establishment of the persistent elevation of kinase activity.

In a final series of studies, Klann and co-workers (1991) determined the effects on the expression of the elevated kinase activity of adding a selective peptide inhibitor of PKC. The inhibitor used for these studies was a synthetic peptide corresponding to an autoinhibitory domain present in PKC, that is, residues 19–36. This peptide, PKC(19–36), significantly attenuated the LTP-associated increase in protein kinase activity, strongly suggesting persistent activation of PKC in the maintenance phase of LTP. Selective peptide inhibitors of CaM-kinase or cAMP-dependent protein kinase had no significant effect on the LTP-associated increase in kinase activity.

These data strongly support the hypothesis that the maintenance phase of LTP is associated with the existence of a Ca^{2+} independent form of PKC, and that the mechanism for the generation of this novel form of PKC is dependent on NMDA receptor activation. Given the fact that activation of PKC is sufficient to produce potentiation of synaptic transmission in the CA1 region of hippocampus, this model is a viable possibility as a mechanism for the maintenance of LTP.

REFERENCES

Abrahams, P. T., and G. G. Schofield. 1992. Norepinephrine-induced Ca^{2+} current inhibition in adult rat sympathetic neurons does not require protein kinase C activation. *Eur. J. Pharmacol. Molec. Pharmacol. Sec.* 227:189–197.

Adams, P. R., D. A. Brown, and A. Constanti. 1982. Pharmacological inhibition of the M-current. *J. Physiol.* 332:223–262.

Adamson, P., I. Hajimohamadreza, M. J. Brammer, and I. C. Campbell. 1989. Intrasynaptosomal free calcium concentration is increased by phorbol esters via a 1,4-dihydropyridine-sensitive (L-type) Ca^{2+} channel. *Eur. J. Pharmacol.* 162:59–66.

Ahlijanian, M. K., J. Striessnig, and W. A. Catterall. 1991. Phosphorylation of an α1-like subunit of an ω-conotoxin-sensitive brain calcium channel by cAMP-dependent protein kinase and protein kinase C. *J. Biol. Chem.* 266:20192–20197.

Akers, R. F., et al. 1986. Translocation of protein kinase C activity may mediate hippocampal long-term potentiation. *Science* 23:587–589.

Alkon, D. L., M. Kubota, J. T. Neary, S. Naito, D. Coulter, and H. Rasmussen. 1986. C-kinase activation prolongs Ca^{2+}-dependent inactivation of K^+ currents. *Biochem. Biophys. Res. Comm.* 134:1245–1253.

Allgaier, C., and G. Hertting. 1986. Polymyxin B, a selective inhibitor of protein kinase C, diminishes the release of noradrenaline and the enhancement of release caused by phorbol 12, 13-butyrate. *Am. J. Physiol.* 251:C833–C840.

Allgaier, C., B. Daschmann, H. Huang, G. Hertting. 1988. Protein kinase C and presynaptic modulation of acetylcholine release in rabbit hippocampus., *Endocrinology.* 122: 2699–2709.

Andersen, P., S. H. Sundberg, O. Sveen, and H. Wigstrom. 1977. Specific long-lasting potentiation of synaptic transmission in hippocampal slices. *Nature* 266:735.

Andreason, T. J., et al. 1983. Purification of a novel calmodulin binding protein from bovine cerebral cortex membranes. *Biochemistry* 22:4615–4618.

Aniksztejn, L., S. Otani, and Y. Ben-Ari. 1992. Quisqualate metabotropic receptors modulate NMDA currents and facilitate induction of long-term potentiation through protein kinase C. *Eur. J. Neurosci.* 4:500–505.

Apel, E. E., et al. 1990. Identification of the protein kinase C phosphorylation site in neuromodulin. *Biochemistry* 29:2330–2335.

Axelrod, J., R. M. Burch, and C. L. Jelsema. 1988. Receptor-mediated activation of phospholipase A_2 via GTP-binding proteins: Arachidonic acid and its metabolites as second messengers. *Trends Neurosci.* 11:117–123.

Baraban, J. M., S. H. Synder, and B. E. Alger. 1985. Protein kinase C regulates ionic conductance in hippocampal pyramidal neurons: Electrophysiological effects of phorbol esters. *Proc. Natl. Acad. Sci. U.S.A.* 82:2538–2542.

Bartmann, P., R. Jackisch, G. Hertting, and C. Allgaier. 1989. A role for protein kinase C in the electrically evoked release of [^{3}H] gamma-aminobutyric acid in rabbit caudate nucleus. *Neuroscience* 29:495–502.

Barrinuevo, G., and T. H. Brown. 1983. Associative long-term potentiation in the hippocampal slices. *Proc. Natl. Acad. Sci. U.S.A.* 80:7347.

Baskys, A., N. K. Bernstein, A. W. Barolet, and P. L. Carlen. 1990. NMDA and quisqualate reduce a Ca^{2+}-dependent K^+ current by a protein kinase-mediated mechanism. *Neurosci. Lett.,* 112:76–81.

Baskys, A., and R. C. Malenka. 1991. Agonists at metabotropic glutamate receptors presynaptically inhibit EPSCs in neonatal rat hippocampus. *J. Physiol.* 444:687–701.

Bazzi, M. D., and G. L. Nelsestuen. 1988a. Association of protein kinase C with phospholipid monolayers: Two-stage irreversible binding. *Biochemistry* 27:6776–6783.

Bazzi, M. D., and G. L. Nelsestuen. 1988b. Constitutive activity of membrane-inserted protein kinase C. *Biochem. Biophys. Res. Comm.* 152:336–343.

Bazzi, M. D., and G. L. Nelsestuen. 1988c. Properties of membrane-inserted protein kinase C. *Biochemistry* 27:7589–7593.

Bekkers, J. M., and C. F. Stevens. 1990. Presynaptic mechanism for long-term potentiation in the hippocampus. *Nature* 346:724–729.

Ben-Ari, Y., L. Aniksztejn, and P. Bregestovski. 1992. Protein kinase C modulation of NMDA currents: An important link for LTP induction. *Trends Neurosci.* 15:333–339.

Bliss, T. V. P., and A. R. Gardner-Medwin. 1973. Long-lasting potentiation of synaptic transmission in the dentate area of the unanesthetized rabbit following stimulation of the perforant path. *J. Physiol.* 232:357.

Bliss, T. V. P., and T. Lomo. 1973. Long-lasting potentiation of synaptic transmission in the dentate area of the anesthetized rabbit following stimulation of the perforant path. *J. Physiol.* 232:331.

Bliss, T., et al. 1986. Correlation between long-term potentiation and release of endogenous amino acids from dentate gyrus of anesthetized rats. *J. Physiol.* 377:391–408.

Boland, L. M., A. C. Allen, and R. Dingledine. 1991. Inhibition by bradykinin of voltage-activated barium current in a rat dorsal root ganglion cell line: Role of protein kinase C. *J. Neurosci.* 11:1140–1149.

Bosma, M. M., and B. Hille. 1989. Protein kinase C is not necessary for peptide-induced suppression of M current or for desensitization of the peptide receptors. *Proc. Natl. Acad. Sci. U.S.A.* 86:2943–2947.

Boss, V., and P. J. Conn. 1992. Metabotropic excitatory amino acid activation stimulates phospholipase D-catalyzed phosphatidylcholine hydrolysis. *J. Neurochem.* 59:2340–2343.

Brown, D. A., and P. R. Adams. 1987. Effects of phorbol dibutyrate on M currents and M current inhibition in bullfrog sympathetic neurons. *Cell. Mol. Biol.* 7:255–269.

Brown, D. A., and H. Higashida. 1988. Inositol 1,4,5-trisphosphate and diacylglycerol mimic bradykinin effects on mouse neuroblastoma x rat glioma hybrid cells. *J. Physiol.* 397:185–207.

Brown, D. A., N. V. Marrion, and T. G. Smart. 1989. On the transduction mechanism for muscarine-induced inhibition of M-current in cultured rat sympathetic neurones. *J. Physiol.* 413:469–488.

Browning, M. D., M. Bureau, E. M. Dudek, and R. W. Olsen. 1990. Protein kinase C and cAMP-dependent protein kinase phosphorylate the β subunit of the purified γ-aminobutyric acid A receptor. *Proc. Natl. Acad. Sci. U.S.A.* 87:1315–1318.

Buchert-Rau, B., and U. Sonnhof. 1982. An analysis of the epileptogenic potency of Co^{2+}—its ability to induce acute convulsive activity in the isolated frog spinal cord. *Pflügers Archiv.* 394:1–11.

Charpak, S., B. H. Gähwiler, K. Q. Do, and T. Knöpfel. 1990. Potassium conductances in hippocampal neurons blocked by excitatory amino-acid transmitters. *Nature* 347:765–767.

Chen, L., and L.-Y. M. Huang. 1991. Sustained potentiation of NMDA receptor-mediated glutamate responses through activation of protein kinase C by a μ opioid. *Neuron* 7:319–326.

Chen, L., and L.-Y. M. Huang. 1992. Protein kinase C reduces Mg^{2+} block of NMDA-receptor channels as a mechanism of modulation. *Nature* 356:521–523.

Chen, S.-J., M. A. Desai, E. Klann, D. G. Winder, J. D. Sweatt, and P. J. Conn. 1992. Amygdala kindling alters protein kinase C activity in dentate gyrus. *J. Neurochemistry* 59:1761–1768.

Colby, K. A., and M. P. Blaustein. 1988. Inhibition of voltage-gated K channels in synaptosomes by *sn*-1,2-dioctanoylglycerol, an activator of protein kinase C. *J. Neurosci.* 8:4685–4692.

Cole, A. E., and R. A. Nicoll. 1983. Acetylcholine mediates a slow synaptic potential in hippocampal pyramidal cells. *Science* 221:1299–1301.

Colley, P., F. Sheu, and A. Routtenberg. 1990. Inhibition of protein kinase C blocks two components of LTP persistence, leaving initial potentiation intact. *J. Neurosci.* 10: 3353–3360.

Collingridge, G. L., S. J. Kehl, and H. McLennan. 1983. Excitatory amino acids in synaptic transmission in the schaffer collateral-commissural pathway of the rat hippocampus. *J. Physiol.* 334:33–46.

Conn, P. J., and L. K. Kaczmarek. 1989. The bag cell neurons of *Aplysia:* A model for the study of the molecular mechanisms involved in the control of prolonged animal behaviors, *Molec. Neurobiol.* 3:1–37.

Conn, P. J., J. A. Strong, E. A. Azhderian, A. Nairn, P. Greengard, and L. Kaczmarek. 1989a. Differential effects of protein kinase inhibitors on cyclic AMP and phorbol ester-induced changes in excitability of *Aplysia* bag cell neurons. *J. Neurosci.* 9: 473–479.

Conn, P. J., J. A. Strong, and L. K. Kaczmarek. 1989b. Inhibitors of protein kinase C prevent enhancement of calcium current and action potentials in peptidergic neurons from *Aplysia. J. Neurosci.* 9:480–487.

Costa, M. R. C., and W. A. Catterall. 1984. Phosphorylation of the α subunit of the sodium channel by protein kinase C. *Cell. Molec. Neurobiol.* 4:291–297.

Critz, S. D., and J. H. Byrne. 1992. Modulation of $I_{K,Ca}$ by phorbol ester-mediated activation of PKC in pleural sensory neurons of *Aplysia. J. Neurophysiol.* 68:1079–1086.

Davies, S., et al. 1989. Temporally distinct pre- and post-synaptic mechanisms maintain long-term potentiation. *Nature* 338:500–503.

De Graan, P. N. E., A. B. Oestreicher, P. Schotman, and L. H. Schrama. 1991. Protein kinase C substrate B-50 (GAP-43) and neurotransmitter release. *Prog. Brain Res.* 89:187–207.

Dekker, L. V., P. N. E. De Graan, D. H. G. Versteeg, A. B. Oestreicher, and W. H. Gispen. 1989. Phosphorylation of B-50 (GAP43) is correlated with neurotransmitter release in rat hippocampal slices. *J. Neurochem.* 52:24–30.

Dekker, L. V., P. N. E. De Graan, M. De Wit, J. J. H. Hens, and W. H. Gispen. 1990. Depolarization-induced phosphorylation of the protein kinase C Substrate B-50 (GAP-43) in rat cortical synaptosomes. *J. Neurochem.* 54:1645–1652.

Dekker L. V., P. N. E. De Graan, and W. H. Gispen. 1991. Transmitter release: Target of regulation by protein kinase C? *Prog. Brain Res.* 89:209–233.

DeReimer, S. A., P. Greengard, and L. K. Kaczmarek. 1985a. Calcium/phosphatidylserine/diacylglycerol-dependent protein phosphorylation in the *Aplysia* nervous system. *J. Neurosci.* 5:2672–2676.

DeReimer, S. A., J. A. Strong, K. A. Albert, P. Greengard, and L. K. Kaczmarek. 1985b. Enhancement of calcium current in *Aplysia* neurones by phorbol ester and protein kinase C. *Nature* 313:313–316.

Desai, M., and P. J. Conn. 1991. Excitatory effects of ACPD receptor activation in the hippocampus are mediated by direct effects on pyramidal cells and blockade of synaptic inhibition. *J. Neurophysiol.* 66:40–52.

Desai, M. A., S. Smith, and P. J. Conn. 1992. Multiple metabotropic glutamiate receptors regulate hippocampal function. *Synapse* 12:206–213.

Doerner, D., and B. E. Alger. 1992. Evidence for hippocampal calcium channel regulation by PKC based on comparison of diacylglycerols and phorbol esters. *Brain Res.* 597:30–40.

Doerner, D., T. A. Pitler, and B. E. Alger. 1988. Protein kinase C activators block specific calcium and potassium current components in isolated hippocampal neurons. *J. Neurosci.* 8:4069–4078.

Doerner, D., M. Abdel-Latif, T. B. Rogers, and B. E. Alger. 1990. Protein kinase C-dependent and independent effects of phorbol esters on hippocampal calcium channel current. *J. Neurosci.* 10:1699–1706.

Doster, S. K., et al. 1991. Expression of the growth-associated protein GAP-43 in adult rat retinal ganglion cells following axon injury. *Neuron* 6:635–647.

Downing, J. E. G., and L. W. Role. 1987. Activators of protein kinase C enhance acetylcholine receptor desensitization in sympathetic ganglion neurons. *Proc. Natl. Acad. Sci. U.S.A.* 84:7739–7743.

Durand, G. M., P. Gregor, X. Zheng, M. V. L. Bennett, G. R. Uhl, and R. S. Zukin. 1992. Cloning of an apparent splice variant of the rat N-methyl-D-aspartate receptor NMDAR1 with altered sensitivity to polyamines and activators of protein kinase C. *Proc. Natl. Acad. Sci. U.S.A.* 89:9359–9363.

Etcheberrigaray, R., L. D. Matzel, I. I. Lederhendler, and D. L. Alkon. 1992. Classical conditioning and protein kinase C activation regulate the same single potassium channel in *Hermissenda crassicornis* photoreceptors. *Proc. Natl. Acad. Sci. U.S.A.* 89:7184–7188.

Etoh, S., A. Baba, and H. Iwata. 1991a. NMDA induces protein kinase C translocation in guinea pig cerebral synaptoneurosomes. *Jap. J. Pharmacol.* 56:287–296.

Etoh, S., A. Baba, and H. Iwata. 1991b. NMDA induces protein kinase C translocation in hippocampal slices of immature rat brain. *Neurosci. Lett.* 126:119–122.

Eusebi, F., M. Molinaro, and B. M. Zani. 1985. Agents that activate protein kinase C reduce acetylcholine sensitivity in cultured myotubes. *J. Cell Biol.* 100:1339–1342.

Eusebi, F., F. Grassi, C. Nervi, C. Caporale, S. Adamo, B. M. Zani, and M. Molinaro. 1987. Acetylcholine may regulate its own nicotinic receptor-channel through the C-kinase system. *Proc. Roy. Soc. Lond.* 230:355–365.

Ewald, D. A., H. J. G. Matthies, T. M. Perney, M. W. Walker, and R. J. Miller. 1988. The effects of down regulation of protein kinase C on the inhibitory modulation of dorsal root ganglion neuron Ca^{2+} currents by neuropeptide Y. *J. Neurosci.* 8:2447–2451.

Farley, J., and S. Auerbach. 1986. Protein kinase C activation induces conductance changes in *Hermissenda* photoreceptors like those seen in associative learning. *Nature* 319:220–223.

Feuerstein, T. J., C. Allgaier, and G. Hertting. 1987. Possible involvement of protein kinase C (PKC) in the regulation of electrically evoked serotonin (5-HT) release from rabbit hippocampal slices. *Eur. J. Pharmacol.* 141:15–21.

Fink, L. A., J. A. Connor, and L. K. Kaczmarek. 1988. Inositol trisphosphate releases intracellularly stored calcium and modulates ion channels in molluscan neurons. *J. Neurosci.* 8:2544–2555.

Fisher, S. K., and B. W. Agranoff. 1987. Receptor activation and inositol lipid hydrolysis in neural tissues. *J. Neurochem.* 48:999–1017.

Forder, J., A. Scriabine, and H. Rasmussen. 1985. Plasma membrane calcium flux, protein kinase C activation and smooth muscle contraction. *J. Pharmacol. Exp. Ther.* 235:267–273.

Freedman, J. E., and G. K. Aghajanian. 1987. Role of phosphoinositide metabolites in the prolongation·of after hyperpolarizations by α_1-adrenoreceptors in rat dorsal raphe neurons. *J. Neurosci.* 7:3897–3906.

Gandhi, V. C., and D. J. Jones. 1992. Protein kinase C modulates the release of [^{3}H]5-hydroxytryptamine in the spinal cord of the rat: The role of L-type voltage-dependent calcium channels. *Neuropharmacology* 31:1101–1109.

Gerber, U., J. A. Sim, and B. H. Gahwiler. 1992. Reduction of potassium conductances mediated by metabotropic glutamate receptors in rat CA3 pyramidal cells does not require protein kinase C or protein kinase A. *Eur. J. Neurosci.* 4:792–797.

Ghirardi, M., et al. 1992. Roles of PKA and PKC in facilitation of voked and spontaneous transmitter release at depressed and nondepressed synapses in aplysia sensory neurons. *Neuron* 9:479–489.

Gianotti, C., et al. 1992. Phosphorylation of the presynaptic protein B-50 (GAP-43) is increased during electrically induced long-term potentiation. *Neuron* 8:843–848.

Girard, P. R., G. J. Mazzei, J. G. Wood, and J. F. Kuo. 1985. Polyclonal antibodies to phospholipid/Ca^{2+}-dependent protein kinase and immunocytochemical localization of the enzymes in rat brain, *Proc. Natl. Acad. Sci. U.S.A.* 82:3030–3034.

Gleason, M. M., and S. F. Flaim. 1986. Phorbol ester contracts rabbit thoracic aorta by increasing intracellular calcium and by activating calcium influx. *Biochem. Biophys. Res. Comm.* 138:1362–1369.

Goslin, D., et al. 1988. Development of neuronal plasticity: GAP-43 distinguishes axonal from dendritic growth cones. *Nature* 336:672–675.

Grafe, P., C. J. Mayer, and J. D. Wood. 1980. Synaptic modulation of calcium-dependent potassium conductance in myenteric neurones in the guinea-pig. *J. Physiol.* 305:235–248.

Grega, D. S., M. A. Werz, and R. L. Macdonald. 1987. Forskolin and phorbol esters reduce the same potassium conductance of mouse neurons in culture. *Science* 235:345–348.

Gross, R. A., and R. L. Macdonald. 1989. Activators of protein kinase C selectively enhance inactivation of a calcium current component of cultured sensory neurons in a pertussis toxin-sensitive manner. *J. Neurophysiol.* 61:1259–1269.

Harris, E. W., and C. W. Cotman. 1986. Long-term potentiation of guinea pig mossy fiber responses is not blocked by the N-methyl-D-aspartate antagonists. *Neurosci. Lett.* 70:132–137.

Harris, K. M., S. Kongsamut, and R. J. Miller. 1986. Protein kinase C mediated regulation of calcium channels in PC-12 pheochromocytoma cells. *FEBS Lett.* 196:365–369.

Hashimoto, T., et al. 1988. Postnatal development of brain-specific subspecies of protein kinase C in rat. *J. Neurosci.* 8:1678–1683.

Hidaka, H., M. Inagaki, S. Kawamoto, and Y. Sasaki. 1984. Isoquinolinesulfonamides, novel and potent inhibitors of cyclic nucleotide dependent protein kinase and protein kinase C. *Biochemistry* 23:5036–5041.

Higashida, H., and D. A. Brown. 1986. Two polyphosphatidylinositide metabolites control two K^+ currents in a neuronal cell. *Nature* 323:333–335.

Hill-Venning, C., and G. A. Cottrell. 1992. Modulation of voltage-dependent calcium current in *Helix aspersa* buccal neurones by serotonin and protein kinase C activators. *Exp. Physiol.* 77:891–901.

Hockberger, P., M. Toselli, D. Swandulla, and H. D. Lux. 1989. A diaclyglycerol analogue reduces neuronal calcium currents independently of protein kinase C activation. *Nature* 338:340–342.

Holz, G. G., S. G. Rane, and K. Dunlap. 1986. GTP-binding proteins mediate transmitter inhibition of voltage-dependent calcium channels. *Nature* 319:670–672.

Hosada, K., et al. 1989. Immunocytochemical localization of the β-I subspecies of protein kinase C in rat brain. *Proc. Nat. Acad. Sci. U.S.A.* 86:1393–1397.

Huganir, R. L. 1987. Regulation of the nicotinic acetylcholine receptor by protein phosphorylation. *J. Receptor Res.* 7:241–256.

Ito, A., et al. 1990. Immunocytochemical localization of the α subspecies of protein kinase C in rat brain. *Proc. Natl. Acad. Sci. U.S.A.* 87:3195–3199.

Jahr, C. E., and C. F. Stevens. 1987. Glutamate activates multiple single channel conductances in hippocampal neurons. *Nature* 325:522–525.

Jones, S. W., and P. R. Adams. 1987. The M-current and other potassium currents of vertebrate neurons. In: *Neuromodulation: The Biochemical Control of Neuronal Excitability* edited by L. Kaczmarek and I. Levitan pp. 159–186. Oxford University Press, New York.

Kandel, E. R., and J. Schwartz. 1982. Molecular mechanisms of memory: Modulation of transmitter release. *Science* 218:433–441.

Kauer, J. A., et al. 1988a. A persistent postsynaptic modification mediates long-term potentiation in the hippocampus. *Neuron* 1:911–917.

Kauer, J. A., et al. 1988b. NMDA application potentiates synaptic transmission in the hippocampus. *Nature* 334:250.

Kelso, S. R., and T. H. Brown. 1984. Differential conditioning of associative synaptic enhancement in hippocampal brain slices. *Science* 232:85.

Kelso, S. R., et al. 1986. Hebbian synapses in the hippocampus. *PNAS* 83:5326.

Keyser, D. O., and B. E. Alger. 1990. Arachidonic acid modulates hippocampal calcium current via protein kinase C and oxygen radicals. *Neuron* 5:545–553.

Kikkawa, U., K. Ogita, M. Go, H. Nomura, T. Kitano, T. Hashimoto, K. Ase, K. Sekiguchi, J. Kuomoto, Y. Nishizuka, N. Saito, and C. Tanaka. 1988. Protein kinase C in membrane signaling. In: *Advances in Second Messenger and Phosphoprotein Research,* Vol. 21, edited by R. Adelstein, C. Klee, and M. Rodbell, pp. 67–74. Raven Press, New York.

Klann, E., et al. 1991. Persistent protein kinase activation in the maintenance phase of long-term potentiation. *J. Biol. Chem.* 266:24253–24256.

Knight, D. E. 1987. Calcium and diacylglycerol control of secretion. *Biosci. Rep.* 7:355–367.

Kofuji, P., J. B. Wang, S. J. Moss, R. L. Huganir, and D. R. Burt. 1991. Generation of two forms of the γ-aminobutyric acid$_A$ receptor γ2-subunit in mice by alternative splicing. *J. Neurochem.* 56:713–715.

Kostyuk, P. G., E. A. Lukyanetz, and A. S. Ter-Markosyan. 1992. Parathyroid hormone enhances calcium current in snail neurons. Stimulation of the effect by phorbol esters. *Pflug. Arch.* 420:146–152.

Kullmann, D., and R. Nicoll. 1992. Long-term potentiation is associated with increases in quantal content and quantal amplitude. *Nature* 357:240–244.

Kuo, J. F., R. G. G. Andersson, B. C. Wise, L. Mackerlova, I. Salomonsson, N. L. Brackett, N. Katoh, M. Shoji, and R. W. Wrenn. 1980. Calcium-dependent protein kinase: Widespread occurrence in various tissues and phyla of the animal kingdom and comparison of effects of phospholipid, calmodulin, and trifluoperazine, *Proc. Natl. Acad. Sci. U.S.A.* 77:7039–7043.

Lacerda, A. E., D. Rampe, and A. M. Brown. 1988. Effects of protein kinase C activators on cardiac Ca^{2+} channels. *Nature* 335:249–251.

Lee, K. S. 1983. Cooperativity among afferents for the induction of long-term potentiation in the CA1 region of hippocampus. *Neuroscience* 3:1369.

Leidenheimer, N. J., S. J. McQuilkin, L. D. Hahner, P. Whiting, and R. A. Harris. 1992. Activation of protein kinase C selectively inhibits the γ-aminobutyric acid$_A$ receptor: Role of desensitization. *Molec. Pharmacol.* 41:1116–1123.

Leonard, J. P., J. Nargeot, T. P. Snutch, N. Davidson, and H. A. Lester. 1987. Ca channels induced in *Xenopus* oocytes by rat brain mRNA. *J. Neurosci.* 7:857–881.

Levitan, I. B., and L. K. Kaczmarek. 1987. Ion currents and ion channels: Substrates for neuromodulation. In: *Neuromodulation: The Biochemical Control of Neuronal Excitability,* edited by L. Kaczmarek and I. Levitan, pp. 18–38. Oxford University Press, New York.

Levy, W. B., and O. Steward. 1979. Synapses as associative memory elements in the hippocampal formation. *Brain Res.* 175:23.

Lewis, D. L., and F. F. Weight. 1988. The protein kinase C activator 1-oleoyl-2-acetylglycerol inhibits voltage-dependent Ca^{2+} current in the pituitary cell line AtT-20. *Neuroendocrinology* 47:169–175.

Lewis, D. V., and W. A. Wilson. 1982. Calcium influx and poststimulus current during early adaptation in *Aplysia* giant neurons. *J. Neurophysiol.* 48:202–216.

Linden, D. J., and A. Routtenberg. 1989. *CIS*-fatty acids, which activate protein kinase C, attenuate Na^+ and Ca^{2+} currents in mouse neuroblastoma cells. *J. Physiol.* 419: 95–119.

Linden, D. J., and J. A. Connor. 1991. Participation of postsynaptic PKC in cerebellar long-term depression in culture. *Science* 254:1656–1659.

Linden, D. J., et al. 1988. NMDA receptor blockade prevents the increase in protein kinase C substrate (protein F1) phosphorylation produced by long-term potentiation. *Brain Res.* 458:142–146.

Linden, D. J., et al. 1992. An electrophysiological correlate of protein kinase C isozyme distribution in cultured cerebellar neurons. *J. Neurosci.* 12:3601–3608.

Llahi, S., and J. N. Fain. 1992. α_1-Adrenergic receptor-mediated activation of phospholipase D in rat cerebral cortex. *J. Biol. Chem.* 267:3679–3685.

Llinas, R., M. Suimorimi, D. E. Hillman, and B. Cherksey. 1992. Distribution and functional significance of the P-type, voltage-dependent Ca^{2+} channels in the mammalian central nervous system. *Trends Neurosci.* 15:351–355.

Loechner, K. J., J. Mattessich-Arrandale, E. M. Azhderian, and L. K. Kaczmarek. 1992. Inhibition of peptide release for invertebrate neurons by the protein kinase inhibitor H-7. *Brain Res.* 581:315–318.

Lotan, I., N. Dascal, Z. Naor, and R. Boton. 1990. Modulation of vertebrate brain Na^+ and K^+ channels by subtypes of protein kinase C. *FEBS Lett.* 267:25–28.

Lovinger, D. M., et al. 1986. Direct relation of long-term synaptic potentiation to phosphorylation of membrane protein F1, a substrate for membrane protein kinase C. *Brain Res.* 399:205–211.

Lynch, G., J. Larson, S. Kelso, G. Barrinuevo, and F. Schottler. 1983. Intracellular injections of EGTA block induction of hippocampal long-term potentiation. *Nature* 305: 719–721.

MacDermott, A. B., M. L. Mayer, G. L. Westbrook, S. J. Smith, and J. L. Barker. 1986. NMDA-receptor activation increases cytoplasmic calcium concentration in cultured spinal cord neurones. *Nature* 321:519–522.

Madison, D. V. 1989. Phorbol esters increase unitary calcium channel activity in cultured hippocampal neurons. *Soc. Neurosci. Abs.* 15:16.

Madison, D. V., and R. A. Nicoll. 1984. Control of repetitive discharge of rat CA1 pyramidal neurones *in vitro. J. Physiol.* 354:319–331.

Madison, D. V., R. C. Malenka, and R. A. Nicoll. 1986. Phorbol esters block a voltage-sensitive chloride current in hippocampal pyramidal cells. *Nature* 321:695–697.

Malenka, R. C., D. V. Madison, R. Andrade, and R. A. Nicoll. 1986a. Phorbol esters mimic some cholinergic actions in hippocampal pyramidal neurons. *J. Neurosci.* 6:475–480.

Malenka, R. C., et al. 1986b. Potentiation of synaptic transmission in the hippocampus by phorbol esters. *Nature* 321:175.

Malenka, R. C., G. S. Ayoub, and R. A. Nicoll. 1987. Phorbol esters enhance transmitter release in rat hippocampal slices. *Brain Res.* 403:198–203.

Malenka, R. C., et al. 1988. Postsynaptic calcium is sufficient for potentiation of hippocampal synaptic transmission. *Science* 242:81–84.

Malgaroli, A., and R. Tsien. 1992. Glutamate-induced long-term potentiation of the frequency of miniature synaptic currents in cultured hippocampal neurons. *Nature* 357:134–139.

Malinow, R., and J. P. Miller. 1988. Postsynaptic hyperpolarization during conditioning reversibly blocks induction of long-term potentiation. *Nature* 320:529.

Malinow, R., et al. 1988. Persistent protein kinase activity underlying long-term potentiation. *Nature* 338:500.

Malinow, R., et al. 1989. Inhibition of postsynaptic PKC or CAMKII blocks induction but not expression of LTP. *Science* 245:862.

Marchetti, C., and A. M. Brown. 1988. Protein kinase activator 1-oleoyl-2-acetyl-*sn*-glycerol inhibits two types of calcium currents in GH3 cells. *Am. J. Physiol.* 254:206–210.

Markram, H., and M. Segal. 1992. Activation of protein kinase C suppresses responses to NMDA in rat CA1 hippocampal neurones. *J. Physiol.* 457:491–501.

Matthies, H. J., H. B. Palfrey, L. D. Hirning, and R. J. Miller. 1987. Down regulation of protein kinase C in neuronal cells: Effects on neurotransmitter release. *Neurosci. Lett.* 75:71–74.

Mayer, M. L., and G. L. Westbrook. 1987. Permeation and block of N-methyl-D-aspartate receptor channels by divalent cations in mouse central neurons. *J. Physiol.* 394:501–528.

McNaughton, B., R. M. Douglas, and G. V. Goddard. 1978. Synaptic enhancement in fascia dentata: Cooperativity among coactive afferents. *Brain Res.* 157:277.

Meiri, K. F., M. Willard, and M. I. Johnson. 1988. Distribution and phosphorylation of the growth-associated protein GAP-43 in regenerating sympathetic neurons in culture. *J. Neurosci.* 8:2571–2581.

Messing, R. O., A. B. Sneade, and B. Savidge. 1990. Protein kinase C participates in up-regulation of dihydropyridine-sensitive calcium channels by ethanol. *J. Neurochem.* 55:1383–1389.

Muller, D., et al. 1988. Phorbol ester-induced synaptic facilitation is different than long-term potentiation. *Proc. Natl. Acad. Sci. U.S.A.* 85:6997.

Murphy, B. J., and W. A. Catterall. 1992. Phosphorylation of purified rat brain Na^+ channel reconstituted into phospholipid vesicles by protein kinase C. *J. Biol. Chem.* 267:16129–16134.

Murray, A. W., A. Fournier, and S. J. Hardy. 1987. Proteolytic activation of protein kinase C: A physiological reaction? *Trends Biochem. Sci.,* 12:53–54.

Nastainczyk, W., A. Röhrkasten, M. Sieber, C. Rudolph, C. Schächtele, D. Marmè, and F. Hofmann. 1987. Phosphorylation of the purified receptor for calcium channel blockers by cAMP kinase and protein kinase C. *Eur. J. Biochem.* 169:137–142.

Navarro, J. 1987. Modulation of [^{3}H]dihydropyridine receptors by activation of protein kinase C in chick muscle cells. *J. Biol. Chem.* 262:4649–4652.

Negro-Vilar, A., D. Conte, and M. Valenca. 1986. Transmembrane signals mediating neural peptide secretion: Role of protein kinase C activators and arachidonic acid metabolites in luteinizing hormone-releasing hormone secretion. *Naunyn-Schmiedeberg's Arch. Pharmacol.* 334:218–221.

Nelson, R., and A. Routtenberg. 1985. Characterization of protein F1: A kinase C substrate directly related to neuronal plasticity. *Exp. Neurol.* 89:213–244.

Nichols, R. A., J. W. Haycock, J. K. Wang, and P. Greengard. 1987. Phorbol ester enhancement of neurotransmitter release from rat brain synaptosomes. *Nature* 325:58–60.

Nielander, H. B., et al. 1990. Mutation of serine 41 in the neuron-specific protein B-50 (GAP-43) prohibits phosphorylation by protein kinase C. *J. Neurochem.* 55:1442–1445.

Nishizuka, Y. 1989. The family of protein kinase C for signal transduction. *JAMA* 262:1826–1833.

Nowak, L., P. Bregestovski, P. Ascher, A. Herbet, and A. Prochiantz. 1984. Magnesium gates glutamate-activated channels in mouse central neurones. *Nature* 307:462–465.

Nowycky, M. C., A. P. Fox, and R. W. Tsien. 1985. Three types of neuronal calcium channel with different calcium agonist sensitivity. *Nature* 316:440–443.

Numann, R., W. A. Catterall, and T. Scheuer. 1991. Functional modulation of brain sodium channels by protein kinase C phosphorylation. *Science* 254:115–118.

O'Callahan, C. M., J. Ptasienski, and M. M. Hosey. 1988. Phosphorylation of the 165-kDa dihydropyridine/phenylalkylamine receptor from skeletal muscle by protein kinase C. *J. Biol. Chem.* 263:17342–17349.

Ottoson, D., and C. Swerup. 1981. Ionic basis of adaptation in crustacean stretch receptors. *J. Physiol.* 317:26–27.

Pfaffinger, P. J., M. D. Leibowitz, W. M. Subers, N. M. Nathanson, W. Almers, and B. Hille. 1988. Agonists that suppress M-current elicit phophoinositide turnover and Ca^{2+} transients, but these events do not explain M-current suppression. *Neuron* 1:477–484.

Plummer, M. R., A. Rittenhouse, M. Kanevsky, and P. Hess. 1991. Neurotransmitter modulation of calcium channels in rat sympathetic neurons. *J. Neurosci.* 11:2339–2348.

Rane, S. G., and K. Dunlap. 1986. Kinase C activator 1,2-oleoacetylglycerol attenuates voltage-dependent calcium current in sensory neurons. *Proc. Natl. Acad. Sci. U.S.A.* 83:184–188.

Rane, S. G., M. P. Walsh, J. R. McDonald, and K. Dunlap. 1989. Specific inhibitors of protein kinase C block transmitter-induced modulation of sensory neuron calcium current. *Neuron* 3:239–245.

Saito, N., et al. 1988. Distribution of protein kinase C-like immunoreactivity neurons in rat brain. *J. Neuroscience* 8:369–382.

Saito, N., et al. 1989. Immunocytochemical localization of β-II subspecies of protein kinase C in rat brain. *Proc. Nat. Acad. Sci. U.S.A.* 86:3409–3413.

Sawada, M., L. J. Cleary, and J. H. Byrne. 1989a. Inositol triphosphate and activators of protein kinase C modulate membrane currents in tail motor neurons of *Aplysia*. *J. Neurophysiol.* 61:302–310.

Sawada, M., M. Ichinose, and T. Maeno. 1989b. Protein kinase C activators reduce the inositol triphosphate-induced outward current and the Ca^{2+}-activated outward current in identified neurons of *Aplysia*. *J. Neurosci. Res.* 22:158–166.

Schroeder, J. E., P. S. Fischbach, and E. W. McCleskey. 1990. T-type calcium channels: Heterogeneous expression in rat sensory neurons and selective modulation by phorbol esters. *J. Neurosci.* 10:947–951.

Scott, R. H., H. A. Pearson, and A. C. Dolphin. 1991. Aspects of vertebrate neuronal voltage-activated calcium currents and their regulation. *Prog. Neurobiol.* 36:485–520.

Shinomura, T., H. Mishima, S. Matsushima, Y. Asoaka, K. Yoshida, M. Oka, and Y. Nishizuka. 1992. Degradation of phospholipids and protein kinase C activation for the control of neuronal functions. *Adv. Exp. Med. Biol.* 318:361–373.

Sigel, E., and R. Baur. 1988. Activation of protein kinase C differentially modulates neuronal Na^+, Ca^+, and γ-aminobutyrate type A channels. *Proc. Natl. Acad. Sci. U.S.A.* 85:6192–6196.

Sigel, E., R. Baur, and P. Malherbe. 1991. Activation of protein kinase C results in down-modulation of different recombinant $GABA_A$-channels. *FEBS Lett.* 291:150–152.

Sim, J. A., U. Gerber, T. Knopfel, and D. A. Brown. 1992. Evidence against a role for protein kinase C in the inhibition of the calcium-activated potassium current I_{AHP} by muscarinic stimulants in rat hippocampal neurons. *Eur. J. Neurosci.* 4:785–791.

Skene, J. H. P. 1989. Axonal growth-associated proteins. *Annu. Rev. Neurosci.* 12:127–156.

Skene, J. H. P., and M. B. Willard. 1981. Axonally transported proteins associated with axon growth in rabbit central nervous systems. *J. Cell. Biol.* 89:96–103.

Smith, M. K., R. J. Colbran, and T. R. Soderling. 1990. Specificities of autoinhibitory domain peptides for four protein kinases. *J. Biol. Chem.* 265:1837.

Sossin, W. S., and J. H. Schwartz. 1992. Selective activation of calcium-activated PKCs in *Aplysia* neurons by 5-HT. *J. Neurosci.* 12:1160–1168.

Spencer, S. A., et al. 1992. GAP-43, a protein associated with axon growth, is phosphorylated at three sites in cultured neurons and rat brain. *J. Biol. Chem.* 267:9059–9064.

Sposi, N. M., et al. 1989. Expression of protein kinase C genes during ontogenic development of the central nervous system. *Molec. Cell. Biol.* 9:2284–2288.

Stabel, S., and P. J. Parker. 1991. Protein kinase C. *Pharm. Ther.* 51:71–95.

Strittmatter, S. M., et al. 1990. Gô is a major growth cone protein subject to regulation by GAP-43. *Nature* 344:836–841.

Strong, J. A., A. P. Fox, R. W. Tsien, and L. K. Kaczmarek. 1987. Stimulation of protein kinase C recruits covert calcium channels in *Aplysia* bag cell neurons. *Nature* 325:714–717.

Swope, S. L., S. J. Moss, C. D. Blackstone, and R. L. Huganir. 1992. Phosphorylation of ligand-gated ion channels: A possible mode of synaptic plasticity. *FASEB J.* 6:2514–2523.

Tanabe, T., H. Takeshima, A. Mikami, V. Flockerzi, H. Takahashi, K. Kangawa, M. Kojima, H. Matsuo, T. Hirose, and S. Numa. 1987. Primary structure of the receptor for calcium channel blockers from skeletal muscle. *Nature* 328:313–318.

Thibault, O., et al. 1989. Long-lasting physiological effects of bath applied N-methyl-D-aspartate. *Brain Res.* 476:170.

Ticku, M. K., and A. K. Mehta. 1990. γ-aminobutyric acid$_a$ receptor desensitization in mice spinal cord cultured neurons: Lack of involvement of protein kinases A and C. *Molec. Pharmacol.* 38:719–724.

Tsien, R. W., D. Lipscombe, D. V. Madison, K. R. Bley, and A. P. Fox. 1988. Multiple types of neuronal calcium channels and their selective modulation. *Trends Neurosci.* 11:431–438.

Tsien, R. W., P. T. Ellinor, and W. A. Horne. 1991. Molecular diversity of voltage-dependent Ca^{2+} channels. *Trends Pharmacol. Sci.* 12:349–353.

Tsuji, S., S. Minota, and K. Kuba. 1987. Regulation of the two ion channels by a common muscarinic receptor-transduction system in a vertebrate neurone. *Neurosci. Lett.* 81:139–145.

Urushihara, H., M. Tohda, and Y. Nomura. 1992. Selective potentiation of N-methyl-D-aspartate-induced current by protein kinase C in *Xenopus* oocytes injected with rat brain RNA. *J. Biol. Chem.* 267:11697–11700.

Vaccarino, F. M., S. Liljequist, and J. F. Tallman. 1991. Modulation of protein kinase C translocation by excitatory and inhibitory amino acids in primary cultures of neurons. *J. Neurochem.* 57:391–396.

Van Hooff, C. O., et al. 1988. B-50 phosphorylation and polyphosphoinositide metabolism in nerve growth cone membranes. *J. Neurosci.* 8:1789–1795.

Vivaudou, M. B., L. H. Clapp, J. V. Walsh Jr., and J. J. Singer. 1988. Regulation of one type of Ca^{2+} current in smooth muscle cells by diacylglycerol and acetylcholine. *FASEB J.* 2:2497–2504.

Wakade, T. D., S. V. Bhave, A. S. Bhave, R. K. Malhotra, and A. R. Wakade. 1991. Depolarizing stimuli and neurotransmitters utilize separate pathways to activate protein kinase C in sympathetic neurons. *J. Biol. Chem.* 266:6424–6428.

Walker, M. W., D. A. Ewald, T. M. Perney, and R. J. Miller. 1988. Neuropeptide Y modulates neurotransmitter release and Ca^{2+} currents in rat sensory neurons. *J. Neurosci.* 8:2438–2446.

Walsh, K. B., and R. S. Kass. 1988. Regulation of a heart potassium channel by protein kinase A and C. *Science* 242:67–69.

Wang, H. Y., and E. Friedman. 1987. Protein kinase C: Regulation of serotonin release from rat brain cortical slices. *Brain Res.* 419:364–368.

Wang, J., and D. Feng. 1992. Postsynaptic protein kinase C essential to induction and maintenance of long-term potentiation in the hippocampal CA1 region. *Proc. Nat. Acad. Sci. U.S.A.* 89:2576–2580.

Weiss, S., J. Ellis, D. Hendley, and R. H. Lenox. 1989. Translocation and activation of protein kinase C in striatal neurons in primary culture: Relationship to phorbol dibutyrate actions on the inositol phosphate generating system and neurotransmitter release. *J. Neurochem.* 52:530–536.

Werz, M. A., and R. L. Macdonald. 1987a. Dual actions of phorbol esters to decrease calcium and potassium conductances of mouse neurons. *Neurosci. Lett.* 78:101–106.

Werz, M. A., and R. L. Macdonald. 1987b. Phorbol esters: Voltage-dependent effects on calcium-dependent action potentials in mouse central and peripheral neurons in cell culture. *J. Neurosci.* 7:1639–1637.

West, J. W., R. Numann, B. J. Murphy, T. Schever, and W. A. Catterall. 1991. A Phosphorylation site in the Na^+ Channel required for modulation by protein kinase C. *Science* 254:866–868.

Whiting, P., R. M. McKernan, and L. L. Iversen. 1990. Another mechanism for creating diversity in γ-aminobutyrate type A receptors: RNA splicing directs expression of two forms of γ2 subunit, one of which contains a protein kinase C phosphorylation site. *Proc. Natl. Acad. Sci. U.S.A.* 87:9966–9970.

Wood, J. G., P. R. Girard, G. J. Mazzei, and J. F. Kuo. 1986. Immunocytochemical localization of protein kinase C in identified neuronal compartments of rat brain. *J. Neurosci.* 6:2571–2577.

Zuber, M. C., et al. 1989. The neuronal growth-cone associated protein GAP-43 induces filipodia in non-neuronal cells. *Science* 244:1193–1195.

Zucker, R. S. 1987. Neurotransmitter release and its modulation. In: *Neuromodulation: The Biochemical Control of Neuronal Excitability,* edited by L. Kaczmarek and I. Levitan, pp. 243–263. Oxford University Press, New York.

8

Protein Kinase C in Smooth Muscle

MASAKATSU NISHIKAWA
HIROYOSHI HIDAKA

Protein kinase C is thought to play important roles in the control of smooth-muscle functions. The activation of this enzyme is often closely linked to changes in phosphoinositide metabolism (Nishizuka, 1992). When agonists such as endothelin, arginine-vasopressin, histamine, and angiotensin II activate specific receptors, an immediate receptor-linked event is the activation of a specific phospholipase C that catalyzes the rapid hydrolysis of phosphatidylinositol 4,5-bisphosphate (PIP_2) to two intracellular messengers, inositol 1,4,5 triphosphate (IP_3) and diacylglycerol (DG) (Griendling et al., 1986; Denthuluri and Brock, 1990). It has been shown that IP_3 triggers the release of Ca^{2+} from intracellular stores in smooth muscle. Likewise, it has been demonstrated that agonists induced rapid and sustained increase in the DG content of cultured smooth-muscle cells, intact tracheal and carotid artery smooth muscle (Griendling et al., 1986; Denthuluri and Brock, 1990). Ca^{2+} regulates smooth-muscle actomyosin ATPase by mechanisms associated with both thick and thin filaments (Adelstein and Eisenberg, 1980; Walsh and Hartshorne, 1982; Sellers, 1991). The increase in DG concentration induces the activation of protein kinase C. The translocation of protein kinase C from the cytosol to the membrane induced by phorbol ester or agonists was demonstrated in intact smooth muscle (Griendling et al., 1989; Langlands and Diamond, 1992), and the temporal pattern of the protein kinase C translocation depends on the agonist used. There seemed to be no direct correlation between changes in the distribution of protein kinase C and the changes in tension. Although activation of protein kinase C is postulated to be involved in smooth-muscle contraction, the role of this kinase is controversial and the precise mechanism is still obscure. In this chapter, some general aspects of protein kinase C in smooth muscle and the relationship to contraction will be discussed (Figure 8.1). Indeed, the vast literature that exists concerning protein kinase C makes it impossible to discuss all aspects of this topics in a short chapter.

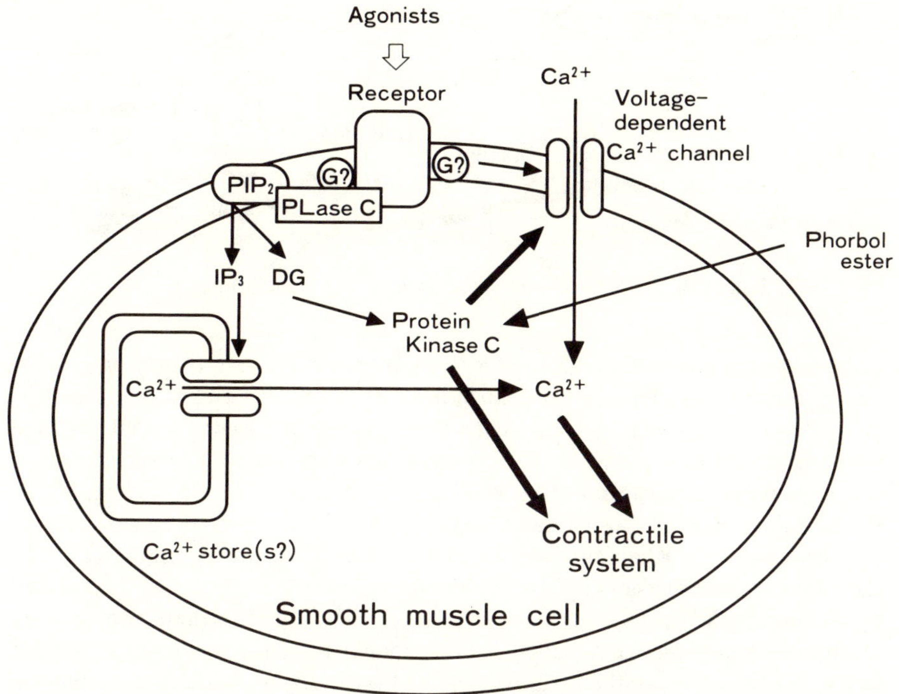

Figure 8.1 Possible roles of protein kinase C in smooth muscle. G: G protein; PIP$_2$: phosphatidylinositol 4,5-bisphosphate; PLase C: phospholipase C; IP$_3$: inositol 1,4,5 triphosphate; DG: diacylglycerol.

ROLES OF PROTEIN KINASE C IN CA^{2+} CHANNEL

Incubation of smooth-muscle strips with phorbol ester, a direct stimulator of protein kinase C, can result in the slow development of tonic force, which is accompanied by increments in Ca^{2+} (Khalil and Van Breemen, 1988). It has been proposed that protein kinase C plays a role in the tonic phase of smooth muscle contraction (Rasmussen et al., 1987). Stimulation of protein kinase C by phorbol ester activates voltage-gated Ca^{2+} channels and causes an increase in Ca^{2+} influx in cultured vascular smooth-muscle cells (Kaczmarek, 1987; Fish et al., 1988; Shearman et al., 1989). Endothelin is a potent stimulator of vascular smooth-muscle contraction, and the contractile response to endothelin is slow in onset and usually long-lasting, like that induced by phorbol ester (Yanagisawa et al., 1988). Endothelin stimulates biphasic accumulation of DG and activates protein kinase C, and also induces a sustained translocation of protein kinase C to the membrane (Griendling et al., 1989; Langlands and Diamond, 1992). Endothelin induces a Ca^{2+} current via opening of L-type channels in freshly dissociated smooth-muscle cells from porcine coronary artery (Goto et al., 1989). Furthermore, the presence of Ca^{2+} channel blockers attenuates both endothelin- and phorbol ester-induced contractions by about 60 percent (Yanagisawa et al., 1988).

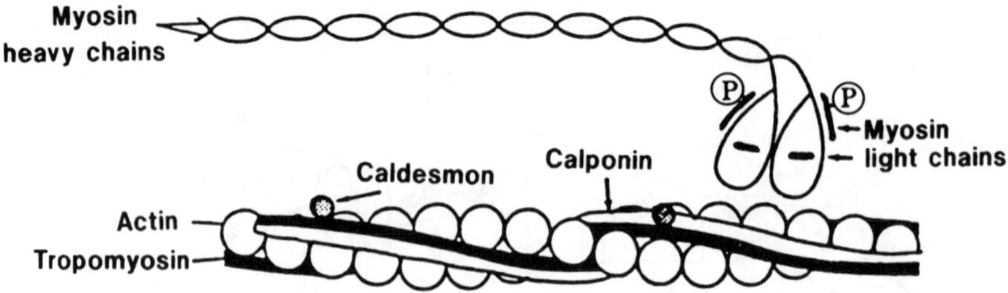

Figure 8.2 Schematic diagram of contractile apparatus in smooth muscle.

Thus, modification of Ca^{2+} channel activity by protein kinase C may be considered a plausible mechanism for controlling Ca^{2+} entry. Several different groups have investigated this possibility, but in some types of vascular smooth muscle no increment was observed in $[Ca^{2+}]_i$ measured with aequorin or fura 2 during the phorbol ester-induced contraction (Jiang and Morgan, 1989; Mori et al., 1990). Furthermore, phorbol esters induce sustained contraction even in the absence of extracellular Ca^{2+} (Khali and Van Breemen, 1988; Jiang and Morgan, 1989). In these cases, phorbol esters induce contraction without changing $[Ca^{2+}]_i$, possibly by increasing Ca^{2+} sensitivity of contractile elements. The discrepancy between the experimental results may reflect the differential distribution of multiple protein kinase C subspecies, or the heterogeneity of L-type channels. Further studies are necessary to elucidate the Ca^{2+} regulatory mechanism of the sustained phase of contraction in smooth muscle.

ROLES OF PROTEIN KINASE C IN CONTRACTILE APPARATUS (Figure 8.2)

Myosin Light-Chain Phosphorylation

Myosin is directly involved in smooth-muscle contraction and is responsible for key functions in non-muscle cells including cytokinesis, capping, phagocytosis, and cell motility (Adelstein and Eisenberg, 1980; Walsh and Hartshorne, 1982; Sellers 1991). Phosphorylation of the 20-kDa light chain (LC_{20}) of myosin by Ca^{2+}/calmodulin-dependent myosin light-chain kinase (MLCK) plays a primary role in the regulation of actomyosin interaction in smooth muscle (Adelstein and Eisenberg, 1980; Walsh and Hartshorne, 1982; Kamm and Stull, 1989; Sellers, 1991; Stull et al., 1991). The regulation of the actomyosin Mg^{2+} ATPase activity of smooth muscle occurs via phosphorylation by MLCK of serine-19 on LC_{20} of myosin. This is the only site on the light chain that phosphorylated when smooth-muscle fibers were stimulated with physiological agonists (Kamm et al., 1989; Singer et al., 1989). The importance of this regulation was clearly demonstrated when single relaxed smooth-muscle cells were microinjected with a Ca^{2+}-independent tryptic fragment of MLCK. The injected cell shortened but there was no increase in sarcoplasmic Ca^{2+} levels, suggesting the phosphorylation of myosin sufficient for contractile activity (Itoh et al., 1989). It is now generally accepted that phosphorylation of myosin LC_{20} by MLCK leads to interaction of actin with myosin (cross-bridge phosphorylation), and the contractile response is initiated. The increases in the levels of phosphorylated LC_{20} are transient, and a low level of LC_{20} phosphorylation is sufficient for steady-state force maintenance in response to contractile stimulus. The force maintenance after LC_{20} dephosphorylation is due to the formation of slowly cycling cross bridges (Hai and Murphy, 1989). Ca^{2+}-stimulated cross-bridge phosphorylation catalyzed by MLCK pathway is both necessary and sufficient for stress development, and maintenance with reduced cross-bridge cycling rates ''latch'' in smooth muscle.

Phosphorylation of LC_{20} at sites other than serine-19 can also modulate the enzymatic activity of myosin. MLCK also catalyzes, at a much slower rate, the phosphorylation of threonine-18 (Ikebe and Hartshorne, 1985; Ikebe et al., 1986). This phosphorylation was shown to increase further the actin-activated Mg^{2+}-ATPase activity of smooth-muscle heavy meromyosin (HMM), the soluble two-headed proteolytic fragment of myosin. Bovine platelet myosin can similarly be phosphorylated at both serine-19 and threonine-18 by MLCK, and phosphorylation also increases actomyosin Mg^{2+}-ATPase activity (Ikebe, 1989). Protein kinase C can phosphorylate serine-1, serine-2, and threonine-9 of LC_{20} of myosin (Sellers, 1991). This phosphorylation has little effect on the Mg^{2+}-ATPase activity of HMM, but does decrease the actin-activated Mg^{2+}-ATPase activity of smooth-muscle HMM that has been phosphorylated on serine-19 by MLCK (Nishikawa et al., 1983, 1984a). The decrease is due to an effect on the K_{app} of the myosin for actin with no effect on the V_{max} (Nishikawa et al., 1984a). Protein kinase C phosphorylation of bovine platelet myosin has an effect on the actin-activated Mg^{2+}-ATPase activity similar to that reported for smooth-muscle HMM (Ikebe and Reardon, 1990a). Protein kinase C phosphorylation also inhibits the conformational change induced by MLCK phosphorylation (Umekawa et al., 1985). In contrast to these findings, the translocation velocity of myosin-coated beads along actin cables in *Nitella* was the same for myosin phosphorylated by MLCK and myosin phosphorylated by both MLCK and protein kinase C (Umemoto et al., 1989). In all the *in vitro* studies of protein kinase C phosphorylation of smooth muscle and non-muscle myosin, both serine-1 (or serine-2) and threonine-19 are phosphorylated. When myosin is phosphorylated by protein kinase C in intact human platelets (Naka et al., 1983; Kawamoto et al., 1989), rat basophilic leukemic 2H3 cells (Ludowyke et al., 1989), or smooth-muscle tissues (Ikebe et al., 1986), only the serine sites are labeled. There is no significant incorporation of phosphate into threonine-9, probably because threonine-9 is readily dephosphorylated by a site-specific phosphatase inside the cell (Ikebe et al., 1986). The phosphatase that selectively dephosphorylates the threonine site of LC_{20} was reported in bovine smooth muscle (Erdodi et al., 1989). This difference in incorporation sites raises the question of whether myosin phosphorylation only at serine-1 or serine-2 possesses the same enzymatic properties as that phosphorylated at both the serine and threonine sites.

The physiological significance of LC_{20} phosphorylation by protein kinase C for contraction of smooth muscle remains unclear. Phosphorylation of LC_{20} by protein kinase C has been demonstrated in certain non-muscle cells in response to antigen and phorbol ester stimulation (Naka et al., 1983; Erdodi et al., 1989; Kawamoto et al., 1989; Ludowyke et al., 1989). The addition of purified protein kinase C, phorbol ester, and phosphatidylserine to skinned contracted vascular smooth muscle resulted in a slow relaxation associated with phosphorylation of LC_{20} in sites phosphorylated by protein kinase C (Inagaki et al., 1987), although the stoichiometry of LC_{20} phosphorylation was not determined. However, phorbol esters have been shown to cause a slow, sustained contraction of various type of smooth muscle (Kamm and Stull, 1989; Stull et al., 1991), associated with a low level of LC_{20} phosphorylation (Singer and Baker, 1987). Phosphorylated LC_{20} (0.05 mol PO_4/mol LC_{20}) from intact vascular smooth muscle activated by phorbol dibutyrate contains approximately equal amounts of phosphoprotein phosphorylated by protein kinase C and MLCK (Singer and Baker, 1987; Sutton and Maeberle, 1990). Stoichiometric phosphorylation (>1.0 mol PO_4/mol LC_{20}) of LC_{20} (at serine-1 and serine-2 sites) by protein kinase C and oleic acid in glycerinated arterial smooth muscle does not activate nor does it significantly modify the contraction of smooth muscle that occurs in response to LC_{20} phosphorylation at serine-19 by MLCK (Sutton and Haeberle, 1990). These data suggest that the low levels of LC_{20} phosphorylation by protein kinase C measured in the intact smooth-muscle tissues may not play a role in the regulation of the phorbol ester-induced contraction. Protein kinase C activation may increase sensitivity of the contractile apparatus to Ca^{2+}. A second regulatory process, other than LC_{20} phosphorylation, may play a role in the regulation of contraction, particularly in phorbol ester-activated muscles.

Multiple sites of phosphorylation of LC_{20} in response to agonists have been recently investigated in smooth- and non-muscle cells (Colburn et al., 1988; Sasaki et al., 1990; Seto et al., 1990a, b). Multiple-site phosphorylation in contracting smooth-muscle tissues or cells is the result of both MLCK and protein kinase C activities or MLCK activity alone. Colburn and co-workers reported that stimulation of tracheal smooth muscle with a high concentration of carbachol results in the formation of both monophosphorylated ($\sim$60% of total LC_{20}) and diphosphorylated ($\sim$10% of total LC_{20}) LC_{20} of myosin, and that LC_{20} of myosin may be diphosphorylated by MLCK at the serine and threonine residues but not by protein kinase (Colburn et al., 1988). Monophosphorylated LC_{20} contained phosphate in serine-19, whereas diphosphorylated LC_{20} was phosphorylated in both serine-19 and threonine-18 (Colburn et al., 1988). Diphosphorylation of LC_{20} of myosin was reported in arterial smooth-muscle tissues and cells, and the amount of diphosphorylated forms is also substantially less than the amount of monophosphorylated forms (Sasaki et al., 1990; Seto et al., 1990a). Prostaglandin F_{2a} induces both monophosphorylation and diphosphorylation of LC_{20} of myosin, whereas K^+ and histamine induce only monophosphorylation in rabbit thoracic arterial smooth muscle (Seto et al., 1990b). The prostaglandin F_{2a}-induced diphosphorylation of LC_{20} is significantly decreased by the treatment of the protein kinase C inhibitor staurosporine, while staurosporine has little effect on the monophosphorylation activity (Seto et al., 1990b). Thus, multiple-site phosphorylation of LC_{20} is probably involved in the regulation of smooth-muscle contraction. Although the diphosphorylated LC_{20} form induced by certain agonists may modify the mode of contraction, the physiological significance of low levels of diphosphorylated LC_{20} is not apparent at present.

The pattern and consequence of the protein kinase C-catalyzed phosphorylation of embryonic myosin are remarkably different from those of adult smooth-muscle myosin (deLanerolle and Nishikawa, 1988). Protein kinase C phosphorylates three sites on the fetal myosin 20-kDa light chain including a serine and threonine residue on the same peptide phosphorylated by MLCK. Phosphorylation by protein kinase C stimulates the actomyosin ATPase activity of fetal myosin, unlike adult myosin, indicating that protein kinase C and MLCK phosphorylate a common site. This may suggest the possibility of alternate pathway for activating fetal smooth-muscle myosin.

Myosin Heavy-Chain Phosphorylation

The phosphorylation of myosin heavy chain has been described for myosins from vertebrate and invertebrate non-muscle and smooth-muscle cells (Kuznicki, 1986; Kawamoto and Adelstein, 1988). In some of these cells, phosphorylation of myosin heavy chains has been shown to affect the interaction of myosin with actin as well as the self-assembly of myosin. Phosphorylation of myosin heavy chain is catalyzed by casein kinase II, Ca^{2+}/calmodulin-dependent protein kinase, and protein kinase C (Conti et al., 1991). In addition to phosphorylating LC_{20} of myosin in vertebrate non-muscle and smooth-muscle cells, protein kinase C has also been shown to phosphorylate the 196-kDa myosin heavy chain in non-muscle cells both *in situ* and *in vitro* (Kawamoto et al., 1989; Conti et al., 1991). The stoichiometry of phosphorylation was 1.0 mol of phosphate per 1 mol of heavy chain. Vertebrate non-muscle myosin contains a single serine residue that can be phosphorylated by protein kinase C near the carboxy terminal end of the α-helix of the myosin rod, and this residue is notably absent in all vertebrate smooth-muscle heavy chain (both 204 and 200 kDa) (Conti et al., 1991). It has been reported that the phosphorylation of myosin heavy chain as well as the 20-kDa light chain of myosin by protein kinase C occurs in intact platelets in response to phorbol ester (Kawamoto et al., 1989). Thus, protein kinase C phosphorylates a serine residue in a concensus sequence present in the vertebrate non-muscle myosin heavy chains, but not in the 204- or 200-kDa smooth-muscle heavy chains. Casein kinase II can phosphorylate both vertebrate non-muscle and smooth-muscle heavy chains. It is not yet clear whether phosphorylation of myosin heavy chain is linked to specific contractile properties in smooth-muscle tissues.

Phosphorylation of Myosin Light-Chain Kinase

MLCK from smooth-muscle and non-muscle cells can be phosphorylated by a number of protein kinases including cAMP- and cGMP-dependent protein kinases (Adelstein and Klee, 1980; Conti and Adelstein, 1980; deLanerolle et al., 1984; Nishikawa et al., 1984b), protein kinase C, (Ikebe et al., 1985; Nishikawa et al., 1985), Ca^{2+}/calmodulin-dependent protein kinase II (Ikebe and Reardon, 1990b;

Hashimoto and Soderling 1990), *in vitro*. Phosphorylation of MLCK by cAMP-dependent protein kinase was originally reported by Adelstein and colleagues. (Adelstein and Klee, 1980; Conti and Adelstein, 1980; deLanerolle et al., 1984; Nishikawa et al., 1984b). In the absence of Ca^{2+}/calmodulin, cAMP-dependent protein kinase phosphorylates purified MLCK at two sites (A and B), whereas in the presence of Ca^{2+}/calmodulin, only one site (B) is phosphorylated by this kinase (Adelstein and Klee, 1980; Conti and Adelstein, 1980; deLanerolle et al., 1984; Nishikawa et al., 1984b). Phosphorylation of site A, which is near the calmodulin-binding domain of MLCK, decreases the affinity of MLCK for Ca^{2+}/calmodulin required for MLCK activity. The cellular effect of this phosphorylation is to increase the concentration of Ca^{2+} required for MLCK activity in intact smooth-muscle cells. This system down-modulates the Ca^{2+} signal, and thus may play an important role in the regulation of the Ca^{2+} sensitivity of myosin light-chain phosphorylation. Purified MLCK is phosphorylated at site A by protein kinase C (Ikebe et al., 1985; Nishikawa et al., 1985). Physiological importance of MLCK phosphorylation by protein kinase C remains unknown in intact smooth muscle. However, a report (Stull et al., 1990) has shown that treatment of tracheal strips with a phorbol ester does not result in significant phosphorylation of site A.

Regulation by Thin-Filament Binding Proteins

A great deal of evidence exists supporting the central role of myosin phosphorylation in the regulation of smooth-muscle contraction (Adelstein and Eisenberg, 1980; Walsh and Hartshorne, 1982; Kamm and Stull, 1989). However, regulation by the phosphorylation of LC_{20} does not explain all aspects of the contractile functions of smooth muscle, in particular, the tonic contractile responses. Therefore, additional regulatory mechanisms have been postulated. The regulation of actomyosin interaction may also involve thin-filament binding proteins such as tropomyosin, caldesmon, and calponin (Sobue et al., 1981; Takahashi et al., 1986; Winder and Walsh, 1990; Sobue and Sellers, 1991). A physiological role for Ca^{2+} control of the thin filaments in smooth muscle has not yet been demonstrated. Nevertheless, *in vitro,* both native thin filaments and synthetic systems consisting of actin, tropomyosin, caldesmon, and calmodulin at ratios similar to native thin filaments are effectively regulated by Ca^{2+} in a manner formally analogous to the regulatory function of troponin in striate muscle. Caldesmon and calponin may function to regulate the contractile state of smooth muscle. Both proteins inhibit the actomyosin ATPase and bind F-actin and tropomyosin in a Ca^{2+}-independent manner and calmodulin in a Ca^{2+}-dependent manner. The two proteins are substrates of protein kinase C and lose their activity to inhibit the actomyosin ATPase on phosphorylation. The tissue content of calponin, equal to that of tropomyosin, is significantly higher than that of caldesmon. However, the functional relationship between caldesmon and calponin is yet unknown.

Caldesmon was originally purified from chicken gizzard smooth muscle and identified as a calmodulin-binding protein that interacts with actin (Sobue et al., 1981). Sobue and colleagues found that caldesmon bound calmodulin in the presence of Ca^{2+} and bound to filamentous actin (F-actin) in the absence of Ca^{2+}, and thereby proposed a flip-flop mechanism for Ca^{2+} regulation of caldesmon binding to thin filaments (Sobue et al., 1981; Sobue and Sellers, 1991). Caldesmon has been classified into two categories on the basis of differences in apparent molecular weight; the higher molecular mass form (h-Caldesmon, 120–150kDa) appears to be found predominantly in smooth-muscle tissue, while the lower molecular mass form (1-caldesmon, 70–80 kDa) is present in non-muscle tissue. There are considerable differences in the caldesmon content of different types of smooth muscle, with the phasic muscle, in general, containing higher concentrations than tonic smooth muscle (Haeberle et al., 1992). The calmodulin- and actin-binding peptide GS17C, which contains the residues from Gly^{651} to Ser^{667} of the caldesmon sequence plus added cysteine, has been reported to induce contraction in vascular smooth-muscle cells of ferret aorta (Katsuyama et al., 1992). The contraction appears to be evoked by the blockage of the inhibitory action of endogenous caldesmon. Caldesmon is a very potent inhibitor of actomyosin superprecipitation and actomyosin ATPase activity, and Ca^{2+}/calmodulin can relieve this inhibition. Tropomyosin does not affect the stoichiometry of caldesmon to actin but does increase the affinity about two- to four-fold. Several investigators have suggested that caldesmon may be involved in latch-bridge formation—the formation of slowly cycling or noncycling cross bridges that are responsible for force maintenance at low levels of ATP consumption and in the presence of intermediate Ca^{2+} concentrations (Hai and Murphy, 1989; Haeberle et al., 1992). Caldesmon phosphorylation may provide an additional step in the regulation of thin-filament function. Caldesmon is phosphorylated by at least five protein kinases: protein kinase C, Ca^{2+}/calmodulin-dependent protein kinase II, casein kinase II, the cdc2 protein kinase, and cAMP-dependent protein kinase (Sobue and Sellers, 1991). Protein kinase C incorporates more than 3 mol of phosphate per mol of caldesmon into the carboxyl terminal 35-kDa domain (Sobue and Sellers, 1991, Katsuyama et al., 1992). This phosphorylation reduces affinity of caldesmon or its fragment to both calmodulin and F-actin, and subsequently the inhibition of the actin-activated Mg^{2+}-ATPase activity of myosin by caldesmon is reduced (Tanaka et al., 1990). Caldesmon is also phosphorylated in vivo. Treatment of intact platelets with phorbol ester results in an increased incorporation of phosphate into caldesmon (Hettasch and Sellers, 1991). However, platelet activation induced by the physiological agonists, thrombin and collagen, is not correlated with an increase in caldesmon phosphorylation (Hettasch and Sellers, 1991). Caldesmon is phosphorylated when porcine carotid arteries are stimulated with various agonists, but the increase responsible for this phosphorylation is not known (Adam et al., 1989). Moreover, it is known that in smooth muscles, caldesmon binds to the myosin as well as to actin; this bind results in cross-linking of thick and thin filaments (Hemic and Chalovich, 1990) and thus increases the probability that myosin and actin remain associated when inhibited by caldesmon.

Another smooth-muscle protein, calponin, has been described that may function to regulate the contractile state of the muscle (Takahashi et al., 1986; Winder and Walsh, 1990). Calponin, Ca^{2+}- and calmodulin-binding troponin-T-like protein, is a heat-stable, basic, 34-kDa protein that interacts with F-actin and tropomyosin in a Ca^{2+}-independent and with calmodulin in a Ca^{2+}-dependent manner. It is present in smooth muscle at the same molar concentration as tropomyosin (Takahashi et al., 1986). Calponin is clearly a distinct protein from caldesmon and MLCK, but it is antigenically related to the C-terminal half of rabbit skeletal and bovine cardiac troponin T. Calponin inhibits the actin-activated Mg^{2+} ATPase activity of fully phosphorylated myosin of smooth muscle. This inhibition does not require Ca^{2+}, although calponin is a Ca^{2+}-binding protein (Takahashi et al., 1986; Winder and Walsh, 1990). Calponin is phosphorylated *in vitro* by protein kinase C and Ca^{2+}/calmodulin-dependent protein kinase II (Winder and Walsh, 1990; Haeberle et al., 1992). Phosphorylation of calponin by either kinase results in loss of its ability to inhibit the actomyosin ATPase due to the reduction of actin-binding capacity. F-actin and tropomyosin as well as calmodulin decrease the rate of phosphorylation of calponin by protein kinase C (Winder and Walsh, 1990; Nakamura et al., 1993). Calponin phosphorylation in response to the contraction of smooth muscle by carbachol has been reported in tracheal smooth muscle (Pohl et al., 1991).

THE ROLE OF PROTEIN KINASE C IN INTERMEDIATE FILAMENTS

Cytoskeletal and contractile proteins in smooth-muscle cells appear to be distributed in two domains: (1) longitudinal intermediate filaments free from myosin, but containing filamin, actin, α-actinin, vinculin, and desmin, and (2) contractile protein elements containing actin, myosin, tropomyosin, caldesmon, and calponin (Small et al., 1986). It has been suggested that the intermediate filament domain may be involved in stress maintenance or in tonic contractions, whereas the contractile protein domain would be involved in initiation of contraction (Small et al., 1986). One way of modulating the intermediate filament domain is through protein phosphorylation/dephosphorylation reactions. Phosphorylation of purified intermediate filament proteins affects assembly properties. Desmin phosphorylated by cAMP-dependent protein kinase or protein kinase C is not able to polimerize (Geisler and Weber, 1988; Inagaki et al., 1988). Carbachol causes a transient phosphorylation of LC_{20} of myosin in tracheal smooth muscle, whereas desmin, synemin, and caldesmon are phosphorylated at a slower rate and maintained for a long time (Park and Rasmussen, 1986). Phorbol ester also causes a slow sustained contraction that was associated with phosphorylation of desmin and synemin as well as caldesmon (Rasmussen et al., 1987). Thus, phosphorylation of these proteins could allow for dynamic regulation of assembly and turnover of intermediate filaments. It is not clear whether intermediate filaments play a functional role in smooth-muscle contraction.

CONCLUDING REMARKS

Ca^{2+} regulation of the contractile elements of smooth muscle is generally agreed to be via specific phosphorylation of LC_{20} of myosin by Ca^{2+}/calmodulin-dependent MLCK. Although protein kinase C may also have regulatory roles in the function of smooth muscle through the phosphorylation of multiple substrates, the primary mechanism is yet unknown.

REFERENCES

Adam, L. P., J. R. Haeberle, and D. R. Hathaway. 1989. Phosphorylation of caldesmon in arterial smooth muscle. *J. Biol. Chem.* 264:7698–7703.

Adelstein, R. S., and E. Eisenberg. 1980. Regulation and kinetics of the actin-myosin-ATP interaction. *Annu. Rev. Biochem.* 49:921–956.

Adelstein, R. S., and C. B. Klee. 1980. Smooth muscle myosin light chain kinase. *Calc. Cell Func.* 1:167–182.

Colburn, J. C., C. H. Michnoff, L. -C. Hsu, C. A. Slaughter, K. E. Kamm, and J. T. Stull. 1988. Sites phosphorylated in myosin light chain in contracting smooth muscle. *J. Biol. Chem.* 263:19166–19173.

Conti, M. A., and R. S. Adelstein. 1980. The relationship between calmodulin binding and phosphorylation of smooth muscle myosin kinase by the catalytic subunit of 3':5' cAMP-dependent protein kinase. *J. Biol. Chem.* 256:3178–3181.

Conti, M. A., J. R. Sellers, R. S. Adelstein, and M. Elzinga. 1991. Identification of the serine residue phosphorylated by protein kinase C in vertebrate nonmuscle myosin heavy chains. *Biochemistry* 30:966–970.

Denthuluri, N. R., and T. A. Brock. 1990. Endothelin receptor-coupling mechanisms in vascular smooth muscle: A role for protein kinase C. *J. Pharmacol. Exp. Ther.* 254:393–399.

Erdodi, F., A. Rokolya, M. Barany, and K. Barany. 1989. Dephosphorylation of distinct sites in myosin light chain by two types of phosphatase in aortic smooth muscle. *Biochem. Biophys. Acta* 1011:67–74.

Fish, D. R., G. Sperti, W. S. Colucci, and D. E. Clapham. 1988. Phorbol ester increases the dihydropyridine-sensitive calcium conductance in vascular smooth muscle cell line. *Circ. Res.* 62:1049–1054.

Geisler, N., and K. Weber. 1988. Phosphorylation of desmin in vitro inhibits formation of intermediate filaments: Identification of three kinase A sites in the aminoterminal head domain. *EMBO J.* 7:15–20.

Goto, K., Y. Kusuya, N. Matsuki, Y. Takuwa, H. Kurihara, T. Ishikawa, S. Kimura, M. Yanagisawa, and T. Masaki. 1989. Endothelin activates the dihydropyridine-sensitive, voltage-dependent Ca^{2+} channel in vascular smooth muscle. *Proc. Natl. Acad. Sci. U.S.A.* 86:3915–3918.

Griendling, K. K., S. E. Rittenhouse, T. A. Brock, L. S. Ekstein, Jr M. A. Gimbrone, and R. W. Alexander. 1986. Sustained diacylglycerol formation from inositol phospholipids in angiotensin II-stimulated vascular smooth muscle cells. *J. Biol. Chem.* 261:5901–5906.

Griendling, K. K., T. Tsuda, and R. W. Alexander. 1989. Endothelin stimulates diacylglycerol accumulation and activates/protein kinase C in cultured vascular smooth muscle cells. *J. Biol. Chem.* 264:8237–8240.

Haeberle, J. R., D. R. Hathaway, and C. L. Smith. 1992. Caldesmon content of mammalian smooth muscles. *J. Mus. Res. Cell Motil.* 13:81–89.

Hai, C.-M., and R. A. Murphy. 1989. Ca^{2+}, cross-bridge phosphorylation, and contraction. *Annu. Rev. Physiol.* 51:285–298.

Hashimoto, Y., and T. R. Soderling. 1990. Phosphorylation of smooth muscle myosin light chain kinase by Ca^{2+}/calmodulin-dependent protein kinase II. *Arch. Biochem. Biophys.* 278:41–45.

Hemic, M. E., and J. M. Chalovich. 1990. Characterization of caldesmon binding to myosin. *J. Biol. Chem.* 265:19672–19678.

Hettasch, J. M., and J. R. Sellers. 1991. Caldesmon phosphorylation in intact human platelets by cAMP-dependent protein kinase and protein kinase C. *J. Biol. Chem.* 266:11876–11881.

Inagaki, M., H. Yokokura, T. Itoh, Y. Kanmura, H. Kuriyama, and H. Hidaka. 1987. Purified rabbit protein kinase C relaxed skinned vascular smooth muscle and phosphorylated myosin light chain. *Arch. Biochem. Biophys.* 254:136–141.

Inagaki, M., Y. Gonda, M. Matsuyama, K. Nishikawa, Y. Nishi, and C. Sato. 1988. Intermediate filament reconstitution in vitro. The role of phosphorylation on the assembly-disassembly of desmin. *J. Biol. Chem.* 263:5970–5978.

Ikebe, M. 1989. Phosphorylation of a second site for myosin light chain kinase on platelet myosin. *Biochemistry* 28:8750–8755.

Ikebe, M., and D. J. Hartshorne. 1985. Phosphorylation of smooth muscle myosin at two distinct sites by myosin light chain kinase. *J. Biol. Chem.* 260:10027–10031.

Ikebe, M., and S. Reardon. 1990a. Phosphorylation of bovine platelet myosin by protein kinase C. *Biochemistry* 29:2713–2720.

Ikebe, M., and S. Reardon. 1990b. Phosphorylation of smooth muscle light chain kinase by smooth muscle Ca^{2+}/calmodulin-dependent multifunctional protein kinase. *J. Biol. Chem.* 265:8975–8978.

Ikebe, M., M. Inagaki, K. Kanamaru, and H. Hidaka. 1985. Phosphorylation of smooth muscle myosin light chain kinase by Ca^{2+}-activated, phospholipid-dependent protein kinase. *J. Biol. Chem.* 260:4547–4550.

Ikebe, M., D. J. Hartshorne, and M. Elzinga. 1986. Identification, phosphorylation, dephosphorylation of a second site for myosin light chain kinase on the 20,000-dalton light chain of smooth muscle myosin. *J. Biol. Chem.* 261:36–39.

Itoh, I., M. Ikebe, G. J. Kargacin, D. J. Hartshorne, B. E. Kemp, and F. S. Fay. 1989. Effects of modulators of myosin light chain kinase activity in single smooth muscle cells. *Nature* 338:164–167.

Jiang, M. J., and K. G. Morgan. 1989. Agonist-specific myosin phosphorylation and intracellular calcium during isometric contractions of arterial smooth muscle. *Eur. J. Pharmacol.* 413:637–643.

Kaczmarek, L. K. 1987. The role of protein kinase C in the regulation of ion channels and neurotransmitter release. *Trends Neurosci.* 10:30–34.

Kamm, K. E., and J. T. Stull. 1989. Regulation of smooth muscle contractile elements by second messengers. *Annu. Rev. Physiol.* 51:299–313.

Kamm, K. E., L.-C. Hsu, Y. Kubota, and J. T. Stull. 1989. Phosphorylation of smooth muscle myosin heavy and light chains. Effects of phorbol dibutyrate and agonists. *J. Biol. Chem.* 264:21223–21229.

Katsuyama, H., C.-L. A. Wang, and K. G. Morgan. 1992. Regulation of vascular smooth muscle tone by caldesmon. *J. Biol. Chem.* 267:14555–14558.

Kawamoto, S., and R. S. Adelstein. 1988. The heavy chain of smooth muscle myosin is phosphorylated in aorta cells. *J. Biol. Chem.* 263:1099–1102.

Kawamoto, S., A. R. Bengur, J. R. Sellers, and R. S. Adelstein. 1989. In situ phosphorylation of human platelet myosin heavy and light chains by protein kinase C. *J. Biol. Chem.* 264:2258–2265.

Khalil, R., and C. Van Breemen. 1988. Sustained contraction of vascular smooth muscle: Calcium influx or C kinase activation? *J. Pharmacol. Exp. Ther.* 244:537–542.

Kuznicki, J. 1986. Phosphorylation of myosin in nonmuscle and smooth muscle cells. *FEBS Lett.* 204:169–176.

deLanerolle, P., and M. Nishikawa. 1988. Regulation of embryonic smooth muscle myosin by protein kinase C. *J. Biol. Chem.* 263:9071–9074.

deLanerolle, P., M. Nishikawa, D. A. Yost, and R. S. Adelstein. 1984. Increased phosphorylation of myosin light chain kinase after an increase in cyclic AMP in intact smooth muscle. *Science* 223:1415–1417.

Langlands, J. M., and J. Diamond. 1992. Translocation of protein kinase C in bovine tracheal smooth muscle strips: The effect of methacholine and isoprenaline. *Eur. J. Pharmacol.* 227:131–138.

Ludowyke, R. I., I. Peleg, M. A. Beaven, and R. S. Adelstein. 1989. Antigen-induced secretion of histamine and the phosphorylation of myosin by protein kinase C in rat basophilic leukemia cells. *J. Biol. Chem.* 264:12492–12501.

Mori, T., T. Yanagisawa, and N. Taira. 1990. Phorbol 12, 13-dibutyrate increases vascular tone but has a dual action on intracellular calcium levels in porcine coronary artery. *Naunyn-Schmiedeberg's Arch. Pharmacol.* 341:251–255.

Naka, M., M. Nishikawa, R. S. Adelstein, and H. Hidaka. 1983. Phorbol ester-induced activation of human platelets is associated with protein kinase C phosphorylation of myosin light chains. *Nature* 306:490–492.

Nakamura, F., T. Mino, J. Yamamoto, M. Naka, and T. Tanaka. 1993. Identification of the regulatory site in smooth muscle calponin that is phosphorylated by protein kinase C. *J. Biol. Chem,* in press.

Nishizuka, Y. 1992. Intracellular signaling by hydrolysis of phospholipids and activation of protein kinase C. *Science* 258:607–614.

Nishikawa, M., H. Hidaka, and R. S. Adelstein. 1983. Phosphorylation of smooth muscle heavy meromyosin by calcium-activated phospholipid-dependent protein kinase. *J. Biol. Chem.* 258:14069–14072.

Nishikawa, M., J. R. Sellers, R. S. Adelstein, and H. Hidaka. 1984a. Protein kinase C modulates in vitro phosphorylation of the smooth muscle heavy meromyosin by myosin light chain kinase. *J. Biol. Chem.* 259:8808–8814.

Nishikawa, M., P. deLanerolle, T. M. Lincoln, and R. S. Adelstein. 1984b. Phosphorylation of mammalian myosin light chain kinases by the catalytic subunit of cAMP-dependent protein kinase and cGMP-dependent protein kinase. *J. Biol. Chem.* 259:8429–8436.

Nishikawa, M., S. Shirakawa, and R. S. Adelstein. 1985. Phosphorylation of smooth muscle myosin light chain kinase by protein kinase C. Comparative study of the phosphorylated site. *J. Biol. Chem.* 260:8978–8983.

Park, S., and H. Rasmussen. 1986. Carbachol-induced phosphorylation changes in bovine tracheal smooth muscle. *J. Biol. Chem.* 261:15734–15739.

Pohl, J., M. P. Walsh, and W. T. Gerthoffer. 1991. Calponin and caldesmon phosphorylation in canine tracheal smooth muscle. *Biophys. J.* 59:58a.

Rasmussen, H., Y. Takuwa, and S. Park. 1987. Protein kinase C in the regulation of smooth muscle contraction. *FASEB J.* 1:177–185.

Sasaki, Y., M. Seto, and K. Komatsu. 1990. Diphosphorylation of myosin light chain in smooth muscle cells in culture. Possible involvement of protein kinase C. *FEBS Lett.* 276:161–164.

Sellers, J. R. 1991. Regulation of cytoplasmic and smooth myosin. *Curr. Opin. Cell Biol.* 3:98–104.

Seto, M., Y. Sasaki, and Y. Sasaki. 1990a. Alternation in the myosin phosphorylation pattern of smooth muscle by phorbol ester. *Am. J. Pharmacol.* 259:C769–C774.

Seto, M., Y. Sasaki, and Y. Sasaki. 1990b. Stimulus-specific patterns of myosin light chain phosphorylation in smooth muscle of rabbit thoracic artery. *Eur. J. Physiol.* 415: 484–489.

Shearman, M. S., K. Sekiguchi, and Y. Nishizuka. 1989. Modulation of ion channel activity: A key function of the protein kinase C enzyme family. *Pharmacol. Rev.* 41: 210–237.

Singer, H. A., and K. M. Baker. 1987. Calcium dependence of phorbol 12,13-dibutyrate-induced force and myosin light chain phosphorylation in arterial smooth muscle. *J. Pharmacol. Exp. Ther.* 243:814–821.

Singer, H. A., J. W. Oren, and H. A. Benscoter. 1989. Myosin light chain phosphorylation in ^{32}P-labelled rabbit aorta stimulated by phorbol 12,13-dibutyrate and phenylephrine. *J. Biol. Chem.* 264:21215–21222.

Small, J. V., D. O. Furst, and J. DeMay. 1986. Localization of filamin in smooth muscle. *J. Cell. Biol.* 102:210–220.

Sobue, K., and J. R. Sellers. 1991. Caldesmon, a novel regulatory protein in smooth muscle and nonmuscle actomyosin system. *J. Biol. Chem.* 266:12115–12118.

Sobue, K., Y. Muramoto, M. Fujita, and S. Kakiuchi. 1981. Purification of a calmodulin-binding protein from chicken gizzard that interacts with F-actin. *Proc. Natl. Acad. Sci. U.S.A.* 78:5652–5655.

Stull, J. T., L. -C. Hsu, M. G. Tansey, and K. E. Kamm. 1990. Myosin light chain kinase phosphorylation in trachea smooth muscle. *J. Biol. Chem.* 265:16683–16690.

Stull, J. T., P. J. Gallagher, B. P. Herring, and K. E. Kamm. 1991. Vascular smooth muscle contractile elements. *Hypertension* 17:723–732.

Sutton, T. A., and J. R. Haeberle. 1990. Phosphorylation by protein kinase C of the 20,000-dalton light chain of myosin in intact and chemically skinned vascular smooth muscle. *J. Biol. Chem.* 265:2749–2754.

Takahashi, K., K. Hiwada, and T. Kokubo. 1986. Isolation and characterization of a 34000-dalton calmodulin- and F-actin-binding protein from chicken gizzard smooth muscle. *Biochem. Biophys. Res. Commun.* 141:20–26.

Tanaka, T., H. Ohta, K. Kanda, T. Tanaka, H. Hidaka, and K. Sobue. 1990. Phosphorylation of high-Mr caldesmon by protein kinase C modulates the regulation of this protein on the interaction between actin and myosin. *Eur. J. Biochem.* 188:490–500.

Umekawa, H., M. Naka, M. Inagaki, H. Onishi, T. Wakabayashi, and H. Hidaka. 1985. Conformational studies of myosin phosphorylated by protein kinase C. *J. Biol. Chem.* 260:9833–9837.

Umemoto, S., A. R. Bengur, and J. R. Sellers. 1989. Effect of multiple phosphorylation of smooth muscle and cytoplasmic myosins on movement in an in vitro motility assay. *J. Biol. Chem.* 264:1431–1436.

Walsh, M. P., and D. J. Hartshorne. 1982. Actomyosin of smooth muscle. *Calc. Cell Func.* 3:223–269.

Winder, S. J., and M. P. Walsh. 1990. Smooth muscle calponin. *J. Biol. Chem.* 265:10148–10155.

Yanagisawa, M., H. Kurihara, S. Kimura, Y. Tomobe, M. Kobayashi, Y. Mitsui, Y. Yazaki, K. Goto, and T. Masaki. 1988. A novel potent vasoconstrictor peptide produced by vascular endothelial cells. *Nature* 332:411–415.

9

Protein Kinase C in the Heart

MICHEL PUCÉAT
JOAN HELLER BROWN

Extracellular hormones or neurotransmitters exert their cellular effects by eliciting the formation of intracellular second messengers, such as cyclic AMP, cyclic GMP, and diacylglycerol (DAG). These molecules directly or indirectly activate protein kinases, leading to the phosphorylation of specific cellular proteins. In cardiac tissue, phosphorylations induced by extracellular stimuli have been shown to regulate various steps in excitation-contraction coupling and in turn to affect cardiac rhythm and contractility. More recently, phosphorylation events have been implicated in the control of cardiac gene expression.

Agonists that activate the cyclic AMP-dependent protein kinase (PKA), for example β-adrenergic receptor agonists and glucagon, exert a positive inotropic effect. An extensive study of this signal transduction pathway has identified the main cardiac targets of the cyclic AMP-dependent protein kinase. There is now general agreement that the positive inotropic effect is mainly due to an increase in intracellular calcium following the PKA-mediated phosphorylation of the voltage-sensitive Ca^{2+} channel. Agonists that stimulate phosphoinositide turnover, for example α_1-adrenergic, purinergic receptor agonists or endothelin, also trigger a positive inotropic effect. In addition to their acute effects, these agonists can induce cardiac cell hypertrophy. Protein kinase C has been suggested to play a role in both these responses, but the biochemical basis for this response is unclear despite considerable advances in knowledge over the last decade. The physiological substrates for PKC-mediated phosphorylation in cardiac muscle are still not known. The presence of PKC isozymes adds further complexity to analysis.

This chapter summarizes progress and provides an update concerning the potential regulatory roles of PKC in cardiac function.

PRESENCE AND CHARACTERIZATION OF DIFFERENT PKC ISOFORMS

In 1979, Takai and colleagues described the presence of a novel protein kinase in brain, the activity of which was dependent both on Ca^{2+} and phospholipids.

Subsequently, Kuo and co-workers (1980) found that this kinase activity occurred in most tissues including the heart, and throughout the animal kingdom. Katoh and colleagues (1981) further characterized this protein kinase in cardiac tissue. The activity of the Ca^{2+}_ and phospholipid-dependent kinase is low in heart compared with its activity in other tissues like lung and brain, but bovine heart displays twice the activity of myocardial preparations from other species. Wise and co-workers (1982) chose the bovine heart to analyze in more detail the biochemical properties of the kinase. The bovine cardiac enzyme migrates in SDS-PAGE (sodium dodecyl sulfonate polyacrylamide gel electrophoresis) with a molecular mass of 83.5 kDa and in isoelectrofocusing with an isoelectric point of 5.2 to 5.8. Its activation depends both on Ca^{2+} (K_a 35 μM) and phospholipids (K_a 10 μM) and it is highly dependent on Mg^{2+}. The kinase has a K_m for ATP of 4 μM and a pH optimum of 6.5.

The enzyme that had initially been purified as a single protein was later shown to exist as multiple isoforms. First described in brain, the different isoforms could be separated by chromatography into types I, II, and III, which were subsequently shown to be encoded by different genes (γ, β, α respectively; see Chapters 1, 2, and 3). Using specific antibodies raised against the different brain isoforms, Kosaka and co-workers (1988) investigated the presence of these isoforms in various rat tissues including the heart. These authors found that whole cardiac homogenate contained both the type II and type III enzyme. Type III was the most abundant cardiac isoform. The cardiac type II enzyme appeared to be immunologically and biochemically different from brain type II since it was not recognized by an antibody raised against brain type II PKC and was only slightly activated by diacylglycerol in the presence of phosphatidylserine. The authors did not exclude the possibility that this isoform was one of the novel Ca^{2+}-insensitive isoforms.

Allen and Katz (1991) further extended these findings by investigating the biochemical characteristics of the different bovine cardiac isoforms. Using hydroxylapatite column chromatography, they were able to separate two peaks of kinase activity that were immunologically identified as the β_2 (II) and α (III) subtypes. Type III was the most abundant isoform. Although both peaks of kinase activity were dependent on calcium, the activation of type II displayed a lesser dependence on this ion than type III, when assayed in the presence of both phosphatidylserine and diacylglycerol. The authors did not totally exclude the possibility that co-purification of a Ca^{2+}-insensitive PKC isoform, for example PKC ϵ, might account for the lower Ca^{2+} dependency of what appeared to be type II kinase. Both isoforms were activated by DAG in the presence of phospholipids, phosphatidylserine being the most effective lipid ($K_{0.5}$ 10 μg/ml at 10 μM Ca^{2+}; 40 μg/ml at 1 μM Ca^{2+}). Both cardiac isoforms were also stimulated by unsaturated fatty acids. In this case, phospholipids were no longer required but the kinase required a much higher calcium concentration to be activated. It is interesting to note that arachidonic acid was almost as potent as oleic acid, the most potent fatty acid activator of the kinase. This could be of physiological significance, since arachidonic acid is generated by activation of phospholipase A_2, (PLA$_2$), which is often induced by the same agonists that stimulate phosphoinositol (PI) turnover in various tissues (for review see Billah and Anthes, 1990). Thus an interesting possibility is that arachidonic acid might directly activate a cytosolic PKC leading to the phosphorylation of its cytosolic substrates.

Qu and colleagues (1991) also purified by hydroxylapatite chromatography three PKC isozymes from rat heart recognized by specific monoclonal antibodies. By contrast to previous studies (Kosaka et al., 1988; Allen and Katz, 1991), they found that cardiac tissue contained type I as well as types II and III. Moreover, type II appeared as the most abundant isoform. The discrepancy between this and other published work could be related to the different enzyme extraction and purification methods, as well as to differences in the specific antibodies used by each laboratory.

In addition to the studies done on whole heart, several laboratories have recently confirmed the presence of the α and β isoforms of PKC in purified preparations of cardiac myocytes. Using neonatal rat ventricular cells, Mochly-Rosen and co-workers (1990) detected both the α and β isoforms by immuno-cytochemistry and Western analysis. The β isoform was detected using an anti-body suggested to be specific for this isoform in brain (see Mochly-Rosen et al., 1987). In our recent studies examining neonatal and adult rat ventricular cells (Pucéat et al., 1993a), we also found a significant amount of PKC α in both cell types. However, using an antibody raised against brain PKC β, we could not detect any PKC β in either of these cardiac myocyte preparations. Bogoyevitch and colleagues (1993) were also unable to detect the β isoform of PKC in their studies in adult rat ventricular cells.

Besides the Ca^{2+}-dependent PKC isoforms (cPKC), Ca^{2+}-independent isoforms that lack the C2 conserved domain (nPKC) have been found in most tissues. In cardiac tissue, Northern blot analysis had suggested the presence of at least one of these isozymes, namely PKC ϵ (Schaap et al., 1990). Koide and colleagues (1992) detected by Western blot analysis a small amount of this isoform in heart. Very recently, Westel and co-workers (1992) extensively investigated the presence of the different PKC isoforms including the nPKCs in numerous tissues. Using both Western blot analysis and an immunocytochemical approach, they found in whole heart the Ca^{2+}-dependent PKC isoforms α, β_1, and β_2 and the Ca^{2+}-independent isoforms ϵ, δ, and ζ the ζ appearing in this study as the most abundant isoform, at least on the basis of immunoreactivity. Since these studies used whole heart, however, one cannot conclude that these isoforms are all present in cardiac myocytes. Using adult rat ventricular myocytes, Bogoyevitch and col-leagues (1993) found that PKC ϵ was the major isoform but failed to detect any δ or ζ isoform. Pucéat and co-workers (1993a) also showed that both neonatal and adult ventricular myocytes contained a high amount of PKC ϵ isoform. These authors also found a significant PKC δ immunoreactivity in both cell types (manu-script in preparation). Smith and colleagues (pers. comm.) also reported that neo-natal ventricular cells contained the nPKCs ϵ, δ, and ζ. The identification of these more recently described Ca^{2+}-independent isoforms in cardiac preparations indi-cates that the two peaks of PKC activity separated chromatographically and orig-inally thought to represent α and δ isoforms must also contain ϵ and δ. Which forms are in cardiomyocytes is less clear. There is general agreement that isolated cardiomyocytes express PKC ϵ and the consensus is that they also express PKC α and δ but not γ. Whether a genuine PKC β isoform is expressed in these cells is still controversial.

Few data are currently available concerning the localization of the kinase in the cardiac cell. In their pioneering work, Katoh and colleagues (1981) demonstrated that the kinase was heterogeneously distributed. The cytosol, sarcolemmal, microsomal, mitochondrial, and nuclear fractions contained 72, 9, 18, and 1 percent of total phospholipid-, Ca^{2+}-dependent kinase activity, respectively. Presti and co-workers (1985) found that purified sarcolemma contained an endogenous PKC activity that could be activated by phorbol esters. A PKC immunoreactivity was also found in the cytosol, sarcolemma, and sarcoplasmic reticulum as well as in the myofilaments of neonatal ventricular myocytes (Liu et al., 1989). Mochly-Rosen and co-workers (1990) also found a putative PKC immunoreactivity (identified using a monoclonal antibody CK 1.4; see Mochly-Rosen et al., 1987) in the myofilaments of neonatal cells.

So far, little is known about PKC ontogeny in heart. One study reported that the enzymatic activity is higher in neonate than in adult rats and that the enzyme distribution between cytosol and membrane is not different in 20-day-old fetal versus adult hearts (Nogushi et al., 1988).

REGULATION AND KINASE ACTIVATION

PKC can be activated *in vivo* by two different pathways: (1) by phorbol ester or analogs of DAG like dioctanoylglycerol (DiC_8) that are able to cross the membrane and directly activate the kinase; (2) by a receptor-mediated pathway in which neurohormones stimulate phospholipid turnover and the generation of endogenous DAG, which then activates the kinase.

Perfusion of the heart with the phorbol ester phorbol dibutyrate (PDBu 1 μM) has been shown to increase PKC activity in the membrane and to decrease cytosolic activity, indicating that the kinase has been redistributed from the cytosol to the membrane. A maximal effect occurs within 5 minutes (Yuan et al., 1987). Capogrossi and co-workers (1990) confirmed these findings in ventricular cells isolated from adult rat heart. In this cardiac preparation, phorbol 12-myristate 13-acetate (PMA) triggers the redistribution of the kinase from the cytosol to the membrane; a maximal effect is observed at 10 nM and occurs within 5 minutes. DiC_8 (100 μM) also induces the redistribution of the cytosolic kinase to the membrane but is much less effective than PMA. PMA also induces PKC redistribution from the cytosol to the membrane in neonatal rat heart cells (Henrich and Simpson, 1988). Indeed, these authors showed that the phorbol ester increased particulate PKC activity by tenfold within 5 minutes, an effect that was sustained up to 20 minutes. None of these studies has attempted to identify the specific protein kinase C isoforms affected by PMA. However, recently Mochly-Rosen and co-workers (1990) used an *in situ* immunofluorescence technique to investigate the redistribution of specific isoforms of the kinase in neonatal cells. They showed that PMA triggered the redistribution of the β isoform of PKC to the membrane and nucleus and of another putative PKC isoform recognized by the antibody CK 1.4 (see above) to the cytoskeleton or myofilaments. More recently, Bogoyevitch and colleagues (1993) reported that PMA triggered the redistribution of the ε isoform to the membrane in adult ventricular cells. Pucéat and colleagues (1993a) found that PMA was able to increase both membrane-associated ε and α isoforms in neonatal and adult ventricular cells. PKC δ was also redistributed to the membrane in both cell types stimulated with PMA (Pucéat et al., manuscript in preparation). Smith and colleagues (pers. comm.) have shown that the phorbol ester induced the redistribution of β, δ, and ζ as well as α and ε isoforms in neonatal rat heart cells.

Numerous agonists (e.g., α_1-adrenergic, purinergic, muscarinic, endothelin, angiotensin II, thrombin) stimulate PI turnover in cardiac tissue. In most studies the neurohormonal modulation of PI turnover has been demonstrated by measuring the formation of ^{3}H-inositol phosphates in ^{3}H-inositol-labeled cells stimulated with agonists. While an increase in inositol phospholipid hydrolysis suggests an increase in DAG formation, this has been shown in relatively few instances. Similarly, there is rather limited evidence that PKC is activated following neurohormonal stimulation.

Most available data concern the response to α_1-adrenergic stimulation. In this regard, Henrich and Simpson (1988) observed that norepinephrine increased the particulate PKC activity of neonatal cells by threefold within 30 seconds. The particulate PKC activity returned toward control values by 5 minutes. Using adult rat cardiomyocytes, Kaku and co-workers (1991) showed that norepinephrine increased membrane-associated PKC activity. A maximal effect was observed at 5 minutes and was sustained up to 10 minutes, when cells were stimulated with 100 μM of the α_1-adrenoceptor agonist. Talosi and Kranias (1992) also have reported that the perfusion of the isolated heart with 10 μM phenylephrine for 4 minutes increased both the total and specific membrane-associated PKC activities. Prazosin, a specific α_1-adrenoceptor antagonist, fully prevented this effect. Regarding hormonal activation of specific isoforms in neonatal cells, phenylephrine (2 μM) triggers within 1 minute the redistribution of an as yet uncharacterized isoform recognized by the antibody CK 1.4 to the myofilaments (Mochly-Rosen et al., 1990). It is noteworthy that receptors for activated PKC have been identified in a detergent insoluble fraction that may correspond to myofilaments from neonatal rat cardiac cells (Mochly-Rosen et al., 1991).

Pucéat and colleagues (1993a) found that phenylephrine and ATP increased membrane-associated PKC ε in neonatal cells. Endothelin and carbachol as well as phenylephrine and ATP exerted a similar effect in adult ventricular cells. However, none of these agonists increased membrane-associated PKC α either in neonatal cells or in adult cells. Bogoyevitch et al. (1993) also reported that endothelin and epinephrine induced the redistribution of cytosolic PKC ε to the membrane in adult ventricular cells, and Smith and co-workers (pers. comm.) also found that phenylephrine and carbachol triggered the redistribution of cytosolic PKC ε to the membrane in neonatal cells; phenylephrine also slightly increased membrane-associated PKC δ.

CARDIAC SUBSTRATES OF PKC

Katoh and co-workers (1981) were the first to investigate the putative endogenous PKC substrates in different subcellular fractions prepared from guinea pig heart. In the cytosolic fraction, PKC phosphorylated two major proteins that appeared in SDS-PAGE with molecular masses of 49 and 38 kDa. These authors failed to find other substrates in fractions enriched in nuclei, mitochondria, microsomes, or plasma membrane. A few years later, Presti and colleagues (1985) demonstrated the presence of an endogenous sarcolemmal substrate for PKC based on evidence that the addition of phorbol ester to the sarcolemmal fraction purified from canine heart triggers the phosphorylation of a 15-kDa protein. This protein could also be phosphorylated by exogenous PKC or PKA. The phosphorylations induced by both kinases are additive, suggesting that they occur at different sites. These findings were further confirmed using a neonatal cardiac cell microsomal fraction (Meij et al., 1991). The subsequent cloning of this PKC substrate indicated a protein having a predicted molecular mass of 8.4 kDa rather than 15 kDa; on the basis of a reasonable degree of homology to phospholamban, the 15-kDa substrate was named phospholemman (Palmer et al., 1991). This study also confirmed that PKA and PKC phosphorylate the protein at different sites. The physiological significance of the phosphorylation of this protein remains unclear since its function is still not known. A recent study suggested that the 15-kDa protein could be a chloride channel (Moorman et al., 1992).

Qu and co-workers (1992) investigated the sarcolemmal substrates for different PKC isoforms using purified PKC isozymes (see Qu et al., 1991). They found that types II (β) and III (α) phosphorylated four proteins having molecular masses of 19, 21, 35, and 95 kDa. There was apparently no 15-kDa protein phosphorylation. The phosphorylation induced by type II and type III isoforms was additive. Type III phosphorylated the 21-kDa protein to a greater extent than type II. So far, little is known about the physiological significance of these substrates.

A recent study reported that the muscarinic receptor purified from chick heart ventricles was a good substrate for PKC (Richardson and Hosey, 1990). Phosphorylation by PKC occurred only in the absence of the associated G protein, but when the phosphorylated receptor was reconstituted into lipid vesicles with a G protein it induced less GTPγS binding and agonist-stimulated GTPase activity (G_o) (Richardson et al, 1992). This could be of physiological relevance if phosphorylation diminishes the capacity of the receptor to transduce the hormonal signal.

PKC can also phosphorylate the sarcoplasmic reticular protein, phospholamban (Iwasa and Hosey, 1984; Movsesian et al., 1984). This phosphorylation increases the sarcoplasmic reticular Ca^{2+} ATPase activity and thus Ca^{2+} uptake from the cytosol (Movsesian et al., 1984). However, this phosphorylation appears not to occur *in vivo*. Indeed, Edes and Kranias (1990) and Hartmann and Schrader (1992) did not observe any phospholamban phosphorylation following perfusion of guinea pig hearts or incubation of isolated rat myocytes with PMA.

Besides membrane proteins, PKC is able to phosphorylate contractile proteins. Four such proteins are substrates for the Ca^{2+}- and phospholipid-dependent kinase: two thick-filament proteins (the C protein and the myosin light chain 2 [MLC$_2$]) and two thin-filament proteins (troponins I and T). The C protein is phosphorylated by PKC *in vitro* at the same sites phosphorylated by the cAMP-dependent kinase (Lim et al., 1985; Venema and Kuo, 1993). PMA also induces the phosphorylation of the C protein in isolated adult rat myocytes (Venema and Kuo, 1993), whereas the perfusion of isolated heart with PMA does not induce any phosphorylation of this contractile protein (Edes and Kranias, 1990). MLC$_2$ has recently been shown to be a substrate for PKC *in vitro* (Venema and Kuo, 1993). However, using this same experimental model (i.e., myofibrils isolated from adult rat ventricular myocytes), Clement and co-workers (1992) did not observe phosphorylation of MLC$_2$ by PKC. The reason for this discrepancy is unclear. Possible explanations are the different PKC or myofilament preparations used by the two laboratories. The stimulation of adult rat ventricular myocytes with PMA also induces the phosphorylation of MLC$_2$ but to a smaller extent than *in vitro* (Venema and Kuo, 1993).

Both the inhibitory subunit and the tropomyosin binding subunit of the troponin complex (TNI and TNT, respectively) are substrates for PKC *in vitro*. Katoh and colleagues (1983) found that TNT was a better substrate (K_m 0.3 μM) than TNI (K_m 3.4 μM) and that PKC could phosphorylate both the purified proteins and the proteins in the tropomyosin complex. By contrast, Bazzi and co-workers (1987) reported that TNI could not be phosphorylated by PKC in the presence of the Ca^{2+}-binding subunit of troponin complex, TNC. Purified PKC is also able to phosphorylate both TNI and TNT when added to intact myofilaments (Clement et al., 1992; Noland and Kuo, 1993). Furthermore, Liu and co-workers (1989) found that the stimulation of intact neonatal cells with PMA induces the phosphorylation of TNT; TNI is phosphorylated to a lesser extent. By contrast, Venema and Kuo (1993) did not observe any TNT phosphorylation following PMA stimulation of adult ventricular cells, whereas TNI was phosphorylated to a significant extent. Noland and Kuo (1991) analyzed the consequences of TNT and TNI phosphorylations on the Ca^{2+}-activated actomyosin ATPase. They found that PKC-dependent phosphorylation of either TNT or TNI, either purified or in the troponin-tropomyosin complex, induced a decrease in the V_{max} of the Ca^{2+}-ATPase of the reconstituted actomyosin complex. This effect was attributed to a decreased interaction of TNT with the other components of the thin filament (Noland and Kuo, 1992) as well as to an alteration in the interaction between the thin and the thick filament (Noland and Kuo, 1993). Phosphorylation of TNI and TNT together in intact myofilaments does not change the Ca^{2+} sensitivity of the actomyosin ATPase (Clement et al., 1992; Noland and Kuo, 1993). The enzyme V_{max} has been reported to be unchanged (Clement et al., 1992) or decreased (Noland and Kuo, 1993) following PKC phosphorylation of the myofilaments. PKC phosphorylation sites of TNI and TNT were examined in more detail by two groups. Using two-dimensional phosphopeptide maps, Noland and colleagues (1989) showed that PKC phosphorylated TNI at serines 43 and/or 45 and 78, and threonine 114, and TNT at threonines 190, 199, and 280. Using ^{31}P nuclear magnetic resonance and phosphopeptide isolation by Swiderek and co-workers (1990) obtained different results; they found that serines 23 and/or 24 of TNI and serine

194 of TNT were phosphorylated by PKC. It is interesting to note from this study that PKA and PKC share the same TNI phosphorylation sites. However, Venema and Kuo (1993) did not confirm this result. On the contrary, these authors found that PKC and PKA phosphorylated TNI at different sites.

Finally, Dosemeci and co-workers (1988) reported that brief PMA stimulation of rat neonatal ventricular myocytes induced the phosphorylation of two proteins having molecular masses of 32 and 83 kDa. Using two-dimensional electrophoresis, they were able to show that the 83-kDa protein was an acidic protein, probably the rodent form of bovine myristoylated alanine-rich C kinase substrate (MARCKS). Irons and colleagues (1991) also found, using atrial cells, that a protein that migrates in SDS-PAGE with an apparent molecular mass of 80 kDa was phosphorylated in response to PDBu.

EFFECTS OF PKC ACTIVATION ON ION CHANNELS

PKC modulates the activity of at least two ion channels that participate in the cardiac action potential plateau and in the repolarization phase, namely the calcium and potassium channels, respectively. Most data concerning effects of PKC on these channels were obtained in studies where phorbol esters were used to activate the kinase.

Ca^{2+} currents are increased, decreased, or unchanged by cell treatment with phorbol ester. For example, the treatment of embryonic chick heart cells with PMA (10 nM–1 μM) inhibited the Ca^{2+} current as indicated by a decrease in $^{45}Ca^{2+}$ influx into the cells (Levitan, 1985). By contrast, using patch clamp recording in neonatal rat cardiomyocytes, Dosemeci and co-workers (1988) observed an increase in Ca^{2+} current in response to PMA (10nM–85 nM). In adult cells, phorbol esters were first reported not to affect I_{ca} either in rat (Tohse et al., 1987) or guinea pig (Apkon and Nerbonne, 1988) ventricular cells. Nevertheless, using adult rat ventricular myocytes, Scamps and colleagues (1992) observed a decrease in I_{CaL} that occurred as soon as PMA (100 nM) was applied to the cell. Tseng and Boyden (1991) analyzed the effects of phorbol ester on the two types of Ca^{2+} currents, the slow inward I_{caL} and the transient I_{caT}. In canine ventricular cells, PMA (100 nM) exerted a biphasic effect on I_{caL}; it first increased and then decreased the ionic current. The phorbol ester decreased or abolished I_{caT} in Purkinje and ventricular cells, respectively. The effects on both I_{caL} and I_{caT} were prevented by the protein kinase inhibitor H8. These authors have interpreted the decrease in I_{caT} as a Ca^{2+}-dependent inactivation of the channel following the early PKC-dependent increase in I_{caL} activity.

Using single-channel analysis, Lacerda and co-workers (1988) had also reported that PMA effects were dependent on the application duration. Indeed, they observed that a brief application of the phorbol ester (5 s) to neonatal cells increased the opening probability of Ca^{2+} channels, whereas a much longer treatment (20 min) decreased this probability. The authors failed to observe the inhibitory effect when DiC_8 was used instead of PMA and therefore suggested that a decrease in PKC (i.e., down-regulation of the enzyme) following the longer PMA exposure might explain the inhibition of I_{ca}. Singer-Lahat and colleagues (1992) recently expressed the cardiac L-type calcium channel in *Xenopus* oocytes in order to study its regulation by PKC. PMA (10 nM) was used to activate the kinase, and Ba current (I_{Ba}) served as an index of Ca^{2+} channel activity. The L-type cardiac Ca^{2+} channel consists of five subunits; the α_1 subunit, which forms the pore, and the α_2, δ, β_1 and β_2 subunits. In oocytes injected with α_1 subunit RNA, PMA first increased I_{Ba} within 7 to 12 minutes and then decreased it. Similar PMA effects were observed in oocytes expressing α_1 together with α_2, δ, and/or β subunit. However, expression of the β subunit together with α_1 appeared to diminish the early activating effect of PMA. By contrast, in oocytes that did not express the α_1 subunit, PMA decreased I_{Ba} only for up to 40 minutes. Neither the kinetics nor the voltage dependence of I_{Ba} was affected by PMA. PMA effects were significantly inhibited by staurosporine (1 μM). These authors suggested from their data that the α_1 subunit is the main target of PKC in the cardiac Ca^{2+} channel and the β subunit could play a modulatory role.

PKC is also able to regulate potassium channels. As in the case of Ca^{2+} channels, the findings are somewhat inconsistent. Apkon and Nerbonne (1988) reported that a Ca^{2+}-independent voltage-activated potassium outward current in rat ventricular myocytes is decreased by superfusion with PMA (10 nM–1 μM) or with 1,2 oleylacetylglycerol (OAG) (60 μM). In guinea pig ventricular cells, Tohse and colleagues (1987) observed an increase in the delayed outward potassium current. This effect was temperature dependent and was prevented by the kinase inhibitor H7. Similarly, Walsh and Kass (1988) observed a temperature-dependent increase in the delayed rectifier potassium current when guinea pig ventricular cells were superfused with PDBu. Finally, a recent study suggests that a chloride conductance can be activated by PKC (Walsh, 1991). Indeed, Walsh found that PMA, an α_1- adrenoceptor agonist (norepinephrine in the presence of propranolol), or dialysis of the cardiomyocyte with partially purified PKC, increased the conductance. Additional investigations assessing more directly the effect of PKC on ionic conductances are required to further understand the role of the kinase in the modulation of the ionic channels.

EFFECTS OF PKC ACTIVATION ON INTRACELLULAR CALCIUM AND PH

Several studies have shown that PKC regulates cellular Ca^{2+} homeostasis in cardiac cells. Using the Ca^{2+}-sensitive probe Fura-2, Leatherman and co-workers (1987) reported a decrease in Ca^{2+}_i in embryonic chick heart cells treated with PMA (1 μM). There was no change in Ca^{2+}_i in neonatal rat cardiac cells treated for short or long times with PMA (Lacerda et al., 1988), although the Ca^{2+} uptake capacity of the SR has been reported to decrease in the same permeabilized cells treated with PMA (Rogers et al., 1990). In electrically driven adult cardiac cells, both diastolic Ca^{2+}_i and the magnitude of the Ca^{2+} transient are decreased following PMA (100 nM) stimulation (Capogrossi et al., 1990). These effects are mimicked by DiC_8 (10 μM). Although the authors have not further investigated the origin of this effect, it can be postulated that a decrease in SR Ca^{2+} uptake (Rogers et al., 1990) as well as a decrease in I_{caL} (Tseng and Boyden, 1991) account for the decreased Ca^{2+} transient observed in the presence of the phorbol ester. However, as discussed by the authors, the fact that diastolic Ca^{2+}_i is also decreased suggests that PMA additionally affects sarcolemmal Ca^{2+} extruding mechanisms, such as the Ca^{2+} pump and/or the Na^+/Ca^{2+} exchanger. In agreement with such possibilities, Gwathmey and Hajjar observed in human permeabilized trabeculae a decrease in sarco plasmic reticulum (SR) Ca^{2+} loading after treatment of the muscle with the phorbol ester 12-deoxyphorbol 13-isobutyrate 20-acetate (DPBA). The decrease in SR Ca^{2+} loading cannot be attributed to an alteration of SR function since the kinetics of Ca^{2+} transients, which usually reflect SR Ca^{2+} release and uptake, were not changed after DPBA treatment of the muscle.

Numerous studies have suggested that PKC is involved in intracellular pH regulation. More specifically, phorbol ester has been reported to stimulate the Na/H antiport, the main alkalinizing pH_i regulatory mechanism in various tissues (Frelin et al., 1988). In cardiac tissue, PKC has also been implicated in Na/H antiport modulation. Using isolated ventricular myocytes, McLeod and Harding (1992) showed that PMA weakly accelerated pH_i recovery following an imposed acidosis, a situation where the Na/H antiport is the main albeit not the sole pH_i regulatory mechanism. By contrast, using an experimental condition where only the Na/H antiport is functional, Pucéat and co-workers (1993b) observed that PMA (100 nM) did not activate the antiport. These last results are not surprising given the fact that cardiac adult antiport lacks phosphorylation sites for the Ca^{2+} phospholipid-dependent kinase (Fliegel et al., 1991) and cannot be phosphorylated *in vitro* by PKC (Fliegel et al., 1992).

EFFECTS OF PKC ON CARDIAC CONTRACTILITY

In cardiac muscle, contractile force can be regulated both by the amount of Ca^{2+} mobilized during the action potential and by the myofilament Ca^{2+} sensitivity. PKC has been shown to regulate both Ca^{2+} (see above) and the Ca^{2+} responsiveness of the myofilaments. Gwathmey and Hajjar (1990) reported that DPBA (1 μM) decreased myofilament Ca^{2+} sensitivity of human trabeculae permeabilized with saponin. Using the relationship between cell shortening and the magnitude of the Ca^{2+} transient as an index of the Ca^{2+} sensitivity of ventricular cell myofilaments, Capogrossi and colleagues (1990) observed an increase or no change in myofilament Ca^{2+} responsiveness following PMA or DiC8 stimulation. Pucéat and co-workers (1990) presented evidence in favor of a Ca^{2+} sensitization of the myofilaments following PKC stimulation. These authors measured the force developed by a single-skinned cell superfused with different Ca^{2+}-containing solutions to establish a tension/pCa^{2+} relationship. Following PMA or DiC_8 stimulation, the tension/pCa^{2+} relationship established after skinning the cells was shifted toward lower Ca^{2+} concentrations, indicating an increase in the Ca^{2+} sensitivity of the myofilaments. This effect was reversed by treating the myofilaments with alkaline phosphatase, suggesting a PKC-dependent phosphorylation pathway. Since PKC has also been shown to exert a synergistic effect on Ca^{2+} sensitization induced by myosin light-chain kinase (Clement et al., 1992), it remains to be established whether PKC directly affects *in vivo* the Ca^{2+} sensitivity of the myofilaments. It should be pointed out that such a Ca^{2+} sensitization of the myofilaments, by altering the Ca^{2+} binding to troponin C, could also account for the decrease in diastolic Ca^{2+}_i observed by Capogrossi and colleagues (1990).

Although PKC activation may cause a Ca^{2+} sensitization of the myofilaments, most studies examining the effects of phorbol esters on contractile force have shown that these agents exert a negative rather than a positive inotropic effect. In isolated perfused heart, PDBu and PMA (100 nM–1 μM) induced a decrease in contractile force and cardiac rhythm as well as in coronary flow (Yuan et al., 1987). These same responses are elicted with DiC_8 (20–60 μM). Leatherman and co-workers (1987) also observed negative inotropic and chronotropic effects of phorbol ester in spontaneously beating embryo chick cells. In permeabilized human trabeculae, Gwathmey and Hajjar (1990) showed a decrease in contractile force following the addition of DPBA (1 μM). They suggested that the negative inotropic effect was due to a decrease in both the magnitude of the Ca^{2+} transient measured with the Ca^{2+} indicator aequorin and the Ca^{2+} sensitivity of the myofilaments. PMA and DiC_8 also decreased the contractility of electrically - stimulated isolated ventricular cells estimated by cell length recording; this negative inotropic response was enhanced by bathing the cell in a high potassium extracellular buffer to increase Ca^{2+}_i (Capogrossi et al., 1990), which suggests that the negative inotropic effect is mainly due to an alteration of Ca^{2+} movements in the cell. Teutsch and colleagues (1987) observed a biphasic effect of PMA (30 nM) on electrically driven guinea pig atria. The phorbol ester first increased the contractile force and then decreased it. PDBu (30 nM) induced only a negative inotropic effect, whereas DiC_8 (10–100 μM) exerted a positive inotropic effect on the same preparation. Kushida (1988) and Otani (1988) and their co-workers failed to observe any effect of phorbol esters on contractile force in rabbit or rat papillary muscle respectively. Similarly, Endou and colleagues (1991) were unable to reproduce the positive inotropic effect induced by α_1-adrenoceptor stimulation with phorbol esters.

Since the contractile response is modulated by both changes in Ca^{2+} and in myofilament Ca^{2+} sensitivity, and since PMA appears to affect these parameters in opposite directions, one cannot use PMAs effect to draw definitive conclusions about the mechanism by which PKC regulates cardiac contractility. It should be kept in mind that most neurohormones (e.g., α_1-adrenoceptor-, purinoceptor-agonists, endothelin) that activate PI turnover in myocardium induce a positive inotropic effect (Endoh, 1986; Legssyer et al., 1988; Kramer et al., 1991). Contrary to PMA, the agonists that activate PI turnover do not decrease Ca^{2+}_i. Thus, PKC-mediated myofilament Ca^{2+} sensitization may well participate in the positive inotropic response to the neurohormones. A recent study by Otani and co-workers (1992) provides a highly informative comparison of the effects of the α_1-adrenoceptor agonist, phenylephrine, and the phorbol ester, PDBu, on both contractility and PKC activity in rat papillary muscle. These authors confirmed that phenylephrine exerted a positive inotropic effect whereas PDBu induced a negative inotropic effect. Both phenylephrine and PDBu were shown to increase membrane-associated PKC activity. However, only PDBu decreased total (cytosolic and membrane-associated) PKC activity and increased the Ca^{2+}/phospholipid-independent activity of this protein kinase. They also observed that leupeptin and E64, two calpain inhibitors, attenuated the PDBu-induced negative inotropic effect. As proposed by the authors, these data suggest that the unique ability of phorbol ester to trigger a protease-mediated cleavage of PKC may account for the opposite inotropic effect of phorbol ester and agonists. Furthermore, Watson and Karmazin (1991) reported that the negative inotropic effect induced by the β-phorbol esters PMA and PDBu could be reproduced by the α-phorbol ester α-PDD, which does not activate PKC. In addition, their study showed that phorbol esters triggered a number of cellular events that occur simultaneously, such as an increase in intracellular lactate concentration or a decrease in ATP levels. These data suggest caution in using the cellular effects of phorbol ester to draw inferences about the involvement of PKC in hormonal responses.

ROLE OF PKC IN GENE EXPRESSION AND CARDIAC HYPERTROPHY

Myocardial hypertrophy occurs mainly in response to hemodynamic overload but can also be induced by hormonal, electrical, or genetic stimuli. Since adult cardiac cells do not retain the ability to proliferate, hypertrophy results not from an increase in cell number but rather from an increase in protein content and thus in cell size. Myocardial cell hypertrophy is associated with a sequence of genetic responses, among the earliest of which is induction of immediate early gene expression (c-*myc*, c-*fos*, c-*jun*, Egr1), which occurs within 15 to 60 minutes after the appropriate intervention. Reactivation of the expression of embryonic genes (skeletal α-actin, β-myosin heavy chain, ANF) and an up-regulation of constitutively expressed contractile protein genes (myosin light-chain 2, cardiac α-actin) are later responses. In a neonatal rat ventricular cell model of hypertrophy, the expression of these genes is maximally induced by 24 to 48 hours of treatment with phenylephrine or endothelin (Dunnmon et al., 1990; Sei et al., 1991). In *in vivo* models of pressure overload, the ANF gene is fully expressed by 7 days (Rockman et al., 1991). PKC has been implicated in these early and late responses.

The phorbol ester PMA was first shown by Starksen and colleagues (1986) to increase the level of c-*myc* RNA, a gene involved in cell proliferation and transformation and to induce cardiac hypertrophy in cultured cardiomyocytes. A few years later, Dunnmon and colleagues (1990) reported that phorbol ester triggered a rapid (5–30 min) expression of the protooncogenes c-*fos* and c-*jun* and of the inducible zinc finger gene Egr-1 in neonatal rat ventricular myocytes. These protooncogenes encode transcription factors that have been implicated in the control of gene expression. Allo and co-workers (1991, 1992) investigated the mechanism underlying phorbol ester-induced increase in protein synthesis. They showed that PMA triggered a rapid (1 min) redistribution of PKC to the membrane and nucleus. Membrane-associated and nuclear PKC activity remained elevated up to

20 minutes in the presence of PMA and was at control levels or below after 24 to 48 hours. The transcriptional activity of nuclei isolated from cells stimulated for 48 hours with PMA remained elevated as indicated both by a rise in ribosomal DNA transcription rate and RNA polymerase activity. These results suggest that PMA has effects at the transcriptional level that may be initiated by early PKC redistribution to nuclei.

Besides immediate early gene expression, PKC plays a role in the reactivation of embryonic genes and up-regulation of cardiac-specific genes during myocardial hypertrophy. The atrial natriuretic factor (ANF) gene is expressed in all cardiac cells at early embryonic stages. In the adult myocardium, this gene is normally expressed in atrial but not in ventricular myocytes, except during agonist- or overload-induced hypertrophy. The α_1-adrenoceptor-mediated reactivation of the ANF gene in ventricular neonatal cells can be partially inhibited by the protein kinase inhibitor H7 (Sei et al., 1991) suggesting that PKC plays a role in the inducible expression of this gene. Shubeita and co-workers (1992) presented further evidence in this regard, demonstrating that PMA increased the level of ANF mRNA by 12-fold in neonatal cells and, like phenylephrine, exerted its effect at the transcriptional level as assessed by increased expression of an ANF-luciferase reporter gene. These authors also demonstrated that expression of a constitutively activated form of PKC leads to transcriptional activation of an ANF promoter-luciferase reporter gene.

PKC is also involved in the up-regulation of β-MHC that occurs in certain types of hypertrophy. Using neonatal cells co-transfected with β-MHC/chloramphenicol acyltransferase (CAT) reporter construct and a plasmid containing constitutively active α-or β-PKC cDNA, Kariya and colleagues (1991) demonstrated that β-MHC gene transcription was stimulated by PKC. β-PKC appeared to be more effective than α-PKC, suggesting differential effects of isoforms.

The expression of the MLC-2 gene and the organization of contractile proteins into myofilaments is also increased by PMA in neonatal ventricular myocytes (Dunnmon et al., 1990). Shubeita and co-workers (1992) demonstrated that PMA, like phenylephrine, induced a four- to fivefold stimulation increase in MLC_2 transcription using an MLC_2-luciferase reporter gene. Co-transfection of vectors encoding constitutively activated α- or β-PKC also enhanced the expression of the co-transfected MLC-2 promoter/luciferase gene. The expression of an activator protein (AP-1)-sensitive reporter gene was also increased following PMA stimulation or transfection of constitutively activated PKC cDNA into neonatal ventricular myocytes, indicating that the DNA binding activity of this transcription factor is increased. This raises the possibility that AP-1, and its constituent Jun and Fos proteins, serve as transcriptional activators of the MLC-2 and ANF genes, and that the effects of PKC are mediated through activation of this transcription factor. Indeed, the ANF gene has been shown to contain AP-1-like binding sequences and to be activated by Fos and Jun expression under some conditions (Kovacic-Milijevic and Gardner, 1992). However, more recent analysis of the ANF promoter indicates that a binding factor other than AP-1 regulates ANF gene expression in response to alpha receptor stimulation (A. Sprenkle and C. Glembotski, personal communication).

PATHOPHYSIOLOGICAL CHANGES IN PKC

Several diseases (or pathological situations) have been associated with changes in PKC activity in cardiac tissue. Chambers and Eilon (1988) investigated cardiac PKC activity in cardiomyopathic hamsters. They found that particulate kinase activity was increased by twofold in 200-day-old myopathic hamsters compared with control animals. In agreement with this finding, an increase in the phosphorylation of a number of particulate PKC substrates (e.g., 26-, 31-, 45-, 53-, 69-, 98-, and 105-kDa proteins) was also observed in the cardiomyopathic animals. Curiously, PKC activity increases were observed only in 150 to 300-day-old animals; no significant difference between control and cardiomyopathic hamsters was observed before this age.

In spontaneously hypertensive rats, cardiac cytosolic PKC was shown to be increased at 20 weeks of age whereas particulate PKC activity did not change (Makita and Yasuda, 1990). The authors failed to observe any change in the phosphorylation of the cytosolic PKC substrates, namely, proteins of molecular masses of 26, 32, 43, and 95 kDa.

In isolated ischemic hearts (30 min ischemia), membrane PKC activity is markedly enhanced (15-fold) compared with normoxic perfused hearts. Cytosolic kinase activity is not altered (Prasad and Jones, 1992). The increased kinase activity is sustained during the following 30-minute reperfusion. Moreover, the phosphorylation of endogenous membrane proteins (e.g., <29, 52, 63, and 92 kDa) is also enhanced in ischemic hearts. The increased Ca^{2+}_i often observed in this pathological situation could play a major role in PKC redistribution to the membrane and activation. Strasser and colleagues (1992) also recently observed in ischemic hearts a two-fold increase in particulate PKC activity associated with a concomitant decrease in cytosolic kinase activity. They also suggested that PKC was involved in the sensitization of forskolin-stimulated adenylyl cyclase, which also occurs in ischemic hearts. In support of this hypothesis, the kinase inhibitors polymyxin B, H7, and staurosporine abolished or prevented adenylyl cyclase sensitization while perfusion of isolated hearts for 5 to 10 minutes with PMA reproduced the sensitization of the adenylyl cyclase.

Far more detailed investigations are required to understand the role of PKC in the progression of these diseases. It is noteworthy that in most of the pathological situations so far investigated, an increase in particulate PKC activity was observed. The development of specific PKC inhibitors is necessary before one can adequately test the involvement of PKC in the pathophysiology and potentially in the treatment of these diseases.

FUTURE DIRECTIONS

Since the discovery of the novel Ca^{2+} and phospholipid-dependent protein kinase, considerable progress has been made toward understanding the regulation and biochemical properties of this enzyme in the cardiac system. However, numerous questions concerning its function still remain to be answered. Specifically, although several putative PKC substrates have been found in cardiac tissue, further investigation is required to determine whether the phosphorylation of these proteins occurs *in vivo*. Moreover, little is known about the physiological significance of these phosphorylations. Much additional work is also needed to determine whether different PKC isoforms expressed in the cardiac myocyte display specific cellular substrates or whether these isoforms have synergistic effects on the same targets.

Along the same line, the regulation of the different isoforms by external agonists is poorly understood. Certain neurohormones that activate PI turnover in cardiac tissue do not increase $Ca^{2+}{}_i$. This suggests the possibility that in these cases, Ca^{2+} independent PKC isoforms might be preferentially activated. The successive biochemical steps that follow the agonist-induced hydrolysis of the phosphoinositides and lead to kinase activation need to be further investigated *in vivo*. Indeed, it is not yet clearly established whether the redistribution of the kinase to the membrane is always mandatory for its activation. For example, second messengers such as arachidonic acid could activate or autophosphorylate the kinase in cytosol without phospholipid environment. This point should be addressed in more detail in cardiac tissue. A better understanding of kinase regulatory mechanism would suggest means of preventing kinase activation at different steps, which would be helpful in elucidating the physiological functions of the kinase.

It is noteworthy that a large number of studies investigating the role of PKC in cardiac tissue used phorbol esters to activate the kinase. As mentioned in this chapter, these compounds may exert some PKC-independent effects, which means that PKC is probably not their only target. More important, they are unlikely to have any selectivity for the various isoforms of PKC. It thus becomes increasingly necessary to use other experimental approaches to address the role of PKC in cardiac function. For example, the overexpression of particular isoforms by transfection of cDNA into cardiac cells or the selective inhibition of particular isoforms by microinjection of peptide inhibitors could be a useful means of elucidating the role of protein kinase C in cardiac cell regulation.

REFERENCES

Allen, B. G., and S. Katz. 1991. Isolation and characterization of the calcium- and phospholipid-dependent protein kinase (protein kinase C) subtypes from bovine heart. *Biochemistry* 30:4334–4343.

Allo, S. M., P. J. McDermott, L. L. Carl, and H. E. Morgan. 1991. Phorbol ester stimulation of protein kinase C activity and ribosomal DNA transcription. Role in hypertrophic growth of cultured cardiomyocytes. *J. Biol. Chem.* 266:22003–22009.

Allo, S. N., L. L. Carl, and H. E. Morgan. 1992. Acceleration of growth of cultured cardiomyocytes and translocation of protein kinase C. *Am. J. Physiol.* 263:C319–C325.

Apkon, M., and J. Nerbonne. 1988. α_1-Adrenergic agonists selectively suppress voltage dependent K^+ currents in rat ventricular myocytes. *Proc. Natl. Acad. Sci. U.S.A.* 85:8756–8760.

Bazzi, M., P. D. Lampe, G. M. Strasburg, and G. L. Nelsestuen. 1987. Phosphorylation of troponin I by protein kinase C: Mechanism of inhibition by calmodulin and troponin C. *Biochim. Biophys. Acta* 931:339–346.

Billah, M. M., and J. C. Anthes. 1990. The regulation and cellular functions of phosphatidylcholine hydrolysis. *Biochem. J.* 269:281–291.

Bogoyevitch, M. A., P. J. Parker, and P. H. Sugden. 1993. Characterization of protein kinase C isotype expression in adult rat heart. Protein kinase C-ϵ is a major isotype present and is activated by phorbol esters, epinephrine and endothelin. *Circ. Res.,* 72:757–767.

Capogrossi, M. C., T. Kaku, C. R. Filburn, D. J. Pelto, R. G. Hansford, H. A. Spurgeon, and E. G. Lakatta. 1990. Phorbol ester and dioctanoyl glycerol stimulate membrane association of protein kinase C and have a negative inotropic effect mediated by changes in cytosolic Ca2+ in adult rat cardiac myocytes. *Circ. Res.* 66:1143–1155.

Chambers, T. C., and G. Eilon. 1988. Heart protein kinase C activity increases during progression of disease in the cardiomyopathic hamster. *Biochem. Biophys. Res. Comm.* 157:507–514.

Clément, O., M. Pucéat, M. Walsh, and G. Vassort. 1992. Protein kinase C enhances myosin light-chain kinase effects on force development and ATPase activity in rat single skinned cardiac cells. *Biochem. J.* 285:311–317.

Dosemeci, A., R. S. Dhallan, M. M. Cohen, W. J. Lederer, and T. B. Rogers. 1988. Phorbol ester increases calcium current and stimulates the effect of angiotensin II in cultured neonatal rat heart myocytes. *Circ. Res.* 62:347–357.

Dunnmon, P. M., K. Iwaki, S. A. Henderson, A. Sen, and K. R. Chien. 1990. Phorbol esters induce immediate-early genes and activate cardiac gene transcription in neonatal rat myocardial cells. *J. Molec. Cell. Cardiol.* 22:901–910.

Edes, I., and E. G. Kranias. 1990. Phospholamban and troponin I are substrates for protein kinase C in vitro but not in intact beating guinea pig hearts. *Circ. Res.* 67:394–400.

Endoh, M. 1986. Regulation of myocardial contractility via adrenoceptors; differential mechanisms of α- and β-adrenoceptor-mediated actions. In: *New aspects of the Role of Adrenoceptor in the Cardiovascular System,* edited by H. Grobecker, A. Philippu, and K. Starke, pp. 78–105. Springer Verlag, Berlin.

Endou, M., Y. Hattori, N. Tohse, and M. Kanno. 1991. Protein kinase C is not involved in α1-adrenoceptor-mediated effect. *Am. J. Physiol.* 260:H27–H36.

Fliegel, L., C. Sardet, J. Pouyssegur, and A. Barr. 1991. Identification of the protein and cDNA of the cardiac Na+/H+ exchanger. *FEBS* Lett. 279:25–29.

Fliegel, L., M. P. Walsh, D. Singh, C. Wong, and A. Barr. 1992. Phosphorylation of the C-terminal domain of the Na+/H+ exchanger by Ca2+/calmodulin-dependent protein kinase II. *Biochem. J.* 282:139–145.

Frelin, C., P. Vigne, A. Ladoux, and M. Ladunski. 1988. The regulation of the intracellular pH in cells from vertebrates. *Eur. Biochem. J.* 174:3–14.

Gwathmey, J. K., and R. J. Hajjar. 1990. Effect of protein kinase C activation on sarcoplasmic reticulum function and apparent myofibrillar Ca^{++} sensitivity in intact and skinned muscles from normal and diseased human myocardium. *Circ. Res.* 67:744–752.

Hartmann, M., and J. Schrader. 1992. Protein kinase C phosphorylates a 15 kDa protein but not phospholamban in intact rat cardiac myocytes. *Eur. J. Pharmacol.* 226:225–231.

Henrich, C. J., and P. C. Simpson. 1988. Differential acute and chronic response of protein kinase C in cultured neonatal rat heart myocytes to α_1-adrenergic and phorbol ester stimulation. *J. Molec. Cell. Cardiol.* 20:1081–1085.

Irons, C. E., C. A. Sei, H. Hidaka, and C. C. Glembotski. 1992. Protein kinase C and calmodulin kinase are required for endothelin-stimulated atrial natriuretic factor secretion fron primary atrial myocytes. *J. Biol. Chem.* 267:5211–5216.

Iwasa, Y., and M. M. Hosey. 1984. Phosphorylation of cardiac sarcolemma proteins by the calcium-activated phospholipid-dependent protein kinase. *J. Biol. Chem.* 259: 534–540.

Kaku, T., E. Lakatta, and C. Filburn. 1991. α-Adrenergic regulation of phosphoinositide metabolism and protein kinase C in isolated cardiac myocytes. *Am. J. Physiol.* 260: C635–C642.

Kariya, K.-I., L. R. Karns, and P. C. Simpson. 1991. Expression of a constitutively activated mutant of the β-isozyme of protein kinase C in cardiac myocytes stimulates the promoter of the β-myosin heavy chain isogene. *J. Biol. Chem.* 266:10023–10026.

Katoh, N., R. W. Wrenn, B. C. Wise, M. Shoji, and J. F. Kuo. 1981. Substrate proteins for calmodulin-sensitive and phospholipid-sensitive Ca-dependent protein kinases in heart, and inhibition of their phosphorylation by palmitoylcarnitine. *Proc. Natl. Acad. Sci. U.S.A.* 78:4813–4817.

Katoh, N., B. C. Wise, and J. F. Kuo. 1983. Phosphorylation of cardiac troponin inhibitory subunit (troponin I) and tropomyosin-binding subunit (troponin T) by cardiac phospholipid Ca2+-dependent protein kinase. *Biochem J.* 209:189–195.

Koide, H., K., Ogita, U. Kikkawa, and Y. Nishizuka. 1992. Isolation and characterization of the ε subspecies of protein kinase C from rat brain. *Proc. Natl. Acad. Sci U.S.A.* 89:1149–1153.

Kosaka, Y., K. Ogita, K. Ase, H. Nomura, U. Kikkawa, and Y. Nishizuka. 1988. The heterogeneity of protein kinase C in various rat tissues. *Biochem. Biophys. Res. Comm.* 151:973–981.

Kovacic-Milijevic, B., and D. G. Gardner. 1992. Divergent regulation of the human atrial natriuretic peptide gene by c-jun and c-fos. *Molec. Cell. Biol.* 12:292–301.

Kramer, B. K., T. W. Smith, and R. A. Kelly. 1991. Endothelin and increased contractility in adult rat ventricular myocytes. Role of intracellular alkolosis induced by activation of the protein kinase C-dependent Na^+-H^+ exchanger. *Circ. Res.* 68:269–279.

Kuo J. F., R. G. G. Andersson, B. C. Wise, L. Mackerlova, I. Salomonsson, N. L. Brackett, N. Katoh, M. Shoji, and R. W. Wrenn. 1980. Calcium-dependent protein kinase: Widespread occurrence in various tissues and phyla of the animal kingdom and comparison of effects of phospholipid, calmodulin, and trifluoperazine. *Proc. Natl. Acad. Sci. U.S.A.* 77:7039–7043.

Kushida, H., T. Hiramoto, H. Satoh, and M. Endoh. 1988. Phorbol ester does not mimic but antagonizes the α-adrenoceptor mediated positive inotropic effect in the rabbit papillary muscle. *Naunyn-Schmiedeberg's Arch. Pharmacol.* 337:169–176.

Lacerda, A. E., D. Rampe, and A. Brown. 1988. Effects of protein kinase C activators on cardiac Ca2+ channels. *Nature* 335:249–251.

Leatherman, G. F., D. Kim, and T. W. Smith. 1987. Effect of phorbol esters on contractile state and calcium flux in cultured chick heart cells. *Am. J. Physiol.* 253:205–209.

Legssyer, A., J. Poggioli, D. Renard, and G. Vassort. 1988. ATP and other adenine compounds increase mechanical activity and inositol trisphosphate production in rat heart. *J. Physiol.* 401:185–189.

Levitan, I. B. 1985. Phosphorylation of ion channels. *J. Memb. Biol.* 87:177–180.

Lim, M. S., C. Sutherland, and M. P. Walsh. 1985. Phosphorylation of bovine cardiac C-protein by protein kinase C. *Biochem. Biophys. Res. Comm.* 132:1187–1195.

Liu, J. D., J. D. Wood, R. L. Raynor, Yi-Chong Wang, A. Noland, A. A. Ansari, and J. F. Kuo. 1989. Subcellular distribution and immunocytochemical localization of protein kinase C in myocardium, and phosphorylation of troponin in isolated myocytes stimulated by isoproterenol or phorbol ester. *Biochem. Biophys. Res. Comm.* 162:1105–1110.

Makita, N., and H. Yasuda. 1990. Alterations of phosphoinositide-specific phospholipase C and protein kinase C in the myocardium of spontaneously hypertensive rats. *Bas. Res. Cardiol.* 85:435–443.

McLeod, K. T., and S. E. Harding. 1991. Effects of phorbol ester on contraction, intracellular pH and Ca2+ in isolated mammalian ventricular myocytes. *J. Physiol.* 444:481–498.

Meij, J. T. A., K. Bezstarosti, V. Panagia, and J. M. J. Lamers. 1991. Phorbol ester and the actions of phosphatidylinositol 4,5-biphosphate specific phospholipase C and protein kinase C in microsomes prepared from cultured myocytes. *Molec. Cell. Biochem.* 105:37–47.

Mochly-Rosen, D., A. I. Babaum, and Jr. D. E. Koshland. 1987. Distinct cellular and regional localization of immunoreactive protein kinase C in rat brain. *Proc. Natl. Acad. Sci. U.S.A.* 84:4660–4664.

Mochly-Rosen, D., C. J. Henrich, L. Cheever, H. Khaner, and P. C. Simpson. 1990. A protein kinase C isozyme is translocated to cytoskeletal elements on activation. *Cell Reg.* 1:693–706.

Mochly-Rosen, D., H. Khaner, J. Lopez, and B. Smith. 1991. Intracellular receptors for activated protein kinase C. *Proc. Natl. Acad. Sci. U.S.A.* 266:14866–14868.

Moorman, J. R., C. J. Palmer, J. E. John III, M. E. Durieux, and L. Jones. 1992. Phospholemman expression induces a hyperpolarization-activated chloride current in *Xenopus* oocytes. *J. Biol. Chem.* 267:14551–14554.

Movsesian, M. A., M. Nishikawa, and R. S. Adelstein. 1984. Phosphorylation of phospholamban by calcium activated, phospholipid dependent protein kinase. *J. Biol. Chem.* 259:8029–8032.

Nogushi, A., J. DeGuire, and P. Zanaboni. 1988. Protein kinase C in the developing rat liver, heart and brain. *Dev. Pharmacol. Ther.* 11:37–43.

Noland, T. A., and J. F. Kuo. 1991. Protein kinase C phosphorylation of cardiac troponin I or troponin T inhibits Ca^{++}-stimulated actomyosin MgATPase activity. *J. Biol. Chem.* 266:4974–4978.

Noland, T. A., and J. F. Kuo. 1992. Protein kinase C phosphorylation of cardiac troponin decreases Ca^{++}-dependent actomyosin MgATPase activity and troponin T binding to tropomyosin-F-actin complex. *Biochem. J.* 288:123–129.

Noland, T. A., and J. F. Kuo. 1993. Protein kinase C phosphorylation of cardiac troponin I and troponin T inhibits Ca^{++}-stimulated MgATPase activity in reconstituted actomyosin and isolated myofibrils, and decreases actin-myosin interactions. *J. Molec. Cell. Cardiol.* 25:53–65.

Noland, T. A., R. L. Raynor, and J. F. Kuo. 1989. Identification of sites phosphorylated in bovine cardiac troponin T by protein kinase C and comparative substrate activity of synthetic peptides containing the phosphorylation sites. *J. Biol. Chem.* 264:20778–20785.

Otani, H., H. Otani, and D. H. Das. 1988. α_1-Adrenoceptor mediated phosphoinositide breakdown and inotropic response in rat left ventricular papillary muscle. *Circ. Res.* 62:8–17.

Otani, H., H. Mitsuyoshi, Z. Xun-Tsing, K. Omori, and C. Inagaki. 1992. Different patterns of protein kinase C redistribution mediated by α_1-adrenoceptor stimulation and phorbol ester in rat isolated left ventricular papillary muscle. *Br. J. Pharmacol.* 107:22–26.

Palmer, C. J., B. T. Scott, and L. R. Jones. 1991. Purification and complete sequence determination of the major plasma membrane substrate for cAMP-dependent protein kinase and protein kinase C in myocardium. *J. Biol. Chem.* 266:11126–11130.

Prasad, M. R., and R. M. Jones. 1992. Enhanced membrane protein kinase C activity in myocardial ischemia. *Bas. Res. Cardiol.* 87:19–26.

Presti, C. F., B. T. Scott, and L. R. Jones. 1985. Identification of an endogenous protein kinase C activity and its intrinsic 15-kilodalton substrate in purified canine cardiac sarcolemmal vesicles. *J. Biol. Chem.* 260:13879–13889.

Pucéat, M., O. Clément, P. Lechene, J. M. Pélosin, R. Ventura-Clapier, and G. Vassort. 1990. Neurohormonal control of calcium sensitivity of myofilaments in rat single heart cells. *Circ. Res.* 67:517–524.

Pucéat, M., O. Clément-Chomienne, A. Terzic, and G. Vassort. 1993a. α1-Adrenoceptor and purinoceptor agonists modulate the Na/H antiport in single cardiac cells. *Am. J. Physiol.* 264:H310–H319.

Pucéat, M., R. Hilal Dandan, L. L. Brunton, and J. H. Brown. (1993b) Neurohormonal regulation of PKC isozymes in isolated cardiac cells. *Biophys. J.* 64:A76.

Qu, Y., J. Torchia, T. Duc Phan, and A. M. Sen. 1991. Purification and characterization of protein kinase C isozymes from rat heart. *Molec. Cell. Biochem.* 103:171–180.

Qu, Y., J. Torchia, T. D. Phan, P. H. Wu, and A. K. Sen. 1992. Endogenous substrates of rat heart protein kinase C type I, II, and III isozymic forms in cardiac sarcolemma. *Biochem. Cell. Biol.* 70:81–85.

Richardson, R. M., and M. M. Hosey. 1990. Agonist-independent phosphorylation of purified cardiac muscarinic cholinergic receptors by PKC. *Biochemistry* 29:8555–8561.

Richardson, R. M., J. Ptasienski, and M. Hosey. 1992. Functional effects of protein kinase C-mediated phosphorylation of chick heart muscarinic cholinergic receptors. *J. Biol. Chem.* 267:10127–10132.

Rockman, H. A., R. S. Ross, A. N. Harris, K. U. Knowlton, M. E. Steinhelper, L. J. Field, Jr. J. Ross and K. R. Chien. 1991. Segregation of atrial-specific and inducible expression of an atrial natriuretic factor transgene in an in vivo murine model of cardiac hypertrophy. *Proc. Natl. Acad. Sci. U.S.A.* 88:8277–8281.

Rogers, T. B., S. T. Gaa, C. Massey, and A. Dosemeci. 1990. Protein kinase C inhibits Ca^{++} accumulation in cardiac sarcoplamic reticulum. *J. Biol. Chem.* 265:4302–4308.

Scamps, F., V. Rybin, M. Pucéat, V. Tkachuk, and G. Vassort. 1992. A Gs-protein couples P2-purinergic stimulation to cardiac Ca channels without cyclic AMP production. *J. Gen. Physiol.* 100:675–701.

Schaap, D., J. Hsuan, N. Totty, and P. J. Parker. 1990. Proteolytic activation of protein kinase C-ε. *Eur. J. Biochem.* 191:431–435.

Sei, C. A., C. E. Irons, A. B. Sprenkle, P. M. McDonough, J. H. Brown, and C. C. Glembotski. 1991. The α-adrenergic stimulation of atrial natriuretic factor expression in cardiac myocytes requires calcium influx, protein kinase C and calmodulin regulated pathways. *J. Biol. Chem.* 266:15910–15916.

Shubeita, H. E., E. A. Martinson, M. Van Bilsen, K. R. Chien, and J. H. Brown. 1992. Transcriptional activation of the cardiac myosin light chain 2 and atrial natriuretic factor genes by protein kinase C in neonatal rat ventricular myocytes. *Proc. Natl. Acad. Sci U.S.A.* 89:1305–1309.

Singer-Lahat, D., E. Gershon, H. Lotan, R. Hullin, M. Biel, V. Flockerzi, F. Hofmann, and N. Dascal. 1992. Modulation of cardiac Ca2+ channels in xenopus oocytes by protein kinase C. *FEBS Lett.* 306:113–118.

Starksen, N. F., P. C. Simpson, N. Bishopric, S. R. Coughlin, W. M. F. Lee, J. A. Escobedo, and L. T. Williams. 1986. Cardiac myocyte hypertrophy is associated with c-myc protooncogene expression. *Proc. Natl. Acad. Sci. U.S.A.* 83:8343–8350.

Strasser, R. H., R. Braun-Dullaeus, H. Walendzik, and R. Marquetant. 1992. α_1-Receptor-independent activation of protein kinase C in acute myocardial ischemia. Mechanisms for sensitization of the adenylyl cyclase system. *Circ. Res.* 70:1304–1312.

Swiderek, K., K. Jaquet, H. E. Meyer, C. Schachtele, F. Hofmann, and L. M. Heilmeyer. 1990. Sites phosphorylated in bovine cardiac troponin T and I. Characterization by ^{31}P-NMR spectroscopy and phosphorylation by protein kinases. *Eur. J. Biochem.* 190:575–582.

Takai Y., A. Kishimoto, Y. Iwasa, Y. Kawahara, T. Mori, and Y. Nishizuka. 1979. Calcium dependent activation of a multifunctional protein kinase by membrane phospholipids. *J. Biol. Chem.* 254:3692–3695.

Talosi, L., and E. G. Kranias. 1992. Effect of α-adrenergic stimulation on activation of protein kinase C and phosphorylation of proteins in intact rabbit hearts. *Circ. Res.* 70:670–678.

Teutsch, I., A. Weible, and M. Siess. 1987. Differential inotropic and chronotropic effects of various protein kinase C activators on isolated guinea pig atria. *Eur. J. Pharmacol.* 144:363–367.

Tohse, N., M. Kameyama, and H. Irisawa. 1987. Intracellular Ca^{2+} and protein kinase C modulate K^+ current in guinea pig heart cell. *Am. J. Physiol.* 252:1321–1324.

Tseng, G.-N., and P. A. Boyden. 1991. Different effects of intracellular Ca and protein kinase C on cardiac T and L Ca currents. *Am. J. Physiol.* 261:H364–H379.

Venema, R. C., and J. F. Kuo. 1993. Protein kinase C-mediated phosphorylation of troponin I and C-protein in isolated myocardial cells is associated with inhibition of myofibrillar actomyosin Mg ATPase. *J. Biol. Chem.* 268:2705–2711.

Walsh, K. B. 1991. Activation of a heart chloride conductance during stimulation of protein kinase C. *Molec. Pharmacol.* 40:342–346.

Walsh, K. B., and R. S. Kass. 1988. Regulation of a heart potassium channel by protein kinase A and C. *Science* 242:67–69.

Watson, J. E., and M. Karmazyn. 1991. Concentration-dependent effects of protein kinase C-activating and nonactivating phorbol esters on myocardial contractility, coronary resistance, energy metabolism, prostacyclin synthesis, and ultrastructure in isolated rat hearts. *Circ. Res.* 69:1114–1131.

Westel, W. C., W. A. Khan, I. Merchenthaler, H. Rivera, A. E. Halpern, H. M. Phung, A. Negro-Vilar, and Y. A. Hannun. 1992. Tissue and cellular distribution of the extended family of protein kinase C isoenzymes. *J. Cell. Biol.* 117:121–133.

Wise, B. C., R. L. Raynor, and J. F. Kuo. 1982. Phospholipid-sensitive Ca2+ dependent protein kinase from heart. I. Purification and general properties. *J. Biol. Chem.* 257:8481–8488.

Yuan, S., F. A. Sunahara, and A. K. Sen. 1987. Tumor-promoting phorbol esters inhibit cardiac functions and induce redistribution of protein kinase C in perfused beating rat heart. *Circ. Res.* 61:372–378.

10

The Role of Protein Kinase C in Exocytosis

RONALD W. HOLZ
MARY A. BITTNER

The role of protein kinase C in mediating the response to extracellular stimuli was first demonstrated in studies of serotonin secretion from platelets and was then found in many other secretory systems (see Nishizuka, 1986, 1992, for reviews). This effect of protein kinase C was one of the initial indications of the importance of the enzyme in signaling pathways. The studies in platelets demonstrated a dramatic synergism between Ca^{2+} mobilization and activation of protein kinase C (Kaibuchi et al., 1983; Sano et al., 1983; Yamanishi et al., 1983; Knight and Scrutton, 1986), and suggested a role for protein kinase C in modulating the intracellular pathway for exocytosis. Physiological stimuli can mobilize Ca^{2+} and activate protein kinase C through phosphatidylinositol 4,5-bisphosphate hydrolysis. The production of inositol 1,4,5 trisphosphate (IP_3) causes release of Ca^{2+} from internal stores; the production of diacylglycerol together with Ca^{2+} activates protein kinase C (Nishizuka, 1986). The different arms of the pathway can be activated by Ca^{2+} ionophore and an exogenous activator of protein kinase C (e.g., phorbol ester, synthetic diacylglycerol).

One of the complications in determining the mechanisms underlying the enhancement of secretion caused by protein kinase C is that protein kinase C-induced phosphorylation modulates a multitude of systems. Protein kinase C activation can alter secretory responses by affecting the intracellular secretory pathway or by modulating extracellular signaling through changes in function of plasma membrane receptors and ion channels (e.g., Wakade et al., 1986; Barrie et al., 1991; Knox et al., 1992; Swartz et al., 1993). This chapter will focus on

This work was supported by grants from the National Institutes of Health, RO1 DK27959 (R.W.H.) and from the National Science Foundation, BNS-9112391 (R.W.H.) and BNS 9008685 (M.A.B.).

recent work that attempts to identify the mechanisms by which protein kinase C enhances the *intracellular* pathway of regulated secretion and the possible protein substrates of the enzyme that are responsible for the effects. These mechanistic studies required the development and use of permeabilized cells. The following will emphasize studies in permeabilized platelets, mast cells, and chromaffin cells.

STUDIES WITH PLATELETS

Platelets contain three types of secretory vesicles. α Granules are the most numerous and contain a variety of proteins including von Willebrand factor and β-thromboglobulin. Dense granules contain 5-hydroxytryptamine (5HT) and nucleotides. Lysosomes contain acid hydrolases. All three types of the granules undergo exocytosis in electropermeabilized cells in the presence of 1 to 10 μM Ca^{2+} (Haslam and Davidson, 1984b; Knight et al., 1984; Coorssen et al., 1990), and the exocytosis of each is enhanced by activation of protein kinase C by phorbol esters or diacylglycerol. Activation of protein kinase C causes an increase in sensitivity to Ca^{2+} of secretion of 5HT from dense granules (Knight et al., 1984; Coorssen et al., 1990) and β-thromboglobulin from α granules (Coorssen et al., 1990). Half-maximal secretion is shifted from approximately 3 μM to less than 1 μM Ca^{2+}. Protein kinase C activation causes a small amount of exocytosis of dense granules but not of α granules in the virtual absence of Ca^{2+} (Coorssen et al., 1990). Protein kinase C activation does not alter the Ca^{2+} sensitivity of lysosome exocytosis (measured by β-N-acetylglucosaminidase secretion), but instead increases the extent of exocytosis (Knight et al., 1984). Thus, different granules in the same cell respond differently to phosphorylation caused by protein kinase C. It is likely that the secretory vesicle membrane, either directly through different substrates on its surface or indirectly through interactions with specific targets, determines the characteristics of the response to protein kinase C-induced phosphorylation.

In intact platelets, thrombin activates phospholipase C to cause hydrolysis of phosphatidylinositol 4,5-bisphosphate (Agranoff et al., 1983) and the production of IP_3 and diacylglycerol (Brass et al., 1986). Protein kinase C is activated and pleckstrin, a specific PKC substrate (see below), is phosphorylated. In electro-permeabilized platelets, thrombin causes a leftward shift in the Ca^{2+}-activation curve for 5HT (dense granule) secretion (Haslam and Davidson, 1984a; Knight et al., 1984) and an increase in the maximal extent of β-N-acetylglucosaminidase (lysosome) secretion (Knight et al., 1984). Because these effects are similar to those of phorbol esters, it is plausible that thrombin acts in large part by increasing intracellular Ca^{2+} and activating protein kinase C through the production of diacylglycerol.

Activation of PKC is unlikely to be the common final pathway for secretion in platelets. Ca^{2+}-induced secretion is not well correlated with phosphorylation of pleckstrin, a measure of protein kinase C activation. Little or no secretion of serotonin or β-thromboglobulin occurs upon significant phosphorylation of pleckstrin in the presence of submicromolar Ca^{2+} in permeabilized platelets (Haslam and Davidson, 1984a; Coorssen et al., 1990). Although phorbol ester and diacylglycerol can induce secretion in intact platelets without a change in cytosolic Ca^{2+}, the rate of secretion is slow compared with secretion stimulated by the combination of phorbol ester or diacylglycerol and Ca^{2+} ionophore (Rink et al., 1983). These data suggest that Ca^{2+} plays a role in the secretory pathway in platelets independent of its ability to activate protein kinase C. Further, there is evidence that a guanine nucleotide-dependent effect in addition to activation of polyphosphoinositide hydrolysis is important in stimulating secretion in platelets. The effects on secretion of phorbol ester and GTPγS in the virtual absence of Ca^{2+} ($< 10^{-9}\ M$) (when there is no stimulation of inositol phosphate production) are greater than additive (Coorssen et al., 1990). Thus, a GTP-binding protein may be involved in secretion from platelets in a role unrelated to the activation of phospholipase C, a topic that is being extensively investigated in many secretory cells (see Gomperts, 1990, for review).

STUDIES WITH CHROMAFFIN CELLS

The Ca^{2+}-activated secretory pathway in chromaffin cells can be separated into distinct temporal and biochemical steps that include a distinct ATP-dependent reaction that precedes the final triggering of secretion by Ca^{2+} (Knight and Baker, 1982; Dunn and Holz, 1983; Holz et al., 1989; Bittner and Holz, 1992a,b). The following indicates that activation of PKC modulates secretion in an ATP-dependent manner but is not responsible for ATP-dependent secretion in the absence of exogenous activators of protein kinase C.

Synthetic diacylglycerols and phorbol esters stimulate slow, Ca^{2+}-dependent release of catecholamine from intact cells that is strongly synergistic with low concentrations of Ca^{2+} ionophore (Brocklehurst et al., 1985; Pocotte et al., 1985). Similar effects are observed in PC12 cells (Pozzan et al., 1984), a cloned rat pheochromocytoma cell line. Surprisingly, secretion induced by phorbol ester in the absence of ionophore occurs predominantly from the norepinephrine-containing cells in primary, bovine chromaffin cell cultures (Cahill and Perlman, 1992). However, Ca^{2+}-dependent secretion induced by phorbol esters in subsequently permeabilized cells occurs from both epinephrine- and norepinephrine-containing cells (M. A. Bittner and R. W. Holz, unpublished observations; A. L. Cahill and R. L. Perlman, unpublished observations). The specificity of effects in intact cells for norepinephrine- rather than epinephrine-containing cells may be a consequence of differences in Ca^{2+} metabolism in the cells. Norepinephrine-containing cells may have higher cytosolic Ca^{2+} concentrations in the absence of Ca^{2+} ionophore.

Knight and Baker found that treatment of electropermeabilized chromaffin cells with tumor-promoting phorbol esters enhances catecholamine secretion (Knight and Baker, 1983). Activation of protein kinase C increased the sensitivity to Ca^{2+} of the exocytotic triggering mechanisms. Subsequent studies with digitonin-permeabilized cells also demonstrated that phorbol ester-pretreated cells respond more vigorously to Ca^{2+} with increases in both Ca^{2+} sensitivity and maximal extent of secretion (Brocklehurst and Pollard, 1985; Pocotte et al., 1985).[1] In chromaffin cells permeabilized with staphylococcus α toxin (which are impermeant to proteins), protein kinase C activation stimulated by incubation with guanine nucleotide also enhances Ca^{2+}-dependent secretion (Bader et al. 1989).[2]

Ninety to ninety-five percent of the total PKC in unstimulated chromaffin cells is soluble when cells are homogenized in isoosmotic solutions containing EGTA (TerBush and Holz, 1986). Bovine chromaffin cells contain mainly the α subtype of protein kinase C, with small amounts of β and ϵ subtypes (B. Strulovici, M. A. Bittner, and R. W. Holz, unpublished observations). For unexplained reasons, homogenization in hypoosmotic solutions decreases the proportion of soluble enzyme to 60 to 80 percent (TerBush and Holz, 1986). Dioctanoylglycerol and phorbol esters that activate protein kinase C increase the percentage of membrane-bound PKC activity in bovine adrenal chromaffin cells to 20 to 50 percent and decrease the rate of efflux of PKC from digitonin-permeabilized cells (TerBush and Holz, 1986). The translocation of protein kinase C to membranes is associated with a large increase in PKC activity measured *in situ* in permeabilized cells with a specific peptide substrate (TerBush and Holz, 1990).

The quantitative relationship between membrane-bound PKC and Ca^{2+}-dependent secretion was determined in cells rendered leaky by digitonin treatment. Intact cells were incubated with various concentrations of TPA to activate and cause translocation of protein kinase C to membrane before permeabilization in the presence of Ca^{2+}. Translocation of as little as 2 to 3 percent of the cellular PKC to the membrane enhanced Ca^{2+}-dependent secretion 25 to 30 percent (TerBush et al., 1988). These results indicate that the secretory pathway in permeabilized cells is exquisitely sensitive to modulation by PKC.

Associated with the increase in membrane-bound protein kinase C is an increase in protein phosphorylation in intact and permeabilized cells (Lee and Holz, 1986). Because many proteins are phosphorylated, such studies do not reveal the PKC substrate that is responsible for the enhancement of secretion. Recent experiments with digitonin-permeabilized cells indicate that phosphorylation induced by membrane-bound protein kinase C, which leads to enhanced secretion, occurs no earlier than 6 minutes before the Ca^{2+} stimulus (M. A. Bittner, D. R. TerBush, and R. W. Holz, manuscript in preparation).

Ca^{2+}-Influx Causes Rapid Translocation of Protein Kinase C to Membranes in Bovine Chromaffin Cells, Which May Enhance the Secretory Response

Nicotinic stimulation or a depolarizing concentration of K^+ causes a rapid (within 2 seconds), transient translocation to membranes of as much as 14 percent of the total cellular PKC activity (TerBush et al., 1988). The translocation of PKC is likely caused by Ca^{2+} influx and a rise in cytosolic Ca^{2+}, since it is entirely dependent on Ca^{2+} in the extracellular medium and occurs at the time of most rapid influx of Ca^{2+} (Holz et al., 1982). The magnitude of the translocation is sufficient to enhance secretion (see above).

The activation of PKC by nicotinic agonist or elevated K^+ may play a role in regulating the intracellular machinery of exocytosis. Staurosporine, a membrane-permeant inhibitor of protein kinase C, at concentrations that inhibit TPA-induced enhancements of elevated K^+-stimulated secretion, inhibits by 20 to 30 percent elevated K^+-stimulated secretion in the absence of TPA. The inhibition did not result from a reduced Ca^{2+} signal, since staurosporine under conditions of the experiment actually increased Ca^{2+} influx[3] (TerBush and Holz, 1990). These experiments provide further evidence that the activation of PKC in intact cells enhances secretion at a step after Ca^{2+} entry.

Protein Kinase C Is Not Required for Exocytosis

Experiments with permeabilized chromaffin cells indicate that activation of PKC modulates but is not necessary for Ca^{2+}-activated secretion. The most direct experiments used a specific pseudosubstrate inhibitory peptide (PKC [19–31]). In permeabilized cells, PKC(19–31) inhibited the phorbol ester-mediated enhancement of Ca^{2+}-dependent secretion as much as 90 percent but had no effect on Ca^{2+}-dependent secretion in the absence of phorbol ester (TerBush and Holz, 1990). The inhibition of the phorbol ester-induced enhancement of secretion by PKC(19–31) was closely correlated with the ability of the peptide to inhibit *in situ* phorbol ester-stimulated PKC activity. PKC(19–31) also blocked TPA-induced phosphorylation of numerous endogenous proteins in permeabilized cells but had no effect on Ca^{2+}-stimulated phosphorylation of tyrosine hydroxylase (TerBush and Holz, 1990).

The use of the peptide, Ca/CaM kinase II(291–317), derived from the calmodulin-binding region of Ca/calmodulin kinase II, demonstrates the specificity of the effects of PKC(19–31). Ca/CaM kinase II(291–317) had no effect on Ca^{2+}-dependent secretion in the presence or absence of phorbol ester. However, the peptide completely blocked the Ca^{2+}-dependent increase in tyrosine hydroxylase phosphorylation while having no effect on TPA-induced phosphorylation of endogenous proteins. The experiments also indicate that Ca/CaM kinase II is not necessary for secretion and suggest that other calmodulin-dependent steps are not important in the final stages of exocytosis.

Experiments with less specific PKC inhibitors also indicate that protein kinase C enhances but is not necessary for secretion (Tachikawa et al., 1990). Polymixin B, sphingosine, H-7, and staurosporine inhibited almost completely the TPA-induced enhancement of Ca^{2+}-evoked secretion from digitonin-permeabilized cells. At the same concentrations, these inhibitors had virtually no effect on Ca^{2+}-induced secretion in the absence of exogenous protein kinase C activators. Similar results were obtained with trypsin, which caused loss of activated PKC from digitonin-permeabilized cells (Holz and Senter, 1988). Low concentrations of trypsin inhibited the TPA-induced enhancement of secretion but did not alter secretion that was stimulated in the absence of TPA.

A number of agents including H-7 and sphingosine inhibited Ca^{2+}-dependent secretion from electropermeabilized cells in the absence of exogenous protein kinase C activators (Knight et al., 1988). Although the drugs inhibit PKC activity *in vitro,* their ability to inhibit phorbol ester-induced enhancements of Ca^{2+}-

dependent secretion under the conditions of the experiments was not investigated. It is, therefore, uncertain whether the reported effects of the drugs in electropermeabilized cells are related to inhibition of PKC.

It is likely that protein kinase C-mediated phosphorylation that is responsible for the enhancement of secretion occurs only minutes before the Ca^{2+} stimulus. Despite the TPA-induced phosphorylation of a host of proteins in the intact cell, subsequent incubation of permeabilized cells with the pseudosubstrate inhibitor PKC(19–31) almost completely inhibits the secretory effects caused by prior activation of the enzyme. In addition, secretion from permeabilized cells is not enhanced by prior treatment of intact cells with TPA if cells are permeabilized for several minutes in the absence of ATP (to inhibit phosphorylation) before stimulation with Ca^{2+} (D. R. TerBush and R. W., Holz unpublished observations). Thus, not only does the effect of activated protein kinase C require ATP (Lee and Holz, 1986), but the ATP must be present within minutes of the Ca^{2+} trigger of exocytosis. These data also make it unlikely that basal levels of activated PKC in intact cells result in a long-lived phosphoprotein that is necessary for secretion. This conclusion is supported by the finding that inhibition of basal protein kinase C activity in intact cells for 25 minutes with staurosporine has little effect on subsequent secretion from permeabilized cells not preincubated with TPA (TerBush and Holz, 1990).

Down-regulation of protein kinase C in chromaffin cells by prolonged incubation with phorbol esters inhibits nicotinic secretion from intact cells (Burgoyne et al., 1988; Wilson, 1990) and can alter secretion from permeabilized cells. In one study, despite continued reduction of protein kinase C, nicotinic agonist-induced secretion recovered several days later (Wilson, 1990). Thus it seems unlikely that the loss of secretory response reflects an essential role for PKC in the intracellular pathway of exocytosis. The effects of down-regulation of protein kinase C on secretion from intact cells may reflect effects of PKC on cholinergic mechanisms, Ca^{2+} channels (TerBush and Holz, 1990), or Na^+ channels (Wilson, 1990).

We found that prolonged incubation with TPA results in loss of the TPA-induced enhancement of secretory response in subsequently permeabilized cells that correlated with the loss of protein kinase C. However, secretion was virtually the same as secretion in the complete absence of phorbol ester (Bittner and Holz, 1986). In another study, down-regulation of PKC resulted in a 42- to 70-percent decrease in Ca^{2+}-dependent secretion from digitonin-permeabilized cells (Burgoyne et al., 1988).

Effects of down-regulation of protein kinase C are difficult to interpret. TPA, 1 μM, after 24 hours causes loss of 85 to 90 percent of the total cellular protein kinase C, but all of the remaining activity is membrane bound (TerBush et al., 1988). This represents a two- to three-fold *increase* in membrane-bound protein kinase C compared with untreated cells. Thus, changes in secretion could be caused by either a loss of cytosolic PKC or a prolonged increase in membrane-bound, active PKC.

In PC12 cells, prolonged incubation with phorbol ester results in complete loss of protein kinase C activity (Matthies et al., 1988). Secretion is similar in these cells devoid of PKC and in cells not treated with phorbol ester. Thus, protein kinase C is not necessary for secretion in PC12 cells.

STUDIES WITH MAST CELLS

The role of protein kinase C in secretion from mast cells is complex and probably involves modulation of a number of different processes. One of the complications in interpreting experiments with populations of mast cells is the finding that fractional release of granule contents from the population does not necessarily reflect the response of individual mast cells (Tatham and Gomperts, 1991; discussed in Lindau and Gomperts, 1991). Instead, some mast cells respond to stimuli by undergoing exocytosis of all of the secretory vesicles, some undergo partial degranulation, and some do not undergo any degranulation. The measured population response reflects the average of these disparate cell responses. This is unlike the situation in chromaffin cells in which the response in the population of cells reflects individual cell responses (Wightman et al., 1991; Jankowski et al., 1992).

The difficulty in relating the course of a secretory event in a single cell to secretion in a population is demonstrated by comparing results in patch clamp studies (which monitor single-cell responses by measuring the increased capacitance associated with increases in plasma membrane area) to results from studies of permeabilized cells (which report the release of vesicular contents of a large number of cells). When mast cells are permeabilized with GTPγS and low Ca^{2+} (50 nM) in chloride solution, no secretion is observed. However, when mast cells are patched with GTPγS and very low Ca^{2+} in chloride solution, degranulation occurs in about half of the cells, but the onset is slow (8–20 min) (Lindau and Gomperts, 1991). In those cells that do initiate secretion, degranulation is rapid and is complete within 1 to 2 minutes. Secretion is not detected in the permeabilized cells, probably because the latency period at this low Ca^{2+} concentration extends beyond the usual time course of such experiments. Again, with a glutamate rather than chloride solution[4] in the patch pipette, the same concentrations of Ca^{2+} and GTPγS stimulate complete degranulation after a delay of 60 to 90 seconds, but at a slower rate. When a similar experiment is done with permeabilized cells, this result is manifest as a partial secretory response.

In experiments with permeabilized mast cells, a given manipulation could alter the probability of a given cell undergoing complete exocytosis, the latency before exocytosis commences, or the extent of secretion. It is difficult to interpret apparent changes in affinity for different effectors (Ca^{2+} or GTPγS). The following section discusses possible roles for protein kinase C in light of these issues.

Activation of Protein Kinase C Can Reduce or Eliminate the Requirement for the Presence of Both Ca^{2+} and GTPγS.

Gomperts and colleagues, using streptolysin-O permeabilized mast cells (pretreated with metabolic inhibitors) (Howell et al. 1987), and Koopman and Jackson, using digitonin-permeabilized mast cells (Koopman and Jackson, 1990), found that the combination of both Ca^{2+} and GTPγS is necessary for histamine secretion in chloride solution. Exogenous ATP is not necessary. Koopman and Jackson found that pretreatment of intact cells with TPA for 5 minutes permitted either Ca^{2+} alone or GTPγS alone to stimulate secretion during a 5-minute incubation of subsequently digitonin-permeabilized cells in the absence of ATP. The effects of TPA could be blocked by the kinase inhibitor staurosporine and by prolonged incubation with TPA, which reduced protein kinase C to undetectable levels. However, secretion in response to the combination of Ca^{2+} and GTPγS was vigorous in the presence of staurosporine, and was substantial after downregulation of PKC.

For reasons that are unclear, Gomperts and colleagues did not observe similar effects in most experiments (Howell et al., 1989) (but see Figure 6 in Koffer et al., 1990). Their experiments were done in chloride solution with streptolysin-O-permeabilized cells that had been pretreated with metabolic inhibitors. Treatment of intact cells with TPA before permeabilization resulted in increased apparent affinities for Ca^{2+} and GTPγS, but both effectors were still necessary. However, the ability of protein kinase C to permit Ca^{2+} alone to trigger secretion is suggested in experiments by Gomperts and colleagues done under conditions in which secretion requires ATP (after the cells have run down). In these experiments PKC inhibitors block secretion in the absence of phorbol esters (Churcher et al., 1990).

These studies considered together allow three conclusions to be drawn. (1) Protein kinase C is not necessary for secretion from mast cells (Koopman and Jackson, 1990). (2) Neither Ca^{2+} nor GTPγS acts through activation of protein kinase C in the absence of ATP (Koopman and Jackson, 1990). (3) Protein kinase C activation can mimic either arm of the dual effector system in mast cells (Churcher and Gomperts, 1990; Churcher et al., 1990; Koffer et al., 1990; Koopman and Jackson, 1990). The data suggest that phosphorylation by protein kinase C activates effectors normally activated by Ca^{2+} and GTPγS. Alternatively, protein kinase C activates another secretory pathway.

Activation of Protein Kinase C Reduces Delay in the Secretory Response in the Presence of ATP

As mentioned above, ATP is not necessary for secretion stimulated by the combination of Ca^{2+} and GTPγS shortly after streptolysin-O permeabilization (Tatham and Gomperts, 1989; Koopman and Jackson, 1990). Tatham and Gomperts found that the presence of ATP during stimulation with the combination of Ca^{2+} and GTPγS in chloride solution profoundly affects the kinetics of hexosaminidase secretion in streptolysin-O-permeabilized cells (Tatham and Gomperts, 1989). In the absence of ATP, secretion begins within a few seconds on introduction of Ca^{2+} and GTPγS. ATP delays the onset of secretion by as much as 1 minute. It is unclear whether the delay requires the hydrolysis of ATP (Tatham and Gomperts, 1989; Lillie et al., 1991). In cells not pretreated with metabolic inhibitors, pretreatment with TPA for 5 minutes greatly reduces the delay (Tatham and Gomperts, 1989). The data suggest that when cells are stimulated soon after permeabilization in the presence of ATP, protein kinase C-mediated phosphorylation speeds processes that are necessary to initiate secretion. This may represent an effect of PKC that is distinct from its ability to restore secretion after the cells have completely lost their secretory response in the absence of ATP (Churcher et al., 1990).

Phorbol Ester Can Restore a Requirement for GTPγS in ATP-dependent Secretion

Secretion that is *dependent* on exogenous ATP after cells have been permeabilized for several minutes can be stimulated by Ca^{2+} alone (Churcher et al., 1990). A requirement for GTPγS in addition to Ca^{2+} is restored in ATP-dependent secretion if cells are pretreated with phorbol ester or okadaic acid (a specific inhibitor of protein phosphatases types 1 and 2A) before permeabilization (Churcher et al., 1990). By increasing the level of phosphorylation, a role for guanine nucleotides is revealed on addition of ATP. Gomperts and colleagues hypothesize that guanine nucleotides activate a protein phosphatase that dephosphorylates and inactivates a protein that inhibits secretion. Thus there may be two substrates for protein kinase C. One is readily phosphorylated by basal PKC activity and enables the cells to respond to Ca^{2+} alone. If phosphorylation is enhanced further by phorbol ester or okadaic acid, then an inhibitor of secretion is activated, an effect that can be reversed by a guanine nucleotide-activated protein phosphatase.

RESTORATION OF SECRETION BY ADDITION OF PROTEIN KINASE C TO PERMEABILIZED CELLS

Protein kinase C addition to permeabilized chromaffin cells (Morgan and Burgoyne, 1992b), PC12 cells (Nishizaki et al., 1992) or pituitary cells (Naor et al., 1989) causes a small enhancement of secretion. In pituitary cells, the α and β but not the γ subtypes of PKC are effective.

In a recent report, a striking reconstitution of Ca^{2+}-dependent, antigen-stimulated secretion from streptolysin-O permeabilized RBL-2H3 cells was accomplished by addition of purified protein kinase C (Ozawa et al., 1993). There was no stimulated secretion from permeabilized cells preincubated for 15 minutes without PKC. The β and δ subtypes were especially effective in maintaining secretion and permitted secretion comparable to that observed in intact cells. The α subtype was also able to reconstitute secretion. Since the δ subtype of protein kinase C does not require Ca^{2+}, the study suggests an action for Ca^{2+} independent of activation of PKC. In these cells, down-regulation of protein kinase C subtypes α and β with prolonged incubation with phorbol ester was associated with a 70-percent reduction of the secretory response. The remaining secretory response may have been supported by the δ subtype, which was not lost from the cells. It therefore appears that in RBL-2H3 cells, protein kinase C is necessary for antigen-stimulated secretion and that different PKC subtypes have different capacities to support secretion.

PROTEINS THAT MAY MEDIATE THE PROTEIN KINASE C-INDUCED ENHANCEMENT OF SECRETION

Exo1 and Protein 14-3-3

Digitonin-permeabilized chromaffin cells lose proteins and lose their ability to secrete in response to Ca^{2+}. Two groups simultaneously purified similar proteins of 27 to 30 or 44 kDa based on their ability to maintain or restore the secretory response (Morgan and Burgoyne, 1992a; Wu et al., 1992). Exo1 (44 kDa) was purified from brain cytosol (Morgan and Burgoyne, 1992a) and the 27 to 30-kDa proteins were purified from adrenal medullary cytosol (Wu et al., 1992). The effects of exogenous exo1 are enhanced by prior activation of protein kinase C before permeabilization (Morgan and Burgoyne, 1992a) or by preincubation of permeabilized chromaffin cells together with partially purified protein kinase C (Morgan and Burgoyne, 1992b). Half-maximal effects of exo1 were obtained at approximately 100 μg/ml purified protein. Because the protein is *not* a substrate for protein kinase C (Morgan and Burgoyne, 1992b), it must in some manner modulate the action of the relevant protein kinase C substrate or modulate phosphorylation by protein kinase C.

Both the 27 to 30-kDa proteins (Wu et al., 1992) and exo1 (Morgan and Burgoyne, 1992a) were shown by protein sequencing to be members of the 14-3-3 family of proteins. The stimulatory effects of adrenal cytosol on secretion from digitonin-permeabilized chromaffin cells could be inhibited by antibody to 14-3-3 (Wu et al., 1992). The inhibition could be prevented by the addition of purified protein 14-3-3 to the cytosol. Other proteins in cytosol also maintain the secretory response, including exo2 (Morgan and Burgoyne, 1992a).

Protein 14-3-3 was first isolated from brain as an acidic protein without known function (Moore and Perez, 1967). It is an abundant protein comprising 1 percent of the total cytosolic protein in brain (Boston et al., 1982). It was recently identified as the protein that is required by Ca^{2+}/calmodulin kinase II (Yamauchi et al., 1981; Ichimura et al., 1988) to activate brain tyrosine hydroxylase and tryptophan hydroxylase. The activating effect on the hydroxylases is similar to the effect in exocytosis in that the 14-3-3 enables a Ca^{2+}-activated protein kinase to have a physiological effect. Indeed, it is reported that protein 14-3-3 activates protein kinase C *in vitro* by stimulating Ca^{2+}/phospholipid-dependent activity approximately twofold (Isobe et al., 1992). However, activation was not observed in another study (Morgan and Burgoyne, 1992b). Exo1 may enhance secretion by enhancing basal and phorbol ester-induced PKC activity.

Another member of the 14-3-3 family that has recently been cloned has arachidonate-specific phospholipase A2 activity for phosphatidylcholine and phosphatidylethanolamine (Zupan et al., 1992). The 14-3-3-type proteins may, therefore, be regulators of enzymes involved in intracellular signaling.

p145, A soluble Rat Brain Protein

Martin and colleagues developed a method using a "cell cracker" that has proven to be the most effective means for depleting cells of soluble cytosolic constituents necessary for secretion (Martin and Walent, 1989; Walent et al., 1992). Using the semiintact PC12 (and GH$_3$) cells, they have identified a 145-kDa cytosolic protein that is necessary for reconstitution of Ca^{2+}-dependent secretion (Walent et al., 1992).

Martin and colleagues have recently found that cells and cytosol depleted of protein kinase C are able to sustain vigorous exocytosis (Nishizaki et al., 1992). The results confirm that PKC is not necessary for secretion (Tachikawa et al., 1990; TerBush and Holz, 1990). Secretion from semiintact cells is further stimulated by 20 to 25 percent when purified PKC is added to PKC-deficient cytosol. The enhancement of secretion is strongest with fractions enriched with p145, although other fractions also support a protein kinase C-dependent enhancement of secretion. Most interestingly, p145 is a substrate for PKC and is phosphorylated under conditions in which protein kinase C enhances secretion (Nishizaki et al., 1992). These studies are promising and suggest that p145 may be a substrate that is responsible for the ability of PKC to enhance secretion. However, the stimulation of secretion was small and the results do not rule out that other proteins may also be substrates that contribute to the enhancement of secretion caused by protein kinase C.

Annexin II or Calpactin I

The annexins are a class of proteins that bind to acidic phospholipids in a Ca^{2+}-dependent manner. The first annexin, synexin, was purified from adrenal chromaffin cells (Creutz et al., 1978) and was postulated to play a role in exocytosis. More recently annexin II or calpactin I has been reported to maintain secretion in digitonin-permeabilized cells (Ali et al., 1989), an effect that may require protein kinase C (Sarafian et al., 1991).

Annexin II tetramer ($M_r = 90$ kDa; also called calpactin I) consists of two heavy chains (p36) linked by two light chains (p11). The light chains bind to the 3-kDa tail at the amino terminus of each p36. The light chains of the dimers interact to form the tetramer. p36 is isolated both as part of the tetramer and as a monomer. p36 was first identified as a substrate for the Rous sarcoma virus tyrosine kinase (Erikson and Erikson, 1980) and was subsequently demonstrated to associate with actin. Because the association with actin requires millimolar Ca^{2+}, the association may not be of physiological importance. Glenney (1986) demonstrated that p36 and the core protein p33 (p36 without the 3-kDa amino terminus) bind to acidic phospholipids in response to micromolar Ca^{2+} with two binding sites for Ca^{2+} on each heavy chain or core. It is likely that the core protein has two binding sites for lipid, since p33 aggregates phosphatidylserine (PS) vesicles (Glenney et al., 1987). Creutz and colleagues demonstrated that p36 was one of a number of proteins that bound in a Ca^{2+}-dependent manner to chromatography columns of chromaffin granule membranes linked to Sepharose (Creutz et al., 1987).

Phosphorylation of the N-terminus of the annexins has variable effects. Tyrosine phosphorylation of the N-terminus of p36 reduces its affinity for lipids (Powell and Glenney, 1987). Serine phosphorylation of the N-terminus of p36 reduces its association with light chain (Johnsson et al., 1986). Recently it was demonstrated that protein kinase C-induced phosphorylation *reduced* the ability of annexin II tetramer to aggregate lipid vesicles but not to bind to vesicles (Johnstone et al., 1992).

The first direct evidence for a role of annexin II and p36 in secretion came from experiments by Ali and colleagues (Ali et al., 1989; Ali and Burgoyne, 1990). Ca^{2+}-dependent secretion from digitonin-permeabilized chromaffin cells decreases with time after permeabilization, possibly because proteins (including calpactin I) exit from the cells. The annexin II tetramer (90 kDa) and p36 similarly maintain the secretory response at 10 μM Ca^{2+}. A subsequent series of experiments provided additional support for a role of annexin II in secretion (Sarafian et al., 1991). In this study annexin II tetramer was much more effective than p36 in maintaining secretion in permeabilized chromaffin cells. Most importantly, the effects required phosphorylation by protein kinase C. In chromaffin cells preincubated with TPA for 24 hours to deplete cells of PKC, annexin II needed to be phosphorylated by PKC *in vitro* in order to enhance secretion. Annexin II is localized in the periphery of the cell in close proximity to chromaffin granules and the plasma membrane (Nakata et al., 1990) and is phosphorylated on serine residues in intact cells upon nicotinic stimulation (Creutz et al., 1987). Since protein kinase C is activated on nicotinic stimulation (TerBush et al., 1988), the effects of phosphorylated annexin II observed in permeabilized cells may reflect a function for annexin II and PKC in modulating secretion in intact cells. However, it is difficult to reconcile these results and ideas with the recent finding that protein kinase C-induced phosphorylation of annexin II tetramer greatly *reduces* the ability of annexin II tetramer to aggregate lipid vesicles (Johnstone et al., 1992). It is likely that important aspects of the role of the annexins in secretion remain to be elucidated.

Although annexin II may play a modulatory role in secretion, there is compelling evidence that cytosolic proteins other than annexin II are required to maintain secretion. Annexin II-depleted cytosol enhances secretion (Wu and Wagner, 1991). As discussed above, p145 from cytosol is required for secretion from PC12 ghosts (Walent et al., 1992). Furthermore, the effects of annexin II are not apparent in all preparations. Annexin II with or without protein kinase C failed to stimulate secretion in PC12 ghosts (Nishizaki et al., 1992).

Chromobindin 9

Chromobindin 9 is a 37-kDa substrate for protein kinase C that binds in a Ca^{2+}-dependent manner to immobilized chromaffin granule membranes (Summers and Creutz, 1985). It is phosphorylated in chromaffin cells when secretion is stimulated by a variety of secretogogues (Michener et al., 1986). Because treatment of chromaffin cells with phorbol ester resulted in phosphorylation of chromobindin 9 that was only 30 percent of the phosphorylation obtained with nicotinic stimulation, it was suggested that the protein may also be a substrate for other protein kinases.

Pleckstrin

In platelets, agonists that stimulate polyphosphoinositide synthesis cause the rapid phosphorylation of a 40 to 47-kDa protein through the activation of protein kinase C (Lyons et al., 1975; Haslam and Lynham, 1977; Kaibuchi et al., 1983; Nishizuka, 1986). The phosphorylation is associated with platelet activation that includes shape changes and secretion. The protein, pleckstrin, is expressed in platelets and various differentiated white blood cells and was the first PKC substrate postulated to mediate the effects of protein kinase C on secretion. However, detailed studies reveal a poor correlation between the phosphorylation of pleckstrin and serotonin secretion (Haslam and Davidson, 1984a; Lapetina et al., 1985), making it unlikely that pleckstrin is the critical protein kinase C substrate in the secretory pathway. cDNA cloning reveals several likely phosphorylation sites but no obvious function for the protein (Tyers et al., 1988, 1989).

Protein Kinase C Reduces Cytoskeletal Actin in Secretory Cells

It has been suggested that a dense cytoskeletal network underlying the plasma membrane prevents secretory vesicle access to the plasma membrane and thereby reduces exocytosis (Orci et al., 1972; Perrin and Aunis, 1985; Trifaro et al., 1985; Sontag et al., 1988). Indeed, reorganization of the cortical cytoskeleton sometimes occurs on secretion and may contribute to the secretory response (Perrin and Aunis, 1985; Cheek and Burgoyne, 1986; Burgoyne et al., 1989; Koffer et al., 1990; Vitale et al., 1991). A decrease in the cytoskeleton is not in itself the trigger for exocytosis. A nicotinic agonist-induced decrease in the cytoskeleton in bovine chromaffin cells is not Ca^{2+} dependent, whereas secretion stimulated by nicotinic agonist is Ca^{2+} dependent (Cheek and Burgoyne, 1986; Burgoyne et al., 1989).

Ca^{2+} regulates the cytoskeleton. In bovine chromaffin cells, elevated K^+ induces a decrease in the f-actin that is Ca^{2+} dependent (Burgoyne et al., 1989). There is no explanation for the difference in Ca^{2+} dependencies of cytoskeletal disruption between nicotinic agonist and elevated K^+. Micromolar Ca^{2+} causes a decrease in cytoskeletal actin in permeabilized chromaffin (Burgoyne et al., 1989) and mast cells (Koffer et al., 1990). In permeabilized chromaffin cells, the Ca^{2+}-dependent decrease in cytoskeletal actin is transient and is not associated with an obvious effect on the rate of secretion (Burgoyne et al., 1989).

Protein kinase C could be one of the modulators of cytoskeletal changes. Protein kinase C associates with cytoskeletal elements *in situ* (Jaken et al., 1989), binds to specific proteins of the cytoskeleton *in vitro* (Matsuuchi et al., 1988), and phosphorylates cytoskeletal proteins (Naka et al., 1983; Werth et al., 1983; Umekawa and Hidaka, 1985; Burgoyne et al., 1986; Gould et al., 1986). Treatment of chromaffin cells with TPA results in a decrease in cytoskeletal actin (Burgoyne et al., 1989), although probably not in the release of g-actin (Cheek and Burgoyne, 1986). Phorbol ester also decreases rhodamine-phalloidin staining of cortical actin in mast cells (Koffer et al., 1990), a likely indication of the partial dissolution of the cortical cytoskeletal network. Thus, physiological activation of protein kinase C may alter the putative cytoskeletal barrier to secretion.

CONCLUSIONS

There is compelling evidence that the activation of protein kinase C results in enhanced secretory response because of alterations in the intracellular pathway of regulated exocytosis. Phosphorylation by PKC can increase both the apparent affinity for Ca^{2+} or other activators of secretion and the extent of secretion. Effects of PKC on secretion may vary with enzyme subtype. Despite a wealth of information concerning the effects of PKC activation on secretion, the substrates responsible for these effects are uncertain. The major limitation is the lack of understanding of the biochemical machinery that is required for exocytosis. However, there is substantial progress in defining cytosolic proteins that maintain or modulate secretion (see above) and in defining secretory vesicle membrane components that may play a role in secretion (De Camilli and Jahn, 1990). In addition, a kinetic description of Ca^{2+}-activated secretion has revealed discrete steps in the pathway (Bittner and Holz, 1992a,b). These advances are likely to lead to the identification of the relevant protein kinase C substrates and the elucidation of their role in secretion.

NOTES

1. Although in bovine chromaffin cells, phorbol ester causes little secretion in the virtual absence of Ca^{2+}, phorbol ester enhances Ca^{2+}-independent as well as Ca^{2+}-dependent secretion in RINm5F (Vallar et al., 1987) and PC12 cells (Peppers and Holz, 1986). Addition of purified protein kinase C to permeabilized anterior pituitary cells enhances secretion in 0.1 μM Ca^{2+} (Naor et al., 1989).

2. Related effects are observed with streptolysin-O-permeabilized acinar cells (Kitagawa et al., 1990).

3. In early experiments it was difficult to demonstrate that activation of protein kinase C with exogenous activators of the enzyme increased secretion stimulated by nicotinic agonists or elevated K^+. Prolonged activation of PKC inhibits elevated K^+-induced Ca^{2+} influx, thereby confounding stimulatory effects of protein kinase C on the secretory machinery (TerBush and Holz, 1990). However, short incubations (5 min or less) with TPA enhanced by 25 percent elevated K^+-induced secretion in spite of a 20-percent inhibition of $^{45}Ca^{2+}$ influx.

4. The characteristics of exocytosis from permeabilized mast cells are strongly dependent on experimental protocol and have been recently reviewed by Lindau and Gomperts (1991). The principle anion in the solutions has important effects on the secretory response. Most experiments with permeabilized mast cells have been conducted in chloride-based solutions in which the combination of Ca^{2+} and a nucleotide capable of activating GTP-binding proteins is necessary for secretion of histamine (Howell et al., 1987; Koopman

and Jackson, 1990) and N-acetylglucosaminidase (hexosaminidase) (Churcher and Gomperts, 1990). Chloride is used rather than glutamate (which gives optimal secretion in chromaffin cells) because glutamate interferes with the determination of histamine (Churcher and Gomperts, 1990). In solutions containing glutamate or other organic anions, either Ca^{2+} or GTPγS alone stimulates hexosaminidase secretion, although there is still synergy between these two effectors (Churcher and Gomperts, 1990). Differences in results in chloride and glutamate solutions may be related to the observation that chloride- but not glutamate-based solutions cause marked morphological alterations during patch clamp experiments, possibly because of changes in the cytoskeleton (Almers and Neher, 1987).

REFERENCES

Agranoff, B. W., P. Murthy, and E. B. Seguin. 1983. Thrombin-induced phosphodiesteratic cleavage of phosphatidylinositol bisphosphate in human platelets. *J. Biol. Chem.* 258:2076–2078.

Ali, S. M., and R. D. Burgoyne. 1990. The stimulatory effect of calpactin (annexin II) on calcium-dependent exocytosis in chromaffin cells: Requirement for both the N-terminal and core domains of p36 and ATP. *Cell Signal.* 2:265–276.

Ali, S. M., M. J. Geisow, and R. D. Burgoyne. 1989. A role for calpactin in calcium-dependent exocytosis in adrenal chromaffin cells. *Nature* 340:313–315.

Almers, W., and E. Neher. 1987. Gradual and stepwise changes in the membrane capacitance of rat peritoneal mast cells. *J. Physiol. (Lond.)* 386:205–217.

Bader, M.-F., J.-M. Sontag, D. Thierse, and D. Aunis. 1989. A reassessment of guanine nucleotide effects on catecholamine secretion from permeabilized adrenal chromaffin cells. *J. Biol. Chem.* 264:16426–16434.

Barrie, A. P., D. G. Nicholls, J. Sanchez-Prieto, and T. S. Sihra. 1991. An ion channel locus for the protein kinase C potentiation of transmitter glutamate release from guinea pig cerebrocortical synaptosomes. *J. Neurochem.* 57:1398–1404.

Bittner, M. A., and R. W. Holz. 1986. Evidence that protein kinase C modulates but is not necessary for Ca^{2+}-dependent catecholamine secretion in digitonin-permeabilized adrenal chromaffin cells. *Soc. Neurosci. Abstr.* 12:998.

Bittner, M. A., and R. W. Holz. 1992a. Kinetic analysis of secretion from permeabilized adrenal chromaffin cells reveals distinct components. *J. Biol. Chem.* 267:16219–16225.

Bittner, M. A., and R. W. Holz. 1992b. A temperature-sensitive step in exocytosis. *J. Biol. Chem.* 265:16226–16229.

Boston, P. F., P. Jackson, and R. J. Thompson. 1982. Human 14-3-3 protein: Radioimmunoassay, tissue distribution and cerebrospinal fluid levels in patients with neurological disorders. *J. Neurochem.* 38:1475–1482.

Brass, L. R., M. Laposata, H. S. Banga, and S. E. Rittenhouse. 1986. Regulation of the phosphoinositide hydrolysis pathway in thrombin-stimulated platelets by a pertussis toxin-sensitive guanine nuclotide-binding protein. *J. Biol. Chem.* 261:16838–16847.

Brocklehurst, K. W., and H. B. Pollard. 1985. Enhancement of Ca^{2+}-induced catecholamine release by the phorbol ester TPA in digitonin-permeabilized cultured bovine adrenal chromaffin cells. *FEBS Lett.* 183:107–110.

Brocklehurst, K. W., K. Morita, and H. B. Pollard. 1985. Characterization of protein kinase

C and its role in catecholamine secretion from bovine adrenal-medullary cells. *Biochem. J.* 228:35–42.

Burgoyne, R. D., T. R. Cheek, and K.-M. Norman. 1986. Identification of a secretory granule-binding protein as caldesmon. *Nature* 319:68–70.

Burgoyne, R. D., A. Morgan, and A. J. O'Sullivan. 1988. A major role for protein kinase C in calcium-activated exocytosis in permeabilised adrenal chromaffin cells. *FEBS. Lett.* 238:151–155.

Burgoyne, R. D., A. Morgan, and A. J. O'Sullivan. 1989. The control of cytoskeletal actin and exocytosis in intact and permeabilized adrenal chromaffin cells: Role of calcium and protein kinase C. *Cell. Signal.* 1:323–334.

Cahill, A. L., and R. L. Perlman. 1992. Phorbol esters cause preferential secretion of norepinephrine from bovine chromaffin cells. *J. Neurochem.* 58:768–771.

Cheek, T. R., and R. D. Burgoyne. 1986. Nicotine-evoked disassembly of cortical actin filaments in adrenal chromaffin cells. *FEBS Lett.* 207:110–114.

Churcher, Y., and B. D. Gomperts. 1990. ATP-dependent and ATP-independent pathways of exocytosis revealed by interchanging glutamate and chloride as the major anion in permeabilized mast cells. *Cell Regul.* 1:337–346.

Churcher, Y., I. M. Kramer, and B. D. Gomperts. 1990. Evidence for protein dephosphorylation as a permissive step in GTP-gamma-S-induced exocytosis from permeabilized mast cells. *Cell Regul.* 1:523–530.

Coorssen, J. R., M. M. Davidson, and R. J. Haslam. 1990. Factors affecting dense and alpha-granule secretion from electropermeabilized human platelets: Ca(2+)-independent actions of phorbol ester and GTP gamma S. *Cell Regul.* 1:1027–1041.

Creutz, C. E., C. J. Pazoles, and H. B. Pollard. 1978. Identification and purification of an adrenal medullary protein (synexin) that causes calcium-dependent aggregation of isolated chromaffin granules. *J. Biol. Chem.* 253:2858–2866.

Creutz, C. E., W. J. Zaks, H. C. Hamman, S. Crane, W. H. Martin. K. L. Gould, K. M. Oddie, and S. J. Parsons. 1987. Identification of chromaffin granule-binding proteins. Relationship of the chromobindins to calelectrin, synhibin, and the tyrosine kinase substrates p35 and p36. *J. Biol. Chem.* 262:1860–1868.

De Camilli, P., and R. Jahn. 1990. Pathways to regulated exocytosis in neurons. *Annu. Rev. Physiol.* 52:625–645.

Dunn, L. A., and R. W. Holz. 1983. Catecholamine secretion from digitonin-treated adrenal medullary chromaffin cells. *J. Biol. Chem.* 258:4989–4993.

Erikson, E., and R. L. Erikson. 1980. Identification of a cellular protein substrate phosphorylated by the avian sarcoma virus-transforming gene product. *Cell* 21:829–836.

Glenney, J. 1986. Phospholipid-dependent Ca^{2+} binding by the 36-kDa tyrosine kinase substrate (calpactin) and its 33-kDa core. *J. Biol. Chem.* 261:7247–7252.

Glenney, J. R., B. Tack, and M. A. Powell. 1987. Calpactins: Two distinct Ca^{2+}-regulated phospholipid- and actin-binding proteins isolated from lung and placenta. *J. Cell. Biol.* 104:503–511.

Gomperts, B. D. 1990. GE: A GTP-binding protein mediating exocytosis. *Annu. Rev. Physiol.* 52:591–606.

Gould, K. L., J. R. Woodgett, C. M. Isack, and T. Hunter. 1986. The protein-tyrosine kinase substrate (p36) is a substrate for protein kinase C in vitro and in vivo. *Molec. Cell Biol.* 6:2738–2744.

Haslam, R. J., and M. M. Davidson. 1984a. Potentiation by thrombin of the secretion of serotonin from permeabilized platelets equilibrated with Ca2+ buffers. Relationship to protein phosphorylation and diacylglycerol formation. *Biochem. J.* 222: 351–361.

Haslam, R. J., and M. M. Davidson. 1984b. Guanine nucleotides decrease the free [Ca2+] required for secretion of serotonin from permeabilized blood platelets. Evidence of a role for a GTP-binding protein in platelet activation. *FEBS Lett.* 174:90–95.

Haslam, R. J., and J. A. Lynham. 1977. Relationship between phosphorylation of blood platelet proteins and secretion of platelet granule constituents. I. Effects of different aggregating agents. *Biochem. Biophys. Res. Commun.* 77:714–722.

Holz, R. W., and R. A. Senter. 1988. Effects of trypsin on secretion stimulated by micromolar Ca2+ and phorbol ester in digitonin-permeabilized adrenal chromaffin cells. *Cell Molec. Neurobiol.* 8:115–128.

Holz, R. W., R. A. Senter, and R. A. Frye. 1982. Relationship between Ca2+ uptake and catecholamine secretion in primary dissociated cultures of adrenal medulla. *J. Neurochem.* 39:635–646.

Holz, R. W., M. A. Bittner, S. C. Peppers, R. A. Senter, and D. A. Eberhard. 1989. MgATP-independent and MgATP-dependent exocytosis. Evidence that MgATP primes adrenal chromaffin cells to undergo exocytosis. *J. Biol. Chem.* 264:5412–5419.

Howell, T. W., S. Cockcroft, and B. D. Gomperts. 1987. Essential synergy between Ca^{2+} and guanine nucleotides in exocytotic secretion from permeabilized rat mast cells. *J. Cell Biol.* 105:191–197.

Howell, T. W., I. M. Kramer, and B. D. Gomperts. 1989. Protein phosphorylation and the dependence on Ca^{2+} and GTP-gamma-S for exocytosis from permeabilised mast cells. *Cell. Signal.* 1:157–163.

Ichimura, T., T. Isobe, T. Okuyama, N. Takahashi, K. Araki, R. Kuwano, and Y. Takahashi. 1988. Molecular cloning of cDNA coding for brain-specific 14-3-3 protein, a protein kinase-dependent activator of tyrosine and tryptophan hydroxylases. *Proc. Natl. Acad. Sci U.S.A.* 85:7084–7088.

Isobe, T., Y. Hiyane, T. Ichimura, T. Okuyama, N. Takahashi, S. Nakajo, and K. Nakaya. 1992. Activation of protein kinase C by the 14-3-3 proteins homologous with exo1 protein that stimulates calcium-dependent exocytosis. *FEBS Lett.* 308:121–124.

Jaken, S., K. Leach, and T. Klauck. 1989. Association of type 3 protein kinase C with focal contacts in rat embryo fibroblasts. *J. Cell Biol.* 109:697–704.

Jankowski, J. A., T. J. Schroeder, R. W. Holz, and R. M. Wightman. 1992. Quantal secretion of catecholamines measured from individual bovine adrenal medullary cells permeabilized with digitonin. *J. Biol. Chem.* 267:18329–18335.

Johnsson, N., P. N. Van, H.-D. Soling, and K. Weber. 1986. Functionally distinct serine phosphorylation sites of p36, the cellular substrate of retroviral protein kinase: Differential inhibition of reassociation with p11. *EMBO J.* 5:3455–3460.

Johnstone, S. A., I. Hubaishy, and D. M. Waisman. 1992. Phosphorylation of annexin II tetramer by protein kinase C inhibits aggregation of lipid vesicles by the protein. *J. Biol. Chem.* 267:25976–25981.

Kaibuchi, K., Y. Takai, M. Sawamura, M. Hoshijima, T. Fujikura, and Y. Nishizuka. 1983. Synergistic functions of protein phosphorylation and calcium mobilization in platelet activation. *J. Biol. Chem.* 258:6701–6704.

Kitagawa, M., J. A. Williams, and R. C. De Lisle. 1990. Amylase release from streptolysin O-permeabilized pancreatic acini. *Am. J. Physiol.* 259:G157–G164.

Knight, D. E., and P. F. Baker. 1982. Calcium-dependence of catecholamine release from bovine adrenal medullary cells after exposure to intense electric fields. *J. Membr. Biol.* 68:107–140.

Knight, D. E., and P. F. Baker. 1983. The phorbol ester TPA increases the affinity of exocytosis for calcium in 'leaky' adrenal medullary cells. *FEBS. Lett.* 160:98–100.

Knight, D. E., and M. C. Scrutton. 1986. Effects of guanine nucleotides on the properties of 5-hydroxytryptamine secretion from electropermeabilised human platelets. *Eur. J. Biochem.* 160:183–190.

Knight, D. E., V. Niggli, and M. C. Scrutton. 1984. Thrombin and activators of protein kinase C modulate secretory responses of permeabilised human platelets induced by Ca2+. *Eur. J. Biochem.* 143:437–446.

Knight, D. E., D. Sugden, and P. F. Baker. 1988. Evidence implicating protein kinase C in exocytosis from electropermeabilized bovine chromaffin cells. *J. Membr. Biol.* 104:21–34.

Knox, R. J., E. A. Quattrocki, J. A. Connor, and L. K. Kaczmarek. 1992. Recruitment of Ca^{2+} channels by protein kinase C during rapid formation of putative neuropeptide release sites in isolated *Aplysia* neurons. *Neuron* 8:883–889.

Koffer, A., P. E. Tatham, and B. D. Gomperts. 1990. Changes in the state of actin during the exocytotic reaction of permeabilized rat mast cells. *J. Cell Biol.* 111:919–927.

Koopman, W. R., and R. C. Jackson. 1990. Calcium- and guanine-nucleotide-dependent exocytosis in permeabilized rat mast cells. *Biochem. J.* 265:365–373.

Lapetina, E. G., J. Silio, and M. Ruggiero. 1985. Thrombin induces serotonin secretion and aggregation independently of inositol phospholipids hydrolysis and protein phosphorylation in human platelets permeabilized with saponin. *J. Biol. Chem.* 260:7078–7083.

Lee, S. A., and R. W. Holz. 1986. Protein phosphorylation and secretion in digitonin-permeabilized adrenal chromaffin cells. Effects of micromolar Ca^{2+}, phorbol esters, and diacylglycerol. *J. Biol. Chem.* 261:17089–17098.

Lillie, T. H. W., T. D. Whalley, and B. D. Gomperts. 1991. Modulation of the exocytotic reaction of permeabilized rat mast cells by ATP, other nucleotides and Mg^{2+}. *Biochim. Biophys. Acta* 1094:355–363.

Lindau, M., and B. D. Gomperts. 1991. Techniques and concepts in exocytosis: Focus on mast cells. *Biochim. Biophys. Acta* 1071:429–471.

Lyons, R. M., N. Stanford, and P. W. Majerus. 1975. Thrombin-induced protein phosphorylation in human platelets. *J. Clin. Invest.* 56:924–936.

Martin, T. F. J., and J. H. Walent. 1989. A new method for cell permeabilization reveals a cytosolic protein requirement for Ca^{2+}-activated secretion in GH_3 pituitary cells. *J. Biol. Chem.* 264:10299–10308.

Matsuuchi, L., K. M. Buckley, A. W. Lowe, and R. B. Kelly. 1988. Targeting of secretory vesicles to cytoplasmic domains in AtT-20 and PC-12 cells. *J. Cell Biol.* 106:239–251.

Matthies, H. J. G., H. C. Palfrey, and R. J. Miller. 1988. Calmodulin- and protein phosphorylation-independent release of catecholamines from P-12 cells. *FEBS. Lett.* 229:238–242.

Michener, M. L., W. B. Dawson, and C. E. Creutz. 1986. Phosphorylation of a chromaffin granule-binding protein in stimulated chromaffin cells. *J. Biol. Chem.* 261:6548–6555.

Moore, B. W., and V. J. Perez. 1967. Specific acidic proteins of the nervous system. In: *Physiological and Biochemical Aspects of Nervous Intergration,* edited by F. D. Carlson, pp. 343–359. Prentice-Hall, Englewood Cliffs, N.J.

Morgan, A., and R. D. Burgoyne. 1992a. Exo1 and Exo2 proteins stimulate calcium-dependent exocytosis in permeabilized adrenal chromaffin cells. *Nature* 355:833–836.

Morgan, A., and R. D. Burgoyne. 1992b. Interaction between protein kinase C and exo1 (14-3-3 protein) and its relevance to exocytosis in permeabilized adrenal chromaffin cells. *Biochem. J.* 286:807–811.

Naka, M., M. Nishikawa, R. S. Adelstein, and H. Hidaka. 1983. Phorbol ester-induced activation of human platelets is associated with protein kinase C phosphorylation of myosin light chains. *Nature* 306:490–492.

Nakata, T., K. Sobue, and N. Hirokawa. 1990. Conformational change and localization of calpactin I complex involved in exocytosis as revealed by quick-freeze, deep-etched electron microscopy and immunocytochemistry. *J. Cell Biol.* 110:13–25.

Naor, Z., H. Dan-Cohen, J. Hermon, and R. Limor. 1989. Induction of exocytosis in permeabilized pituitary cells by alpha- and beta-type protein kinase C. *Proc. Natl. Acad. Sci. U.S.A.* 86:4501–4504.

Nishizaki, T., J. H. Walent, J. A. Kowalchyk, and T. F. J. Martin. 1992. A key role for a 145-kDa cytosolic protein in the stimulation of Ca^{2+}-dependent secretion by protein kinase C. *J. Biol. Chem.* 267:23972–23981.

Nishizuka, Y. 1986. Studies and perspectives of protein kinase C. *Science* 233:305–312.

Nishizuka, Y. 1992. Intracellular signaling by hydrolysis of phospholipids and activation of protein kinase C. *Science* 258:607–614.

Orci, L., K. H. Gabbay, and W. J. Malaisse. 1972. Pancreatic beta-cell web: Its possible role in insulin secretion. *Science* 175:1128–1130.

Ozawa, K., Z. Szallasi, M. G. Kazanietz, P. M. Blumberg, H. Mischak, J. F. Mushinski, and M. A. Beaven. 1993. Ca^{2+}-dependent and Ca^{2+}-independent isozymes of protein kinase C mediate exocytosis in antigen-stimulated rat basophilic RBL-2H3 cells. *J. Biol. Chem.* 268:1749–1756.

Peppers, S. C., and R. W. Holz. 1986. Catecholamine secretion from digitonin-treated PC12 cells. Effects of Ca2+, ATP, and protein kinase C activators. *J. Biol. Chem.* 261:14665–14669.

Perrin, D., and D. Aunis. 1985. Reorganization of alpha-fodrin induced by stimulation of secretory cells. *Nature* 315:589–592.

Pocotte, S. L., R. A. Frye, R. A. Senter, D. R. TerBush, S. A. Lee, and R. W. Holz. 1985. Effects of phorbol ester on catecholamine secretion and protein phosphorylation in adrenal medullary cell cultures. *Proc. Natl. Acad. Sci. U.S.A.* 82:930–934.

Powell, M. A., and J. R. Glenney. 1987. Regulation of calpactin I phospholipid binding by calpactin I light-chain binding and phosphorylation by p60v-src. *Biochem. J.* 247:321–328.

Pozzan, T., G. Gatti, N. Dozio, L. M. Vicentini, and J. Meldolesi. 1984. Ca^{2+}-dependent and -independent release of neurotransmitters from PC12 cells: A role for protein kinase C activation? *J. Cell Biol.* 99:628–638.

Rink, T. J., A. Sanchez, and T. J. Hallam. 1983. Diacylglycerol and phorbol ester stimulate secretion without raising cytoplasmic free calcium in human platelets. *Nature* 305: 317–319.

Sano, K., Y. Takai, J. Yamanishi, and Y. Nishizuka. 1983. A role of calcium-activated phospholipid-dependent protein kinase in human platelet activation. *J. Biol. Chem.* 258:2010–2013.

Sarafian, T., L.-A. Pradel, J.-P. Henry, D. Aunis, and M.-F. Bader. 1991. The participation of annexin II (calpactin I) in calcium-evoked exocytosis requires protein kinase C. *J. Cell Biol.* 114:1135–1147.

Sontag, J. M., D. Aunis, and M. F. Bader. 1988. Peripheral actin filaments control calcium-mediated catecholamine release from streptolysin-O-permeabilized chromaffin cells. *Eur. J. Cell Biol.* 46:316–326.

Summers, T. A., and C. E. Creutz. 1985. Phosphorylation of a chromaffin granule-binding protein by protein kinase C. *J. Biol. Chem.* 260:2437–2443.

Swartz, K. J., A. Merrit, B. P. Bean, and D. M. Lovinger. 1993. Protein kinase C modulates glutamate receptor inhibition of Ca^{2+} channels and synaptic transmission. *Nature* 361:165–168.

Tachikawa, E., S. Takahashi, T. Kashimoto, and Y. Kondo. 1990. Role of a Ca^{2+}/phospholipid-dependent protein kinase in catecholamine secretion from bovine adrenal medullary chromaffin cells. *Biochem. Pharmacol.* 40:1505–1513.

Tatham, P. E., and B. D. Gomperts. 1989. ATP inhibits onset of exocytosis in permeabilised mast cells. *Biosci. Rep.* 9:99–109.

Tatham, P. E., and B. D. Gomperts. 1991. Rat mast cells degranulate in response to microinjection of guanine nucleotide. *J. Cell Sci.* 98, Pt. 2:217–224.

TerBush, D. R., and R. W. Holz. 1986. Effects of phorbol esters, diglyceride, and cholinergic agonists on the subcellular distribution of protein kinase C in intact or digitonin-permeabilized adrenal chromaffin cells. *J. Biol. Chem.* 261:17099–17106.

TerBush, D. R., and R. W. Holz. 1990. Activation of protein kinase C is not required for exocytosis from bovine adrenal chromaffin cells: The effects of protein kinase C (19–31), Ca/CaM kinase II (291–317), and staurosporin. *J. Biol. Chem.* 265:21179–21184.

TerBush, D. R., M. A. Bittner, and R. W. Holz. 1988. Ca2+ influx causes rapid translocation of protein kinase C to membranes. Studies of the effects of secretagogues in adrenal chromaffin cells. *J. Biol. Chem.* 263:18873–18879.

Trifaro, J.-M., M.-F. Bader, and J.-P. Doucet. 1985. Chromaffin cell cytoskeleton: Its possible role in secretion. *Can. J. Biochem. Cell Biol.* 63:661–679.

Tyers, M., R. A. Rachubinski, M. I. Stewart, A. M. Varrichio, R. G. Shorr, R. J. Haslam, and C. B. Harley. 1988. Molecular cloning and expression of the major protein kinase C substrate of platelets. *Nature* 333:470–473.

Tyers, M., R. J. Haslam, R. A. Rachubinski, and C. B. Harley. 1989. Molecular analysis of pleckstrin: The major protein kinase C substrate of platelets. *J. Cell Biochem.* 40:133–145.

Umekawa, H., and H. Hidaka. 1985. Phosphorylation of caldesmon by protein kinase C. *Biochem. Biophys. Res. Commun.* 132:56–62.

Vallar, L., T. J. Biden, and C. B. Wollheim. 1987. Guanine nucleotides induce Ca2+-independent insulin secretion from permeabilized RINm5F cells. *J. Biol. Chem.* 262:5049–5056.

Vitale, M. L., R. Del Castillo, L. Tchakarov, and J.-M. Trifaro. 1991. Cortical filamentous actin disassembly and scinderin redistribution during chromaffin cell stimulation precede exocytosis, a phenomenon not exhibited by gelsolin. *J. Cell Biol.* 113:1057–1067.

Wakade, A. R., R. K. Malhotra, and T. Wakade. 1986. Phorbol ester facilitates ^{45}Ca accumulation and catecholamine secretion by nicotine and excess K^+ but not by muscarine in rat adrenal medulla. *Nature* 321:698–700.

Walent, J. H., B. W. Porter, and T. F. J. Martin. 1992. A novel 145 kd brain cytosolic protein reconstitutes Ca^{2+}-regulated secretion in permeable neuroendocrine cells. *Cell* 70:765–775.

Werth, D. K., J. E. Niedel, and I. Pastan. 1983. Vinculin, a cytoskeletal substrate of protein kinase C. *J. Biol. Chem.* 258:11423–11426.

Wightman, R. M., J. A. Jankowski, R. T. Kennedy, K. T. Kawagoe, T. J. Schroeder, D. J. Leszczyszyn, J. A. Near, E. J. Diliberto Jr., and O. H. Viveros. 1991. Temporally resolved catecholamine spikes correspond to single vesicle release from individual chromaffin cells. *Proc. Natl. Acad. Sci. U.S.A.* 88:10754–10758.

Wilson, S. P. 1990. Regulation of chromaffin cell secretion and protein kinase C activity by chronic phorbol ester treatment. *J. Biol. Chem.* 265:648–651.

Wu, Y. N., and P. D. Wagner. 1991. Calpactin-depleted cytosolic proteins restore Ca^{2+}-dependent secretion to digitonin-permeabilized bovine chromaffin cells. *FEBS Lett.* 282:197–199.

Wu, Y. N., N. D. Vu, and P. D. Wagner. 1992. Anti-(14-3-3 protein) antibody inhibits stimulation of noradrenaline (norepinephrine) secretion by chromaffin-cell cytosolic proteins. *Biochem. J.* 285:697–700.

Yamanishi, J., Y. Takai, K. Kaibuchi, K. Sano, M. Castagna, and Y. Nishizuka. 1983. Synergistic functions of phorbol ester and calcium in serotonin release from human platelets. *Biochem. Biophys. Res. Commun.* 112:778–786.

Yamauchi, T., H. Nakata, and H. Fujisawa. 1981. A new activator protein that activates tryptophan 5-monooxygenase and tyrosine 3-monooxygenase in the presence of Ca^{2+}-, calmodulin-dependent protein kinase. *J. Biol. Chem.* 256:5404–5409.

Zupan, L. A., D. L. Steffens, C. A. Berry, M. Landt, and R. W. Gross. 1992. Cloning and expression of a human 14-3-3 protein mediating phospholipolysis. *J. Biol. Chem.* 267:8707–8710.

11

Protein Kinase C: A Key Enzyme Determining Cell Fate and Apoptosis?

MICHAEL J. MCCABE, JR., STEN ORRENIUS

The expansion of functional cell lineages depends on a biased ratio of cell births (mitosis) to cell deaths (apoptosis). In the past, the focus of many cell biologists and molecular oncologists has been on understanding the control of cell division; however, compelling evidence has emerged that the regulation of cell killing also contributes to morphogenesis and tumorigenesis. Thus, the capacity of a select cell population to perform a differentiative biological function depends on the balanced control of cell fate from both sides of the equation (Figure 11.1).

Physiological cell death occurs by an active "suicide" mechanism known as apoptosis (pronounced ăp-ō-tō-sis). As indicated, an active process of cell death provides an additional way to precisely regulate cell numbers, and so also biological function. Furthermore, active participation by the dying cell affords the possibility that cell death can be suppressed before it reaches a point of no return. Controlled suppression of the cell death pathway is viewed as a key mechanism of cell selection during organogenesis. Additionally, aberrant cell survival by inhibition of cell death may contribute to oncogenesis, and many tumor promoters (e.g., phorbol esters) may act by preventing apoptosis.

Because of its diverse functions in cellular signal transduction and tumor promotion, research on the protein kinase C pathway has spread over many fields of biomedicine. In keeping with the relative novelty of cell death as a means of cell population control, a limited yet compelling body of evidence points to a regulatory role for PKC in apoptosis. This evidence will be discussed in the following pages. Our challenge has been to present various points of view in a thorough, clear, and unbiased fashion despite the limited array of carefully tested hypotheses. Thus, although the literature does support that PKC functions both in inducing and inhibiting apoptoses, how it does so is largely a matter of speculation. In addition, any hypothesis regarding the role of PKC in apoptosis must also consider the rapidly changing notions on the control of cell death by apoptosis.

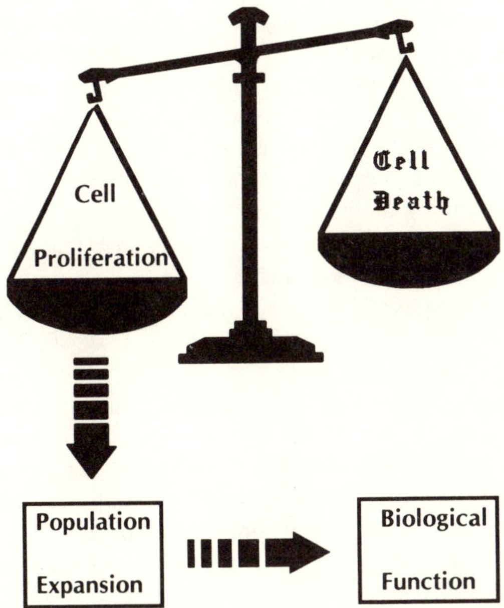

Figure 11.1 Biological function is controlled by the balance of cell death and cell proliferation.

APOPTOSIS, WHAT IS IT?

Apoptosis is a physiological form of cell deletion that plays an essential role in development at all levels of complexity (i.e., cell lineages, tissues, organs). Apoptosis refers to a metabolic cascade that, when activated by a variety of physiological and environmental stimuli, leads to a process of gene-directed cell death in several well-studied model systems (*vide infra*). Programmed cell death, a term often mistakenly used synonymously with apoptosis, refers to a cell autonomous, intrinsic pathway that leads to cell death during normal development. However, not all developmental cell death follows an intrinsic program. In some instances of developmental cell death the initiating stimulus may not be known. Furthermore, a genetic program is not clearly evident in all forms of apoptosis. For these reasons, purists caution against the usage of the two terms, apoptosis and programmed cell death, synonymously (Alles et al., 1991); however, both terms imply that the cell is an active participant in its own death by a mechanism that has evolved by the host to kill its own cells.

In 1972, Kerr, Wyllie, and Currie defined apoptosis on the basis of peculiar morphology distinct from the morphology attributed to necrosis or ''accidental cell death'' (Kerr et al., 1972). The morphological features associated with apoptosis include condensation of chromatin into crescentic caps at the nuclear periphery, nucleolar disintegration, and reduction in nuclear size (Arends et al., 1990). These nuclear changes are accompanied by shrinkage of total cell volume, increased cell density, compaction of cytoplasmic organelles without overt changes in the architecture of the organelles proper, and dilation of the endo-plasmic reticulum. In the final phases of apoptosis, there is budding and separation of both nucleus and cytoplasm into multiple, small, membrane-bound apoptotic-bodies as well as marked plasma membrane budding. Apoptosis is associated with death of individual cells rather than of contiguous patches of tissue; there is no inflammatory infiltrate because dying cells are phagocytosed by neighboring cells rather than by immigrant professional phagocytes.

Accompanying the nuclear collapse described above is a rapid degradation of chromatin. When chromatin degradation is analyzed biochemically, the kinetics are similar to those for the appearance of nuclear condensation, suggesting that the two reflect a single or closely coupled event. In 1980 Wyllie showed that chromatin isolated from thymocytes that had been exposed to glucocorticoid hor-mones, to which these cells are lethally sensitive, was fragmented into single and multiple nucleosomes (Wyllie, 1980). Nucleosomes are spaced regularly along chromatin at approximately 180 base pair (bp) intervals; hence, an electrophoretic separation of chromatin from cells undergoing apoptosis reveals a characteristic "ladder" pattern of oligosomal-sized bands (i.e., 180 bp, 360 bp, 540 bp, etc.). The formation of this ladder pattern of degraded chromatin is believed to be enzymatic; an endonuclease gains access to the internucleosomal linker region of chromatin, where it is rather weakly associated with histone H1. Chromatin in a transcriptionally active conformation, which is depleted in H1 and may allow better access to enzymes in the nucleoplasm, appears to be even more susceptible to degradation. If the conformation of chromatin plays a determinant role in endo-nuclease accessibility, PKC activation may modulate apoptosis by affecting chro-matin structure by mechanisms that may include modulation of intranuclear con-centrations of spermine (Brüne et al., 1991) or by inducing the phosphorylation of histones or transcription factors.

In many cell types (e.g., thymocytes, myelogenous leukemia cells), chromatin degradation is clearly a premortem event (extensive fragmentation is observed before loss of plasma membrane integrity), and the endonuclease appears to be a constitutive enzyme awaiting the appropriate activation stimuli. More recent information using glucocorticoid-treated thymocytes indicates that chromatin is cleaved into larger high molecular weight fragments by an unknown enzyme prior to internucleosomal fragmentation (Brown et al., 1993). The nature of the enzyme(s) responsible for this initial chromatin cleavage into large fragments, its involvement in other instances of apoptosis, and its positive and negative (e.g., PKC?) regulation remain to be determined.

Possible initiators of endonuclease activity (reviewed in McCabe et al., 1992; Orrenius et al., 1992) include changes in the intracellular localization/concentra-tion of second messengers (e.g., Ca^{2+} and cAMP), fluctuations in the concentra-tions of other ions (e.g., H^+ and Zn^{2+}), altered kinase or phosphatase activities, and altered expression of cellular oncogenes such as p53 (Yonish-Rouach et al., 1991), c-*myc* (Shi et al., 1992; Evan et al., 1992), and *bcl*-2 (Hockenberyy et al., 1990, 1991; Bissonnette et al., 1992; Garcia et al., 1992; Miyashita, and Reed, 1992; Campana et al., 1993). These initiators may trigger changes in endonuclease activity, chromatin conformation, or both. Hence, both enzyme activity and substrate conformation appear to be inextricably linked in the initiation of chromatin degradation serving as potential distal sites of regulation by PKC isozymes in the apoptotic program.

In some instances, apoptosis appears to be dependent on protein synthesis and can be abrogated by application of inhibitors shortly after the lethal stimulus. Little is known, however, of the precise intracellular effector mechanisms controlling apoptosis and in particular the regulation and significance of chromatin degradation. Although not a defining principle, the association of macromolecular synthesis with apoptosis helps to conceptually and mechanistically set apart apoptosis (i.e., suicide) from other forms of cell death. Also, macromolecular synthesis implies that particular cell death genes may regulate cell fate. Indeed, identification of genes such as the nematode, *Caenorhabditis elegans, ced-9* (Hengartner et al., 1992), and the mammalian protooncogene *bcl-2,* whose expressions block cell death, has led to the notion that cell fate is regulated by the expression of genes that rescue cells targeted for destruction from the death pathway. How are these genes switched on? Although the PKC signaling pathway interacts with several other pathways including the Ca^{2+}- and the cAMP-signaling pathways and their associated kinases in regulating gene expression (i.e., IL-2 production in T-lymphocytes), a direct link between the PKC pathway and the ''gene rescue'' pathway has not been demonstrated. However, the induction of apoptosis in human fibroblasts by the protein kinase inhibitor staurosporine can be prevented by the overexpression of *bcl-2,* indicating that the target for *bcl-2* is distal to PKC activation (Jacobsen et al., 1993). On the other hand, *bcl-2* and PKC may not always be linked, since PKC activation can block CTLL-2 cell death after IL-2 withdrawal (Rodriguez-Tarduchy and Lopez-Rivas, 1989), but *bcl-2* fails to block the IL-2 deprivation-induced death of CTLL-2 cells (Nunez et al., 1990).

THE THYMOCYTE SELECTION MODEL SYSTEM

In the immune system, clonal deletion of developing T cells and B cells occurs by apoptosis. To be able to recognize and destroy the multitude of foreign antigens that it is exposed to in its lifetime, the vertebrate organism is equipped with a powerful genetic mechanism (DNA rearrangement) to diversify its repertoire of antigen-receptors. Due to the random nature of antigen-receptor diversification, this mechanism inevitably generates immune cells that can respond to the organism's own antigens, potentially leading to autoimmune disease. These self-reactive lymphocytes are usually deleted at an immature stage of their development.

A variety of experimental approaches have collectively established a sequence of maturation steps occurring during T-cell ontogeny (reviewed in Boyd and Hugo, 1991). According to this scheme (Figure 11.2), development toward mature medullary thymocytes that express high levels of the $\alpha\beta$ T-cell receptor ($\alpha\beta$TcR) in association with the CD3 complex and either CD4 or CD8 proceeds through a $\alpha\beta$TcR/CD3^{lo} CD4$^+$CD8$^+$ stage (i.e., ''double positive'' thymocytes). These double positive precursors to the mature helper ($\alpha\beta$TcR^{hi} CD4$^+$CD8$^-$) and suppressor/cytotoxic ($\alpha\beta$TcR^{hi} CD4$^-$CD8$^+$) T cells make up the majority of thy-

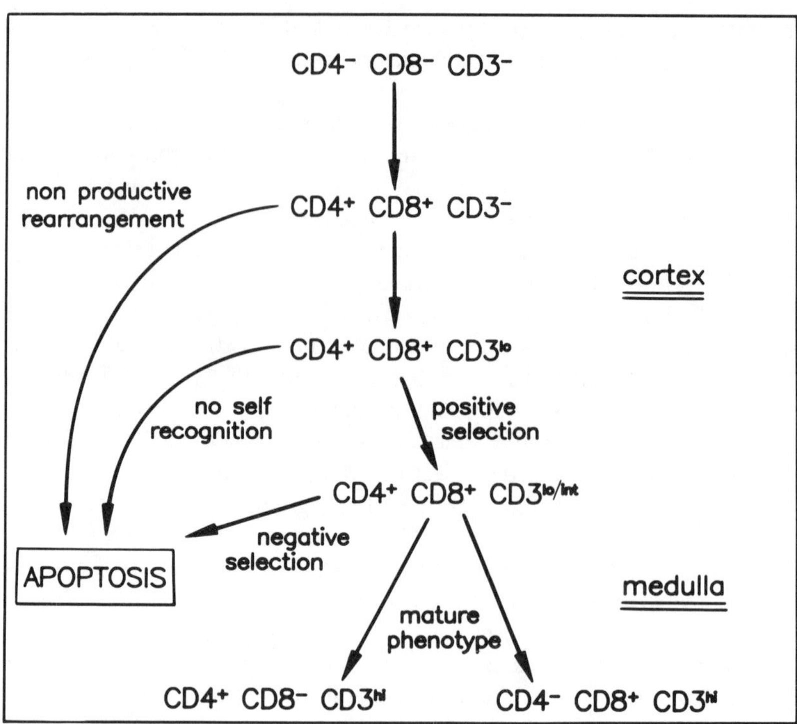

Figure 11.2 Apoptosis during thymocyte maturation.

mocytes (65–80%) and undergo rigorous selection events. Positive selection ensures the maturation of thymocytes bearing $\alpha\beta$TcRs that are able to recognize foreign antigen fragments presented by self major histocompatibility complex proteins. In addition, thymocytes bearing receptors able to react with self-antigens are killed at this stage of thymic maturation by the process of negative selection or clonal deletion; negative selection appears to occur by the mechanism of apoptosis.

Direct evidence that clonal deletion occurs *in vivo* by the mechanism of apoptosis was obtained through experiments using mice transgenic for a TcR recognizing a peptide antigen that induced the deletion of $\alpha\beta$TcR^{lo} CD4$^+$CD8$^+$ thymocytes (Murphy et al., 1990). Deletion of thymocytes expressing this transgene was accompanied by both chromatin degradation and morphological changes characteristic of apoptosis. In addition to studies using transgenic mice, *in vivo* administration of monoclonal antibodies to the $\alpha\beta$TcR/CD3 complex was shown to induce apoptosis, by the same criteria, in immature thymocytes (Shi et al., 1991). Furthermore, in many *in vitro* studies it has been demonstrated that double positive thymocytes are sensitive to apoptosis in response to various stimuli including glucocorticoid hormones (Wyllie, 1980; Cohen and Duke, 1984), γ-irradiation (Sellins and Cohen, 1987), Ca^{2+} elevation (Kizaki et al., 1989; McConkey et al., 1989), topoisomerase inhibitors (Walker et al., 1991; Onishi et al., 1993), and antibodies to the $\alpha\beta$TcR/CD3 complex (Smith et al., 1989; Tadakuma et al., 1990), and thymocytes have been useful as a model system for studying the biochemical mechanisms underlying apoptosis. In most of the described models of apoptosis the biochemical events leading to the point of no return are poorly understood. In addition, thymocyte subpopulations other than the $CD4^+CD8^+$ thymocytes may die by apoptosis, and PKC activation may function differently in other thymocyte subpopulations. For instance, chelation of intracellular zinc induces apoptosis in phenotypically mature thymocytes (McCabe et al., 1993).

Calcium

Investigations from numerous laboratories have established that thymocytes from both rodents and humans undergo apoptosis in response to stimuli that cause a sustained elevation in the cytosolic free calcium concentration ($[Ca^{2+}]_i$). A key role for calcium in regulating apoptosis has come from studies that demonstrate that known Ca^{2+}-elevating agents trigger thymocyte apoptosis (reviewed in McCabe et al., 1992). Furthermore, intracellular Ca^{2+}-chelating agents prevent apoptosis in many instances, as does the removal of Ca^{2+} from the external medium. Recent studies showing that overexpression of the intracellular Ca^{2+}-binding protein calbindin-D_{28K} prevents the induction of apoptosis in a thymoma cell line exposed to glucocorticoid hormone or calcium ionophore further strengthen these arguments, because calbindin serves as an endogenous Ca^{2+}-chelator in these studies (Dowd et al., 1992).

$\alpha\beta$TcR^{lo} $CD4^+CD8^+$ thymocytes appear to exist as at least two separate populations with respect to antigen receptor-stimulated $[Ca^{2+}]_i$ elevation (Finkel et al., 1989). One population does not increase $[Ca^{2+}]_i$ or die on engagement of the $\alpha\beta$TcR; however, direct stimulation of CD3 does result in elevated $[Ca^{2+}]_i$ and cell death. The second population of thymocytes responds to engagement of either the $\alpha\beta$TcR or CD3 with an increase in $[Ca^{2+}]_i$ and subsequent cell death. The former, nondeletable population in which the $\alpha\beta$TcR and CD3 are apparently uncoupled may be the one that undergoes positive selection, while the latter population of thymocytes is susceptible to clonal deletion. These two populations are related to each other in that thymocytes first pass through the nondeletable stage, where they are protected from apoptosis, and then through the deletable stage as they mature (Finkel et al., 1992). In addition to differences in Ca^{2+} signaling between uncoupled and coupled $\alpha\beta$TcRs and CD3 molecules, activation of protein kinase C enzymes may differ between the two populations. Additionally, the expression of select PKC isozymes may differ between these two thymocyte subpopulations. It remains to be determined whether the expression of distinct PKC isozymes is differentially regulated during thymocyte maturation and what their physiological substrates in thymocytes are. Using affinity-purified anti-PKC antisera, high levels of PKC α, PKC β, and PKC ϵ have been found in $CD4^+CD8^+$ thymocytes (Strulovici et al., 1991). Additional studies have confirmed the presence of mRNA for PKC α, β, ϵ, θ, and ζ but not γ, δ or η in immature thymocytes (Freire-Moar et al., 1991; Baier et al., 1993). Hence, during the nondeletable stage of thymocyte maturation, signaling through the $\alpha\beta$TcR involving the activation of distinct PKC isozymes may be associated with positive selection, whereas signaling during the deletable stage solely involves changes in $[Ca^{2+}]_i$. Bypassing the $\alpha\beta$TcR and CD3 with agents that directly elevate Ca^{2+} such as calcium ionophores or thapsigargin (S. A. Jiang and S. Orrenius, unpublished observation), a tumor promoter that releases Ca^{2+} from the intracellular IP_3-sensitive Ca^{2+} storage pool, results in endonuclease activation and apoptotic cell death. Although the release of the intracellular Ca^{2+} pool may be causally linked to further sustained Ca^{2+} elevation through stimulation of capacitative Ca^{2+} entry through plasma membrane Ca^{2+} channels, release of the intracellular Ca^{2+} pool alone is not sufficient for endonuclease activation. As such, PKC may function to impair the influx of Ca^{2+} from the extracellular space. Alternatively, PKC activation, which is known to decrease plasma membrane-associated CD3 (Cantrell et al., 1989), may block the apoptotic pathway by down-regulating the signal-transducing complex from the cell surface. Since CD3-mediated signaling for apoptosis is associated with Ca^{2+} elevation (in fact, both Ca^{2+} ionophore and anti-CD3 appear to cause apoptosis in the same population of thymocytes), experiments addressing whether the activation of PKC can block thapsigargin-induced thymocyte apoptosis may shed light on the question of PKC-mediated antigen-receptor down-regulation as a mechanism for inhibition of the initiation of the apoptotic program in thymocytes. The PKC-mediated down-regulation of other thymocyte membrane glycoproteins, such as CD4 (Wang et al., 1987), may also be involved, since CD4 ligation has been linked to the regulation of apoptosis in thymocytes (MacDonald et al., 1988; Teh et al., 1991), mature T cells (Newell et al., 1990), and HIV-infected T cells (Groux et al., 1992; Meyaard et al., 1992).

Glucocorticoids

The best characterized model of apoptosis in the immune system is the glucocorticoid induction of thymocyte death. *In vivo* and *in vitro* administration of glucocorticoids has been shown to trigger the program of apoptotic self-destruction of predominantly the $CD4^+CD8^+$ population of thymocytes. Human thymocytes exposed *in vitro* to relatively low concentrations ($\sim 10^{-7}$ *M*) of methylprednisolone or dexamethasone display marked nuclear condensation and chromatin degradation within 4 hours, which continues to increase until 16 hours (McCabe and Orrenius, 1993). When added early (i.e., $\leq$ 4 hr.), PKC-activating phorbol esters (i.e., PdB_2 or PMA) completely prevent glucocorticoid-induced endonuclease activation in human thymocyte cultures. The events distal to PKC activation that lead to this blockade of endonuclease activation by glucocorticoid are not known; however, 4 hours after hormone treatment seems to be a point of no return in the cell death pathway. Surprisingly, activation of PKC by phorbol esters does not prevent glucocorticoid-induced apoptosis in rat thymocyte cultures. Moreover, numerous reports with mouse thymocytes have implicated PKC in the induction of apoptosis (Kizaki et al., 1989; Tadakuma et al., 1990; Onishi et al., 1993); phorbol esters, alone, trigger apoptosis but they also potentiate the actions of cAMP in inducing chromatin cleavage (Suzuki et al., 1990). The explanation for these species differences for PKC-mediated influences on apoptosis are probably not related to differences in proliferative potential, but they may be due to differences in PKC-mediated phosphorylation of the glucocorticoid receptor or they may relate to differences in hormone receptor complex translocation to the nucleus. Alternatively, the events downstream of PKC activation, such as increased c-*fos* expression and AP-1 activation, may differ between species after phorbol ester treatment. It has been shown in many experimental systems that the glucocorticoid receptor and the AP-1 transcription factor can inhibit each other by direct protein-protein interaction (Diamond et al., 1990). Such interaction could repress the expression of cell death genes required for the initiation of apoptosis. Inhibition of glucocorticoid-induced apoptosis by phorbol esters suggests that putative cell death genes triggered by glucocorticoid treatment in human thymocytes may be under the control of a composite glucocorticoid response element.

Thymocyte Cell Death Genes

Inhibitors of protein synthesis (e.g., cycloheximide, emetine) block glucocorticoid-induced thymocyte apoptosis in all species, suggesting that *de novo* expression of cell death genes is required in the cascade of events leading to endonuclease activation. Interestingly, inhibition of glucocorticoid-induced apoptosis by cycloheximide follows the same time course as inhibition by phorbol esters (i.e., ≤ 4 hr). A number of novel gene products have been associated with thymocytes undergoing apoptosis, including two mRNAs, RP-2 and RP-8, that were isolated by subtractive hybridization from cDNA libraries obtained from control and dexamethasone-treated rat thymocytes (Owens et al., 1991). The deduced amino acid sequences of these gene products suggest that RP-2 is a plasma membrane-spanning glycoprotein, whereas RP-8, which contains a zinc finger domain, may be a DNA-binding regulatory protein. Neither RP-2 nor RP-8 appears to contain any consensus sequences for PKC phosphorylation; however, RP-2 shows some homology at tyrosine 182 with published tyrosine phosphorylation consensus sequences. The precise roles, if any, that RP-2 and RP-8 play in thymocyte programmed cell death remain to be determined. Using a more complicated subtractive hybridization strategy involving several cell lines known to undergo programmed cell death, Ishida and colleagues isolated and characterized another novel cell death-associated gene product, which they called PD-1 (Ishida et al., 1992). PD-1 shares homology to plasma membrane-spanning glycoproteins of the immunoglobulin gene superfamily. *In vivo,* PD-1 expression is restricted to the thymus, and PD-1 mRNA increases in thymocytes stimulated to undergo apoptosis with anti-CD3. Interestingly, PD-1 contains a potential PKC phosphorylation site at Thr^{210}, as the sequence KDDT^{210}LK shares some homology to the PKC phosphorylation consensus sequence RXXSXR (Hsieh et al., 1991). The human homologues of RP-2, RP-8, and PD-1 remain to be characterized.

Blockade of Thymocyte Apoptosis by Interleukins

PKC activation, IL-2 production, and IL-2R expression are finely tuned during the ontogenic development of T cells. The effect of PKC activation induced by phorbol esters on cell proliferation of human thymic populations was studied by Gonzalez-Fernandez and co-workers 1991. Proliferative responses to phorbol esters alone and in combination with IL-2 were observed in prethymocytes and medullary thymocytes, respectively, whereas cortical thymocytes (i.e., $CD4^+CD8^+$) did not proliferate in the presence of phorbol ester or phorbol ester plus IL-2. In this same study, phorbol ester treatment increased the expression of the low affinity IL-2R on prethymocyte and medullary thymocytes but not cortical thymocytes; phorbol ester treatment induced IL-2 secretion in prethymocyte cultures but not in the cortical or medullary thymocyte cultures. Since IL-2 has been shown to prevent anti-CD3-induced apoptosis of medullary thymocytes (Nieto et al., 1990; Lawetzky et al., 1991), PKC-mediated expression of the IL-2 receptor on medullary thymocytes may function to rescue medullary thymocytes from apoptotic cell death. Refractoriness to phorbol ester-mediated proliferation, IL-2 production, and IL-2R expression occurs exclusively during the cortical stage of development. Additional investigations with murine thymocytes have confirmed that $CD4^+CD8^+$ thymocytes do not proliferate or produce cytokines in response to phorbol esters alone or in combination with calcium ionophores (Fischer et al., 1991; Montgomery and Dallman, 1991). $CD4^+CD8^+$ thymocytes are not, however, completely refractory to phorbol esters, as phenotypic changes can be seen in phorbol ester-treated cortical thymocytes (Gratiot-Deans et al., 1992), implying that the PKC system is operative at this stage.

Kizaki and colleagues (1989) showed that in mouse thymocyte cultures PKC-activating phorbol esters induced apoptosis in immature thymocytes. Reminiscent of the unbalanced signaling hypothesis of thymocyte cell death previously proposed (McConkey et al., 1990; Rothenberg, 1990), calcium ionophore or phorbol ester, alone, induced significant chromatin fragmentation, whereas the addition of phorbol esters at low concentration blocked the calcium ionophore-induced chromatin fragmentation in unfractionated mouse thymocyte cultures. Inhibition of apoptosis under these conditions was accompanied by an increase in DNA synthesis. The interpretation of these investigators was that phorbol esters (i.e., PKC activation) switched a suicide process induced by elevated intracellular Ca^{2+} concentration to an opposite process: stimulation of proliferation. In hindsight, given the information that cortical thymocytes are refractory to the proliferative and growth factor-promoting effects of phorbol esters, a more likely explanation is that phorbol ester stimulation of PKC blocked the apoptotic pathway in cortical thymocytes and stimulated DNA synthesis in another population(s) of thymocytes. Although the concept that stimulation of cell death and cell birth are opposite linked processes appears to hold in some instances (*vide infra*), for $CD4^+CD8^+$ thymocytes the evidence for this on a single-cell basis is weak.

GROWTH FACTOR WITHDRAWAL

Apoptosis can be triggered in many cell culture systems after withdrawal of growth and differentiation factors. The *in vivo* correlate to this growth factor withdrawal may be competition between cells for limiting trophic factors (i.e., survival molecules that are produced at or localized to specific regions) that control the size or specificity of the surviving population. In these experimental models, cell birth (i.e., mitosis) and cell death (i.e., apoptosis) are viewed clearly as opposite events. In these model systems, the induction of apoptosis follows growth factor deprivation or interruption of the cell growth pathway (McCabe et al., 1992; Campana et al., 1993), and agents that block the growth-factor withdrawal-induced apoptotic pathway, such as phorbol esters, are associated with replacing the growth factor stimulus.

A case in point is the induction of apoptosis that occurs in IL-2-dependent T-cell lines (e.g., CTLL-2) and cultured concanavalin A blasts on withdrawal of interleukin-2 (Rodriguez-Tarduchy and Lopez-Rivas, 1989). Inclusion of phorbol ester to these cultures inhibits both endonuclease activity and the decline in viable cell numbers. Although the CTLL cells have been suggested as models for studying the biochemical regulation of thymocyte apoptosis, it cannot be stressed enough that the influence of phorbol esters on apoptosis in these cells is quite unrelated to the observations noted with primary thymocyte cultures; namely, lack of a PdB_2-induced proliferative response in the population of thymocytes rescued from the apoptotic program. Phorbol esters can replace IL-2 as a growth stimulus for IL-2-dependent T-cell lines. In fact, since CTLL-2 cells are commonly used in bioassays for supernatants containing IL-2, it is necessary to charcoal filter supernatants elicited by phorbol ester treatment to remove the phorbol esters that may be carried over to the bioassay.

Human peripheral blood monocytes progressively undergo apoptosis when cultured in the absence of serum, cytokines, or other stimuli. Certain growth factor-independent clones of myeloid leukemic cells can regain a growth factor-dependent state during differentiation into mature macrophages or granulocytes with interleukin-3 (IL-3), interleukin-6 (IL-6), or granulocyte/macrophage-colony-stimulating factor (GM-CSF). Loss of function and viability of these differentiating myeloid leukemia cells in the absence of these cytokines has been associated with chromatin degradation and nuclear condensation typical of apoptosis (Lotem and Sachs, 1992). Tumor-promoting (i.e., PKC-activating) phorbol esters can rescue these differentiating leukemic cells as well as normal myeloid precursors from apoptosis (Lotem et al., 1991). Furthermore, the rescue of myeloid cells from apoptosis by phorbol esters could be blocked by amiloride, an inhibitor of the Na^+/H^+ antiporter that is a major substrate for PKC. In IL-6-dependent myeloid cells, apoptosis triggered by IL-6 removal could be prevented by IL-3 as well as by phorbol esters, although IL-3 and phorbol esters rescued differentiating myeloid cells by apparently different pathways. This also appears to be the case for fibroblasts deprived of FGF or EGF; serum deprivation results in spontaneous apoptosis that can be suppressed either by FGF, EGF, or phorbol esters (Araki et al., 1990). Yet despite this similarity the initiation of the FGF and EGF protective pathways does not appear to be mediated through the activation of PKC (Kanter et al., 1984). PKC inhibitors do not block EGF prevention of apoptosis in serum-deprived fibroblasts, and EGF and FGF appear to involve a tyrosine kinase-dependent mechanism (Tilly et al., 1992).

It should be noted that in the case of myeloid cells described above, the protection from apoptosis afforded by PKC activation more closely resembles the thymocyte selection model, in that the rescued cell population is nonproliferating. As is the case with CD4$^+$CD8$^+$ thymocytes, activation of PKC in the myeloid cell systems is more closely associated with promoting differentiation of the rescued cell population. Induction of cell death in leukemic cells by apoptosis and inhibition of this cell death are obviously of importance in cancer therapy strategies. Hematopoietic cytokines and tumor-promoting phorbol esters can prevent both spontaneous or TGF-β-induced apoptosis in myeloid leukemic cells (Lotem and Sachs, 1992). Two important implications can be gleaned from these experiments. First, many tumor promoters may act by inhibiting apoptosis rather than by stimulating cell proliferation. Second, cytokine therapy may inadvertently be deleterious to cancer patients by preventing apoptosis in malignant cell populations.

CONCLUDING REMARKS

Given the multitude of roles that PKC plays in controlling cellular proliferation and differentiation, it is not surprising that this family of enzymes has been implicated in the control of cell death, particularly in the control of cell selection by apoptosis. In this chapter we have attempted to point out that the notion of physiological cell death by apoptosis or by any other mechanism is a relatively novel concept, and that the regulation of apoptosis in general is poorly understood. Emerging evidence suggests that the PKC family may regulate cell decisions that shape biological functions ranging from the selection of lymphocyte repertoires to cancer cell growth. A variety of hypotheses testing the mechanisms of regulation of apoptosis by these kinases, which no doubt will shed light on many basic biological and clinical problems, need to be addressed.

REFERENCES

Alles, A., K. Alley, J. C. Barrett, R. Buttyan, et al. 1991. Apoptosis: A general comment. *FASEB J.* 5:2127–2128.

Araki, S., Y. Simada, K. Kaji, and H. Hayashi. 1990. Role of protein kinase C in the inhibition by fibroblast growth factor of apoptosis in serum-depleted endothelial cells. *Biochem. Biophys. Res. Comm.* 172:1081–1085.

Arends, M. J., R. G. Morris, and A. H. Wyllie. 1990. Apoptosis: The role of the endonuclease. *Am. J. Pathol.* 136:593–608.

Baier, G., D. Telford, L. Giampa, K. M. Coggeshall,, G. Baier-Bitterlich, N. Isakov, and A. Altman. 1993. Molecular cloning and characterization of PKCθ, a novel member of the protein kinase C (PKC) gene family expressed predominantly in hematopoietic cells. *J. Biol. Chem.* 268:4997–5004.

Bissonnette, R. P., F. Echerverri, A. Mahboubi, and D. R. Green. 1992. Apoptotic cell death induced by c-*myc* is inhibited by *bcl*-2. *Nature* 359:552–554.

Boyd, R. L., and P. Hugo. 1991. Towards an integrated view of thymopoiesis. *Immunol. Today* 12:71–77.

Brown, D. G., X.-M. Sun, and G. M. Cohen. 1993. Dexamethasone-induced apoptosis involves cleavage of DNA to large fragments prior to internucleosomal fragmentation. *J. Biol. Chem.* 268:3037–3039.

Brüne, B. P. Hartzell, P. Nicotera, and S. Orrenius. 1991. Spermine prevents endonuclease activation and apoptosis in thymocytes. *Exp. Cell Res.* 195:323–329.

Campana, D., E. Coustan-Smith, A. Manabe, M. Buschle, S. C. Raimondi, F. G. Behm, R. Ashmun, M. Aricò, A. Biondi, and C.-H. Pui. 1993. Prolonged survival of B-lineage acute lymphoblastic leukemia cells is accompanied by overexpression of *bcl*-2 protein. *Blood* 81:1025–1031.

Cantrell, D. A., B. Friedrich, A. A. Davies, M. Gullberg, and M. J. Crumpton. 1989. Evidence that a kinase distinct from protein kinase C induces CD3 γ-subunit phosphorylation without a concomitant down-regulation in CD3 antigen expression. *J. Immunol.* 142:1626–1630.

Cohen, J. J., and R. C. Duke. 1984. Glucocorticoid activation of a calcium-dependent endonuclease in thymocyte nuclei leads to cell death. *J. Immunol.* 132:38–42.

Diamond, M. I., J. N. Miner, S. K. Yoshinaga, and K. R. Yamamoto. 1990. Transcription factor interactions: Selectors of positive or negative regulation from a single DNA element. *Science* 249:1266–1272.

Dowd, D. R., P. N. MacDonald, B. S. Komm, M. R. Haussler, and R. L. Miesfeld. 1992. Stable expression of the calbindin-D28K complementary DNA interferes with the apoptotic pathway in lymphocytes. *Molec. Endocrin.* 6:1843–1848.

Evan, G. I., A. H. Wyllie, C. S. Gilbert, T. D. Littlewood, H. Land, M. Brooks, C. M Waters, L. Z. Penn, and D. C. Hancock. 1992. Induction of apoptosis in fibroblasts by c-myc protein. *Cell* 69:119–128.

Finkel, T. H., J. C. Cambier, R. T. Kubo, W. K. Born, P. Marrack, and J. W. Kappler. 1989. The thymus has two functionally distinct populations of immature $\alpha\beta^+$T cells: One population is deleted by ligation of $\alpha\beta$TCR. *Cell* 58:1047–1054.

Finkel, T. H., J. W. Kappler, and P. C. Marrack. 1992. Immature thymocytes are protected from deletion early in ontogeny. *Proc. Natl. Acad. Sci.* U.S.A. 89:3372–3374.

Fischer, M., I. MacNeil, T. Suda, J. E. Cupp, K. Shortman, and A. Zlotnik. 1991. Cytokine production by mature and immature thymocytes. *J. Immunol.* 146:3452–3456.

Freire-Moar, J., H. Cherwinski, F. Hwang, J. Ransom, and D. Webb. 1991. Expression of protein kinase C isoenzymes in thymocyte subpopulations and their differential regulation. *J. Immunol.* 147:405–409.

Garcia, I., I. Martinou, Y. Tsujimoto, and J.-C. Martinou. 1992. Prevention of programmed cell death of sympathetic neurons by the *bcl*-2 proto-oncogene. *Science* 258:302–304.

González-Fernández, A., F. Diaz-Espada, M. Kreisler, and F. G. Deza. 1991. Proliferative responses induced by the activation of protein kinase C during the development of human T lymphocytes. *Eur. J. Immunol.* 21:115–121.

Gratiot-Deans, J., D. Keim, J. R. Strahler, L. A. Turka, and S. Hanash. 1992. Differential expression of Op18 phosphoprotein during human thymocyte maturation. *J. Clin. Invest.* 90:1576–1581.

Groux, H., G. Torpier, D. Monté, Y. Mouton, A. Capron, and J. C. Ameisen. 1992. Activation-induced death by apoptosis in CD4$^+$ T cells from human immunodeficiency virus-infected asymptomatic individuals. *J. Exp. Med.* 175:331–340.

Hengartner, M. O., R. E. Ellis, and H. R. Horvitz. 1992. *Caenorhabditis elegans* gene ced-9 protects cells from programmed cell death. *Nature* 356:494–499.

Hockenbery, D. M., G. Nuñez, C. Milliman, R. D. Schreiber, and S. J. Korsmeyer. 1990. Bcl-2 is an inner mitochondrial membrane protein that blocks programmed cell death. *Nature* 348:3334–337.

Hockenbery, D. M., M. Zutter, W. Hickey, M. Nahm, and S. J. Korsmeyer. 1991. BCL2 protein is topographically restricted in tissues characterized by apoptotic cell death. *Proc. Natl. Acad. Sci.* U.S.A. 88:6961–6965.

Hsieh, J.-C., P. W. Jurutka, M. A. Galligan, C. M. Terpening, C. A. Haussler, D. S. Samuels, Y. Shimizu, N. Shimizu, and M. R. Haussler. 1991. Human vitamin D receptor is selectively phosphorylated by protein kinase C on serine 51, a residue crucial to its transactivation function. *Proc. Natl. Acad. Sci.* U.S.A. 88:9315–9319.

Ishida, I., Y. Agata, K. Shibahara, and T. Honjo. 1992. Induced expression of PD-1, a novel member of the immunoglobulin gene superfamily, upon programmed cell death. *EMBO J.* 11:3887–3895.

Jacobson, M. D., J. F. Burne, M. P. King, T. Miyashita, J. C. Reed, and M. C. Raff. 1993. Bcl-2 blocks apoptosis in cells lacking mitochondrial DNA. *Nature* 361:365–369.

Kanter, P., K. J. Leister, L. D. Tomei, P. A. Wenner, and C. E. Wenner. 1984. Epidermal growth factor and tumor promoters prevent DNA fragmentation by different mechanisms. *Biochem. Biophys. Res. Comm.* 118:392–399.

Kerr, J. F. R., A. H. Wyllie, and A. R. Currie. 1972. Apoptosis: A basic biological phenomenon with wide-ranging implications in tissue kinetics. *Br. J. Cancer* 26:239–257.

Kizaki, H., T. Tadakuma, C. Odaka, J. Muramatsu, and Y. Ishimura. 1989. Activation of a suicide process of thymocytes through DNA fragmentation by calcium ionophores and phorbol esters. *J. Immunol.* 143:1790–1794.

Lawetzky, A., M. Kubbies, and T. Hünig. 1991. Rat ''first-wave'' mature thymocytes: Cycling lymphoblasts that are sensitive to activation-induced cell death but rescued by interleukin 2. *Eur. J. Immunol.* 21:2599–2604.

Lotem, J., and L. Sachs. 1992. Hematopoietic cytokines inhibit apoptosis induced by transforming growth factor β1 and cancer chemotherapy compounds in myeloid leukemic cells. *Blood* 80:1750–1757.

Lotem, J., E. J. Cragoe Jr., and L. Sachs. 1991. Rescue from programmed cell death in leukemic and normal myeloid cells. *Blood* 78:953–960.

MacDonald, H. R., H. Hengartner, and T. Pedrazzini. 1988. Intrathymic deletion of self-reactive cells prevented by neonatal anti-CD4 antibody treatment. *Nature* 335:174–176.

McCabe Jr., M. J., and S. Orrenius. 1993. Protein kinase C activating phorbol esters prevent glucocorticoid-induced apoptosis in human thymocyte cultures. In preparation.

McCabe Jr., M. J., P. Nicotera, and S. Orrenius. 1992. Calcium-dependent cell death: Role of the endonuclease, protein kinase C, and chromatin conformation. *Ann. N.Y. Acad. Sci.* 663:269–278.

McCabe Jr., M. J., S. A. Jiang, and S. Orrenius. 1993. Chelation of intracellular zinc induces apoptosis in mature thymocytes. *Lab. Invest.* 69:101–110.

McConkey, D. J., P. Hartzell, P. Nicotera, and S. Orrenius. 1989. Calcium-activated DNA fragmentation kills immature thymocytes. *FASEB J.* 3:1843–1849.

McConkey, D. J., S. Orrenius, and M. Jondal. 1990. Cellular signalling in programmed cell death (apoptosis). *Immunol. Today* 11:120–121.

Meyaard, L., S. A. Otto, R. R. Jonker, M. Mijnster, R. P. M. Keet, and F. Miedema. 1992. Programmed death of T cells in HIV-1 infection. *Science* 257:217–219.

Miyashita, T., and J. C. Reed. 1992. *bcl-2* Gene transfer increases relative resistance of

S49.1 and WEH17.2 lymphoid cells to cell death and DNA fragmentation induced by glucocorticoids and multiple chemotherapeutic drugs. *Cancer Res.* 52:5407–5411.

Montgomery, R. A., and M. J. Dallman. 1991. Analysis of cytokine gene expression during fetal thymic ontogeny using the polymerase chain reaction. *J. Immunol.* 147:554–560.

Murphy, K. M., A. B. Heimberger, and D. Y. Loh. 1990. Induction by antigen of intrathymic apoptosis of $CD4^+CD8^+TCR^{lo}$ thymocytes *in vivo*. *Science* 250:1720–1722.

Newell, M. K., L. J. Haughn, C. R. Maroun, and M. H. Julius. 1990. Death of mature T cells by separate ligation of CD4 and the T-cell receptor for antigen. *Nature* 347:286–289.

Nieto, M. A., A. González, A. López-Rivas, F. Diaz-Espada, and F. Gambón. 1990. IL-2 protects against anti-CD3-induced cell death in human medullary thymocytes. *J. Immunol.* 145:1364–1368.

Nunez, G., L. London, D. Hockenbery, M. Alexander, J. P. McKearn, and S. J. Korsmyer. 1990. Deregulated *bcl*-2 gene expression selectively prolongs survival of growth factor-deprived hemopoietic cell lines. *J. Immunol.* 144:3602–3610.

Onishi, Y., Y. Azuma, Y. Sato, Y. Mizuno, T. Tadakuma, and H. Kizaki. 1993. Topoisomerase inhibitors induce apoptosis in thymocytes. *Biochim. Biophys. Acta* 1175:147–154.

Orrenius, S., M. J. McCabe Jr., and P. Nicotera. 1992. Ca^{2+}-dependent mechanisms of cytotoxicity and programmed cell death. *Toxicol. Lett.* 64, 65:357–364.

Owens, G. P., W. E. Hahn, and J. J. Cohen. 1991. Identification of mRNAs associated with programmed cell death in immature thymocytes. *Molec. Cell. Biol.* 11:4177–4188.

Rodríguez-Tarduchy, G., and A. López-Rivas. 1989. Phorbol esters inhibit apoptosis in IL-2-dependent T lymphocytes. *Biochem. Biophys. Res. Comm.* 164:1069–1075.

Rothenberg, E. V. 1990. Death and transfiguration of cortical thymocytes: A reconsideration. *Immunol. Today* 11:116–119.

Sellins, K. S., and J. J. Cohen. 1987. Gene induction by γ-irradiation leads to DNA fragmentation in lymphocytes. *J. Immunol.* 139:3199–3206.

Shi, Y., R. P. Bissonnette, N. Parfrey, M. Szalay, R. T. Kubo, and D. R. Green. 1991. *In vivo* administration of monoclonal antibodies to the CD3 T cell receptor complex induces cell death (apoptosis) in immature thymocytes. *J. Immunol.* 146:3340–3346.

Shi, Y., J. M. Glynn, L. J. Guilbert, T. G. Cotter, R. P. Bissonnette, and D. R. Green. 1992. Role for c-myc in activation-induced apoptotic cell death in T cell hybridomas. *Science* 257:212–214.

Smith, C. A., G. T. Williams, R. Kingston, E. J. Jenkinson, and J. J. T. Owen. 1989. Antibodies to CD3/T-cell receptor complex induce death by apoptosis in immature T cells in thymic cultures. *Nature* 337:181–184.

Strulovici, B., S. Daniel-Issakani, G. Baxter, J. Knopf, L. Sultzman, H. Cherwinski, J. Nestor Jr., D. R. Webb, and J. Ransom. 1991. Distinct mechanisms of regulation of protein kinase C by hormones and phorbol diesters. *J. Biol. Chem.* 266:168–173.

Suzuki, K., H. Kizaki, T. Tadakuma, and Y. Ishimura. 1990. 12-O-Tetradecanoylphorbol 13-acetate potentiates the action of cAMP in inducing DNA cleavage in thymocytes. *Biochem. Biophys. Res. Comm.* 171:827–831.

Tadakuma, T., H. Kizaki, C. Odaka, R. Kubota, Y. Ishimura, H. Yagita, and K. Okumura.

1990. CD4$^+$CD8$^+$ thymocytes are susceptible to DNA fragmentation induced by phorbol ester, calcium ionophore and anti-CD3 antibody. *Eur. J. Immunol.* 20: 779–784.

Teh, H.-S., A. M. Garvin, K. A. Forbush, D. A. Carlow, C. B. Davis, D. R. Littman, and R. M. Perimutter. 1991. Participation of CD4 coreceptor molecules in T-cell repertoire selection. *Nature* 349:241–243.

Tilly, J. L., H. Billig, K. I. Kowalski, and A. J. W. Hsueh. 1992. Epidermal growth factor and basic fibroblast growth factor suppress the spontaneous onset of apoptosis in cultured rat ovarian granulosa cells and follicles by a tyrosine kinase-dependent mechanism. *Molec. Endocrinol.* 6:1942–1950.

Walker, P. R., C. Smith, T. Youdale, J. Leblanc, J. F. Whitfield, and M. Sikorska. 1991. Topoisomerase II-reactive chemotherapeutic drugs induce apoptosis in thymocytes. *Cancer Res.* 51:1078–1085.

Wang, P. T. H., M. Bigby, and M.-S. SY. 1987. Selective down modulation of L3T4 molecules on murine thymocytes by the tumor promoter, phorbol 12-myristate 13-acetate. *J. Immunol.* 139:2157–2165.

Wyllie, A. H. 1980. Glucocorticoid-induced thymocyte apoptosis is associated with endogenous endonuclease activation. *Nature* 284:555–556.

Yonish-Rouach, E., D. Resnitzky, J. Lotem, L. Sachs, A. Kimchi, and M. Oren. 1991. Wild-type p53 induces apoptosis of myeloid leukaemic cells that is inhibited by interleukin-6. *Nature* 352:345–347.

12

Serine/Threonine Protein Phosphatases and Their Inhibitors

RICHARD E. HONKANEN
ALTON L. BOYNTON

The reversible phosphorylation of biologically active proteins is an important mechanism of regulation for a wide variety of cellular events in eukaryotic cells. Therefore, the regulation of protein kinases has been and continues to be an area under extensive investigation. However, protein phosphorylation is a dynamic process, with the level of phosphorylation at any instant reflecting the relative activities of both protein kinases and protein phosphatases. Like protein kinases, recent data from several systems indicate that protein phosphatases are also highly regulated and potentially responsive to the concentration of intracellular second messengers. Perhaps not as abundant as protein kinases, cDNA cloning has uncovered a wealth of serine/threonine (ser/thr) protein phosphatases. In this chapter the proteins catalyzing the dephosphorylation of biologically active proteins will be discussed, with concentration on ser/thr protein phosphatases, the enzymes most likely responsible for dephosphorylating the substrates of protein kinase C. Several excellent reviews on ser/thr protein phosphatases have been published in the last few years (Cohen, 1989; Shenolikar and Narin, 1991; Sim, 1992). Thus, this chapter will focus on recent progress in the field and minimize reiteration.

CHARACTERIZATION OF SER/THR PROTEIN PHOSPHATASES

Before 1983, although many protein phosphatases were described in the literature, little was known concerning the relationship between these enzymes, and their classification was controversial. Ingebritsen and Cohen (1983) introduced a simple classification based on the suggestion that most, if not all, of the ser/thr protein

Table 12.1 Distinguishing Features of SER/THR Protein Phosphatases

	PP1	PP2A	PP2B	PP2C	PP3
Substrates					
Phosphorylase kinase a	β	α	α	α	β
Phosphorylase a	Yes	Yes	No	No	Yes
Phosphohistone	Yes	Yes	No	Yes	Yes
Myosin light chain	Yes	Yes	Yes	Yes	ND[a]
Modulators					
I-1	Inhibit	NE[b]	NE	NE	ND
I-2	Inhibit	NE	NE	NE	Activate
DARPP-32	Inhibit	NE	NE	NE	ND
NIPP-1	Inhibit	NE	NE	NE	ND
Heparin (phos a substrate)	Inhibit	Activate/NE[c]	NE	NE	NE
Heparin (phosphohistone substrate)	Inhibit	NE	NE	NE	Activate/ Inhibit[c]
Protamine (phos a substrate)	Inhibit	NE	NE	NE	Inhibit
Toxins			$IC^{50}s \ (nM)^d$		
Okadaic acid	49.0	0.28	5000	NE	3.91
Methyl okadaate	NE	~10,000	ND	ND	~100
Okadaol	NE	>10,000	ND	ND	~1000
1-Nor-okadaone	NE	NE	ND	ND	NE
Okadaic acid tetraacetate	NE	NE	ND	ND	NE
Microcystin-LR	2.03	0.04	200	NE	0.18
Microcystin-LA	2.01	0.04	ND	ND	0.19
Nodularin	2.37	0.03	ND	NE	1.39
Calyculin	0.40	0.25	ND	ND	0.25
Tautomycin	23.1	7.51	ND	ND	29.7

[a]ND = not determined.

[b]NE = no effect.

[c]Concentration dependent.

[d]Concentration of toxin required to inhibit 50 percent of activity. IC_{50}s determined at the titration end point for each enzyme.

phosphatases described in the literature could be explained by the activity of four principal catalytic subunits (PP1, PP2A, PP2B, and PP2C) with broad and overlapping substrate specificities. Recently, a fifth catalytic subunit was added, named PP3 (Honkanen et al., 1991b), based on the Ingebritsen and Cohen classification system. On the basis of several criteria, most ser/thr protein phosphatases can be placed into one of three major groups—type 1, type 2, or type 3—depending on whether they preferentially dephosphorylate the α or β subunit of phosphorylase kinase and whether they are sensitive to two heat- and acid-stable proteins, termed inhibitor 1 (I-1) and inhibitor 2 (I-2). Type 2 protein phosphatases (PP1) specifically dephosphorylate the β subunit of phosphorylase kinase and are inhibited by nanomolar concentrations of I-1 and I-2. Type I protein phosphatases (PP2) preferentially dephosphorylate the α subunit of phosphorylase kinase and are not inhibited by I-1 and I-2. PP2 is further divided into three subtypes, PP2A,PP2B (e.g., calcineurin), and PP2C, which can be distinguished by a number of criteria. PP2B and PP2C have an absolute requirement for Ca^{2+}/calmodulin and Mg^{2+}, respectively, while PP2A, like PP1, does not require divalent cations. Type 3, the newest biochemically characterized member of the ser/thr protein phosphatase family, dephosphorylates the β subunit of phosphorylase kinase, like PP1. However, nanomolar concentrations of I-2 stimulate its activity, unlike PP1 and PP2A. PP3 catalytic activity is also, like that of PP1 and PP2A, independent of divalent cations. In addition, PP1 and PP3 prefer phosphorylase over phosphohistone as a substrate *in vitro,* whereas PP2A prefers phosphohistone (Honkanen et al., 1991b). Toxins, such as okadaic acid, microcystin-LR, nodularin, and calyculin A, are also emerging as powerful tools to classify ser/thr protein phosphatases and will be discussed in detail later in this chapter (see Table 12.1 for a summary of the biochemical features of the known ser/thr protein phosphatases).

Recent cDNA cloning studies have greatly advanced our knowledge of the ser/thr protein phosphatases and suggest that the 1983 classification system of Ingebritsen and Cohen may soon require updating. The amino acid sequence for the catalytic subunits of PP1, PP2A, PP2B, and PP2C were deduced from screening of cDNA libraries, indicating that PP1, PP2A, and PP2B belong to a single gene family to which PP2C is completely unrelated (Cohen et al., 1990). Structurally, PP1 and PP2A are more similar than PP2A and PP2B. The cDNA's encoding of the catalytic subunits of PP1 and PP2A (predicted molecular masses of $\sim$37 kDa and $\sim$36 kDa, respectively) indicate that these enzymes share about 50 percent identity in their primary amino acid sequences (Berndt et al., 1987; Green et al., 1987; Stone et al., 1987; Arino et al., 1988; Bai et al., 1988; Cohen, 1988; Silva et al., 1988; Dombradi et al., 1989; Silva et al., 1991). The catalytic subunit of PP2B is slightly larger, M_r $\sim$58 kDa, and PP2B shares about 40 percent identity with both PP1 and PP2A. The "core" of PP2B is very similar to PP1 and PP2A, but PP2B has an extended C-terminal region that appears to interact with Ca^{2+}/calmodulin, as well as a region that suppresses activity in the absence of calcium (Berndt et al., 1987; Cohen, 1988). In fact, it has been speculated that PP2B was derived from the fusion of a PP1-PP2A precursor gene and the calmodulin gene (Kincaid et al., 1988; Ito, et al., 1989). The least studied ser/thr protein phosphatase is PP2C and it shows no significant structural similarity with PP1, PP2A, or PP2B. PP3 has yet to be cloned and structurally described.

Sequencing studies have revealed that PP1, PP2A, and PP2B exist in several isoforms that are present in most, if not all, tissues and share considerable sequence homology. For example, two full-length mammalian cDNAs encoding isoforms of PP1 (PP1α and PP1β) encode proteins that share about 95 percent identity at the amino acid level and differ primarily in their amino-terminal domains (Berndt et al., 1987; Bai et al., 1988; Cohen, 1988). Drosophila have at least three isozymes of PP1 (α_1, α_2, and β) with α_1 and α_2 having 95 percent amino acid sequence identity and α_1, α_2, and β having 85 percent amino acid sequence identity (Dombradi et al., 1990). Full-length cDNAs encoding two isoforms of PP2A having about 98 percent identity (Green et al., 1987; Stone et al., 1987; Arino et al., 1988) and two isoforms of PP2B having about 89 percent identity (Guerini and Klee, 1989; Ito et al., 1989; Kuno et al., 1989; Cohen et al.,

1990; Kincaid et al., 1990; Cyert et al., 1991; Liu et al., 1991; McPartlin et al., 1991; Silva et al., 1991; Chen et al., 1992) have also been reported.

Studies using PCR suggest that even more isoforms of PP1, PP2A, and PP2B exist. PCR employing primers that encode regions conserved in PP1, PP2A, and PP2B identified cDNAs encoding both the α and β isoforms of PP2A, four cDNAs encoding PP1-like sequences, and three cDNAs encoding PP2B-like sequences (Wadzinski et al., 1990). A similar study employing genomic DNA as a template identified several isoforms of PP1, PP2A, and PP2B in humans, yeast, and drosophila (Chen et al., 1992). However, since the PCR products obtained in these studies do not contain overlapping regions, it is not possible to determine if they identified the same isoforms of PP1, PP2A, or PP2B (Wadzinski et al., 1990; Chen et al., 1992). Knowledge of PP2C isoforms is much less. However, recent studies indicate that PP2C is also encoded by two different genes that may prove to be isoforms of this enzyme.

Molecular cloning studies in lower eukaryotes have demonstrated that ser/thr protein phosphatases are very highly conserved through evolution. To date, PP1 has received the most attention, and genes encoding the catalytic subunits of PP1 in humans, drosophila, and fungi reveal that the primary amino acid structure of PP1 is about 86 percent identical in these diverse organisms (Cohen et al., 1990). Similarly, PP2A in humans and drosophila are about 93 percent identical in structure, and both PP1 and PP2A share about 35 percent identity with bacteriophage λ protein, λorf221, and the homologous gene in Φ 80 (Cohen et al., 1988; Cohen and Cohen, 1989); orf221 encodes a bacteriophage protein phosphatase that may aid in viral replication (Cohen et al., 1988; Cohen and Cohen, 1989). PP2B also appears highly conserved between species. However, at present there are too few studies with PP2B or PP2C to make a definitive comparison.

In addition to the above-mentioned isoforms, recent studies have demonstrated the existence of at least nine ser/thr protein phosphatases that cannot readily be classified as PP1, PP2A, PP2B, PP2C, or PP3 on the basis of sequence similarities or on the biochemical properties of the purified proteins. These phosphatases, designated as PPV, PPX, PPY, PPZ1, PPZ2, PPH3, rdgC, and sit4, have been identified in a variety of tissues. PPX, PPY, PPV, PPZ1, PPZ2, and PPH3 were identified from screening of cDNA or genomic libraries, and although these putative protein phosphatases are similar to the PP1/PP2A/PP2B family of phosphatases, they are sufficiently distinct structurally to suggest that they encode novel proteins and not isoforms of PP1 or PP2A. PPY was originally identified from a drosophila cDNA library and maps on drosphila chromosome 2. PPY encodes a putative protein that shares 63 percent identity with mammalian PP1α and 39 percent identity with mammalian PP2As (Dombradi et al., 1989). The cDNA-encoding PPX, isolated from a rabbit liver cDNA library, has 65 percent identity to PP2A and 45 percent amino acid identity to PP1 (Silva et al., 1988). Only a partial cDNA sequence of PPV has been reported, and this putative enzyme is more similar to PP2A than to PP1 (Cohen et al., 1990). PPZ1 and PPZ2 were originally identified as being from a rabbit brain library (Cohen et al., 1990); however, they were later shown to be yeast genes that presumably were contaminates of the brain library (Silva et al., 1991). Again, PPZ1 and PPZ2 are more similar to PP1 than to PP2A or PP2B. PPH3 is a yeast gene and the predicted PPH3 protein is most closely related to rabbit PPX. However the similarity of PPH3 and PPX is not sufficient to suggest PPH3 is the yeast homologue of PPX (Ronne et al., 1991). The assumption that PPH3 is not a yeast homologue of PPX based on structural similarities assumes that PPX is as highly conserved among species as is PP1 and PP2A.

Studies in drosophila on genes involved in retinal degeneration identified another putative ser/thr protein phosphatase that, on the basis of structural similarity, may also be a member of the PP1/PP2A/PP2B family of phosphatases. Analysis of the retinal degeneration C gene (rdgC), a gene that is required to prevent light-induced retinal degeneration (Steele et al., 1992), indicates that the rdgC transcriptional region encodes a protein that is 30 percent identical to the PP1/PP2A/PP2B family of phosphatases. The presence of "EF hand motifs" in the rdgC gene suggests that this putative phosphatase is capable of binding Ca^{2+} and may be regulated directly by changes in the intracellular concentration of calcium (Steele et al., 1992). In *Drosophila,* rdgC is expressed in visual systems as well as in the mushroom bodies of the central brain. rdgC is the only known or putative protein phosphatase described to date that possesses multiple EF hand structures.

The SIT4 protein phosphatase was originally identified in *Saccharomyces cerevisiae* by mutations that restore transcription to the HIS4 gene in the absence of *trans*-acting DNA binding factors that are normally required for HIS4 transcription (Arndt et al., 1989). The protein encoded by SIT4 has a predicted molecular weight of 35.5 kDa and is 55 percent identical to the catalytic subunit of mammalian PP2As and 40 percent identical to mammalian PP1s (Arndt et al., 1989). Mutational studies in yeast indicate that SIT4 is required for the progression of *S. cerevsiae* into S-phase. SIT4 is required for the normal accumulation of G1 cyclin mRNAs that are needed for DNA synthesis, and SIT4 is required for a still unidentified function during bud emergence or cell cycle progression through late G1 or G1/S (Arndt et al., 1989; Sutton et al., 1991; Fernandez-Sarabia et al., 1992).

PP3 is a protein phosphatase identified in the particulate membrane-containing fraction of a bovine brain extract that, on the basis of biochemical, pharmacological, immunological, and partial sequence data, also appears to be a novel ser/thr protein phosphatase (Honkanen et al., 1991b). Like PP1 and PP2A, PP3 has a broad substrate specificity *in vitro* and is sensitive to okadaic acid, microcystin-LR, and nodularin (Honkanen et al., 1990, 1991a,b). The function and sequence of PP3 are not presently known.

INHIBITORS OF SER/THR PROTEIN PHOSPHATASES

The renewed interest in ser/thr protein phosphatases can, in part, be attributed to the discovery that several highly toxic compounds, including okadaic acid, microcystins, nodularin, calyculin A, and tautomycin (see Figure 12.1 for structures), are potent and specific inhibitors of PP1, PP2A, PP3, and to a much lesser extent

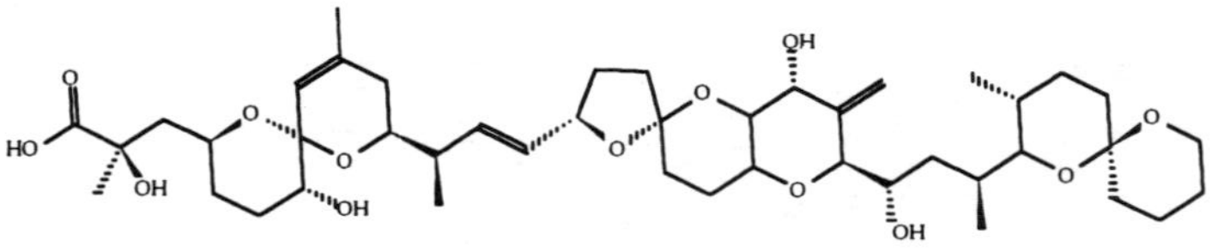

Okadaic acid

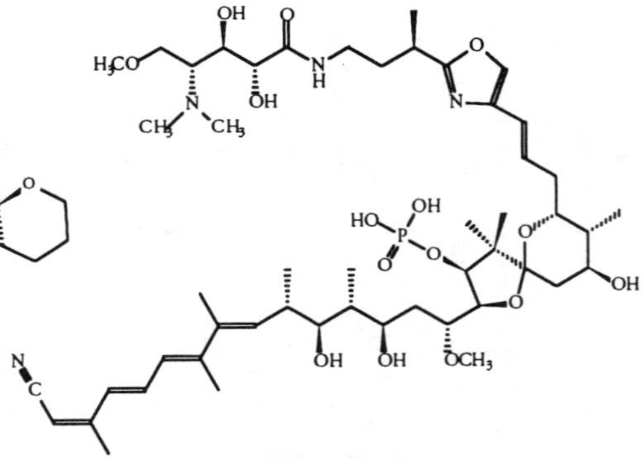

Calyculin A

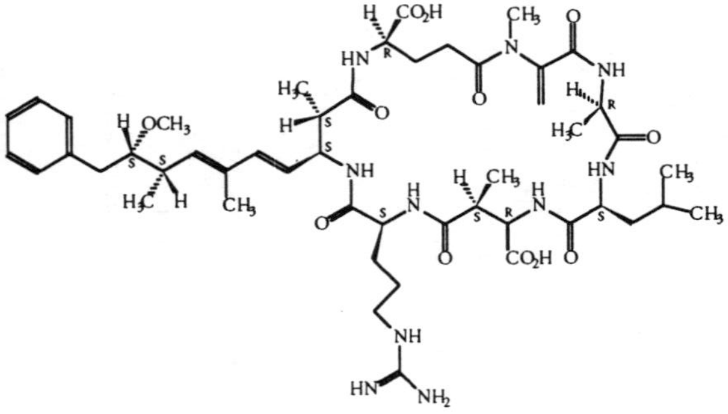

Microcystin-LR

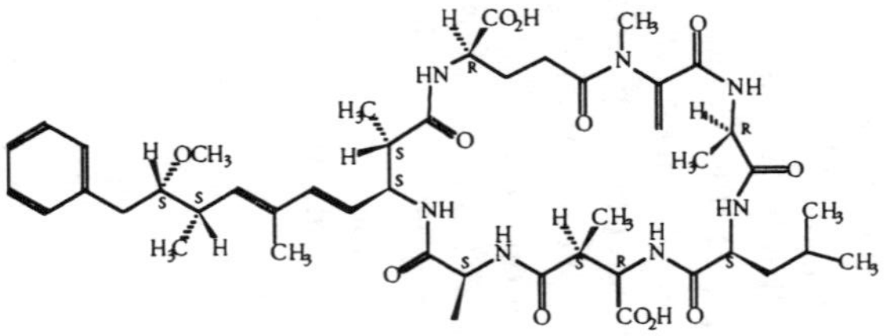

Microcystin-LA

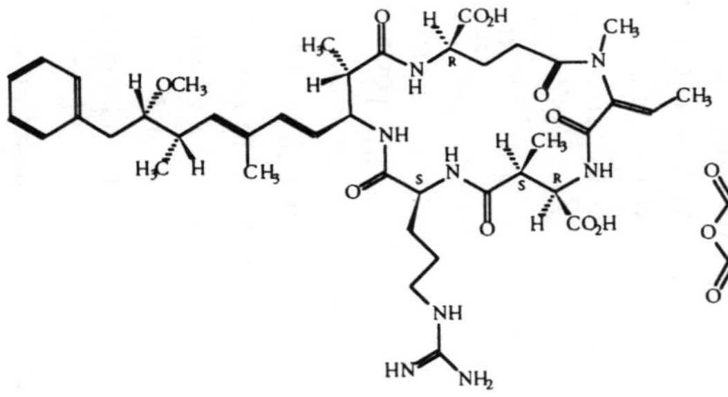

Figure 12.1 Structure of the various known ser/thr protein phosphatase inhibitors.

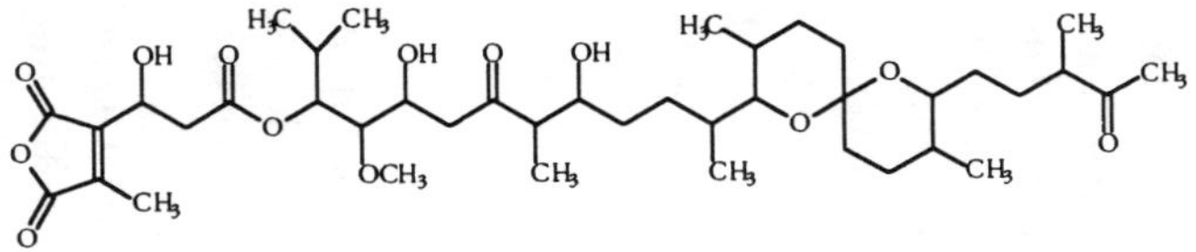

PP2B. In addition, four acid- and heat-stable endogenous modulators of PP1 and PP3 activity have been identified.

Before discussing the inhibition of protein phosphatase catalytic subunits by various toxins and modulator proteins, it is important to note that the high affinity of protein phosphatases for these agents demands that in enzyme assays the concentration of the catalytic subunits be lower than the concentration of the inhibitor employed. Thus, before an IC_{50} or the concentration of specific protein phosphatases in cellular extracts can be estimated, a series of enzyme dilution assays must be carried out to determine the "titration end point." This is defined as the concentration of enzyme after which further dilution no longer affects the IC_{50}. Knowledge of this is essential if the protein phosphatase(s) is not to titrate the inhibitor from the assay mixture (e.g., a concentration of enzyme in the assay mixture greater than okadaic acid or microcystin will give a false IC_{50}; illustrated in Figure 12.2). Failure to consider this unique characteristic of phosphatase/inhibitor interaction may explain the apparent discrepancy of IC_{50}s currently reported in the literature.

Okadaic Acid

The first of the naturally produced toxins identified as having inhibitory activity against ser/thr protein phosphatases was okadaic acid (Bialojan and Takai, 1988). Okadaic acid is a complex ($C_{44}H_{44}O_{13}$) polyether fatty acid (Figure 12.1) produced by several species of marine dinoflagellates of the genus *Dinophysis* and first isolated from a common black sponge, *Halichondria okadaii,* which concentrates the okadaic acid from the dinoflagellate while filter feeding (Murakami et al., 1982). Okadaic acid-producing dinoflagellates are also concentrated by many types of commercially valuable shellfish, and okadaic acid has recently been identified as a major cause of diarrhetic seafood poisoning. Acanthifolicin and dinophysistoxin-1, two toxins with structures very similar to that of okadaic acid, are also associated with diarrhetic seafood poisoning (Kumagi et al., 1986).

The first clue that okadaic acid was an inhibitor of ser/thr protein phosphatase activity came from the observation that okadaic acid caused a prolonged contraction of vascular smooth muscles (Takai et al., 1987). Bialojan and Takai (1988) subsequently demonstrated that okadaic acid very potently inhibited the activity of the purified catalytic subunits of PP1 and PP2A, and we have recently shown that PP3 is also sensitive to okadaic acid (Honkanen et al., 1991b). PP2B is much less sensitive to okadaic acid, requiring more than 1000-fold more okadaic acid for inhibition, and PP2C is apparently not affected by okadaic acid (Bialojan and Takai, 1988; Honkanen et al., 1991b). The apparent IC_{50}s of okadaic acid for PP2A, PP3, and PP1 are 0.28, 3.9, and 49 nM, respectively (Table 12.1). Recently, several derivatives of okadaic acid that demonstrate interesting variations on the potency of these okadaic acid derivatives toward PP1, PP2A, and PP3 have become commercially available (LC Services, Woburn, MA). Esterification of the carboxyl group (C-1) of okadaic acid, producing 1-methyl okadaate, greatly reduces the toxin's ability to inhibit all three enzymes and alters its specificity to favor PP3 (Table 12.1). Thus, 1-methyl okadaate has an apparent IC_{50}

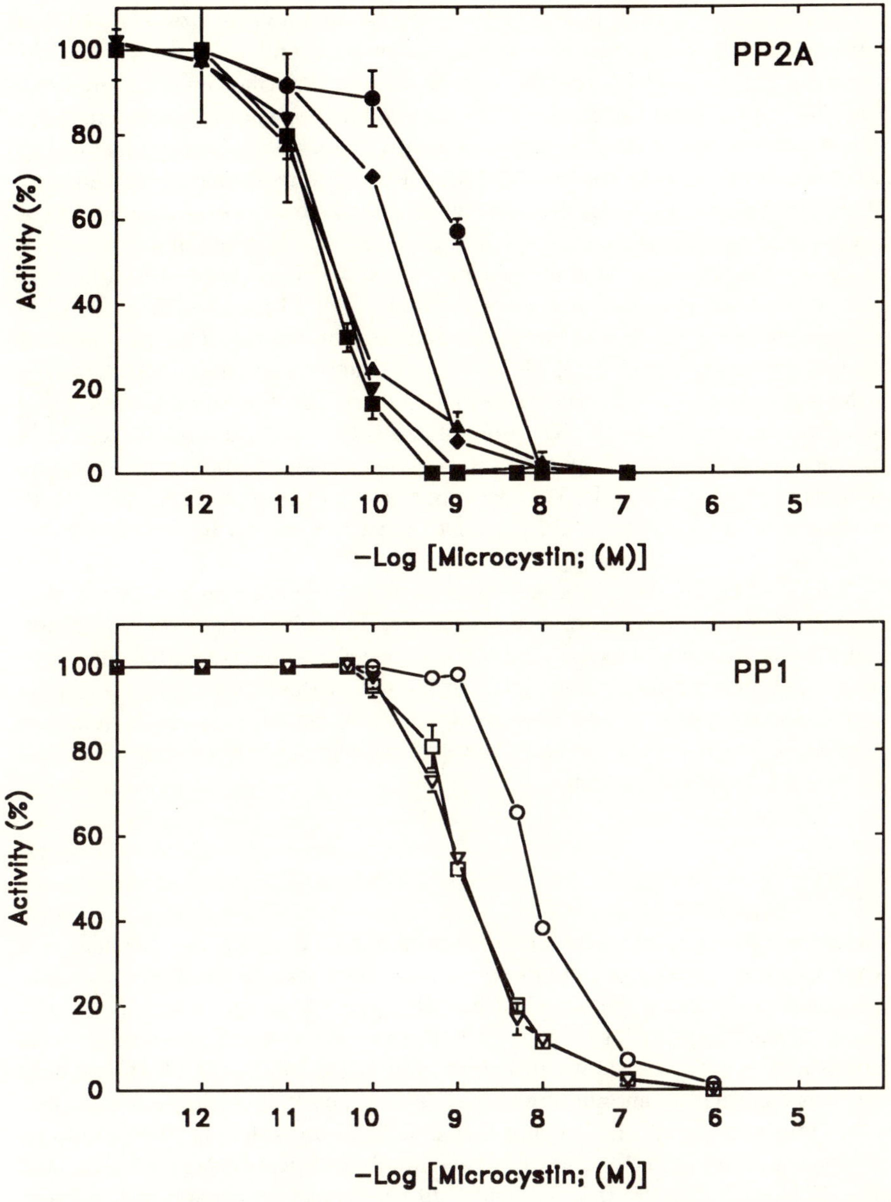

Figure 12.2 Effect of enzyme concentration on the inhibition of purified catalytic subunits of PP1 and PP2A protein phosphatases by microcystin-LR. Phosphatase activity was determined from the amount of ^{32}P released during a 10-minute assay conducted at 30°C in a volume of 80 μl and containing 2 μM phosphohistone as substrate (see Honkanen et al., 1991b, for details). Phosphohistone phosphatase activity before dilution was 25 ± 1.3 and 196 ± 4.1 nmol/min/mg protein for PP1 and PP2A, respectively. Phosphatase activity of undiluted (●), 5- (◆), 20- (▲), 33- (■), and 50- (▼) fold dilution of PP2A and undiluted (○), 10- (□), and 33- (▽) fold dilution of PP1 are shown. The activity is expressed as percent of maximal activity ± S.D. (n = 6). Reproduced with permission of Waverly Press.

of 100 nM for PP3, more than 10 μM for PP2A, and no effect on PP1. At first it would appear that 1-methyl okadaate might be useful in distinguishing the activity of PP3 from PP2A, and this may be the case if purified enzymes are used, providing cost is not a factor. However, with cell homogenates as a protein phosphatase source, the methyl group of 1-methyl okadaate is apparently rapidly removed, and its affinity for PP1, PP2A, and PP3 reverts to that of okadaic acid (R. E. Honkanen and A. L. Boynton, unpublished observations). Reduction of okadaic acid to produce okadaol also had an effect on the affinity of the derivative for the various catalytic subunits of PP1, PP2A, and PP3 (Table 12.1). Like 1-methyl okadaate, okadaol has a greater affinity for PP3 than for PP2A and has no apparent effect on PP1 up to a concentration of 10 μM. The utilization of okadaol for separation of PP3, PP2A, and PP1 activities seems impractical due to the high apparent IC_{50}s and cost factors. Further reduction of okadaol to 1-nor okadaone, or acetylation of C-2 to produce okadaic acid tetraacetate, results in the compounds having no apparent effect on the activities of the purified catalytic subunits of PP1, PP2A, an PP3 at concentrations up to 10 μM (Table 12.1). Both of these two compounds would appear to meet the criteria for a noneffective control for okadaic acid.

Okadaic acid has been reported to be a tumor promoter in the mouse skin (Suganuma et al., 1988) model system as well as in *in vitro* model systems (Katoh et al., 1990). However, other *in vitro* studies have also demonstrated okadaic acid to be a potent inhibitor of the tumor promotion process in a variety of model systems (Mordan et al., 1990; Rivedal et al., 1990), possibly due to its ability to down-regulate growth factor receptors such as platelet-derived and epidermal growth factor receptors (Dean et al., 1991).

Calyculin A

Calyculin A, another recently identified inhibitor of PP1, PP2A, and PP3, is a novel spiro-ketal with a skeleton bearing phosphate, oxazole, nitrile, and amide functionalities. It was originally identified as a cytotoxic component of the marine sponge *Discodermia calyx* (Figure 12.1; Kato et al., 1986). Like okadaic acid, calyculin A was at first shown to induce contraction of muscle fibers and only later was identified as an inhibitor of PP1, PP2A, and PP3 (Ishihara et al., 1989; R. E. Honkanen and A. L. Boynton, unpublished observations). The affinity of calyculin A for PP1, PP2A, and PP3 is essentially identical (Table 12.1; IC_{50}s for PP1, PP2A, and PP3 are 0.4, 0.3, and 0.29 nM, respectively) and thus is not a useful compound in distinguishing among activities of the various phosphatases. The similar IC_{50}s for PP1, PP2A, and PP3 may account for the high cytotoxicity of cells to calyculin A, which would inhibit all three enzymes at essentially the same concentration. In contrast, cells survive for prolonged periods in okadaic acid at a concentration of 10 nM, which inhibits both PP2A and PP3 activities, but they do not survive in slightly higher concentration (50–100 nM), which inhibits all three enzymes. Thus, the use of calyculin A in culture is limited, and interpretations with respect to specific protein phosphatases is complicated. Calyculin A, like okadaic acid, has been demonstrated to act as a tumor promoter *in vivo* in the mouse skin model system (Suganuma et al., 1990).

Microcystins

Studies to determine the molecular nature of several highly toxic monocyclic peptides produced by cyanobacteria have also resulted in the discovery of several potent and specific inhibitors of PP1, PP2A, and PP3. These studies were initiated in 1878 when George Francis published a report showing cyanobacteria to be the toxic component of a ''poisonous Australian lake'' (Francis, 1878). However, the toxins produced by cyanobacteria have been characterized and their structures elucidated only in the last 30 years. To date, cyanobacteria are known to produce both neurotoxins (anatoxins and apphantoxins) and hepatotoxins, initially called fast-death factor (Bishop et al., 1959) then called microcystins (Konst et al., 1965; Carmichael, 1986), cyanoginosin (Botes et al., 1984), and cyanviridin (Kusumi et al., 1987). Further characterization and structural analysis of the hepatotoxic component of many strains of cyanobacteria revealed that microcystin, cyano-ginosin, and cyanviridin are all members of a family of monocyclic heptapeptides now collectively referred to as microcystins. The microcystins, to date, have been isolated from several genera of cyanobacteria including *Microcystis* (Carmichael, 1986; Honkanen et al., 1990), *Oscillatoria and Anabaena* (Krishnamurthy et al., 1986a,b; Kusumi et al., 1987; Eriksson et al., 1988), and *Hapalosiphon* (Prinsep et al., 1992), and the microcystin family consists of over 15 structurally related cyclic heptapeptides (Carmichael et al., 1988b). Microcystins have the general structure cyclo (D-Ala-L-X-D-erythro-β-methylisoAsp-L-Y-Adda-D-isoGlu-*N*-methyldehydroAla), where X and Y are variable L-amino acids (i.e., leucine and arginine in microcystin-LR and leucine and alanine in microcystin-LA) and Adda is a unique β amino acid (2S, 3S, 8S, 9S)-3-amino-9-methoxy-2,6,8-trimethyl-10-phenyl-4(E), 6(E)-decadienoic acid (Figure 12.1).

Although the structure and toxicological properties of several different micro-cystins were known for almost a decade, it was not until the early 1990s that they were shown to be specific inhibitors of ser/thr protein phosphatases. The initial observation of H. Fujiki and co-workers (Fujiki and Moore, 1987; Sassa et al., 1989) that microcystin-LR increased the phosphorylation state of [^{32}P]-labeled cell homogenates in a manner similar to okadaic acid was probably the root of all the studies linking microcystin-LR with the inhibition of protein phosphatases. Microcystin-LR was the first cyanobacteria-produced cyclic peptide found to inhibit the activity of PP1 and PP2A (Honkanen et al., 1990; MacKintosh et al., 1990; Yoshizawa et al., 1990). Microcystin-LA and microcystin LR are potent inhibitors of PP1, PP2A, and PP3 (Honkanen et al., 1990; MacKintosh et al., 1990; Princep et al., 1992). The substitution of alanine (microcystin-LA) for argi-nine (microcystin-LR) has little effect on potency, with apparent IC$_{50}$s for micro-cystin-LR of 0.05, 0.20, and 1.9 nM for PP2A, PP3, and PP1, respectively, and for microcystin-LA of 0.05, 0.33, and 2.3 nM for PP2A, PP3, and PP1, respec-tively, with phosphorylase used as a substrate (Table 12.1). Thus, the variable component of the cyclic peptide does not appear to affect the inhibitory activity of the toxins tested to date (R. E. Honkanen and A. L. Boynton, unpublished observation). Microcystin-LR is known to promote liver tumors *in vivo* (Yoshizawa et al., 1990; Matsushima et al., 1991). Unfortunately, the microcystins are impermeable to most cells *in vitro,* with the notable exception of hepatocytes.

Nodularin

Nodularin is structurally related to microcystins and is a monocyclic pentapeptide recently purified and characterized from *Nodularia spumegina* (Carmichael et al., 1988a; Rinehart et al., 1988; Eriksson et al., 1990). Nodularin also contains Adda and has the structure cyclo (D-β-methyliso-Asp-L-Arg-Adda-D-iso-Glu-N-methyldehydrobutyrine) (Figure 12.1; Rinehart et al., 1988). The inhibition profile for nodularin is slightly different from that obtained with the microcystins and okadaic acid; nodularin inhibits PP1 and PP3 at similar concentrations (IC_{50}s of 2.37 and 1.39 nM, respectively), which is significantly higher than the concentration of toxin required to inhibit the activity of PP2A (IC_{50} = 0.03 nM; Table 12.1; Eriksson et al., 1990; Honkanen et al., 1991a). Like the microcystins, nodularin is hepatotoxic *in vivo* and is impermeable to most cell types *in vitro* with the exception of hepatocytes (Yoshizawa et al., 1990).

Tautomycin

The most recent addition to the toxins that inhibit ser/thr protein phosphatases is tautomycin. Unlike the other toxins described in this chapter, tautomycin is produced by a terrestrial microorganism, *Streptomyces spiroverticillatus* (Figure 12.1; Cheng et al., 1987, 1990) and was originally identified as an antibiotic with strong toxicity against a variety of eukaryotes including fungi, yeast, and animal cells (Cheng et al., 1987). Tautomycin was later found to induce morphological changes in human leukemic cells, which correlated with an apparent increase in protein phosphorylation due to the inhibition of protein phosphatase activity (Magae et al., 1990). Recently, tautomycin was shown to inhibit the activity of PP1, PP2A, and PP3 (Hori et al., 1991; R. E. Honkanen and A. L. Boynton, unpublished observations). However, unlike okadaic acid, microcystin, and nodularin, all of which are very stable compounds, tautomycin is relatively unstable (Cheng et al., 1990). The apparent IC_{50}s for tautomycin are 7.51, 23.1, and 29.7 nM for PP2A, PP1, and PP3, respectively, with phosphohistone used as substrate (Table 12.1).

Inhibitor-1

Inhibitor-1 was first identified as an acid- and heat-stable modulator of PP1 activity (K_i is 1.6 nM), requiring phosphorylation by the cyclic AMP-dependent protein kinase (PKA) for activity against PP1 (Nimmo and Cohen, 1978). Inhibitor-1 has no effect on PP2A, PP2B, and PP2C and its effects on PP3 are not known. Inhibitor 1 has a M_r of 18.7 kDa as determined by its amino acid sequence, yet it has an apparent M_r of 60 kDa and 26 kDa when estimated by gel filtration or

SDS PAGE-electrophoresis, respectively (Aitken et al., 1982). The active fragment of the protein is between amino acids 8–49 with thr 35 identified as the phosphate acceptor of PKA action (Aitken et al., 1982).

Inhibitor-2

Inhibitor-2 is another acid- and heat-stable modulator of PP1 activity. However, unlike I-1, it does not require phosphorylation for activity and has a K_i of 3.1 nM against PP1 (Foulkes et al., 1983). Inhibitor-2 has no effect on PP2A, PP2B, or PP2C; however, I-2 very potently stimulates PP3 activity (i.e., 1 nM stimulates maximal activity when either phosphohistone or phosphorylase is used as a substrate; Honkanen et al., 1991b). Inhibitor-2 has a M_r of 22.8 kDa with an apparent M_r of 32 kDa on SDS-PAGE (Holmes et al., 1986). Because of the inhibitory effect of I-2 on PP1 and its stimulatory effect on PP3, its use in determining the levels of either PP1 or PP3 in cellular extracts is not practical.

DARPP-32

DARPP-32 is a brain-derived **d**opamine and **c**AMP-**r**egulated **p**hospho**p**rotein that inhibits PP1 activity when phosphorylated by PKA (Hemmings et al., 1984). DARPP-32 has no effect on PP2A, PP2B, or PP2C activities and its effects on PP3 are not known. Significant amino acid sequence identity exists in the active domain of these proteins (amino acids 8–48 of DARPP-32 and amino acids 9–49 of I-1); however, little similarity is apparent in the remaining sequence of the two proteins (Shenolikar and Narin, 1991).

NIPP-1

NIPP-1 is a recently described **n**uclear **i**nhibitor of **p**rotein **p**hosphatase 1 isolated from bovine thymus nuclei (Beullens et al., 1992). Two forms, NIPP-1a (M_r 18 kDa) and NIPP-1b (M_r 16 kDa) were isolated and purified and, like I-1, I-2, and DARPP-32, they are acid- and heat-stable proteins. NIPP-1a and NIPP-1b inhibited PP1 with a K_i of about 1 pM, which is about 1000-fold more effective than I-1, I-2, or DARPP-32. NIPP-1a and NIPP-1b had no effect on the activities of PP2A, PP2B, and PP2C, while effects on PP3 have not been determined.

Use of Toxins to Study Protein Phosphatases

The discovery that toxins such as okadaic acid or microcystin are potent and specific inhibitors of PP1, PP2A, and PP3 has greatly improved the ability to measure ser/thr protein phosphatase activities in crude cellular extracts. PP1, PP2A, PP2B, PP2C, and PP3 have broad and overlapping substrate specificities *in vitro* and the identification and quantitation of their activities in crude cell extracts has proven to be difficult. PP1, PP2A, and PP3 activities present in cell extracts can be determined by taking advantage of, first, the different affinities of okadaic acid or microcystin for these three catalytic subunits; second, the intra-

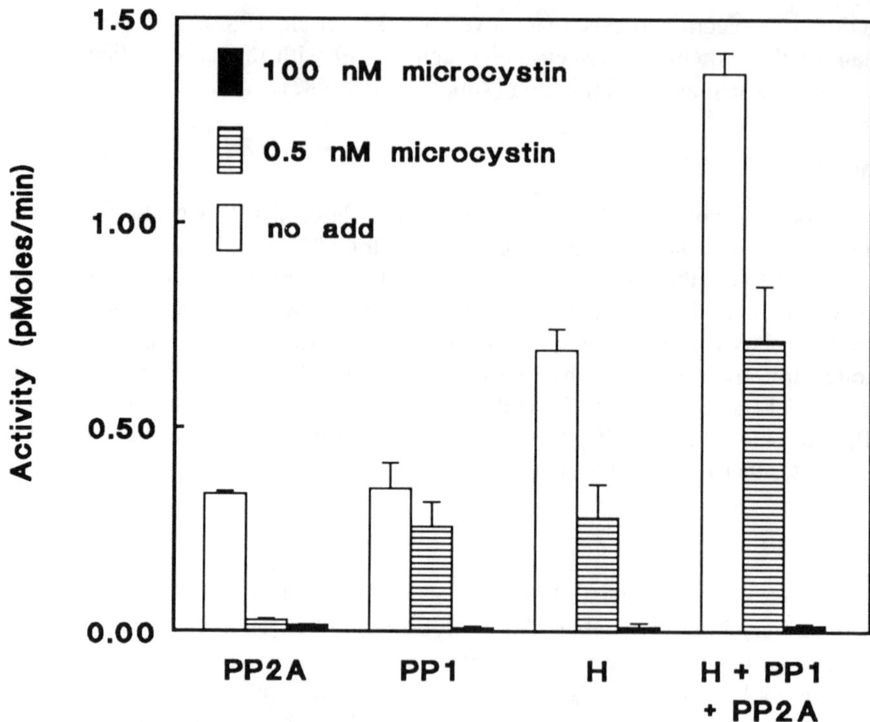

Figure 12.3 Inhibitory effect of microcystin-LR in cell homogenates to which the purified catalytic subunits of PP1 and PP2A protein phosphatases have been added. Assays were performed with dilute solution (1-, 10-, and 10-fold below their titration end points for PP2A, PP1, and T51B cell homogenate (H), respectively), and conducted in the presence or absence of microcystin-LR; □ controls; ▤, 0.5 nM microcystin-LR; ■, 100 nM microcystin-LR. Homogenates of T51B cells were prepared and diluted to the titration end point and assayed for phosphohistone phosphatase activity as described by Honkanen and colleagues (1991b). Reproduced by permission of Waverly Press.

cellular location of these enzymes; and third, the fact that activities due to PP2B and PP2C, which are divalent cation dependent, can be eliminated by inclusion of EGTA/EDTA in the assay mixture. PP3 is found only in the detergent extracts of a crude homogenate or of a high-speed membranous-containing fraction (100,000 $\times$ g pellet), whereas PP2A and PP1 are found in both the supernatant and the pellet fractions after high-speed (100,000 $\times$ g) centrifugation. Thus, in either whole-cell homogenates or in the high-speed supernatant, the difference in phosphatase activity measured in the presence and absence of 0.5 nM microcystin (with EGTA present, which will eliminate activities due to PP2B and PP2C) can be attributed to PP2A activity (0.5 nM microcystin does not affect the activity of PP1), while the difference in activity measured in the presence of 0.5 nM and 100 nM microcystin is PP1 activity. The feasibility of this approach is illustrated in Figure 12.3. Purified catalytic subunits of PP1 and PP2A were added to a homogenate of T51B rat liver epithelial cells (the ''titration end point'' was initially determined), and phosphohistone phosphatase activity was determined. In assays with purified catalytic subunits of PP2A and PP1, 0.5 nM microcystin-LR almost completely inhibited PP2A activity while not affecting PP1 activity, and both enzymes were totally inhibited by 100 nM microcystin-LR. In dilute cellular homogenates, more than 50 percent of the divalent cation-independent (PP1 and PP2A) phosphatase activity was inhibited by 0.5 nM microcystin-LR, and the remainder was inhibited by 100 nM microcystin-LR. Moreover, when diluted PP1 and PP2A catalytic subunits (diluted ten- and one-fold past the "titration end point," respectively) are added to a diluted (ten-fold below the "titration end point") T51B cell homogenate, the amount of phosphatase activity inhibited by 0.5 nM microcystin-LR is similar to the combined activity of the PP2A added and the 0.5-nM sensitive fraction in the original homogenate. One hundred nanomolar microcystin-LR still caused total inhibition of the remaining activity, suggesting that drug titration effects were not occurring. Thus, when appropriately diluted, microcystin-LR may be used as a probe to distinguish PP1 and PP2A activity in a cell homogenate. To assay PP1, PP2A, and PP3 activities in the high-speed particulate fraction, the pellet must first be extracted with dilute detergent such as CHAPS or cholate (Honkanen et al., 1991b). Using okadaic acid, for example, the difference in activity in the presence and absence of 2 nM okadaic acid is defined as PP2A activity (2 nM okadaic acid affects only about 5–10% of the PP3 activity); the difference in activity in the presence of 2 nM and 15 nM okadaic acid is PP3 activity (15 nM okadaic acid inhibits 100% PP3 and only about 10% PP1 activity); and the difference in activity in the presence of 15 nM and 50 nM okadaic acid is the PP1 activity (50 nM okadaic acid inhibits 100% of PP1 activity). Again, it is crucial that "titration end points" be determined when assaying activities of PP1, PP2A, and PP3 in cellular homogenates using okadaic acid or other toxin inhibitors, for reasons previously discussed (see Figure 12.2).

CONCLUSIONS

cDNA cloning has revealed at least 20 ser/thr protein phosphatases. However only five proteins have been identified, purified, and biochemically characterized—PP1, PP2A, PP2B, PP2C, and PP3. While no single criterion is able to biochemically distinguish between these five ser/thr protein phosphatase catalytic subunits, the use of a combination of several criteria (see Table 12.1) can identify specific known ser/thr protein phosphatases. The discovery of marine toxins with high affinities toward PP1, PP2A, and PP3 has improved our ability to study these important enzymes and will aid in uncovering the mysteries of their physiological actions. The challenge for the future is several-fold. First is the identification, purification, and biochemical characterization of additional ser/thr protein phosphatases and identification of their respective cDNAs. Second is the identification of specific inhibitors and/or activators of the activities of the ser/thr protein phosphatases. And last, with use of the tools of recombinant DNA technology, pharmacology, biochemistry, and cell biology, is the determination of the role of specific ser/thr protein phosphatases in cellular function.

REFERENCES

Aitken, A., T. Bilham, and P. Cohen. 1982. Complete primary structure of protein phosphatase inhibitor-1 from rabbit skeletal muscle. *Eur. J. Biochem.* 126:235–246.

Arino, J., C. W. Woon, D. W. Brautigan, T. B. Miller, and G. L. Johnson. 1988. Human liver phosphatase 2A: cDNA and amino acid sequence of two catalytic subunit isotypes. *Proc. Natl. Acad. Sci. U.S.A.* 85:4252–4256.

Arndt, K. T., C. A. Styles, and G. R. Fink. 1989. A suppressor of HIS4 transcriptional defect encodes a protein with homology to the catalytic subunit of protein phosphatases. *Cell* 56:527–537.

Bai, G., Z. Zhang, J. Amin, S. A. Deans-Zirattu, and E. Y. Lee. 1988. Molecular cloning of a cDNA for the catalytic subunit of rabbit muscle phosphorylase phosphatase. *FASEB J.* 2:3010–3016.

Berndt, N., D. G. Campbell, F. B. Caudwell, P. Cohen, E. F. da Cruz e Silva, O. B. da Cruz e Silva, and P. T. W. Cohen. 1987. Isolation and sequence analysis of a cDNA clone encoding a type-1 protein phosphatase catalytic subunit: Homology with protein phosphatase 2A. *FEBS Lett.* 223:340–346.

Beullens, M., A. Van Eynde, W. Stalmans, and M. Bollen. 1992. The isolation of novel inhibitory polypeptides of protein phosphatase 1 from bovine thymus nuclei. *J. Biol. Chem.* 267:16538–16544.

Bialojan, C. and A. Takai. 1988. Inhibitory effect of a marine-sponge toxin, okadaic acid, on protein phosphatases. *Biochem. J.* 256:283–290.

Bishop, C. T., E. F. L. J. Anet, and P. R. Gorham. 1959. Isolation and identification of the fast death factor *Microcystis aeruginosa* NRC-1. *Can J. Biochem. Physiol.* 37: 453–459.

Botes, D. P., A. A. Tuinman, P. L. Wessels, C. C. Viljoin, H. Kruger, D. H. Williams, S. Santikarn, R. J. Smith, and S. J. Hammond. 1984. The structure of cyanoginosin-LA, a cyclic heptapeptide toxin form cyanobacterium *Microcystis aeruginosa*. *J. Chem. Soc. Perkin Trans.* 1:2311–2315.

Carmichael, W. W. 1986. Algal toxins. *Adv. Botan. Res.* 12:47–66.

Carmichael, W. W., J. T. Eschedor, G. M. L. Patterson, and R. E. Moore. 1988a. Toxicity and partial structure of a hepatotoxic peptide produced by the cyanobacterium *Nodularia spumigena* Mertens emend. L575 from New Zealand. *Appl. Environ. Microbiol.* 54:2257–2263.

Carmichael, W. W., V. Beasley, D. L. Bunner, J. N. Eloff, I. Falconer, P. Gorham, K. Harada, T. Krishnamurthy, Y. Juan, R. E. Moore, K. Rinehart, M. Runnegar, O. M. Skulberg, and M. Watanabe. 1988b. Naming of cyclic heptapeptide toxins of Cyanobacteria (blue-green algae). *Toxicon* 26:971–973.

Chen, M. X., Y. H. Chen, and P. T. W. Cohen. 1992. Polymerase chain reaction using *Saccharomyces*, *Drosophila* and human DNA predict a large family of protein serine/threonine phosphatases. *FEBS Lett.* 306:54–58.

Cheng, X. C., T. Kihara, H. Kusakabe, J. Magae, Y. Kobayaswhi, R. Fang, A. Ni, Y. Shen, K. Ko, I. Yamaguchi, and K. Isono. 1987. A new antibiotic, tautomycin. *J. Antibiot.* 40:907–909.

Cheng, X. C., M. Ubukata, and K. Isono. 1990. The structure of tautomycin, a dialkyl-maleic anhydride antibiotic. *J. Antibiot.* 43:809–812.

Cohen, P. 1989. The structure and regulation of protein phosphatases. *Annu. Rev. Biochem.* 58:453–508.

Cohen, P. T. W. 1988. Two isoforms of protein phosphatase 1 may be produced from the same gene. *FEBS Lett.* 232:17–23.

Cohen, P. T. W., and P. Cohen. 1989. Discovery of a protein phosphatase activity encoded in the genome of bacteriophase γ. *Biochem. J.* 260:931–934.

Cohen, P. T. W., J. F. Collins, A. F. W. Coulson, N. Berndt, and O. B. da Cruz e Silva. 1988. Segments of bacteriophage γ (orf221) and Φ are homologous to genes coding for mammalian protein phosphatases. *Gene* 69:131–134.

Cohen, P. T. W., N. D. Brewis, V. Hughes, and D. J. Mann. 1990. Protein serine/threonine phosphatases; an expanding family. *FEBS Lett.* 268:355–359.

Cyert, M. S., R. Kunisawa, D. Kaim, and J. Thorner. 1991. Yeast has homologs (CNA1 and CNA2 gene products) of mammalian calcineurin, a calmodulin-regulated phos-phoprotein phosphatase. *Proc. Natl. Acad. Sci. U.S.A.* 88:7376–7380.

Dean, N. M., L. J. Mordan, K. T. Tse, S. L. Mooberry, and A. L. Boynton. 1991. Okadaic acid inhibits PDGF-induced proliferation and decreases PDGF receptor number in C3H10T1/2 mouse fibroblasts. *Carcinogenesis* 12:665–670.

Dombradi, V., J. M. Axton, D. M. Glover, and P. T. W. Cohen. 1989. Molecular cloning and chromosomal localization of a novel *Drosophila* protein phosphatase. *FEBS Lett.* 247:391–395.

Dombradi, V., J. M. Axton, N. D. Brewis, E. F. da Cruz e Silva, L. Alphey, and P. T. W. Cohen. 1990. *Drosophila* contains three genes that encode distinct isoforms of protein phosphatase 1. *Eur. J. Biochem.* 194:739–745.

Eriksson, J. E., J. A. O. Meriluoto, H. P. Kujari, and O. M. Skulberg. 1988. A comparison of toxins isolated from the cyanobacteria *Oscillatoria agardhii* and *Microcystis aeruginosa. Comp. Biochem. Physiol.* 89C:207–210.

Eriksson, J. E., D. Toivola, J. A. O. Meriluoto, H. Karaki, Y. Han, and D. Hartshorne. 1990. Hepatocyte deformation induced by cyanobacterial toxins reflects inhi-bition of protein phosphatases. *Biochem. Biophys. Res. Commun.* 173:1347–1353.

Fernandez-Sarabia, M. J., A. Sutton, T. Zhong, and K. T. Arndt. 1992. SIT4 protein phosphatase is required for the normal accumulation of SWI4, CLN1, CLN2 and HCS26 RNAs during late G_1. *Genes Dev.* 6:2417–2428.

Foulkes, J. G., S. J. Strada, P. J. Henderson, and P. Cohen. 1983. A kinetic analysis of the effects of inhibitor-1 and inhibitor-2 on the activity of protein phosphatase-1. *Eur. J. Biochem.* 132:309–313.

Francis, G. 1878. Poisonous Australian lake. *Nature (Lond.)* 18:11–16.

Fujiki, H., and R. E. Moore. 1987. Meeting report: US-Japan seminar on natural products and cancer chemoprevention. *Jap. J. Cancer Res. (Gann)* 78:875–878.

Green, D. D., S. Yand, and M. C. Mumby. 1987. Molecular cloning and sequence analysis of the catalytic subunit of bovine type 2A protein phosphatase. *Proc. Natl. Acad. Sci. U.S.A.* 84:4880–4884.

Guerini, D., and C. B. Klee. 1989. Cloning of human calcineurin A: Evidence for two isozymes and identification of a polyproline structural domain. *Proc. Natl. Acad. Sci. U.S.A.* 86:9183–9187.

Hemmings, H. C., P. Greengard, H. Y. Tung, and P. Cohen. 1984. DARPP-32, a dopamine-regulated neuronal phosphoprotein, is a potent inhibitor of protein phosphatase-1. *Nature* 310:503–505.

Holmes, C. F., D. G. Campbell, F. B. Caudwell, A. Aitken, and P. Cohen. 1986. The protein phosphatases involved in cellular regulation. Primary structure of inhibitor-2 from rabbit skeletal muscle. *Eur. J. Biochem.* 155:173–182.

Honkanen, R. E., J. Zwiller, R. E. Moore, S. L. Daily, B. S. Khatra, M. Dukelow, and A. L. Boynton. 1990. Characterization of microcystin-LR, a potent inhibitor of type 1 and type 2A protein phosphatases. *J. Biol. Chem.* 265:19401–19404.

Honkanen, R. E., M. Dukelow, J. Zwiller, R. E. Moore, B. S. Khatra, and A. L. Boynton. 1991a. Cyanobacterial nodularin is a potent inhibitor of type 1 and type 2A protein phosphatases. *Molec. Pharm.* 40:577–583.

Honkanen, R. E., J. Zwiller, S. L. Daily, B. S. Khatra, M. Dukelow, and A. L. Boynton. 1991b. Identification, purification, and characterization of a novel serine/threonine protein phosphatase from bovine brain. *J. Biol. Chem.* 266:6614–6619.

Hori, M., J. Magae, J. G. Han, D. J. Hartshorne, and H. Karaki. 1991. A novel protein phosphatase inhibitor, tautomycin: Effect on smooth muscle. *FEBS Lett.* 285:145–148.

Ingebritsen, T. S., and P. Cohen. 1983. The protein phosphatases involved in cellular regulation. 1. Classification and substrate specificities. *Eur. J. Biochem.* 132:255–261.

Ishihara, H., B. L. Martin, D. L. Brautigan, H. Karaki, H. Ozaki, Y. Kato, N. Fusetani, S. Watabe, K. Hashimoto, D. Uemura, and D. J. Hartshorne. 1989. Calyculin A and okadaic acid: Inhibitors of protein phosphatase activity. *Biochem. Biophys. Res. Commun.* 159:871–877.

Ito, A., T. Hashimoto, M. Hirai, T. Takeda, H. Shuntoh, T. Kuno, and C. Tanaka. 1989. The complete primary structure of calcineurin A, a calmodulin binding protein homologous with protein phosphatases 1 and 2A. *Biochem. Biophys. Res. Commun.* 163:1492–1497.

Kato, Y., N. Fusetani, S. Matsunaga, and K. Hashimoto. 1986. Calyculin A, a novel antitumor metabolite from marine sponge *Discodermia calyx*. *J. Am. Chem. Soc.* 108:2780–2781.

Katoh, F., D. J. Fitzgerald, L. Giroldi, H. Fujiki, T. Sugimura, and J. Yamasaki. 1990. Okadaic acid and phorbol esters: Comparative effects of these tumor promoters on cell transformation, intercellular communication and differentiation in vitro. *Jap. J. Cancer Res.* 81:590–597.

Kincaid, R. L., M. S, Nightingale, and B. M. Martin. 1988. Characterization of a cDNA clone encoding the calmodulin-binding domain of mouse brain calcineurin. *Proc. Natl. Acad. Sci. U.S.A.* 85:8983–8987.

Kincaid, R. L., P. R. Giri, S. Higuchi, J. Tamura, S. C. Dixon, C. A. Marietta, D. A. Amorese, and B. M. Martin. 1990. Cloning and characterization of molecular isoforms of the catalytic subunit of calcineurin using nonisotopic methods. *J. Biol. Chem.* 265:11312–11319.

Konst, H., P. D. McKercher, P. R. Gorham, A. Robertson, and J. Howell. 1965. Symptoms and pathology produced by toxic *Microcystis aeruginosa* NRC-1 in laboratory and domestic animals. *Can. J. Comp. Med. Vet. Sci.* 29:221–228.

Krishnamurthy, T., W. W. Carmichael, and E. W. Sarver. 1986a. Investigations of freshwater cyanobacteria (blue-green algae) toxic peptides. I. Isolation, purification and characterization of peptides from *Microcystis aeruginosa* and *Anabaena flos-aquae*. *Toxicon* 24:865–869.

Krishnamurthy, T., L. Szafraniac, E. W. Sarver, D. F. Hunt, J. Shabonowitz, W. W. Carmichael, S. Missler, O. M. Skulberg, and G. Codd. 1986b. Amino acid analysis of freshwater blue-green algal toxic peptides by faxt atom bombardment tandem mass spectrometric technique. *Proc. 34th Annual Conference on Mass Spectrometry and Allied Topics, Cincinnati,* p. 93a.

Kumagi, M., T. Tanagi, M. Murata, T. Yasumoto, M. Kat, P. Lassus, and J. A. Rodriques-Vazqheaz. 1986. Okadaic acid as the causative toxin of diarrhetic shellfish poisoning in Europe. *Agri. Biol. Chem.* 50:2853–2857.

Kuno, T., T. Takeda, M. Hirai, A. Ito, H. Mukai, and C. Tanaka. 1989. Evidence for a second isoform of the catalytic subunit of calmodulin-dependent protein phosphatase (calcineurin A). *Biochem. Biophys. Res. Commun.* 165:1352–1358.

Kusumi, T., T. Ooi, M. M. Watanabe, H. Takahagh, and H. Kakisawa. 1987. Cyanoviridin-RR, a toxin from the cyanobacterium (blue-green alga) *Microcystis viridis. Tetrahed. Lett.* 28:4695–4697.

Liu, Y., S. Ishii, M. Tokai, H. Tsutsumi, S. Ohki, R. Akada, K. Tanaka, E. Tsuchiya, S. Fukui, and T. Miyakawa. 1991. The *Saccharomyces cerevisiae* genes (CMP1 and CMP2) encoding calmodulin-binding protein homologous to the catalytic subunit of mammalian protein phosphatase 2B. *Molec. Gen. Genet.* 227:52–59.

MacKintosh, C., K. A. Beattie, S. Klumpp, P. Cohen, and G. A. Codd. 1990. Cyanobacterial microcystin-LR is a potent and specific inhibitor of protein phosphatases 1 and 2A from both mammals and higher plants. *FEBS Lett.* 264:187–192.

Magae, J., H. Osada, H. Fujiki, T. Saido, K. Suzuki, K. Nagai, M. Yamasaki, and K. Isono. 1990. Morphological changes of human myeloid leukemia K562 cells by a protein phosphatase inhibitor tautomycin. *Proc. Jap. Acad.* 66:28–33.

Matsushima, R., S. Nishiwaki, T. Ohta, S. Yoshizawa, M. Suganuma, K. Harada, M. R. Watanabe, and H. Fujiki. 1991. Structure-function relationships of microcystins, liver tumor promoters, in interaction with protein phosphatases. *Jap. J. Cancer Res.* 82:993–996.

McPartlin, A. E., H. M. Barker, and P. T. W. Cohen. 1991. Identification of a third alternatively spliced cDNA encoding the catalytic subunit of protein phosphatase 2Bβ. *Biochem. Biophys. Acta* 1088:308–310.

Mordan, L. J., N. M. Dean, R. E. Honkanen, and A. L. Boynton. 1990. Okadaic acid: A reversible inhibitor of neoplastic transformation of mouse fibroblasts. *Cancer Commun.* 2:237–241.

Murakami, Y., Y. Oshima, and T. Yasumoto. 1982. Identification of okadaic acid as a toxic component of a marine dinoflagellate *Prorocentrum lima. Bull. Jap. Soc. Sci. Fish.* 48:69–72.

Nimmo, G. A., and P. Cohen. 1978. The regulation of glycogen metabolism. Purification and characterisation of protein phosphatase inhibitor-1 from rabbit skeletal muscle. *Eur. J. Biochem.* 87:341–351.

Prinsep, M. R., F. Caplan, R. E. Moore, G. M. L. Patterson, R. E. Honkanen, and A. L. Boynton. 1992. Microcystin-LA from blue green alga in the order Stigonematales. *Phytochemistry* 31:1247–1248.

Rinehart, K. L., K. Harada, M. Namikoshi, C. Chen, and C. A. Harvis. 1988. Nodularin, microcystin and the configuration of adda. *J. Am. Chem. Soc.* 110:8557–8558.

Rivedal, E., S. Mikalsen, and T. Sanner. 1990. The non-phorbol ester tumor promoter okadaic acid does not promote morphological transformation or inhibit junctional communication in hamster embryo cells. *Biochem. Biophys. Res. Commun.* 167:1302–1308.

Ronne, H., M. Carlberg, G. Hu, and J. O. Nehlin. 1991. Protein phosphatase 2A in *Saccharomyces cerevisiae:* Effects on cell growth and bud morphogenesis. *Molec. Cell. Biol.* 11:4876–4884.

Sassa, T., W. W. Richter, N. Uda, M. Suganuma, H. Suguri, S. Yoshizawa, M. Hirota, and H. Fujiki. 1989. Apparent "activation" of protein kinases by okadaic acid class tumor promoters. *Biochem. Biophys. Res. Commun.* 159:939–944.

Shenolikar, S., and A. C. Narin. 1991. Protein phosphatases: Recent progress. *Adv. Sec. Mess. Phosphoprot. Res.* 23:1–121.

Silva, O. B., E. F. da Cruz e Silva, and P. T. W. Cohen. 1988. Identification of a novel protein phosphatase catalytic subunit by cDNA cloning. *FEBS Lett.* 242:106–110.

Silva, E. F., V. Hughes, P. McDonald, M. J. R. Stark, and P. T. W. Cohen. 1991. Protein phosphatase $2B_w$ and protein phosphatase Z are *Saccharomyces cerevisiae* enzymes. *Biochem. Biophys. Acta* 1089:269–272.

Sim, A. T. R. 1992. The regulation and function of protein phosphatases in brain. *Molec. Neurobiol.* 5:229–246.

Steele, F. R., T. Washburn, R. Rieger, and J. E. O'Tousa. 1992. Drosophila retinal degeneration C (rdgC) encodes a novel serine/threonine protein phosphatase. *Cell* 69: 669–676.

Stone, S. R., J. Hofsteenge, and B. A. Hemmings. 1987. Molecular cloning of cDNAs encoding two isoforms of the catalytic subunit of protein phosphatase 2A. *Biochemistry* 26:7215–7220.

Suganuma, M., H. Fujiki, H. Suguri, S. Yoshizawa, M. Hirota, M. Nakayasu, M. Ojika, K. Wakamatsu, K. Yamada, and T. Sugimura. 1988. Okadaic acid: An additional non-phorbol-12-tetradecanoate-13 acetate-type tumor promoter. *Proc. Natl. Acad. Sci. U.S.A.* 85:1768–1771.

Suganuma, M., H. Fujiki, F. H. Suguri, S. Yoshizawa, S. Yasumoto, Y. Kato, N. Fusetani, and T. Sugimura. 1990. Calyculin A, an inhibitor of protein phosphatases, a potent tumor promoter on CD-1 mouse skin. *Cancer Res.* 50:3521–3525.

Sutton, A., D. Immanuel, and K. T. Arndt. 1991. The SIT4 protein phosphatase functions in late G_1 for progression into S phase. *Molec. Cell. Biol.* 11:2133–2148.

Takai, A., C. Bialojan, M. Troschka, and J. C. Rueeg. 1987. Smooth muscle myosin phosphatase inhibition and force enhancement by black sponge toxin. *FEBS Lett.* 217:81–84.

Wadzinski, B. E., L. E. Heasley, and G. L. Johnson. 1990. Multiplicity of protein serine-threonine phosphatases in PC12 pheochromocytoma and FTO-2B hepatoma cells. *J. Biol. Chem.* 265:21504–21508.

Yoshizawa, S., R. Matsushima, M. F. Watanabe, K. I. Harada, A. Ichihara, and W. W. Carmichael. 1990. Inhibition of protein phosphatases by microcystin and nodularin associated with hepatotoxicity. *J. Jap. Cancer Res. Clin. Oncol.* 116:609–614.

Index